QUANTITATIVE
CHEMICAL
ANALYSIS

QUANTITATIVE CHEMICAL ANALYSIS

Stanley E. Manahan
University of Missouri

Brooks/Cole Publishing Company
Monterey, California

Brooks/Cole Publishing Company
A Division of Wadsworth, Inc.

Printed in the United States of America

10 9 8 7 6 5 4 3 2 1

Library of Congress Cataloging-in-Publication Data

Manahan, Stanley E.
  Quantitative chemical analysis.

  Includes index.
  1. Chemistry, Analytic—Quantitative. I. Title.
QD101.2.M36 1986   543   86-2590
ISBN 0-534-05538-9

Sponsoring Editor: Sue Ewing
Editorial Assistant: Lorraine McCloud
Production Editor: Penelope Sky
Production Assistant: Dorothy Bell
Permissions Editor: Mary Kay Hancharick
Interior and Cover Design: Victoria A. Van Deventer
Cover Photo: Dennis Yates
Art Coordinator: Michèle Judge
Interior Illustration: John Foster
Photo Editor and Researcher: Sue C. Shepherd
Typesetting: Polyglot Compositors, Singapore
Cover Printing: The Lehigh Press Company, Pennsauken, New Jersey
Printing and Binding: The Maple-Vail Book Manufacturing Group, York, Pennsylvania

Dedicated to
Professor Reynold T. Iwamoto,
The University of Kansas

Preface

This text is designed for an undergraduate laboratory course in quantitative analysis. Chapters 1–12 cover classical quantitative analysis, offering material usually presented in a one-semester or two-quarter course. Chapters 13–23 present simple instrumental aspects of quantitative analysis. Chapter 24 explains the analysis of real samples, and Chapter 25 suggests laboratory experiments.

The book is directed toward a wide range of student abilities and interests, and presents modern quantitative analysis as an essential, dynamic part of present-day chemical science. It includes essential aspects of theory and practice, and some judiciously chosen challenging examples and problems that will increase the interest of chemistry majors. Moreover, the text is intended to serve as a consistently helpful learning partner for the student who is majoring in another discipline. The approach to each principle and concept is therefore gradual and systematic.

Text Organization

Chapter 1 contains introductory material that supplies the essential background for the course. Quantitative analysis is often taken by sophomores who have not already studied organic chemistry. To assist these students, Chapter 1 includes a short section on the fundamentals of organic chemistry; students are referred back to this section in later chapters of the book, when pertinent organic formulas, reactions, and polymers are covered.

The major laboratory techniques of quantitative analysis are introduced in Chapter 2, because quantitative analysis is a laboratory course, and students need this information as they progress through the laboratory sequence.

Chapter 3 covers the quantitative treatment of data at a level appropriate to undergraduates. Emphasis is on the types of errors and data that are commonly encountered in the quantitative analysis laboratory, but the chapter is designed to avoid overwhelming the students by an excessively mathematical treatment of data early in the course.

Chapters 4–14 cover the applications to chemical analysis of acid-base, solubility, complexation, and oxidation-reduction. Acid-base equilibria and reactions are covered first because they are so important to the other phenomena discussed in these chapters. The applicable fundamental equilibria are discussed before each topic is examined: acid-base in Chapter 4, solubility in Chapter 7, complexation in Chapter 10, and oxidation-reduction in Chapter 11.

Depending upon what equipment is available, instructors may choose to use laboratory instrumental techniques in connection with the various areas discussed in Chapters 4–12. Instrumental pH measurement is so frequently required that it is introduced early, in Chapter 5. For other instrumental techniques, the instructor may refer students to material in Chapters 13–23. For example, the action of acid-base indicators (presented in Chapter 5) is demonstrated very clearly in the laboratory by solution spectrophotometric measurements of indicator solutions at various pH values (theory and practice covered in Chapter 21).

Because they are likely to be used for reference in the manner just described, Chapters 13–23 can stand alone. In addition to supplementing a standard quantitative analysis course, these instrumental chapters can be used in a non-majors course that includes simple instrumental analysis. The nature of the material, the depth of coverage, and the degree of sophistication are appropriate to such courses.

Learning Aids

The text includes special features to help students master the course material.

At the beginning of each chapter is a list of learning goals that serves as a guide to the topics to be mastered as the chapter is studied.

For many students, solving problems, particularly those involving chemical equilibria, is the most challenging part of a quantitative analysis course. Excessively specific schemes for the solution of problems in quantitative analysis are often counter-productive, and any approach that encourages students to depend on formulas is particularly deadly to the learning process. Therefore, the problem-solving sections of the text offer examples that will help students achieve an ease in solving problems that will be beneficial in a variety of situations.

At the end of each chapter are programmed summaries in a fill-in-the-blanks format that enables students to check their own mastery of the topics in the chapter. The answers that belong in the blanks are at the back of the book.

Also at the end of each chapter are separate question and problem sections, with selected answers at the end of the book.

The answers and solutions to all problems are in the student solutions manual, available as a desk copy to the instructor who adopts this text.

Acknowledgments

Special thanks are due to Bruce Thrasher of Willard Grant Press and Sue Ewing of Brooks/Cole Publishing Company for their efforts in developing this text. The manuscript was reviewed by John DeVries, California State University at Hayward; John Grove, South Dakota State University; Arlin Gyberg, Augsberg College; and Herbert Hill, Washington State University.

Many individuals devoted their time and talents to preparing this text for publication. Foremost among these was Production Editor Penelope Sky, whose patience, attention to detail, and overall competence are outstanding. Others whose excellent skills enhanced the book are Designer Victoria Van Deventer, Art Coordinator Michèle Judge, Illustrator John Foster, Photo Editor Sue C. Shepherd, Permissions Editor Mary Kay Hancharick, Editorial Assistant Lorraine McCloud, and Proofreader Tricia Cain. Their contributions are greatly appreciated.

Stanley E. Manahan

Contents

12   Oxidation-Reduction Titrations   316

13   Potentiometry   360

14   Voltammetry and Amperometric Titrations   397

15   Electrogravimetric and Coulometric Determinations   427

22   Atomic Spectrophotometric Analysis   590

23   Nuclear and Radiochemical Analysis   610

24   The Analysis of Real Samples   628

Appendix   681

Index   691

1

What Is Analytical Chemistry?

L E A R N I N G   G O A L S

1. Understanding of the chemical analysis process.
2. Distinctions between physical and chemical methods and between classical and instrumental methods of analysis.
3. Concepts of standards and calibration as applied to chemical analysis.
4. Units of measurement commonly used in chemical analysis.
5. Fundamentals of organic chemical structure and functional groups.
6. Basic principles of stoichiometry.
7. Terms and units employed to express solution concentration.
8. Solution equilibrium and the equilibrium constant expression.
9. The major types of solution equilibria—for example, acid-base.

1.1 Introduction

Analytical chemistry is that branch of the chemical sciences used to determine the composition of a sample of material. A **qualitative analysis** is performed to determine *what* is in a sample. The amount, concentration, composition, or percent of a substance present is determined by **quantitative analysis**. Sometimes both qualitative and quantitative analyses are performed as part of the same process. As its title states, this text deals primarily with quantitative analysis.

Analytical chemistry is important in practically all areas of human endeavor. The nutritional value of food is determined largely by chemical analysis. Chemical analyses are performed to assess potential dangers from hazardous waste sites, to tell the amounts of fertilizers needed for good soil productivity, and to assay the percentage of a commercially valuable mineral in an ore. The commonly performed clinical blood chemistry test consists of a number of chemical measurements made by automated instruments to provide a profile of an individual's physiological status.

Analytical chemistry is a dynamic discipline. New chemicals and increasingly sophisticated instruments and computational capabilities are constantly coming into use to improve the ways in which chemical analyses are done. Some of these improvements involve the determination of ever smaller quantities of a substance; others greatly shorten the time required for an analysis; and some enable us to tell with much greater specificity the identities of a large number of compounds in a complex sample. Robotic technology is even coming into play in analytical chemistry, and robots can now do some of the routine work that must otherwise be performed by a chemist working at a laboratory bench.

This chapter is an introduction to analytical chemistry and quantitative analysis. In addition, it gives some basic information needed to begin the study of quantitative analysis. For example, to discuss quantitative analysis, it is necessary to know the basic units of measurement employed. Some of the users of this text will not have had organic chemistry; therefore a short section on this subject is provided to aid in understanding the formulas and structures of organic compounds presented later in the text. Fundamentals of stoichiometry, solution concentration, and equilibrium are also presented.

This chapter has several learning goals. These should be kept in mind as the chapter is read. After the chapter is read, they should be reviewed to see if they have been mastered.

1.2   The Chemical Analysis Process

The steps common to most quantitative chemical analyses are shown schematically in Figure 1.1. Rather than as an individual laboratory operation or "experiment," chemical analysis should be thought of in terms of an overall *chemical analysis process*, each step of which is crucial to obtaining a valid, useful result. The first step is to obtain a representative sample or samples. A **sample** is that portion of matter upon which the analysis is performed. Its composition should be as close as possible to that of the bulk of material that it represents.

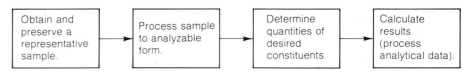

FIGURE 1.1   Schematic representation of the major steps involved in chemical analysis process

When this is accomplished it is a **representative sample**. Generally samples are provided to quantitative analysis students in the laboratory. However, it must be kept in mind that obtaining a good sample is a crucial step in the chemical analysis process—a step where far too many analyses go wrong and are therefore worthless from the very beginning. More details are provided on taking and preserving a representative sample in Chapter 24.

After the best possible sample is obtained, it is subjected to **sample processing** to get it into a form that can be analyzed. This step most commonly consists of putting the sample into solution. It may be as simple as dissolving commercial fertilizer in dilute HCl to enable determination of its potassium content, or as complicated as fusing ground rock with molten alkali salts or digesting fish tissue in hot oxidizing acid (HNO_3, $HClO_4$).

When a sample is oxidized, dissolved in acid, or otherwise greatly altered as part of the analytical process, the chemical analysis is called *destructive*. In some cases, such as those where evidence of a crime is involved, it is important to preserve the sample in an unaltered form. This requires *nondestructive* methods of analysis, such as instrumental neutron activation analysis (Chapter 23).

After sample processing, it is often necessary to eliminate **interferences** from substances in the sample that can cause erroneous results. For example, in the complexometric determination of iron(III) with ethylenediaminetetra-acetic acid (EDTA) (Chapter 10), adjustment of the sample solution to pH 3 will eliminate interferences from other metal ions. In some cases it may be necessary to remove an interfering substance from a sample before the quantity of the desired constituent can be measured.

After any interferences are eliminated, the quantity of the desired sample constituent is determined. At this point, several important terms used in the discussion of quantitative analysis should be defined. The constituent being sought—potassium in fertilizer, sulfur in coal, iron in iron ore—is called an **analyte.** The specific measurement of a constituent such as sulfur in coal is the **determination** of that analyte. The total process that a sample undergoes is called an **analysis** and may involve the determination of more than one analyte. Thus an analytical chemist might say that "I analyzed a sample of paint and determined the chromium and titanium contents of the paint pigments." It would not be correct for the chemist to say that "I analyzed chromium and titanium in a paint sample." In some cases, particularly where modern instruments can be applied, the actual determination of the quantity of analyte in a properly processed sample is the easiest step in the chemical analysis process.

The final step in a chemical analysis is the calculation of results. This step may consist of a few simple arithmetic calculations, or it may involve a complicated data processing operation that calculates and compensates for interferences in the method. In addition to providing a number for the quantity or percentage of analyte in a sample, the calculation of results usually involves an evaluation of the reliability (accuracy and precision, Chapter 3) of the analytical values. In modern analytical laboratories results are calculated and stored by computer.

Major Categories of Chemical Analysis

Both qualitative and quantitative analysis are divided between **wet chemical** or **classical** methods, involving primarily chemical reactions and simple measurements of weight and volume, and **instrumental methods**. The latter use instruments to measure physical manifestations of chemical species and chemical reactions — such as absorption of light, electrical potentials, or small changes in temperature.

A similar division recognizes **chemical** and **physical** methods of analysis. Chemical methods almost always involve the measurement of a mass of a chemical species or volume of a reagent solution produced or consumed by a chemical reaction. As a simple example, the quantity of chloride in a sample may be determined by precipitating it with a solution of silver nitrate,

$$Ag^+ + Cl^- \rightarrow AgCl(s) \tag{1.1}$$

weighing the AgCl produced, and relating that quantity by a simple stoichiometric calculation to the chloride originally present in the sample. This is an example of a **gravimetric** determination. Alternatively, the silver nitrate may be contained in a solution of known Ag^+ concentration, and the volume of the solution required to precipitate all the Cl^- in the sample can be measured. Such a procedure is known as a **titrimetric** determination. Both gravimetric and titrimetric analysis are discussed in more detail in Chapter 2.

Physical methods of analysis normally involve a measurement of a physical parameter other than mass or volume. For example, the quantity of chloride in a sample may be measured by dissolving a weighed quantity of the sample in water, diluting to an accurately measured volume, measuring the electrical potential of a chloride ion–selective electrode versus a reference electrode in the solution (Chapter 13), and calculating the concentration of chloride ion from the measured potential. The concentration of sodium ion in a drinking water sample may be determined by aspirating some of the sample into the flame of an atomic emission (flame photometric) analysis instrument (Chapter 22) and measuring the intensity of the yellow light characteristic of thermally excited sodium atoms. The level of ammonium ion in an irrigation runoff water sample may be determined by treating the sample with Nessler's reagent, an alkaline solution of mercuric iodide in potassium iodide, and measuring the light absorbed and scattered in the 400 to 500 nanometer wavelength range by the dispersion of colloidal particles formed by the orange-brown product of the reaction. This product is formed by the reaction

$$2HgI_4^{2-} + NH_3 + 3OH^- \rightarrow \begin{matrix} Hg{\Large\diagup}^{\displaystyle I} \\ {\LARGE\diagdown}O(s) \\ Hg{\LARGE\diagup} \\ {}_{\displaystyle NH_2} \end{matrix} + 7I^- + 2H_2O \tag{1.2}$$

This determination involves both a chemical reaction and a physical measurement and may be classified as a *physicochemical* method. Physical methods of analysis are almost invariably instrumental methods, because they make use of instruments such as potentiometers for measuring electrical potential, or spectrophotometers for measuring the wavelength and intensity of light.

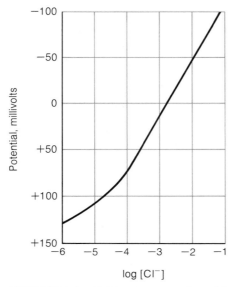

FIGURE 1.2   Calibration curve for a chloride ion–selective electrode

Generally, physical methods of analysis make use of **calibration curves** to relate the physical parameter measured to the quantity or concentration of an analyte. Such a calibration curve is shown in Figure 1.2 relating log [Cl⁻] to the potential of a chloride–reference electrode system. Such curves are obtained by the measurement of physical response obtained from solutions of known concentration called **standard solutions**. Often there is not a straight-line relationship between the concentration, or log of concentration, of an analyte and the physical quantity measured, although the calibration curve may still be useful. In modern practice of routine chemical analysis, the calibration curve is stored on a computer that automatically calculates the concentration corresponding to a measured physical quantity.

1.4   Calibration Standards

Calibration

There are only two absolute quantities that can be related directly to the amount of analyte in a sample without standardization; these are the mass of the product obtained by gravimetric analysis involving stoichiometric reactions and the quantity of electricity consumed by an electrochemical reaction, such as the electrodeposition of Ag^+ ion (Chapter 15):

$$Ag^+(aq) + e^- \rightarrow Ag(s) \tag{1.3}$$

In all other cases the concentration or quantity of analyte C_A is related to a measured physical quantity X by the general relationship

$$C_A = f(X) \tag{1.4}$$

This relationship states that C_A is a function of X, which may be a volume of reagent, electrical current, or other physical quantity. The determination of this relationship is called **calibration**. In the simplest form of this relationship, C_A varies linearly with X, so that

$$C_A = kX \tag{1.5}$$

where k is a proportionality constant or calibration factor.

EXAMPLE

The quantity of strong acid, H^+, in a sample may be determined by measuring the volume of a base solution required to react with all the acid according to the reaction,

$$H^+ + OH^- \rightarrow H_2O$$

If 0.0318 moles of acid require 0.0426 liters of the base solution for neutralization, what is the value of the proportionality constant k?

$$C_A = 0.0318 \text{ moles of acid} \qquad X = 0.0426 \text{ liters of base}$$

$$k = \frac{C_A}{X} = \frac{0.0318 \text{ moles of acid}}{0.0426 \text{ liters of base}} = 0.746 \frac{\text{moles of acid}}{\text{liter of base}}$$

If an unknown sample requires 0.0488 liters of base to neutralize the strong acid in it, the acid content is given by

$$C_A = 0.746 \frac{\text{moles of acid}}{\text{liter of base}} \times 0.0488 \text{ liters of base}$$

$$= 0.0364 \text{ moles of acid}$$

Chemical Standards for Calibration

The calibration process requires the use of chemical standards to provide known values of C_A for the determination of $f(X)$. The best of these are termed **primary standards**. The desirable properties of a primary standard chemical are the following:

1. *Stability*. The primary standard should be stable with respect to the absorption of water, drying, reaction with air, and in solution.
2. *Purity*. The primary standard should have a known, specified purity, preferably of at least 99.5%.
3. *Weighability*. The primary standard should be in a form that is

readily weighed, which implies that it be a chemically stable solid. A high molecular weight is desirable to minimize the relative errors of small absolute errors in weighing.

    4. *Solubility*. The primary standard should be soluble in water or common acids or bases.

In addition, it is desirable that a primary standard, like all reagents used in chemical analysis, be readily available at reasonable cost and not be unduly toxic.

1.5   Units of Measurement in Analytical Chemistry

Normally analytical chemical measurements are made using metric units. Recall that these consist of a basic unit with a prefix used to designate a multiple of the basic unit, as shown in Table 1.1. All analytical measurements are ultimately related back to the determination of **mass**, expressed in grams, g. The basic metric unit of volume is the liter, L. Volumes used in routine chemical analysis are normally most conveniently expressed in units of $\frac{1}{1000}$ L, called milliliters, mL (1 mL = 1 cubic centimeter, cm^3). In clinical analysis, volumes are normally expressed in $\frac{1}{10}$ L, or deciliters, dL. The volumes of small samples injected into a chromatograph (Chapters 18 and 19) or into the graphite furnace of an atomic absorption instrument (Chapter 22) are normally expressed as microliters, μL. The wavelength of visible or ultraviolet light associated with spectrophotometric analysis (Chapters 20 to 22) is in the range of hundreds of nanometers, nm, whereas the wavelength of infrared radiation is expressed in units of micrometers, μm (10^{-6} meters). Temperatures in analytical chemistry are usually expressed in degrees Celsius, °C. Other units are introduced in the text in the chapters where they are used.

    According to the Système Internationale (SI) of measurements, the base unit for quantity of substance is the **mole**, abbreviated mol. The amount of

TABLE 1.1   Prefixes used in the metric system

Prefix and symbol	Multiple of the basic unit
Mega, M	10^6
Kilo, k	10^3
Hecto, h	10^2
Deka, da	10
Deci, d	10^{-1}
Centi, c	10^{-2}
Milli, m	10^{-3}
Micro, μ	10^{-6}
Nano, n	10^{-9}
Pico, p	10^{-12}

substance is distinct from mass. A mole of a specific entity, such as atoms, molecules, ions, or electrons, contains Avogadro's number of those entities, 6.022×10^{23}. The mass of a mole of chemical species in grams is equal to its **gram-formula weight**. Thus, nitrogen gas, N_2, has a formula weight (fw) of 28.0, so a mole of nitrogen gas has a mass of 28.0 g. Similarly, a mole of $Al_2(SO_4)_3$, formula weight 342.15, has a mass of 342.15 g. The number of moles of a chemical species is given by the formula

$$\text{Number of moles} = \frac{\text{mass, g}}{\text{gram-formula weight}} \tag{1.6}$$

On the scale used in analytical chemistry, it is often convenient to deal with units of $\frac{1}{1000}$ mol. This is the **millimole**, mmol. The number of millimoles of a chemical species is given by

$$\text{Number of mmoles} = \frac{\text{mass, mg}}{\text{milligram-formula weight}} \tag{1.7}$$

$$\text{Number of mmoles} = 1000 \times \text{number of mol} \tag{1.8}$$

EXAMPLE

How many moles of the detergent builder sodium tripolyphosphate, $Na_5P_3O_{10}$, fw 367.9, are contained in 5.04 g of this compound?

$$\text{Moles } Na_5P_3O_{10} = \frac{5.04 \text{ g}}{367.9 \text{ g/mol}} = 0.0137 \text{ mol}$$

How many millimoles of $Na_5P_3O_{10}$ are contained in the same mass of this compound?

$$5.04 \text{ g} = 5.04 \times 10^3 \text{ mg}$$

$$\text{mmoles } Na_5P_3O_{10} = \frac{5.04 \times 10^3 \text{ mg}}{367.9 \text{ mg/mmol}} = 13.7 \text{ mmol}$$

Note that

$$\text{mmoles } Na_5P_3O_{10} = 1000 \times \text{moles } Na_5P_3O_{10}$$

1.6   Some Basics of Organic Chemistry

Importance of Organic Chemistry to Quantitative Analysis

Organic chemistry is that branch of chemistry dealing with practically all known carbon compounds, most of which contain carbon atoms covalently bonded to each other. Carbon-containing compounds that are not considered organic chemicals include the carbides, such as calcium carbide, CaC_2; cyanides, containing the CN^- ion; and carbonates, such as calcium carbonate, $CaCO_3$. Most users of this book already have some familiarity with organic chemistry through courses in the subject or by having covered it in beginning chemistry. Since quantitative analysis inevitably involves some use of organic chemistry, a few of the basic principles of this discipline are reviewed here.

Structure of Organic Compounds

Because of the ability of carbon atoms to bond to each other, organic molecules may consist of many atoms. These may be in the form of structures consisting of carbon atoms arranged as straight chains, branched chains, or rings. Such structures are shown in Figure 1.3 for three **hydrocarbons**, organic compounds containing only C and H atoms.

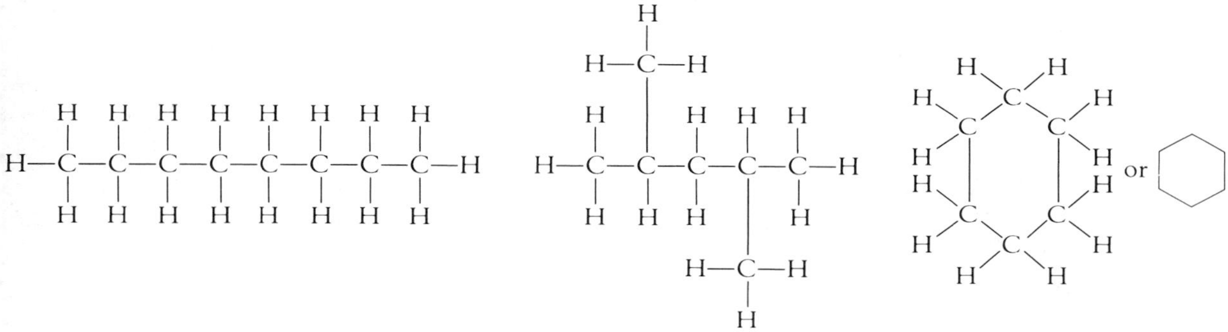

n-Octane, a straight-chain hydrocarbon

2,4-Dimethylpentane, a branched-chain hydrocarbon

Cyclohexane, a ring hydrocarbon

FIGURE 1.3 Organic compounds (hydrocarbons) in straight chain, branched-chain, and ring configurations

Naming Organic Compounds

Organic nomenclature deals with the names of organic compounds. The subject is too large to go into in detail, but it should be mentioned that the names of organic compounds describe their structures. For the examples cited in Figure 1.3, *n-octane* denotes a straight-chain alkane (hydrocarbon with only single bonds between C atoms) containing *eight* carbons in the chain, *2,4-dimethylpentane* has *five* C atoms in a chain with methyl ($-CH_3$) groups attached to the number 2 and 4 carbon atoms of the chain, and *cyclohex*ane has a *ring* of *six* C atoms.

Aromatic Compounds

A large number of organic compounds used in chemical analysis are **aromatic compounds** containing rings of six carbon atoms involving so-called π (pi, pronounced "pie") bonds between carbon atoms. The simplest of the aromatic compounds is the hydrocarbon benzene, C_6H_6. It is an oversimplification, but the structure of benzene can be visualized as resonating between the two structures shown in Figure 1.4. This would imply a ring structure with three double bonds. Actually, the bonds in benzene are much stronger, and the reactions of benzene are much different than this structure implies. In fact, the C atoms in benzene are contained in a planar ring with π electron clouds above and below the plane. Such a structure is shown simply as a hexagon with a circle in it,

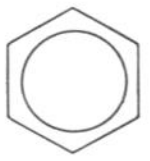

wherein it is implied that carbon atoms are located at each corner of the hexagon, and each C has one H bonded to it.

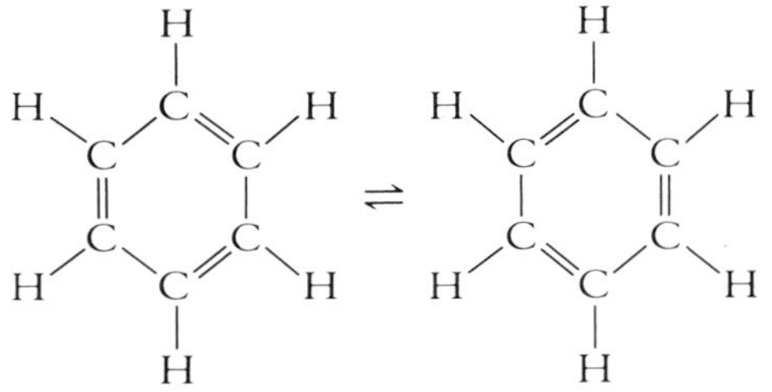

FIGURE 1.4   Benzene illustrated as two ring structures, each with three double bonds

The kind of symbolic structure shown for benzene is used to illustrate more complicated organic structures. For example, the compound consisting of two benzene rings sharing two carbon atoms is naphthalene, $C_{10}H_8$, shown in Figure 1.5. In the structure of naphthalene, there is a C atom at each vertex with one H atom bonded to each of the numbered C positions. The compound 1,2,3,4-tetrahydronaphthalene has one aromatic ring (shown with a circle in it) and one nonaromatic ring (the hexagon without a circle). As its name implies, this compound is naphthalene with one additional H atom at each of the 1, 2, 3, and 4 positions. This gives two H atoms at each of the 1, 2, 3, and 4 positions, one H atom at each of the 5, 6, 7, and 8 positions, and no H atoms on each of the C's shared between the two rings.

Naphthalene   Tetralin, or 1,2,3,4-tetrahydronaphthalene

FIGURE 1.5   Structures of naphthalene and tetralin.

Functional Groups in Organic Chemistry

Many organic compounds contain **functional groups** in which at least one atom other than carbon or hydrogen is present. These groups may play a role in quantitative analysis by giving the compound a desired property, such as the ability to bind to metal ions (see chelation, Chapter 10). Sometimes a functional group itself is the subject of a quantitative determination. Organic compounds are classified on the basis of functional groups, as shown in Table 1.2.

The generalized formula of an organic molecule is often given by showing the important functional group, or groups, and using R or R′ to designate the rest of the molecule, often a hydrocarbon group. A typical group is an alkyl entity, such as the ethyl group,

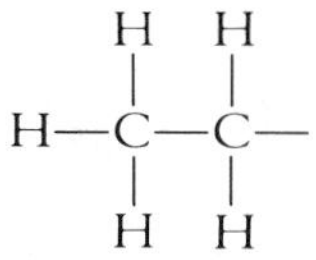

$$
\begin{array}{c}
\quad\; H \quad H \\
\quad\; | \quad\;\; | \\
H-C-C- \\
\quad\; | \quad\;\; | \\
\quad\; H \quad H
\end{array}
$$

Such a group constituting a major portion of an organic molecule is called a **moiety**. Thus, a generalized formula of a ketone can be expressed as,

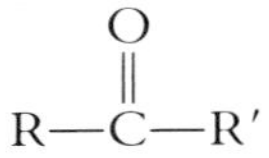

$$
\begin{array}{c}
\quad\;\; O \\
\quad\;\; \| \\
R-C-R'
\end{array}
$$

where R and R′ are hydrocarbon moieties. Aromatic moieties are designated by the symbol ϕ or Ar.

Polymers

One of the most important aspects of organic chemistry is the production of large aggregates of molecules bonded together to form **polymers**. For example, large numbers of styrene molecules can bond together to form long chains of polystyrene polymers as shown below:

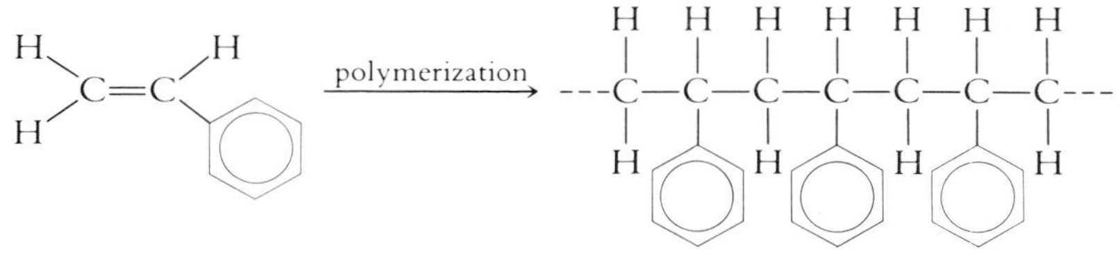

Styrene   Polystyrene

(1.9)

Polymers are widely used in analytical chemistry for the manufacture of laboratory apparatus, filters, membranes, ion-exchange resins (see Chapter 16), and chromatographic packings (see Chapter 18).

TABLE 1.2   Organic compounds classified according to functional group

Class of compound	Name of example compound	Structural formula with functional group shaded
Alcohol	Ethanol	
Phenol	Phenol	(aromatic alcohol)
Ether	Diethyl ether	
Aldehyde	Acetaldehyde	
Ketone	Acetone	
Carboxylic acid	Acetic acid	
Amine	1,2-Diaminoethane	
Amide	Acetamide	
Nitrile	Acetonitrile	
Nitro compounds	Nitrobenzene	
Heterocyclic nitrogen compounds	Pyridine	

TABLE 1.2 (*continued*)

Class of compound	Name of example compound	Structural formula with functional group shaded
Nitroso compounds	α-Nitroso-β-naphthol	
Sulfonic acids	Benzenesulfonic acid	
Organohalides	Dichloromethane	

1.7   Stoichiometry

The quantities of chemical species reacting with each other and the quantities of the products of these reactions are considered under the category of **stoichiometry**. Whereas the chemist measures mass in the laboratory in units like grams and milligrams, chemical reactions occur in multiples of moles and millimoles. Thus it is necessary for the analyst to convert back and forth between grams and moles and milligrams and millimoles. How this is done can be illustrated by several examples. Consider the calculation of the mass of solid Li_2CO_3, fw 73.9, that can be precipitated from a solution containing 5.24 g of LiCl, fw 42.4, by the addition of sodium carbonate, Na_2CO_3.

$$2Li^+ + 2Cl^- + 2Na^+ + CO_3^{2-} \rightarrow Li_2CO_3(s) + 2Na^+ + 2Cl^- \qquad (1.10)$$

$$\text{Mass } Li_2CO_3 = \text{mass } LiCl \times \underbrace{\frac{1 \text{ mol LiCl}}{42.4 \text{ g LiCl}}}_{\substack{\text{Converts mass} \\ \text{of LiCl to moles} \\ \text{of LiCl}}} \times \underbrace{\frac{1 \text{ mol } Li_2CO_3}{2 \text{ mol LiCl}}}_{\substack{\text{Converts} \\ \text{moles of LiCl} \\ \text{to moles of} \\ Li_2CO_3}} \times \underbrace{\frac{73.9 \text{ g } Li_2CO_3}{\text{mol } Li_2CO_3}}_{\substack{\text{Converts} \\ \text{moles of } Li_2CO_3 \\ \text{to grams of} \\ Li_2CO_3}}$$

$$= 5.24 \text{ g LiCl} \times \frac{1 \text{ mol LiCl}}{42.4 \text{ g LiCl}} \times \frac{1 \text{ mol } Li_2CO_3}{2 \text{ mol LiCl}} \times \frac{73.9 \text{ g } Li_2CO_3}{\text{mol } Li_2CO_3}$$

$$= 4.57 \text{ g } Li_2CO_3 \qquad (1.11)$$

Note how the units cancel to give the desired units of grams of Li_2CO_3. Carrying units through a stoichiometric calculation in this manner is a check to ensure that the calculation was carried out properly.

It is just as easy to carry out a stoichiometric calculation using millimoles and milligrams. Suppose that the mass of $Al_2(SO_4)_3$, a chemical used for purifying water, is to be determined in a mixture with inert material. This is accomplished by dissolving the sample in water and precipitating the sulfate as $BaSO_4$ by the addition of a solution of $BaCl_2$. If 3.78 g of $BaSO_4$ are produced, what was the quantity of $Al_2(SO_4)_3$, fw 342.2, in the original sample? The net ionic reaction for the precipitation of $BaSO_4$ is

$$Ba^{2+} + SO_4^{2-} \rightarrow BaSO_4(s) \tag{1.12}$$

and each millimole of $Al_2(SO_4)_3$ produces 3 mmol of $BaSO_4$. The calculation of the mass of aluminum sulfate in the sample is the following:

$$\text{Mass } Al_2(SO_4)_3 = \text{mass } BaSO_4 \times \frac{1 \text{ mmol } BaSO_4}{233.4 \text{ mg } BaSO_4}$$

$$\times \frac{1 \text{ mmol } Al_2(SO_4)_3}{3 \text{ mmol } BaSO_4} \times \frac{342.2 \text{ mg } Al_2(SO_4)_3}{\text{mmol } Al_2(SO_4)_3} \tag{1.13}$$

A mass of 3.78 g of $BaSO_4$ is equal to 3.78×10^3 mg of this compound. Substituting this value into the preceding equation gives

$$\text{Mass } Al_2(SO_4)_3 = 3.78 \times 10^3 \text{ mg } BaSO_4 \times \frac{1 \text{ mmol } BaSO_4}{233.4 \text{ mg } BaSO_4}$$

$$\times \frac{1 \text{ mmol } Al_2(SO_4)_3}{3 \text{ mmol } BaSO_4} \times \frac{342.2 \text{ mg } Al_2(SO_4)_3}{\text{mmol } Al_2(SO_4)_3}$$

$$= 1.85 \times 10^3 \text{ mg } Al_2(SO_4)_3$$

EXAMPLE

How many mg of $AgNO_3$, fw 169.9, are required to precipitate as Ag_2CrO_4 all the CrO_4^{2-} in a solution containing 529 mg Na_2CrO_4, fw 162.0?

$$2Ag^+ + CrO_4^{2-} \rightarrow Ag_2CrO_4(s)$$

$$\text{Mass } AgNO_3 = 529 \text{ mg } Na_2CrO_4 \times \frac{1 \text{ mmol } Na_2CrO_4}{162.0 \text{ mg } Na_2CrO_4}$$

$$\times \frac{2 \text{ mmol } AgNO_3}{1 \text{ mmol } Na_2CrO_4}$$

$$\times \frac{169.9 \text{ mg } AgNO_3}{1 \text{ mmol } AgNO_3}$$

$$= 1110 \text{ mg } AgNO_3$$

Note how the units cancel to give the correct units of mg $AgNO_3$.

1.8   Solution Concentration

Concentration in Moles per Liter

Solution **concentration** refers to the quantity of **solute** dissolved per unit quantity of **solution** or **solvent**. Solution concentrations are expressed in many ways. The most common expression of solution concentration used in this text is **molarity**, M, defined as,

$$M = \frac{\text{moles of solute}}{\text{liter of solution}} = \frac{\text{millimoles of solute}}{\text{milliliter of solution}} \tag{1.14}$$

EXAMPLE

A total of 6.48 g of glucose, $C_6H_{12}O_6$, fw 180.2, is dissolved in 476 mL of solution. What is the molar concentration of glucose in the solution?

fw of glucose = 180.2

$$\text{Moles of glucose} = \frac{6.48 \text{ g}}{180.2 \text{ g/mol}} = 3.60 \times 10^{-2} \text{ mol}$$

$$M = \frac{3.60 \times 10^{-2} \text{ mol}}{0.476 \text{ L}} = 7.56 \times 10^{-2} \text{ mol/L, or } 7.56 \times 10^{-2} M$$

Alternatively,

$$M = \frac{36.0 \text{ mmol}}{476 \text{ mL}} = 7.56 \times 10^{-2} \text{ mmol/mL}$$

Thus the units of molar concentration are moles per liter, mol/L, or millimoles per milliliter, mmol/mL; the numerical value of the concentration is the same in both cases. These units are commonly abbreviated as M; for example, a solution containing 0.138 mol of a solute per liter of solution is said to be 0.138 M in concentration of that solute.

Analytical Concentration

It is important to understand that the **analytical concentration** or **nominal molarity** of a solute designates the number of moles of solute dissolved to make a liter of the solution (or millimoles to make a milliliter), but does not necessarily specify the equilibrium concentration of that species in solution. This is because many solutes dissociate, react with water (hydrolyze), or undergo other reactions in solution. For example, when 0.0500 mol of H_3PO_4

is dissolved in a solution made up to 1.00 L, the analytical concentration of H_3PO_4 is 0.0500 M—that is, $C_{H_3PO_4} = 0.0500\ M$. However, at this concentration H_3PO_4 is 31.2% dissociated to H^+ and $H_2PO_4^-$ according to the reaction

$$H_3PO_4 \rightleftharpoons H^+ + H_2PO_4^- \tag{1.15}$$

The **equilibrium** or **species** concentrations of H_3PO_4, $H_2PO_4^-$, and H^+ in the solution are designated by the formulas of the species enclosed in brackets as follows:

$$[H_3PO_4] = 0.688 \times C_{H_3PO_4} = 0.688 \times 0.0500\ M = 0.0344\ M \tag{1.16}$$

Since H_3PO_4 is 31.2% dissociated, it is 68.8% (0.688) undissociated at equilibrium.

$$[H_2PO_4^-] = [H^+] = 0.312 \times 0.0500\ M = 0.0156\ M \tag{1.17}$$

Whenever a molar concentration is designated by a chemical formula in brackets, the concentration is meant to be that of the particular species so designated. The analytical concentration or nominal molarity is also known as the **formal concentration**, F. Thus, for the example just discussed, the formal concentration of H_3PO_4 is 0.0500 F.

EXAMPLE

Exactly 3.00 mg of acetic acid, $HC_2H_3O_2$ (HAc), fw 60.05, was dissolved in water and the solution brought to a volume of 250 mL. The acetic acid was 25.6% dissociated. Calculate the analytical concentration of acetic acid, designated C_{HAc}, and the species concentrations of undissociated HAc—Ac$^-$ and H^+.

$$\text{mmoles HAc added} = \frac{3.00\ \text{mg}}{60.05\ \text{mg/mmol}} = 5.00 \times 10^{-2}\ \text{mmol HAc}$$

$$C_{HAc} = \frac{\text{mmol HAc added}}{\text{mL solution}} = \frac{5.00 \times 10^{-2}\ \text{mmol}}{250\ \text{mL}}$$

$$= 2.00 \times 10^{-4}\ \text{mmol/mL}$$

$$HAc \rightleftharpoons H^+ + Ac^-$$

(The fraction of HAc dissociated is 0.256, that undissociated is 0.744.)

$$[HAc] = 0.744 \times 2.00 \times 10^{-4}\ \text{mmol/mL} = 1.49 \times 10^{-4}\ \text{mmol/mL}$$

$$[Ac^-] = [H^+] = 0.256 \times 2.00 \times 10^{-4}\ \text{mmol/mL}$$

$$= 5.1 \times 10^{-5}\ \text{mmol/mL}$$

EXAMPLE How many grams of Na_2CO_3, fw 106.0, should be taken to make 250 mL of a solution 0.0600 M in Na^+ ion?

$$\text{mmoles } Na^+ = 0.0600 \text{ mmol/mL} \times 250 \text{ mL} = 15 \text{ mmol}$$

$$\text{Moles } Na^+ = \frac{1 \text{ mol}}{1000 \text{ mmol}} \times 15 \text{ mmol} = 0.015 \text{ mol}$$

Each mole of Na_2CO_3 yields 2 moles of Na^+, therefore,

$$\text{Moles } Na_2CO_3 = \frac{1 \text{ mol } Na_2CO_3}{2 \text{ mol } Na^+} \times 0.015 \text{ mol } Na^+ = 7.50 \times 10^{-3} \text{ mol}$$

$$\text{Mass } Na_2CO_3 = 7.50 \times 10^{-3} \text{ mol} \times 106.0 \text{ g/mol} = 0.795 \text{ g}$$

Normal Concentrations and Titer

Later in the text use will be made of concentrations expressed as **normality** or **normal concentration**. Such concentrations are expressed in **equivalents** of solute per liter, where an equivalent is a mass of a chemical species that will undergo a specified amount of a particular reaction. For example, an equivalent of NaOH will neutralize just as much acid as an equivalent of $Ca(OH)_2$. However, there are two moles (two equivalents) of OH^- ion in one mole of $Ca(OH)_2$, but only one mole of OH^- (one equivalent) per mole of NaOH. Therefore, the normality of a $Ca(OH)_2$ solution, in terms of its acid-neutralizing capacity, is twice its molarity, whereas the normality of a solution of NaOH is the same as its molarity. Normality is defined for specific types of reactions in later chapters.

Titer refers to a specified mass of a substance that will react with a unit volume of a solution. Thus 1 mL of an acid solution with a $Ca(OH)_2$ titer of 2.0 mg/mL will neutralize 2.0 mg of $Ca(OH)_2$.

Other Concentration Units

Concentrations, especially for some solutions sold commercially, may be expressed as **percentage concentration**. Such a concentration must be specified exactly to know what it means. A **weight percent** (w/w) is given by

$$\text{Weight percent} = \frac{\text{weight of solute}}{\text{weight of solution}} \times 100 \tag{1.18}$$

a **volume percent** (v/v) is expressed as

$$\text{Volume percent} = \frac{\text{volume of solute}}{\text{volume of solution}} \times 100 \qquad (1.19)$$

and the formula for a **weight-volume percent** (w/v) is the following:

$$\text{Weight-volume percent} = \frac{\text{weight of solute, g}}{\text{volume of solution, mL}} \times 100 \qquad (1.20)$$

EXAMPLE

Commercial nitric acid, HNO_3, is 69.0 weight percent (w/w) HNO_3. What is the molarity of HNO_3 in such a solution, given that its density is 1.409 g/mL?

$$\text{Mass of 1 L} = 1\ \text{L} \times 1000\ \text{mL/L} \times 1.409\ \text{g/mL} = 1409\ \text{g}$$

$$\text{Mass } HNO_3 \text{ in 1 L} = 1409\ \text{g solution} \times 0.690\ \frac{\text{grams of } HNO_3}{\text{gram of solution}}$$

$$= 972\ \text{g } HNO_3$$

$$\text{Moles } HNO_3 \text{ in 1 L} = \frac{972\ \text{g } HNO_3}{63.0\ \text{g } HNO_3/\text{mol}} = 15.4\ \text{mol}$$

Therefore the molar concentration of HNO_3 in the commercial solution is 15.4 mol/L.

EXAMPLE

Commercial concentrated solutions of ammonia, NH_3, fw 17.0, are 14.8 M in NH_3 and have a density of 0.898 g/mL. What is the weight percent of NH_3 in such a solution?

There are 14.8 mol NH_3 in a liter of the solution

$$\frac{\text{Mass of } NH_3}{\text{Liter of solution}} = 14.8\ \text{mol} \times 17.0\ \text{g/mol} = 252\ \text{g}$$

The mass of 1 L of the solution is given by

$$\text{Mass of 1 L of solution} = 0.898\ \text{g/mL} \times 1000\ \text{mL} = 898\ \text{g}$$

$$\frac{\text{Weight percent}}{\text{concentration}} = \frac{\text{weight } NH_3 \text{ in 1 L of solution}}{\text{weight of 1 L of solution}} \times 100$$

$$\frac{\text{Weight percent}}{\text{concentration}} = \frac{252\ \text{g}}{898\ \text{g}} \times 100 = 28.1\%\ NH_3$$

Chemical analysis is being used increasingly for the determination of very low concentrations of solutes, particularly pollutants in water. Such analysis is commonly called **trace analysis**. The concentrations of the solutes in such solutions and in the standard solutions used to calibrate the instruments are commonly expressed in the following terms:

$$\text{Parts per million, ppm} = \frac{\text{weight of solute}}{\text{weight of solution}} \times 10^6 \tag{1.21}$$

$$\text{mg/L} = \frac{\text{milligrams of solute}}{\text{liters of solution}} \tag{1.22}$$

$$\text{Parts per billion, ppb} = \frac{\text{weight of solute}}{\text{weight of solution}} \times 10^9 \tag{1.23}$$

$$\mu\text{g/L} = \frac{\text{micrograms of solute}}{\text{liters of solution}} \tag{1.24}$$

Since a liter of a dilute solution (density 1.0 g/mL) weighs essentially 1000 g (10^6 mg or 10^9 μg), a concentration in parts per million is essentially that of the same solution in milligrams per liter, and parts per billion is essentially the same as micrograms per liter. Thus for dilute solutions 1 ppm is 1 mg/L and 1 ppb is 1 μg/L.

EXAMPLE

What volume of 1.00×10^{-2} M $Pb(NO_3)_3$, atomic weight (aw) 207.2, should be diluted to 1.00 L to give a standard solution for the atomic absorption analysis of lead containing 10.0 mg/L Pb?

The number of millimoles of Pb in 1.00 L (1000 mL) of 10.0 mg/L Pb solution is

$$\text{mmoles Pb} = \frac{10.0 \text{ mg}/1000 \text{ mL}}{207.2 \text{ mg/mmol}} \times 1000 \text{ mL} = 0.0483 \text{ mmol}$$

The number of milliliters of 1.00×10^{-2} M $Pb(NO_3)_3$ solution that must be diluted to 1000 mL to obtain 1000 mL of a solution containing 0.0483 mmol of Pb is given by

$$\text{Milliliters of } 1.00 \times 10^{-2} M \text{ } Pb(NO_3)_3 = \frac{0.0483 \text{ mmol}}{1.00 \times 10^{-2} \text{ mmol/mL}}$$

$$= 4.83 \text{ mL}$$

Check: $4.83 \text{ mL} \times 1.00 \times 10^{-2} \text{ mmol/mL} \times 207.2 \text{ mg/mmol} = 10.0 \text{ mg Pb}$

It is useful to keep in mind that whenever solution concentrations are expressed in quantity (mass, moles) per unit volume, the equation applying to dilution of the solution is,

$$C_{\text{before}} V_{\text{before}} = C_{\text{after}} V_{\text{after}} \tag{1.25}$$

where C and V are concentration and volume, respectively, and the subscripts apply to before and after dilution. This equation can be used, for example, to calculate the volume of a known solution required for dilution to give a specified volume of a less concentrated solution.

Much of the theory of quantitative analysis has to do with solution equilibrium, which in general terms involves the extent to which acid–base, solubilization (precipitation), complexation, or oxidation-reduction reactions proceed in the direction written. This is important, for example, in determining whether a precipitate is insoluble enough to be used for gravimetric analysis, or whether a reaction is sufficiently complete to form the basis of a titration analysis. Other examples will be cited in later chapters.

A generalized equilibrium reaction may be written in the form,

$$a\text{A} + b\text{B} \rightleftharpoons c\text{C} + d\text{D} \tag{1.26}$$

where the double arrow indicates a **reversible reaction** in which there is no net change in the relative quantities of reactants and products. The equilibrium of such a reaction is expressed by an *equilibrium constant expression*,

$$\frac{[\text{C}]^c[\text{D}]^d}{[\text{A}]^a[\text{B}]^b} = K \tag{1.27}$$

where K is a numerical quantity designated as the **equilibrium constant**.

A reaction may approach a true state of equilibrium from either side; that is, the reaction is reversible. For Reaction 1.26, a state of equilibrium may be attained by mixing A and B, which will react to form some C and D, or by mixing C and D to form A and B. In both cases, at equilibrium the concentrations of A, B, C, and D will be such that the equilibrium constant expression has the value K.

As noted from the **principle of Le Châtelier**, covered in beginning chemistry courses, a stress placed upon a system in equilibrium will shift the equilibrium to relieve the stress. This phenomenon is most commonly used or observed in analytical chemistry by changing the concentration of one of the reaction participants. Thus, if the concentration of product C is increased in Reaction 1.26, the reaction will shift to the left; if A is increased, it will shift to the right. These are examples of the **mass action effect**.

In most cases, this text uses concentrations and pressures in equilibrium constant expressions. This is an approximation in which K is not exactly constant with varying concentrations and pressures, so that it is called an *approximate* equilibrium constant and applies only to limited conditions. An exact form, the **thermodynamic equilibrium constant**, is derived from thermodynamic data, makes use of *activities* in place of concentrations and is applicable over a wide concentration range at constant temperature (K varies with temperature). Activities are discussed in more detail in Chapter 4. It may be noted here that the activity of a solute in solution is a measure of its effectiveness in its interactions with other things, such as other solutes and electrodes (Chapter 13) in solution. The activity of a species X is commonly denoted by the symbol a_X. Activities approach concentrations at low values of concentrations. For ions, activities are decreased from concentrations by the presence of electrolytes (dissolved salts) in solution. Activities of ions from ionic solutes increase with the concentration of the ionic solute. In the absence of other electrolytes, the increase is not linear above about 1×10^{-3} M, the activity increasing less than the concentration. As mentioned, thermodynamic equilibrium constants are expressed in terms of activities. Thus for the reaction in Equation 1.26, the thermodynamic equilibrium constant is expressed as

$$\frac{a_C^c a_D^d}{a_A^a a_B^b} = K \tag{1.28}$$

Units in Equilibrium Constant Expressions

Although units of concentration and pressure may be carried through equilibrium constant expression calculations, they become rather cumbersome in expressions involving a number of terms. The concentrations and pressures used in these expressions may be considered as concentration or pressure ratios. In such a case the concentration or pressure of the species is divided by the **standard state** concentration or pressure. In such a calculation, the standard state concentration of a solute is regarded as exactly 1 M, and the standard state pressure of a gas is defined as exactly 1 atmosphere (atm). Thus for the value of the concentration of a dissolved species A, the following concentration ratio is substituted into an equilibrium constant expression:

$$[A] = \frac{\text{concentration of A, } \text{mol}/\text{L}}{1.00 \ \text{mol}/\text{L}} \tag{1.29}$$

For example, if the molar concentration of A is 1.00×10^{-3} mol/L, the number 1.00×10^{-3} is commonly used in the equilibrium constant expression without units. The following can be used for the pressure of a gas, P_g, substituted into an equilibrium constant expression,

$$P_g = \frac{\text{pressure of gas, } \text{atm}}{1.00 \ \text{atm}} \tag{1.30}$$

As shown for both cases above, the units cancel so that the values of concentrations or pressures substituted into an equilibrium constant expression are numerically equal to the respective concentrations and pressures, but without units.

The concentrations (more exactly, the activities) of solids and pure liquids are constant and have been assigned the value of exactly 1; therefore they do not appear in equilibrium constant expressions. Furthermore, the activity of water is taken as 1 for equilibrium reactions occurring in aqueous solution. Whenever water or solids are participants in a chemical reaction, their formulas do appear in the reaction, even though they are not included in the equilibrium constant expression describing the reaction equilibrium.

Acid-Base Equilibria

Acids in water are sources of the hydrogen ion, H^+. This species is often designated as H_3O^+, called the hydronium ion, to emphasize the fact that H^+ in water is always bound to water molecules (see the discussion at the end of Section 4.1).

The most fundamental acid-base reaction in water is the dissociation of water itself,

$$H_2O \rightleftarrows H^+ + OH^-$$ (1.31)

where OH^- is the hydroxide ion. The equilibrium constant expression for this reaction is

$$K_w = [H^+][OH^-]$$ (1.32)

in which K_w is called the **ion-product constant** for water. It has a value at $25°$ C of 1.01×10^{-14}, usually rounded off to 1.00×10^{-14}.

Acid-base equilibrium calculations typically involve the calculation of the degree of dissociation of a weak acid, such as acetic acid, HAc

$$HAc \rightleftarrows H^+ + Ac^- \qquad K_a = \frac{[H^+][Ac^-]}{[HAc]}$$ (1.33)

Such calculations are covered in Chapter 7.

Since H^+ concentration in aqueous solution may vary over many orders of magnitude, it is convenient to express it in logarithmic form as **pH**, the negative log of $[H^+]$ (or, more rigorously, the negative log of hydrogen ion activity). Similarly, $pOH = -\log [OH^-]$.

Solubility Equilibria

Solubility equilibria deal with reactions such as

$$AgCl(s) \rightleftarrows Ag^+ + Cl^-$$ (1.34)

in which one of the participants is an insoluble (actually, slightly soluble) salt. The equilibrium constant expression for this reaction is,

$$K_{sp} = [Ag^+][Cl^-]$$ (1.35)

where K_{sp} is the **solubility product**. Note that AgCl does not appear in the equilibrium constant expression because it is a solid with an activity defined as exactly 1. Solubility equilibria are discussed in Chapter 7.

Complexation

As discussed in Chapter 10, metal **complexes** or **complex ions** are formed by the reaction of a metal ion in solution with a **complexing agent** or **ligand**, both of which are capable of independent existence in solution. One of the simplest examples of complexation that has been used for chemical analysis is the complex formed by the reaction,

$$\underset{\substack{\text{Ferric metal}\\\text{ion}}}{Fe^{3+}} \quad + \quad \underset{\substack{\text{Thiocyanate}\\\text{ligand}}}{SCN^-} \quad \rightleftarrows \quad \underset{\substack{\text{Complex ion}\\\text{(red color)}}}{FeSCN^{2+}} \tag{1.36}$$

The intensity of the red $FeSCN^{2+}$ ion color in solution can be used as a measure of ferric ion concentration. The equilibrium constant for a complexation reaction is expressed in terms of a **formation constant**, K_f, as shown below for the preceding reaction:

$$K_f = \frac{[FeSCN^{2+}]}{[Fe^{3+}][SCN^-]} = 1.07 \times 10^3 \tag{1.37}$$

Oxidation-Reduction Equilibria

Oxidation-reduction reactions are those involving transfers of electrons between chemical species. An oxidation-reduction reaction that constituted the basis for one of the first means of determining iron in solution involves the following reaction between Fe^{2+} ion and permanganate ion, MnO_4^-,

$$MnO_4^- + 5Fe^{2+} + 8H^+ \rightleftarrows Mn^{2+} + 5Fe^{3+} + 4H_2O \tag{1.38}$$

The equilibrium of this reaction lies very far to the right, so that a stoichiometric oxidation of ferrous ion by permanganate ion occurs.

The equilibrium expression for Reaction 1.38 is the following:

$$K = \frac{[Mn^{2+}][Fe^{3+}]^5}{[MnO_4^-][Fe^{2+}]^5[H^+]^8} \tag{1.39}$$

The value of K at 25° C is 3×10^{62}. This value is calculated from the Nernst equation, as explained in Chapter 11.

Distribution between Phases

One of the more important operations in analytical chemistry is the separation of an analyte or an interfering substance into a phase that is physically distinct and separable from the material (usually an aqueous solution) that the analyte is in. For example, an organic-soluble pesticide in a water sample may be extracted into cyclohexane (see structure in Figure 1.3) for subsequent gas chromatographic determination (Chapter 19). The distribution of a solute X between an aqueous phase and an organic phase may be represented by the equilibrium reaction,

$$X_{aq} \rightleftarrows X_{org} \tag{1.40}$$

where the subscripts aq and org represent the aqueous and organic phases, respectively. This kind of equilibrium is described by the **distribution law** expressed by a **distribution coefficient** or **partition coefficient** in the following form:

$$K_d = \frac{[X]_{org}}{[X]_{aq}}$$

Ion Exchange

The transfer of ions between water and another phase capable of chemically binding ions is called **ion exchange**. Most commonly an **ion-exchange resin** is used that consists of a synthetic organic polymer to which are bonded either negatively or positively charged functional groups capable of bonding reversibly with cations or anions, respectively. (See Section 1.6 for definitions of polymer and functional group.)

The equilibrium of ions between aqueous solution and an ion-exchange resin may be expressed by a reaction such as,

$$Na^+(aq) + \{Res\}^-H^+ \rightleftarrows H^+(aq) + \{Res\}^-Na^+ \tag{1.41}$$

where $\{Res\}^-$ represents one ion-exchanging functional group on a cation-exchange resin. The equilibrium of this reaction is given by the following equilibrium constant expression:

$$K = \frac{[Na^+]_{res}[H^+]_{aq}}{[Na^+]_{aq}[H^+]_{res}} \tag{1.42}$$

In this expression $[H^+]_{aq}$ and $[Na^+]_{aq}$ are the molar concentrations of these ions in aqueous solutions, whereas $[H^+]_{res}$ and $[Na^+]_{res}$ are concentrations of hydrogen and sodium ions, respectively, on the cation exchanger in units of moles per mass of ion exchanger.

Programmed Summary of Chapter 1

The major terms and concepts introduced in this chapter are contained in this summary in a programmed format. To derive the most benefit from the summary, you should fill in the blanks for each question, then check the answers at the end of the book to see if your choices are correct.

The determination of the identities of sample constituents is called (1) _____, whereas the determination of the amounts or percentages of substances present is (2) _____. The first step in the overall chemical analysis process is (3) _____. That portion of material upon which a chemical analysis is performed is called the (4) _____ and a specific constituent determined in it is called (5) _____. Two major purposes of sample processing are (6) _____ and (7) _____. A chemical analysis in which the sample is retained intact is called (8) _____. Some terms that are used to describe chemical analyses that involve primarily chemical reactions and simple measurements of mass or volume are (9) _____, (10) _____, and (11) _____. Analytical methods that involve the measurement of parameters other than mass or volume are termed (12) _____ or (13) _____ methods. A chemical determination that is based upon the measurement of the mass of a product produced from the chemical reaction of an analyte is called a (14) _____ determination, whereas one that involves the measurement of the volume of a reagent required to react with an analyte is called a (15) _____ determination. A plot of a measured parameter versus the known concentration or quantity of a substance to be determined is called a (16) _____. Solutions used to prepare such a plot are called (17) _____. The process of determining the magnitude of a measurable parameter as a function of known analyte concentration is called (18) _____. The desirable properties of a primary standard chemical are (19) _____, (20) _____, (21) _____, and (22) _____. A gram is a metric unit of (23) _____. The basic metric unit of volume is the (24) _____, and that of temperature is the (25) _____. A metric unit that is exactly 1000 times that of a base unit is prefixed by (26) _____, 0.01 is prefixed by (27) _____, and 0.001 is prefixed by (28) _____. In the SI system of measurements, the base unit for quantity of substance is the (29) _____; on the scale of chemical analysis it is usually more convenient to work with (30) _____. The branch of chemistry dealing with the vast majority of known carbon-containing compounds is called (31) _____. The thing in common among benzene, naphthalene, phenol, nitrobenzene, α-nitroso-β-naphthol, and benzene-sulfonic acid is that they are (32) _____ compounds. A group constituting a major portion of an organic molecule is called a (33) _____, and can be designated in general by the letter R. In organic chemistry specific arrangements of atoms involving multiple bonds or at least one atom of an element other than carbon are called (34) _____. When a large number of organic molecules with unsaturated bonds bond together to form an aggregate of these molecules, the product is called a (35) _____. Stoichiometry deals with (36) _____. The units milli-moles of solute per milliliter of solution express concentration in terms of (37) _____. When 0.100 mol of H_3PO_4 is dissolved and diluted to 1 L, it is not correct to say that $[H_3PO_4] = 0.100\ M$, because a large fraction of the H_3PO_4 is dissociated. Instead, it may be said that the (38) _____ of H_3PO_4 is 0.100 M. Normality, or normal concentration is expressed in units of (39) _____ per liter. A concentration expressed as a specified mass of a substance that will react with a unit volume of a solution is known as the solution (40) _____. Units used to express very low concentrations of solutes, such as those of some pollutants, are (41) _____, essentially equal to (42) _____, or ppb, essentially equal to (43) _____. As it refers to the generalized chemical reaction $aA + bB \rightleftarrows cC + dD$, the expression

$$\frac{[C]^c[D]^d}{[A]^a[B]^b} = K$$

is known as (44) _____, and K is the (45) _____. Often concentrations and pressures are used in the preceding expression, so that K is not exactly constant, even at constant temperature; when activities are used instead, K is constant and is called a (46) _____. The equilibrium constant expression for the reaction $H_2O \rightleftarrows H^+ + OH^-$ is (47) _____, and is known as the ion product for water. The negative log of the hydrogen ion concentration is called (48) _____. For the slight dissolution in water of the sparingly soluble silver salt, silver chromate, $Ag_2CrO_4(s) \rightleftarrows 2Ag^+ + CrO_4^{2-}$, the equilibrium constant expression is (49) _____ and is known as the (50) _____. The reaction $Cu^{2+} + NH_3 \rightleftarrows CuNH_3^+$ is an example of (51) _____, for which the equilibrium constant expression is (52) _____, known as a (53) _____. A chemical reaction that involves an exchange of electrons between species is called (54) _____ reaction. An equilibrium constant that describes the relative concentrations of a species in two different phases is called the (55) _____. Two such phases other than water might be (56) _____.

Questions

1. What is wrong with saying that "Lead was analyzed in a blood sample of a person suspected of suffering from lead poisoning"?
2. What is the general term applied to a constituent determined in a sample?
3. Skim through the introduction to instrumental neutron activation analysis in Chapter 23, and explain why this technique is generally considered nondestructive.
4. Match the following:
   a. Representative sample
   b. Sample processing
   c. Interferences
   d. Determination

   1. Must be removed or chemically "tied up" so as not to affect the determination.
   2. Actual process of finding out how much of an analyte is present.
   3. Usually involves putting into solution.
   4. Ideally should have a composition the same as that of the whole mass of material whose composition it is desired to know.

5. Mass and volume are the two parameters most generally measured in a chemical or wet chemical method of analysis. What two broad categories of analysis do these two measurements represent?

6. For a particular method of analysis, the concentration of analyte C_A is a function of a physical measurement of magnitude X which ranges from 0 to 10 on the instrument used for the calibration. The calibration relationship is $C_A = X + X^{1/3}$. Plot the general shape of the calibration curve.

7. From the examples cited in Figure 1.3, Figure 1.5, and Table 1.2, give the structures of 2-methylpropane and 2-nitronaphthalene.

8. Using R and ϕ to designate alkyl and aromatic groups, respectively, give general formulas of alcohols, phenols, ethers, aldehydes, ketones, carboxylic acids, amines, amides, nitriles, nitro compounds, nitroso compounds, sulfonic acids, and organohalides.

9. A sample consisting of hydrated copper (II) orthophosphate, $Cu_3(PO_4)_2 \cdot 3H_2O$, and other materials that contained neither phosphorus nor copper was dissolved and the copper deposited on a platinum electrode by an electrical current. Give the equation from which the percentage of phosphorus in the sample can be calculated from the mass of copper deposited, the mass of the sample, the appropriate atomic and formula weights, and other needed parameters.

10. Write the equilibrium constant expression for K for the reaction

$$Cr_2O_7^{2-} + 6Fe^{2+} + 14H^+ \rightleftharpoons 2Cr^{3+} + 6Fe^{3+} + 7H_2O$$

Problems

1. The heat content (heat released per unit mass) of coal in a 10,000 metric ton unit coal train (1000 kg per metric ton) was determined. A representative sample was obtained by "grab sampling" 64 approximately 1-kg portions of coal from the train. These were reduced to three 1-g samples, for which the heat content was determined and the average taken. What percentage of the total coal in the train was taken for analysis?

2. In an analysis similar to that described in Problem 1, three 0.8-g samples were obtained from a single coal car containing 96 metric tons of coal. What percentage of the coal was taken for analysis?

3. Three 1-g samples were taken from a pile of 960 metric tons of coal scraped from the surface of a power plant coal pile. The coal was suspected of having weathered, with a resulting loss of heat content. What percentage of this pile of coal was sampled?

4. A chloride-selective electrode calibrated with the calibration curve shown in Figure 1.2 gave a reading of 50 mV in an irrigation water sample suspected of saline contamination. Estimate the molar concentration of $[Cl^-]$ in the water sample.

5. Another sample was analyzed according to the procedure outlined in Problem 4 giving a reading of 140 mV. What was $[Cl^-]$?

6. Using the procedure described in Problem 4, what is the value of $[Cl^-]$ corresponding to a potential of 90 mV?

7. How many millimoles are present in 0.836 g of potassium hydrogen phthalate (KHP), $KHC_8H_4O_4$, fw 204.2.

8. How many millimoles are present in

0.946 g of benzoic acid, $HC_7H_5O_2$, fw 122.1?

9. How many millimoles are there in 0.358 g of acetic acid, $HC_2H_3O_2$, fw 60.05?

10. What is the mass of solid $BaSO_4$, fw 233.4, that may be precipitated from a solution containing 1.538 g of $Fe_2(SO_4)_3$, fw 399.9, by the addition of excess $BaCl_2$? First write the balanced precipitation reaction.

11. What mass of Fe_2O_3 can be obtained by the reaction of 0.948 g of pure Fe, aw 55.8, with oxygen to form Fe_2O_3, fw 159.7?

12. Oxidation of Al, aw 27.0, produced 1.382 g of Al_2O_3, fw 102.0. What was the original mass of Al present?

13. The chemical analysis of a lake-water sample for dissolved O_2 showed 0.668 mg O_2 dissolved in a 75.0-mL sample. What was the molar concentration of O_2 in the lake water?

14. As an ingredient of an intravenous feeding solution, 0.884 g of glucose, $C_6H_{12}O_6$, fw 180, was dissolved in 1.00 dL (see Table 1.1) of solution. What was the molarity of the solution?

15. An antifreeze solution was prepared by dissolving 3.00 kg of ethylene glycol, $C_2H_6O_2$, fw 62.1, to make a total of 9.50 daL (see Table 1.1) of solution. What was the molar concentration of ethylene glycol in the solution?

16. A mass of 3.00 mg of acetic acid ($HC_2H_3O_2$, also abbreviated HAc), fw 60.05, was dissolved in water to make 225 mL of acetic acid solution, which at this concentration is 24.5% dissociated. Calculate the analytical concentration of acetic acid (C_{HAc}), and the equilibrium concentrations of HAc, H^+, and Ac^-.

17. A total of 2.28 mg of NH_3 gas, fw 17.0, was bubbled into 500 mL of water making a solution with a total volume of 500 mL. What is the analytical concentration of ammonia, C_{NH_3}, and the equilibrium concentrations of [NH_3], [NH_4^+], and [OH^-]? The reaction of NH_3 with water is $NH_3 + H_2O \rightleftharpoons NH_4^+ + OH^-$, and at a concentration of 2.28 mg of NH_3 in 500 mL of water, this reaction occurs to the extent that 22.6% of the NH_3 is converted to NH_4^+.

18. A solution was prepared by adding 14.9 mg of H_3PO_4, fw 98.0, to water and diluting the solution to a volume of 500 mL. At this concentration, 62.0% of the H_3PO_4 is dissociated according to the reaction $H_3PO_4 \rightleftharpoons H^+ + H_2PO_4^-$. Calculate the analytical concentration of H_3PO_4, $C_{H_3PO_4}$, as well as the equilibrium concentrations of [H^+], [$H_2PO_4^-$], and [H_3PO_4].

19. A standard HCl solution was prepared to analyze a stream of sodium carbonate solution, Na_2CO_3, added continuously to water as part of a municipal water-treatment process. When standardized, it was found that 42.73 mL of the acid neutralized 0.7691 g of pure Na_2CO_3 dissolved in water by the reaction,

$$Na_2CO_3(aq) + 2HCl(aq) \rightarrow$$
$$2NaCl(aq) + H_2O + CO_2(g).$$

What was the Na_2CO_3 titer of the standard HCl solution?

20. A 25.00-mL sample of the water-treatment stream containing Na_2CO_3 required 32.60 mL of the standard HCl for neutralization (Problem 19). What was the Na_2CO_3 content of the water-treatment stream in milligrams of Na_2CO_3 per liter?

21. A solution is 32.0 weight percent (w/w) in compound X, fw 182, and has a density of 1.26 g/mL. What is the molar concentration of X in the solution?

22. A solution of compound Y, fw 206, is 28.0 weight percent (w/w) Y, and has a

density of 1.22 g/mL. What is the molar concentration of Y in the solution?

23. A solution of compound Z, fw 156, is 6.00 w/w% Z and has a density of 1.08 g/mL. What is the molar concentration of Z in the solution?

24. A lead nitrate, $Pb(NO_3)_2$, standard solution contains 12.0 mg/L Pb, aw 207. What is the molar concentration of NO_3^- in the solution?

25. A basic (high-pH) wastewater solution contains 36.2 mg/L NH_3, fw 17. What is the molar concentration of the NH_3?

26. As part of a study of the physiology of joggers, perspiration from a jogger was collected and found to be 64.8 mg/L NaCl. What was the molar concentration of NaCl in the perspiration?

27. An alkaline cleaning solution has $[OH^-] = 3.24 \times 10^{-2} M$. What are the values of $[H^+]$ and pH?

28. Fog droplets collected from a two-day "acid-fog" incident in Los Angeles in December 1982 had a pH of 1.7. What was the value of $[H^+]$ in these droplets?

29. A total of 0.327 g of NaOH was dissolved in 500 mL of water. What was the pH of the solution?

30. The solubility product K_{sp} of AgI is 1.00×10^{-16}. What is the solubility S of AgI in pure water?

31. What is the solubility S of AgI in a solution initially $2.56 \times 10^{-4} M$ in Ag^+?

32. Look up the solubility product of AgCl. What is the solubility of AgCl in a solution initially $2.46 \times 10^{-3} M$ in Ag^+?

33. The formation constant of the complex ion $FeSCN^{2+}$ is 1.07×10^3 (see Equation 1.37). What is $[FeSCN^{2+}]$ in a solution that at equilibrium has $[Fe^{3+}] = 1.26 \times 10^{-4} M$ and $[SCN^-] = 1.00 \times 10^{-2} M$?

34. Referring to the previous problem, a solution at equilibrium has $[FeSCN^{2+}] = 2.52 \times 10^{-3} M$ and $[SCN^-] = 1.27 \times 10^{-2} M$. What is $[Fe^{3+}]$?

35. A solution of $FeSCN^{2+}$ was formed by mixing an equal number of moles of Fe^{3+} and SCN^-. At equilibrium $[FeSCN^{2+}]$ was $2.15 \times 10^{-3} M$. What were the values of $[Fe^{3+}]$ and $[SCN^+]$ at equilibrium?

36. A total of 32.6 mmol of compound X was equilibrated between 25 mL of benzene and 75 mL of water. The benzene phase was isolated and the solvent evaporated leaving a residue of 23.2 mmol of X. Assuming the remainder of the X stayed in the aqueous phase, what is the distribution coefficient K_d?

37. A 50-mL solution of compound Y in water was equilibrated with 75 mL of chloroform. At equilibrium $[Y]_{aq} = 7.53 \times 10^{-4} M$ and $[Y]_{org} = 9.08 \times 10^{-2} M$. What was the value of the distribution coefficient K_d?

38. From the answer to Problem 37, if 36.0 mmol of compound Y are equilibrated with 100 mL of water and 75 mL of chloroform, which is immiscible with water, how many millimoles of Y are in the water solution and how many are in chloroform solution at equilibrium?

2

An Introduction
to the Major Techniques
of Quantitative Analysis

L E A R N I N G   G O A L S

1. The essentials of mass measurement or weighing to the degree of accuracy required in the analytical laboratory.
2. The basic components and fundamentals of operation of the analytical balance.
3. How to dry solid samples, maintain them in a dry state in a desiccator, and weigh out portions of them for chemical analysis.
4. The processes of filtration in the analytical laboratory.
5. The fundamental procedures for drying, ashing, and igniting a precipitate.
6. The measurement of solution volumes by pipet, buret, and volumetric flask.
7. Calibration of volumetric glassware.

2.1 The Importance of Good Laboratory Technique

This course is almost certainly being taken in conjunction with laboratory work, and a substantial part of the course grade probably depends upon obtaining accurate results in the analysis of unknown samples. Quantitative analysis is often the first course encountered by a student in which a good grade *depends upon doing things right in the laboratory*. A properly taught, rigorous quantitative analysis laboratory course is often the best preparation that a student may have for the real world of chemistry in industry, medicine, agriculture, and all the other fields to which chemistry applies.

It is important, therefore, that the student learn and observe *good laboratory technique* from the beginning. Good laboratory technique is simply a question of observing the rules below:

1. Follow directions exactly.
2. Know why the directions are being followed—that is, the reasons for each step of laboratory procedure.
3. Keep a laboratory notebook in which all directions, results, and calculations are carefully recorded.

4. Be well organized
   a. in your thinking
   b. in your notebook
   c. in the laboratory operations
5. Know how to use each item of laboratory equipment and treat it with great care.
6. Know and observe the proper steps for cleaning laboratory glassware.
7. Follow all safety regulations; *proper protective eyewear must be worn at all times in the laboratory.*
8. Know which laboratory chemicals are hazardous; use them in accordance with safety regulations and discard them in accordance with hazardous waste regulations.

There are a number of key items of apparatus and essential laboratory procedures that are used in the quantitative analysis laboratory. These are discussed in the remainder of the chapter.

2.2 Mass Measurements

Mass and Weight

The terms mass and weight are often used interchangeably in analytical chemistry, but the difference between these two terms should be known. **Mass** is a measure of the quantity of matter; the mass of a specific object does not vary. **Weight** is the force of attraction which other objects, principally the earth, exert upon an object. It varies somewhat with latitude; an object weighs more in Anchorage than in Costa Rica because of a slight flattening of the earth at the poles compared to the equator. An object also weighs slightly less in Laramie, Wyoming, at an altitude of about 2400 m than it does in New York City, near sea level. The relationship between mass M and weight W is

$$W = Mg \tag{2.1}$$

where g is the gravitational constant.

Principles of the Balance

The determination of mass, commonly called *weighing*, is the most fundamental operation undertaken in the analytical laboratory. This determination is carried out by comparing the mass of the unknown object to objects with known masses, commonly called *weights*. Since the operation is carried out in the same place, g is the same for both the unknown and standards, so that if both have the same weight, they both have the same mass. This may be understood by examination of a free-swinging, equal-arm balance as shown in

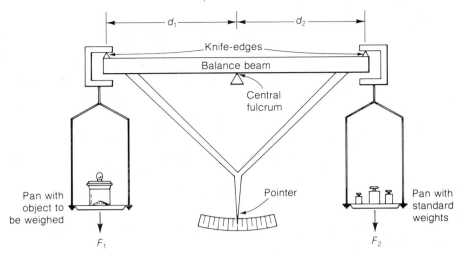

FIGURE 2.1   A simple double-pan balance arrangement

Figure 2.1. The central component of such a balance is a rigid beam suspended in the middle by a fulcrum or knife-edge. At either end are balance pans suspended from the beam by knife-edges that are equidistant from the fulcrum — that is, $d_1 = d_2$. The pointer attached to the beam indicates when the balance is on–center. If an object of mass M_1 is placed on the left pan, the pointer will deflect to the right because of a force F_1 exerted downward by the object. The pointer can be returned to the midpoint by putting standard weights of mass M_2 on the right pan, exerting a force of F_2. At this point, according to the principle of the lever,

$$F_1 \times d_1 = F_2 \times d_2 \tag{2.2}$$

Since $d_1 = d_2$,

$$F_1 = F_2 \quad \text{at balance} \tag{2.3}$$

These forces are due to gravity's attraction for the object being weighed and the standard weights, leading to the relationships

$$F_1 = M_1 g \tag{2.4}$$

$$F_2 = M_2 g \tag{2.5}$$

Therefore,

$$\frac{F_1}{F_2} = \frac{M_1 g}{M_2 g} = \frac{M_1}{M_2} \tag{2.6}$$

Noting the relationship in Equation 2.3, at balance

$$M_1 = M_2 \tag{2.7}$$

In practice it is difficult to have exactly the right combination of weights to exactly match the weight of the object, and a small rider is moved along a graduated scale on the beam to compensate for the difference in weights. The position of the rider on the beam and a slight imbalance of the pointer from the midpoint can be used to measure the exact mass of the object being weighed.

The Analytical Balance

The normal practice of quantitative analysis requires the measurement of mass to 0.1 mg—that is, 0.0001 g. This is accomplished by a very sensitive balance called an *analytical balance*. These used to be **double-pan balances** of the general construction shown in Figure 2.1, but are now almost invariably **single-pan balances** such as that shown in Figure 2.2.

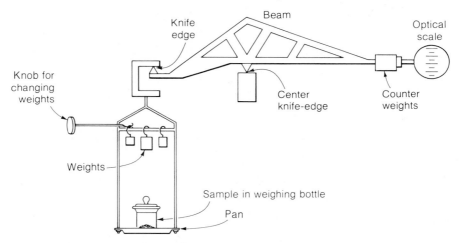

FIGURE 2.2 Single-pan balance (From Robert L. Pecsok (ed.), *Modern Chemical Technology*, Vol. 1, rev. ed., American Chemical Society, Washington, D.C., 1973, p. 82.)

With a single-pan balance, the weights are on the same side of the balance beam as the object to be weighed. A counterweight holds the beam in balance. There are only two knife-edges. When an object to be weighed is placed on the pan, the force on that end of the beam obviously is greater than that exerted by the counterweight. The balance is restored to equilibrium by *removing* weights with a mechanical weight lifter. As a result, the balance always has a *constant load* at any weight of object up to the limit of the balance. This is important because it enables the **sensitivity**—the smallest difference in mass detectable by the balance—to remain the same regardless of the mass on the pan, up to the capacity of the balance. The net effect is an increase in the accuracy of the balance. The typical laboratory balance can be used to determine the mass of a 160-g object to the nearest 0.0001 g.

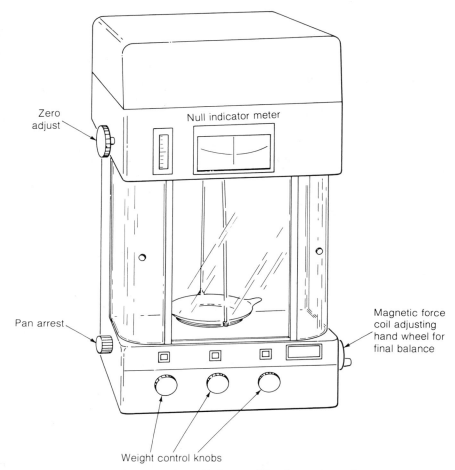

Zero adjust

Null indicator meter

Pan arrest

Magnetic force coil adjusting hand wheel for final balance

Weight control knobs

FIGURE 2.3   A single-pan analytical balance of the electronic torsion type.

Determining Mass with an Analytical Balance

Analytical balances come in a variety of styles that require different operating procedures. Typically, with some of the older single-pan models that many of us still use, the following steps are involved in weighing.

    1. With the pan empty and all weights dialed to zero, turn the *arrest* knob to the full release position so that the beam and pan rest freely on the knife-edges.

    2. Adjust the illuminated scale to zero by rotating the *zero adjustment* knob. This zeroes the balance with nothing on the pan.

3. After arresting the balance beam (locking it in place), place the object to be weighed in the center of the pan.

4. Partially release the beam with the arrest knob.

5. Starting with the heaviest weight scale, turn the *weight control knob* until the instruction to "remove weight" appears or until the illuminated scale shifts position; when this happens move the weight control knob back one notch. Repeat with the control knobs for the lighter weights.

6. Fully release the beam with the arrest knob and allow the illuminated scale to attain a position of equilibrium.

7. The total mass is the sum of the readings on the dials associated with the weight control knobs and the reading on the illuminated scale; normally a vernier is employed to read to the nearest 0.1 g.

An analytical balance is a sensitive instrument and must be treated as such. The knife-edges must be protected from damage by arresting the mechanism when the balance is not in use and when changing the object on the pan or changing the weights. Only proper containers should be placed on the pan; *chemicals must never be placed directly on the pan*. The balance should be kept thoroughly clean; a camel's hair brush is normally employed for this purpose.

An object should be at room temperature when weighed; weighing a warm object results in erratic and inaccurate weights. Bare fingers introduce moisture and oils onto an object and should not be employed to handle anything to be weighed on an analytical balance. Tongs or a strip of lint-free paper should be employed to handle such objects.

Buoyancy Corrections in Weighing

In some cases the effect of the **buoyancy** of air upon objects must be considered in determining their masses. This occurs when the weighed object has a density appreciably different from that of the weights used in the balance. The densities of most solids are sufficiently close to those of the weights that a buoyancy correction is not required, but such a correction is required for low-density solids, as well as liquids or gases. The buoyancy-corrected mass of an object, M_c, may be obtained from the formula,

$$M_c = M_w + M_w \left(\frac{d_{air}}{d_c} - \frac{d_{air}}{d_w} \right) \tag{2.8}$$

where M_w is the mass of the weights and d_{air}, d_c, and d_w are the densities of air, the object to be weighed, and the weights, respectively, in grams per milliliter. The value of d_{air} is 0.0012 g/mL.

As an example of a buoyancy correction, consider the weighing of a liquid, $d_c = 0.87$ g/mL, using a balance with tantalum weights, $d_w = 16.6$ g/mL. The liquid was weighed by first weighing an empty glass weighing bottle, adding a quantity of liquid and weighing the bottle and contents. The difference in these two weights was 1.0376 g, and is the value of M_w. The value of M_c is given by the following substitution into Equation 2.8:

$$M_c = 1.0376 \text{ g} + 1.0376 \text{ g} \left(\frac{0.0012 \text{ g/mL}}{0.87 \text{ g/mL}} - \frac{0.0012 \text{ g/mL}}{16.6 \text{ g/mL}} \right) \tag{2.9}$$

$$= 1.0376 \text{ g} + 1.0376 \text{ g} \times (0.00138 - 0.00007) \tag{2.10}$$

$$= 1.0376 \text{ g} + 0.0014 \text{ g} = 1.0390 \text{ g} \tag{2.11}$$

Auxiliary Balances

Many weighing operations in the analytical laboratory do not require the sensitivity of an analytical balance. In these cases an auxiliary balance is used. Some of the more sensitive of these devices can be used to weigh up to 200 g to within 1 mg. Most such balances are now electronic with a digital readout of mass (Figure 2.4). These balances usually have an automatic taring feature, which sets the balance to zero with the empty container on the pan so that the mass of substance is weighed directly after it is put into the container. The most convenient of these balances are top-loading balances, in which the pan is mounted on top of the balance as shown in Figure 2.4. Auxiliary balances are generally convenient, fast, rugged, and have a relatively large capacity, so that they should be used whenever the accuracy of an analytical balance is not required. (Electronic single-pan balances with digital readout and automatic taring are now readily available. These, like related auxiliary balances, can be interfaced to a computer for the direct entering of weights into computer memory.)

FIGURE 2.4   Left and center: Top-loading auxiliary balance with digital readout and automatic tare. **Right:** Single-pan electronic analytical balance with automatic tare and automatic digital readout. This balance has top as well as side access. It is sensitive to 0.01 mg for weights to 31 g and to 0.1 mg for weights to 162 g.

2.3   **Apparatus and Operations of Weighing**

Handling Solids for Weighing

The previous section discussed the balance, which is involved in any weighing process. However, several important items of apparatus and several important processes are involved with weighing. These are described here.

Solids, such as solid samples to be analyzed or primary standards (see Section 1.4), are normally handled in a granular or finely powdered form. Such solids need to be dried and kept in a dry condition, with minimum exposure to the atmosphere. They are normally handled in weighing bottles (Figure 2.5), light-weight glass containers with close-fitting ground-glass lids. To prevent exposure of solids to atmospheric moisture, the lids may be placed on weighing bottles outside the desiccator whenever material is not being transferred to or from the bottle. To prevent contamination from the hands, with subsequent weighing errors, these bottles may be handled with tongs or with a strip of lint-free paper wrapped around the bottle and pinched snugly with the fingers.

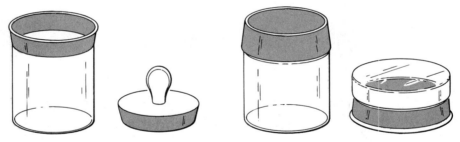

FIGURE 2.5   Typical weighing bottles. The sizes are approximately those shown in the figure.

Weighing by Difference

In removing portions of a sample or primary standard from a weighing bottle for determination of mass, greatest convenience and accuracy as well as minimal chance of sample loss, are attained by *weighing by difference*. This offers the advantage that N accurately weighed portions of material may be obtained with only $N + 1$ weighing operations. Suppose that it is desired to weigh out three portions of a solid sample contained in a weighing bottle. The bottle and its contents should be weighed initially, a portion of solid removed, a second weighing performed, a second portion of solid removed, followed by a third weighing, removal of a third portion, then a fourth and final weighing.

Typical data might appear as follows:

Mass of bottle and contents		Before sample removal	19.0163 g	18.1376 g	16.9805 g
		After sample removal	18.1376 g	16.9805 g	16.0036 g
		Mass of sample	0.8787 g	1.1571 g	0.9769 g

   A solid sample can be removed from a weighing bottle with a stainless steel spatula (Figure 2.6). This introduces the possibility of some error as a result of solid clinging to the spatula. Another method of solid transfer from a weighing bottle to a beaker or conical flask involves holding the weighing bottle horizontally over the other container with a strip of paper wrapped tight around the weighing bottle and tapping some of the contents of the bottle into the flask or beaker. Here, again, a steady hand and care to avoid spillage are required.

FIGURE 2.6   Stainless steel spatula for transferring solids

Weighing Liquids

Weighing liquids can be difficult. Less volatile liquids can be poured into preweighed, stoppered containers, which are then weighed again to get the mass of the liquid. More volatile liquids may have to be drawn into a weighed ampoule by heating the ampoule and immersing its neck in the liquid; the liquid is sucked into the ampoule as it cools. The neck can be sealed off with a flame, the full ampoule and any glass removed in sealing weighed, and the ampoule broken in the container which is to hold the sample. The procedure requires considerable skill and is potentially dangerous.

2.4   Drying Solids

The Drying Oven

Solids are typically dried in a **drying oven** (Figure 2.7). This device maintains a preset temperature within narrow limits and provides limited ventilation to remove moist air. Samples to be dried should be placed in the drying oven with their lids off to allow moisture to escape. The temperature of drying depends upon the material; some substances are unstable when heated and can be dried only at temperatures around 100°C.

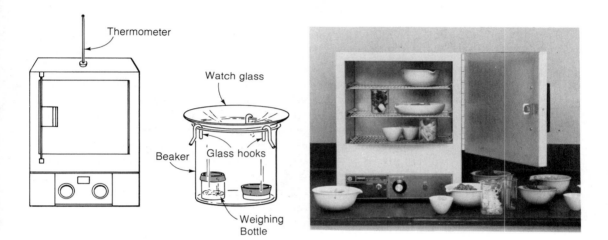

FIGURE 2.7   Left: A drying oven, and a weighing bottle in a beaker covered in a manner that allows air circulation. **Right:** Fisher Isotemp oven for drying samples, chemicals, and apparatus.

The Desiccator

A sample is maintained in a dry state and may even be dried more in a **desiccator** (Figure 2.8). This is a container, commonly consisting of glass or aluminum, which contains a **desiccant**, and which is sealed from the atmosphere. The lid has a ground-glass planar bottom surface which fits tightly over the ground-glass upper surface of the base. These surfaces may be lightly greased to complete the seal; some chemists prefer to leave them dry. Desiccants may consist of a number of substances, the most common of which are anhydrous calcium sulfate, anhydrous calcium chloride, anhydrous magnesium perchlorate, and phosphorus pentoxide. Other desiccants include

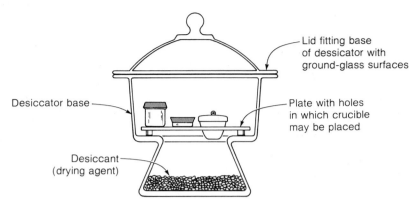

FIGURE 2.8   A laboratory desiccator. A crucible is shown in one of the holes on the plate, and a weighing bottle is on top of the plate.

anhydrous copper sulfate, NaOH sticks, sulfuric acid, silica gel, KOH sticks, Al_2O_3, and BaO. These desiccants differ in their ability to dry air. Anhydrous $CaCl_2$ reduces the moisture content of air to 2.8 mg/L of water in air, whereas P_4O_{10} takes it down to an equilibrium value of only 0.00002 mg/L.

A desiccator lid should be removed by sliding it sideways. When a warm object is first placed in a desiccator, the lid should be left slightly ajar for a few moments to allow the escape of heated air. Then, as the contents of the desiccator cool, the lid should be opened momentarily a few times to prevent the cooled air from forming too much of a vacuum in the desiccator, which would make lid removal difficult.

Drying to Constant Weight

Drying to constant weight is a commonly performed process in which an object or material is alternately heated, cooled in a desiccator, and weighed, with the cycle repeated until a constant weight is obtained. This may be done with an empty container, such as a crucible, a precipitate in a crucible, or a sample in a weighing bottle. It is done to ensure that changes caused by heating—for example, loss of moisture from weighing bottle walls or loss of volatile matter from a precipitate—have occurred to a maximum extent. Usually weight constant to 0.2 to 0.3 mg is attained. In some cases, particularly where very high heating temperatures are involved, the process is called ignition to constant weight.

Weighing Hygroscopic Substances

Special measures are required for the weighing of hygroscopic substances, which strongly absorb water from the atmosphere. One approach that may be used is to dry a weighing bottle and lid to constant mass, add a quantity of the hygroscopic substance to the bottle, dry to constant mass again, then remove the hygroscopic substance quantitatively; its mass is the difference in the two masses measured.

Filtration

Types of Filters

A number of quantitative analysis procedures, especially those involving gravimetric determinations (see Chapter 8), require **filtration**. This is the retention of a solid suspended in a liquid by a porous medium such as filter paper, a glass frit in the bottom of a glass filtering crucible, unglazed porcelain in the bottom of a porcelain filtering crucible, or an asbestos mat deposited as a layer on the perforated bottom of a Gooch crucible.

The simplest method of filtration involves use of filter paper in a glass funnel (Figure 2.9). Before it is placed in the funnel, the circular filter paper is

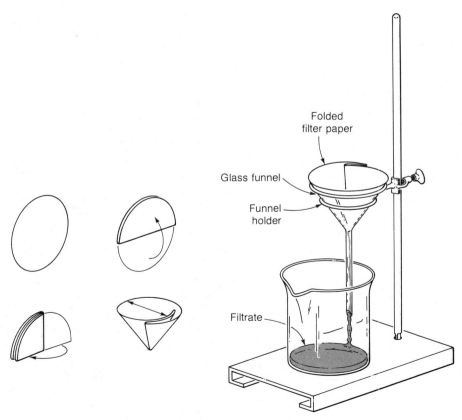

FIGURE 2.9 Gravity filtration with a folded filter paper in a glass funnel

folded once in half, then folded a second time, giving it the appearance of a quarter-circle, and finally opened so that one wall of the cone just formed is composed of three layers of filter paper and the other of just one layer. Thus any liquid poured into the cone has to go through filter paper to escape, and precipitate is retained by the paper.

Filter paper is employed when the precipitate is to be ignited at a high temperature. This burns off the filter paper; the paper should have the lowest possible ash content. To prevent error because of the filter paper ash, ashless filter paper is employed, typically having an ash residue of about 0.1 mg for 9- or 11-cm-diameter filters. Filter paper is relatively low in cost and comes in a number of different porosities. It is the only medium that can be used with some success to filter gelatinous precipitates.

The asbestos filter employed in a Gooch crucible is not widely used anymore because of concerns over the health effects of asbestos. It is especially useful for the ignition of precipitates up to 1200° C. The porosity is difficult to control and it is somewhat inconvenient to use. Sintered-glass filtering crucibles are rapid and convenient to use and come in three major porosities— coarse, medium, and fine. Glass crucibles can be heated to 500° C. Porcelain filter crucibles offer all the conveniences of glass but can be heated to 1100° C. Aluminum oxide crucibles are even better in that they may be heated to 1450° C.

The Filtration Process

Once a filter has been prepared, the first step in the filtration process is decantation of the **supernatant liquid** above the precipitate through the filter. Thus, initially, few particles lodge on the filter, so that it does not clog as rapidly. To avoid spilling the decanted liquid, it is poured down a stirring rod onto the filter (Figure 2.10); the glass rod is used to collect any drops of liquid remaining on the lip of the beaker after pouring and transferring these drops back to the beaker. After decantation of the original liquid is complete, the precipitate must be washed to remove solution containing soluble salts from it. This is accomplished by adding wash liquid to the precipitate, agitating the precipitate with the liquid, then decanting the wash liquid through the filter. This gives more efficient washing than simply pouring the wash liquid over the precipitate on a filter. After the washing process is complete, the precipitate is washed onto the filter by a carefully directed stream of wash liquid. Last traces of the precipitate are dislodged by a *rubber policeman* on the end of a glass rod (Figure 2.11) while washing with a stream of water.

Except when filter paper is used, filtration is normally accomplished with a vacuum. The apparatus for doing this is shown in Figure 2.12. In its simplest form the crucible is tightly fitted into the hole of a rubber ring, which makes a vacuum-tight fitting with the crucible. More elaborate crucible holders for vacuum filtration incorporate a funnel with a stem that directs liquid down into the collection flask.

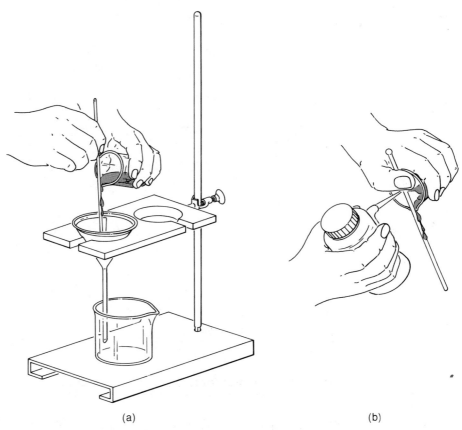

(a)   (b)

FIGURE 2.10 Procedures for (a) pouring a supernatant liquid from a precipitate and (b) rinsing the precipitate into the funnel

Rubber policeman

FIGURE 2.11 Rubber policeman fastened to the end of a glass rod for transfer of precipitates

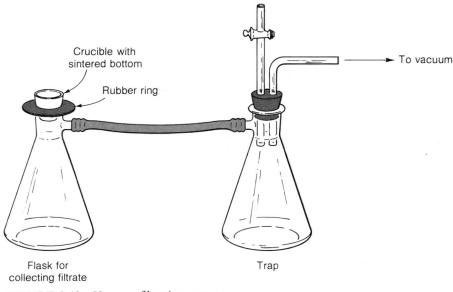

Crucible with
sintered bottom

Rubber ring

To vacuum

Flask for
collecting filtrate

Trap

FIGURE 2.12   Vacuum filtration apparatus

2.6   Drying, Ashing, and Ignition of a Precipitate

Once a precipitate is collected, it certainly must be dried and may have to be
ignited to convert it to a chemically stable weighing form. There is no definite
delineation between drying and ignition. Normally the former involves
heating in a thermostatically controlled electric drying oven with a maximum
temperature of about 250° C. Ignition commonly requires heating to temperatures ranging from 250 to 1200° C. Ignition can be accomplished by heating a
crucible directly in a burner flame or in a temperature-controlled electrical
muffle furnace. Ignition requires the use of porcelain or silica (SiO_2) crucibles,
which are heat-resistant.

Collection of a precipitate collected on a filter paper requires incineration
of the filter paper. The paper containing the precipitate is carefully folded to
enclose the precipitate and is then placed in a crucible that has been ignited to
constant weight (see Section 2.4). The crucible and contents are then placed in
a clay-pipe or silica-tube triangle held on a ring. The crucible is inclined at an
angle, and the lid placed partially on it (Figure 2.13). The crucible and contents
are gradually heated in the following steps:

1. Slow drying at a very low flame taking care that steam formed from
   the wet filter paper does not expel precipitate.
2. Slow carbonization of the paper; if the paper burns, the lid must be
   placed on the crucible immediately to extinguish the flame.

3. Oxidation of residual carbon by heating the bottom of the crucible strongly with the lid partially ajar.
4. Ignition for 30 to 60 min in the hottest portion of the burner flame with the lid on the crucible.

2.7   Measurement of Volume

Volumetric Glassware

Next to weighing, measurement of volume is the most exacting operation in routine analytical practices and is the most common measurement normally made. There are three major devices used for volume measurement in the quantitative analysis laboratory, as shown in Figure 2.14. These are the **pipet**, which delivers a preselected volume of liquid; the **buret**, which is used in titration to deliver a measured variable volume of liquid required to bring about a chemical reaction between a reactant dissolved in the solution in the

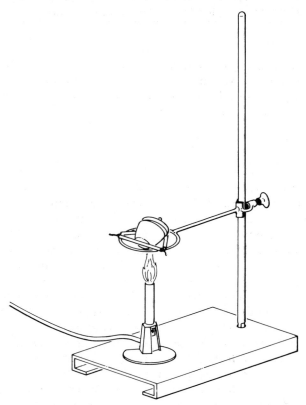

FIGURE 2.13   Heating a crucible containing a precipitate in filter paper

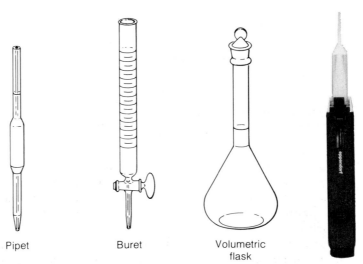

Pipet   Buret   Volumetric
                                        flask

FIGURE 2.14   Left: Three general classes of glassware for measuring volumes of liquids. **Right:** Syringe pipet for rapid pipetting of solutions.

buret and a quantity of another reactant in solution in a beaker or flask; and the **volumetric flask**, which contains a fixed volume of solution. Volumes in the quantitative analysis laboratory are most commonly measured in milliliters; large volumes are expressed in liters.

Pipets

Within each class of volumetric device shown in Figure 2.14, there are numerous variations. This is especially true of pipets. Figure 2.15 illustrates the three kinds of pipets most often used in the quantitative analysis laboratory. The volumetric or transfer pipet is designed to deliver one specified volume of reagent. It is calibrated such that free drainage of the reagent from the pipet delivers the desired volume, and the thin film of reagent remaining on the pipet walls must not be rinsed into the receiving vessel. These pipets are available in volumes ranging from 0.5 to 200 mL and often have a colored strip around the top to aid sorting. The measuring pipet, such as the Mohr pipet shown in Figure 2.15, is designed to deliver any volume of liquid up to the maximum capacity of the pipet. Commercially available measuring pipets range in capacity from 0.1 to 25 mL. A number of automatic pipets are available to deliver single, preselected volumes. Some of these are adjustable, so that a nonstandard volume may be chosen. The simple syringe pipet delivers a volume predetermined by the syringe. Syringes that deliver from 1 μL to 1 mL are available.

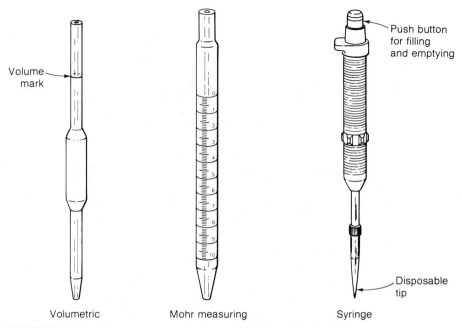

FIGURE 2.15   Common types of pipets used in the analytical laboratory

Burets

A simple laboratory buret consists of a long tube with graduations etched on it calibrated to deliver any volume up to the capacity of the buret. A rotatable stopcock at the bottom of the buret allows reagent to drain from the device. This stopcock may be made of glass and requires a lubricant to maintain a seal. Stopcocks made of chemically resistant Teflon resist chemical attack, have little tendency to "freeze" in place, and do not require lubrication. Buret capacities commonly range from 1 to 50 mL; the 50-mL buret is usually used in the quantitative analysis laboratory.

A 50-mL buret should be read to 0.01 mL. Some care is required to read a buret that closely. The top of a column of liquid in a buret forms a dish-shaped depression known as a **meniscus** as shown in Figure 2.16. The buret volume should be read from the bottom of the meniscus. The meniscus can be made to stand out very clearly as a black dish shape above a clear solution by placing a buret-reading card consisting of a piece of black tape on a white card behind the buret so that the upper edge of the tape is even with the bottom of the meniscus. The eye should be exactly at the level of the meniscus to eliminate parallax error.

Burets come in a variety of more sophisticated forms, including those that refill automatically and that zero automatically. Sophisticated electronic

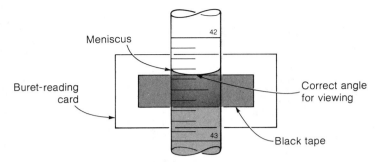

FIGURE 2.16   Reading a buret with a buret-reading card

burets are available in which the reagent is delivered by an electrically driven syringe, and the amount delivered read from a meter.

Volumetric Flasks

Volumetric flasks are calibrated to contain a specific volume of liquid. They are normally used to contain a dissolved solute diluted to a specific volume. For example, if a reagent is to be made up containing a particular mass of solute in 250.0 mL of solution, the weighed solute may first be added to the volumetric flask, dissolved in an adequate amount of water (then cooled if the dissolution is exothermic), water added almost to the mark followed by mixing, and finally water added to the mark followed by additional mixing. The thoroughness of the final mixing should be assured by placing the stopper firmly on the flask and inverting 20 times. Some volumetric flasks have two calibration marks on the neck. When the flask is filled to the upper one of these, it delivers the stated capacity of the flask when the contents are poured out.

From the preceding discussion it is obvious that some volumetric ware is calibrated to *contain* a particular volume, whereas other volumetric ware is calibrated to *deliver* a specified volume. Burets are calibrated to deliver, as are most pipets. An exception to the latter is one form of the lambda pipet, which is to be washed out with a solvent after delivery of reagent, and therefore is calibrated to contain. It has just been noted that most volumetric flasks are calibrated to contain, whereas others can also deliver a particular volume.

Cleaning Volumetric Glassware

One of the tasks facing the analytical chemist is cleaning volumetric ware. Clean glassware is denoted by an unbroken surface film of water upon rinsing; breaks in this surface film as the water drains, or drops of solution clinging to the glassware surface, show a lack of cleanliness. Cleaning is usually accomplished by brief soaking in warm detergent solution accompanied by gentle brushing with a suitable brush. Long-term soaking in detergent solutions is to be avoided because permanent marks can develop at the interface of the detergent solution and air. These marks cause breaks in the water film. In extreme cases, glassware may be cleaned with *dichromate cleaning solution,* composed of approximately 70 g of $Na_2Cr_2O_7 \cdot 2H_2O$ dissolved in 1 L of concentrated sulfuric acid. *Dichromate cleaning solution is an extremely corrosive acidic oxidant with a dehydrating action and causes immediate burns to exposed skin. It should be used only by those who have been properly trained in all precautions involving its use. Both it and the spent solution are regarded as hazardous waste materials and their disposal must be in accordance with institutional hazardous waste regulations.*

Glassware may be allowed to contact dichromate cleaning solution for periods of several minutes to overnight, depending upon how dirty the surface is. The technique for cleaning a buret is shown in Figure 2.17. Cleaning solution may be poured into volumetric flasks, and pipets may be immersed in the solution contained in a glass cylinder. Relatively new cleaning formulations free of acid and dichromate are much safer than dichromate cleaning solution and do not present substantial disposal problems.

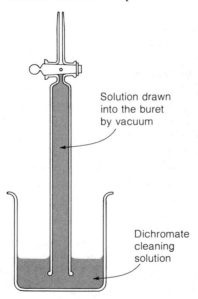

Solution drawn
into the buret
by vacuum

Dichromate
cleaning
solution

FIGURE 2.17   Cleaning the inside surface of a buret in dichromate cleaning solution

2.8   Calibration of Volumetric Glassware

For very exacting work, glassware must be calibrated to see if it contains or delivers nominal values of volumes; if not, corrections can be made. The process involves determination of the masses of quantities of liquids added to or delivered from a volumetric vessel and calculating the true volumes from the masses and densities of the liquids. Variations in temperature affect both the density of the liquid and the volume of the container; the buoyancy effect upon the determination of mass (Section 2.2) must also be taken into account. Water is normally the liquid chosen to calibrate volumetric ware; mercury is used in some special cases.

A pipet is calibrated by weighing the water delivered by the pipet. For example, suppose at 25° C a 10-mL volumetric pipet delivers 9.988 g of water weighed with a balance having stainless steel weights. The density of water at 25° C is 0.9971 g/mL. The mass of water corrected for buoyancy can be calculated with Equation 2.8 using 0.0012 g/mL for the density of air and 7.8 g/mL for the density of stainless steel weights:

$$M_c = 9.988 \text{ g} + 9.988 \text{ g} \left(\frac{0.0012 \text{ g/mL}}{0.9971 \text{ g/mL}} - \frac{0.0012 \text{ g/mL}}{7.8 \text{ g/mL}} \right) = 9.998 \text{ g} \quad (2.12)$$

The actual volume of water delivered is given by dividing the mass of the water by its density:

$$\begin{array}{l} \text{Volume of} \\ \text{water delivered} \end{array} = \frac{9.998 \text{ g}}{0.9971 \text{ g/mL}} = 10.03 \text{ mL true volume delivered}$$

$$(2.13)$$

A buret is calibrated by weighing successive 10-mL portions of water drained from the buret. Initially the buret is filled with water and drained almost down to the 0-mL mark at the top of the graduations etched onto the buret, making sure that all the air is expelled from the buret tip. After 1 min to allow for drainage, the buret is drained to exactly the 0 mark and allowed to stand for about 10 min to make sure there is no leakage around the stopcock. In the meantime, a 125-mL conical flask and rubber stopper are weighed. After that the buret is drained into the flask in 10-mL intervals and the mass of water in the flask is weighed after each drainage. About 1 min should be allowed to drain 10 mL; the buret tip should be touched to the wall of the flask after each drainage to collect any retained drop of water on the tip, and 1 min should be allowed after the buret is drained to read the volume. After the process is completed, six masses will have been recorded—one of the empty flask and stopper and five at 10-mL intervals as the buret was drained. Data are recorded as shown in Table 2.1

With a single-pan balance, calibration of a volumetric flask is accomplished by simply filling the flask to the mark with water, weighing the flask and the water, subtracting the mass of the empty flask, correcting for buoyancy, and calculating the true volume held by the flask. A problem may be encountered in finding a balance with a sufficient capacity to weigh a filled volumetric flask over 100 mL in capacity. In that case, a good option is to calibrate a flask against a specific volumetric pipet. For example, a 50-mL pipet can be used to add five portions of water to a 250-mL volumetric flask, and a

TABLE 2.1 Representative data for the calibration of a 50-mL buret at 25° C measured with a balance with stainless steel weights

Approximate volume interval*	Buret reading	Apparent volume	Mass of flask and water[†]	Apparent mass of water delivered	True mass of water delivered	True volume of water delivered	Correction	Cumulative correction
Initial	0.00		41.283	—	—	—	—	—
0–10	9.99	9.99	51.243	9.960	9.970	10.00	0.01	0.01
10–20	19.99	10.00	61.203	9.960	9.970	10.00	0.00	0.01
20–30	30.00	10.01	71.153	9.950	9.960	9.99	−0.02	−0.01
30–40	40.02	10.02	81.113	9.960	9.970	10.00	−0.02	−0.03
40–50	49.99	9.97	91.013	9.900	9.910	9.94	−0.03	−0.05

[†] All volumes are expressed as mL.
* All masses are expressed as grams.

51

mark made at the meniscus with water-resistant tape. Henceforth, when the flask is filled to the taped mark, withdrawal of an aliquot with the same 50-mL pipet will assure that one-fifth of the solution is taken each time.

Programmed Summary of Chapter 2

This summary contains in a programmed format the major terms and concepts introduced in this chapter. To get the most out of it, you should fill in the blanks for each question, and then check the answers at the end of the book to see if you are correct.

The measure of the quantity of matter is known as (1) _____. The force of attraction of the earth for an object is known as the object's (2) _____. The virtually frictionless device upon which a balance beam rests on the body of a balance, or a balance pan is suspended from the beam, is known as a (3) _____. The device that is used to compensate for the slight mismatch in mass between an object on one pan of a double-pan balance and the weights on the other pan is a (4) _____. The common laboratory analytical balance measures mass to the nearest (5) _____. The sensitivity of a single-pan balance remains the same because it has a (6) _____. The arrest knob on a single-pan balance serves to (7) _____. Weight control knobs on a single-pan balance are used to (8) _____. When a weighed object has a density considerably less than that of the weights used in the balance, a correction must be made for (9) _____. A balance that is used for routine weighing in the analytical laboratory that does not require the accuracy of an analytical balance is called (10) _____. The most convenient of these balances are those that are (11) _____. The feature that enables setting a balance to zero with an empty container on the pan is the (12) _____. The two major physical forms of weighing bottles are (13) _____. A technique that permits weighing N samples with only $N + 1$ weighings is called (14) _____. The four major components of a desiccator prepared to hold samples are (15) _____. An excellent desiccant that takes air down to 0.00002 mg/L of water in air is (16) _____. Alternate heating, cooling, and drying of an object or sample, repeated as many cycles as necessary, is called (17) _____. Four major types of filters or filter materials are (18) _____. In order for a precipitate collected on a filter paper to be ignited successfully the paper must be of the (19) _____ type. The major steps in filtering a precipitate from the solution from which it has formed are (20) _____. When some material stronger than filter paper is employed as a filtering medium, the filtration is assisted by (21) _____. Drying a precipitate generally involves temperatures up to (22) _____, whereas ignition is usually accomplished in the temperature range of (23) _____. The three major types of volumetric glassware commonly used in the analytical laboratory are (24) _____. A pipet that delivers a fixed volume of liquid is called a (25) _____ pipet, whereas one of the types that can deliver a variable volume of liquid is the (26) _____ pipet. A type of automatic pipet that can be obtained in sizes ranging from 1 μL to 1 mL is the (27) _____. The advantages of a Teflon stopcock on a buret are (28) _____. A 50-mL buret should be read to the nearest 0.01 mL. The level of liquid in a buret is read from (29) _____ with which the eye should be level to avoid a (30) _____ error. Reading a buret is facilitated by use of a (31) _____ consisting of a (32) _____. Pipets and burets are normally calibrated to (33) _____ a particular volume, whereas volumetric flasks most commonly are calibrated to (34) _____ a particular volume. Cleaning volumetric glassware is usually accomplished with (35) _____, whereas in extreme cases (36) _____ may be employed. The latter solution is prepared by adding (37) _____ to (38) _____. For most accurate results volumetric ware should be (39) _____ before use. This is usually accomplished by determining the (40) _____ contained in or delivered by an item of volumetric ware.

Questions

1. What is the relationship between weight and mass?
2. What are the two major types of construction of analytical balances?
3. Why does the sensitivity of a single-pan balance not vary with the mass of an object?
4. What are the major controls of a single-pan balance?
5. In what position is the single-pan balance arrest knob when weights are being dialed on or off?
6. What precision part of the balance is most susceptible to damage by not having the mechanism arrested when it should be?

7. For weighing a low-density material, is the buoyancy correction increased or decreased by using more dense weights?
8. What is the buoyancy correction if the weights and the object being weighed have the same density?
9. What is meant by taring?
10. What is the advantage of weighing by difference?
11. What is the function of the chemical placed in the bottom of a desiccator base?
12. What is the purpose of drying to constant weight?

Problems

1. An object was weighed composed of an experimental structural material with a density of 0.518 g/mL, giving an uncorrected mass of 0.13260 g. The balance employed had stainless steel weights ($d_w = 7.8$ g/mL), and the density of air is 0.0012 g/mL. What is the buoyancy-corrected mass of the object M_c?

2. Repeat the preceding problem assuming that the balance had tantalum weights, $d_w = 16.6$ g/mL.

3. An object with a density of 1.578 g/mL yielded an uncorrected mass of 2.0391 g on a balance with platinum-iridium weights, having a density of 21.5 g/mL. What was the buoyancy-corrected mass of the object?

4. The density of water at $25°C$ is 0.9971 g/mL, the density of air at this temperature is 0.0012 g/mL and the density of stainless steel weights is 7.8 g/mL. Derive a formula in the form

$$M_c = X \cdot M_u$$

to give the buoyancy-corrected mass M_c from the uncorrected mass M_u for water weighed at $25°C$ on a balance with stainless steel weights.

5. Using the results of the preceding question, calculate the volume contained by a 25-mL volumetric flask weighing 12.7018 g empty at $25°C$ on an analytical balance with stainless steel weights and 37.6054 g under the same conditions when filled to the mark with water.

6. Using the same conditions as in the preceding question, calculate the volume contained by a 10-mL flask that delivers 9.8987 g of water when drained from the mark.

7. Using the same conditions as in the preceding two questions, calculate the volume contained by a 50-mL volumetric flask that registers a weight gain of 49.988 g when filled to the mark with water.

8. A buret was calibrated by weighing at $25°C$, on an analytical balance with stainless steel weights, successive approximate 10-mL increments of water contained in the buret. Starting at 0 mL the weights of water delivered from each of the preceding buret readings were the following: 10.01 mL, 9.980 g; 19.99 mL, 9.955 g; 29.97 mL, 9.950 g, 40.02 mL, 9.980 g, and 50.00 mL, 9.960 g. Show the volume correction and cumulative correction for each 10-mL interval.

9. A buret calibrated as in the preceding example gave the following values: 9.99 mL, 9.990 g; 20.00 mL, 9.945 g; 30.15 mL, 9.960 g; 39.99 mL, 9.950 g; 49.98 mL, 9.948 g. Give the cumulative corrections at 10-mL intervals.

10. A 10-mL total capacity buret was calibrated as in the two preceding problems, giving the following values: 1.990 mL, 1.903 g; 4.006 mL, 1.965 g; 6.104 mL, 1.987 g; 7.904 mL, 1.885 g; 9.999 mL, 2.001 g. Show the volume correction and cumulative correction for each 2-mL interval.

11. A sample of pond water to be analyzed for nitrate contamination was stored at $4°C$ to minimize bacterial action, and a 25-mL sample was withdrawn at $4°C$ for analysis. The coefficient of expansion for pure water and dilute solutions is 2.5×10^{-4} per deg—that is, in the temperature range of 4 to $25°C$ where aqueous samples are normally handled, the volume of an aliquot of water increases by a fraction of 0.00025 for each degree Celsius increase in temperature. What is the volume V_c of the aliquot withdrawn, corrected to $20°C$, given that the uncorrected volume V_u is 25.00 mL at $4°C$?

12. What is the percentage increase of the volume of a sample in going from 8 to $25°C$?

13. Applying the temperature correction given in the two preceding problems and taking the density of water as 0.9971 g/mL at $25°C$, what is the density of water at $4°C$?

3

The Quantitative Treatment of Data

L E A R N I N G   G O A L S

1. The concept of error, including determinate and indeterminate error.
2. The meaning of accuracy and precision.
3. Significant digits in calculations and rounding numbers properly; the meaning of uncertainty and relative uncertainty of a number.
4. Terms pertinent to quantitative results, such as the mean of a set of values, the true value of a parameter, absolute error, and relative error.
5. Mathematical expressions of precision, such as range and standard deviation.
6. The distribution of random errors and the Gaussian distribution of random errors.
7. Application of confidence limits to analytical results.
8. Use of the Q-test in deciding whether to reject a discordant result.
9. The basics of plotting experimental data.

3.1 Numbers in Chemical Analysis

The quality of an analytical result depends upon the reliability of the numbers used and obtained in the analytical procedure. The analyst must have confidence that these numbers are as correct and reproducible between analyses as possible. Such numbers come from many sources; they may be masses read from an analytical balance; volumes measured from a buret; fundamental constants; or the results of calculations using numbers representing masses, volumes, electrical potentials, light intensities, and other parameters. This chapter deals with the handling of such numbers in quantitative analysis.

3.2 Analytical Results and Error

Determinate and Indeterminate Errors

One of the major objectives of analytical measurements is to obtain *reproducible results*. For example, if three determinations of the percentage of iron in the same iron ore sample gave values of 25.32, 25.30, and 25.33%, the fact that these values are quite close to each other increases the analyst's confidence in the validity of the results. The degree to which numbers in a group are in agreement with each other is the **precision** of the group of numbers. A lack of precision may indicate the presence of **indeterminate**, or **random, errors**. Such errors vary randomly in direction and magnitude and are from sources that cannot be determined.

In addition to indeterminate errors, another major type of error consists of **determinate errors**. These have a definite cause (although it may be unknown to the analyst), and each type of determinate error is always in the same direction. A particular determinate error may be of the same magnitude each time it occurs or may be proportional to the quantity measured. For example, a volumetric pipet nominally of 25.00 mL that actually delivers 25.40 mL of solution each time it is used introduces a determinate error into an analysis.

Accuracy and Precision

Just because reproducible data are obtained in a chemical analysis does not necessarily mean that the data represent the true value. The extent to which the data or the average value of a set of data agree with the true value being determined is the **accuracy** of the data. The relationship between accuracy and precision is shown graphically in Figure 3.1. In addition to the combinations shown in Figure 3.1, it is possible that the average of a number of imprecise values may be quite close to the true value due in part to the central tendency of random errors. Generally, however, when the precision of a set of data is low, the accuracy is likewise low.

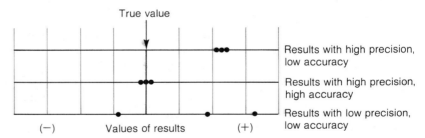

FIGURE 3.1   Representation of precision and accuracy. Each solid circle, ●, stands for an experimentally determined value.

3.3   **Significant Digits**

Absolute Uncertainty in a Number

Before discussing calculations with quantitative data, it is necessary to consider the expression of numbers properly, including the expression of results of a mathematical calculation. This requires that a number of rules be followed; they are given in this section.

> Rule 1. In expressing a number, retain no digits beyond the first uncertain one.

EXAMPLE

As discussed in Chapter 2, the common laboratory analytical balance weighs to the nearest 0.0001 g. Therefore, masses measured on such a balance expressed to the correct number of significant figures are those such as

$$46.2158 \text{ g}$$
$$3.8023 \text{ g}$$
$$0.6714 \text{ g}$$
$$12.3807 \text{ g}$$

Each of these masses has an uncertainty of 0.0001 g; the last digit is the only uncertain one.

Suppose that a laboratory buret were drained from an initial volume of 0.55 mL to 44.15 mL. It would *not be correct* to express the volume of solution delivered as 43.6 mL, because the last digit is *not* uncertain. With a good buret and a careful analyst, the uncertainty in reading a buret is ±0.01 mL. Therefore, the result should be expressed as 43.60 mL, where the last digit is uncertain. It would not be correct to express the volume delivered as 43.602 mL because both of the last two digits are uncertain, and only one uncertain digit should be retained.

The 0.01 mL uncertainty in reading an analytical-quality 50-mL buret is the **absolute uncertainty** of the measurement. If a number has units, milliliters in this case, the absolute uncertainty has the same units.

The absolute uncertainty in reading a volume from a properly calibrated (see Section 2.8) 50-mL buret is 0.01 mL throughout the measurement range of the buret. Thus an accurately observed 4-mL volume reading on a 50-mL buret should be regarded as having the same 0.01-mL absolute uncertainty as a 40-mL volume reading on the same buret.

Relative Uncertainty of a Number

The **relative uncertainty** of a number is defined as

$$\text{Relative uncertainty} = \frac{\text{absolute uncertainty of a number}}{\text{value of the number}} \tag{3.1}$$

Relative uncertainty may be expressed as a ratio or as a decimal. It may be multiplied by 100 to give the relative uncertainty in percent or by 1000 to express relative uncertainty in parts per thousand. Obviously, the relative uncertainty for a small number is greater than that of an appreciably larger number when both have the same absolute uncertainties, as shown by the following examples.

EXAMPLE

A volume of 3.00 mL is known with an absolute uncertainty of 0.02 mL, and a volume of 30.00 mL is also known with an absolute uncertainty of 0.02 mL. What are the relative uncertainties of the two volumes?

For 3.0 mL,

$$\text{Relative uncertainty} = \frac{0.02 \text{ mL}}{3.00 \text{ mL}} = \frac{2}{300} = \frac{1}{150}$$

$$= 0.007 = 0.7\% = 7 \text{ parts per thousand}$$

For 30.0 mL,

$$\text{Relative uncertainty} = \frac{0.02 \text{ mL}}{30.00 \text{ mL}} = \frac{2}{3000} = \frac{1}{1500}$$

$$= 0.0007 = 0.07\% = 0.7 \text{ parts per thousand}$$

The *larger* volume has a much *smaller* relative uncertainty. This is the reason that it is best to measure volumes that are close to the maximum capacity of the buret.

Zeros as Significant Digits

Zeros can cause considerable confusion in the expression of significant figures. Zeros to the left of the first nonzero digit in a number are not significant. Thus, in the number 0.00237,

0.00<u>237</u>

   ↑   ↑
       Three significant digits, 237
  The three zeros are not significant digits.

the 237 portion is significant whereas the three zeros are not; therefore, the number has three significant digits. Zeros to the right of any nonzero digit should be considered significant. Thus, in the number 0.084100 there are five significant digits.

<u>0.084100</u>

  ↑   ↑
     Five significant digits, 84100
  The two zeros are not significant digits.

Confusion arises with a number such as 6200. Does 6200 have four significant digits or only two, the 62 portion? According to some conventions, the number written as 6200 has only two significant digits. If it is written as

6200.

the zeros to the left of the decimal point are considered significant and the number has four significant digits. This kind of confusion is avoided by using exponential notation, as shown in Table 3.1.

TABLE 3.1 Exponential notation and significant digits

Number	Number of significant digits
6.2×10^3	2
6.200×10^3	4
1.2097×10^{-5}	5
9×10^{-6}	1
3.030×10^6	4

EXAMPLE

Give the number of significant digits in each of the numbers on the left.

3.1	2
0.0120	3
7300	Not known with certainty
7300.	4
7300.0	5
5.200×10^5	4

Exact Numbers

Confusion can arise with single-digit numbers that are meant to be exact, but are not followed by any zeros. For example, suppose that a 50-mL pipet was used to withdraw exactly one-fifth of the acid from a 250-mL volumetric flask, and that chemical analysis showed 3.274 mmol of acid in the 50-mL portion. The total number of millimoles of acid in the 250-mL volumetric flask is given by

$$\text{Total millimoles of acid} = 5 \times 3.274 \text{ mmol} = 16.37 \text{ mmol} \tag{3.2}$$

Here the 5 is assumed to be 5.000 . . . and does not limit the number of significant figures in the answer.

Rounding Off Numbers

Rule 2. In rounding off numbers, if the residue (digit to be dropped) is greater than 5, increase the digit to the immediate left of the residue by 1.

EXAMPLE

The number 32.147 correctly rounded off by dropping the last digit is 32.15.

$$32.147 \rightarrow 32.15$$
$$\uparrow$$
Residue greater than 5

Rule 3. If the residue is less than 5, leave the last digit to be retained unchanged.

EXAMPLE

The number 7362 rounded off by dropping the last digit is 7.36×10^3.

$$7362 \rightarrow 7.36 \times 10^3$$
$$\uparrow$$
Residue less than 5

It would be incorrect to express the rounded number as 7360, because that could be taken to mean that the last digit, 0, is significant, which it is not.

Rule 4. In rounding off numbers, if the residue is exactly 5, leave an *even* digit to the left of the residue unchanged and raise an *odd* digit in the same position by 1.

EXAMPLE

The number 4.865 rounded by dropping the last digit is 4.86.

$$4.865 \rightarrow 4.86$$
$$\uparrow$$
Even digit

The number 17,035 properly rounded by dropping the last digit is 1.704×10^4.

$$17,035 \rightarrow 1.704 \times 10^4$$
$$\uparrow$$
Uneven digit

Rounding Off the Results of Calculations

Rule 5. The result of the addition or subtraction of two or more numbers should retain only as many digits to the right of the decimal point as the number having the fewest such digits.

EXAMPLE

$$
\begin{array}{r}
+\ \ 2.7089 \\
-43.213 \\
+50.74 \quad \longleftarrow \text{Number with the fewest digits to the} \\
-\ \ 0.17432 \quad \text{right of the decimal point} \\
\hline
10.06158 \rightarrow 10.06 \\
\uparrow
\end{array}
$$

Result should retain only
two digits to the right of
the decimal point

Rule 6. The product or quotient from multiplication or division should have a relative uncertainty approaching as closely as possible the relative uncertainty of the factor with the greatest relative uncertainty. Therefore, the product or quotient should have a relative uncertainty ranging over the decade of $2x$ down to $\frac{1}{10} 2x$—that is, $2x$ to $\frac{1}{5} x$, where x is the relative uncertainty of the factor with the fewest significant digits.

EXAMPLE

$$
0.48581 \times 0.2013 \times 101 = 9.87715
$$
$$
\downarrow
$$
$$
9.9
$$

The least-certain factor is 101. Its relative uncertainty is $1/101$. Therefore the relative uncertainty of the result should be greater than $1/5 \times 1/101 = 1/505$ and less than $2 \times 1/101 = 1/50.5$. The relative uncertainty in 9.9 is $1/99$, which lies between these two values. Therefore, the correctly rounded answer is 9.9.

EXAMPLE

$$
\frac{200.6 \times 0.29834 \times 0.321704}{0.98 \times 19.167 \times 0.009211685} = 111.27141
$$
$$
\downarrow
$$
$$
111
$$

The least certain factor is 0.98. The relative uncertainty is $1/98$. The relative uncertainty of the result should lie between $1/5 \times 1/98 = 1/490$ and $2 \times 1/98 = 1/49$. The result expressed to the correct number of digits is 111, with a relative uncertainty of $1/111$.

The two preceding examples illustrate a major pitfall in expressing the result of multiplication or division to the correct number of significant digits. In these cases it is the relative uncertainty that must be taken into account. It is not necessarily correct to have the same number of digits in the result as are in the least-certain factor (although this is a commonly followed "rule of thumb"). In the first example above, the least-certain factor, 101, has three significant digits and a relative uncertainty of essentially 1 part in 100. If the result were expressed to three digits, 9.98, it would have a relative uncertainty of essentially 1 part per 1000 (that is, 1/998, which is essentially 1/1000), much less relative uncertainty than is justified from the least-certain digit. In the second example, the least-certain factor, 0.98, has a relative uncertainty of essentially 1 part in 100 (1/98). If the result were expressed to only two digits, 1.1×10^2, the uncertainty would be about 1 part in 10, far greater than the relative uncertainty implied by the least-certain factor.

Rounding Logarithms

> Rule 7. The **logarithm** of a number should contain as many significant figures in its mantissa as there are in the number.

EXAMPLE

Log $2.71 \times 10^4 = 4.\underline{433}$

　　　　　　↑　↑
　　　　　　Characteristic   Mantissa with three significant digits

Log $\underline{9475} = 3.\underline{9766}$

　　　　↑　　　　　↑
　　　　|　　　　Mantissa with four significant digits
　　Number with four significant digits

3.4   Some Terms Pertaining to Quantitative Results

Absolute Error

To ensure minimum error, it is common for several determinations to be made in the analysis of the sample for a specific constituent. The result of each determination is commonly designated as x_i, where i indicates the order in which the particular measurement was made—for example, 1 for the first, 5 for the fifth, and so on. The **mean**, or **average**, of a set of measurements is given by $\bar{x}$, where

$$\bar{x} = \frac{x_1 + x_2 + \cdots + x_n}{n} \tag{3.3}$$

Expressed more concisely, this relationship is

$$\bar{x} = \frac{\sum_{i=1}^{n} x_i}{n} = \frac{1}{n} \sum_{i=1}^{n} x_i \tag{3.4}$$

This equation may be abbreviated simply to

$$\bar{x} = \frac{\sum x_i}{n} \tag{3.5}$$

EXAMPLE

Four determinations of the percentage of sulfate in a sample gave values of 9.35, 9.29, 9.38, and 9.27%. What is $\bar{x}$?

$$\bar{x} = \frac{9.35 + 9.29 + 9.38 + 9.27}{4} = 9.32\%$$

Note that in the above calculation the 4 is considered to be an exact number (see Section 3.3), and does not enter into the consideration of significant figures.

The **true value** of a parameter, such as the percentage of sulfate in a sample, is given by the symbol μ. If the true value is known, the **absolute error** of an individual value is given by

$$\text{Absolute error} = x_i - \mu \tag{3.6}$$

and the absolute error of the mean of a set of values is

$$\text{Absolute error} = \bar{x} - \mu \tag{3.7}$$

EXAMPLE

If the true value of the percentage of sulfate in the example just cited was 9.30% sulfate, what were the absolute errors of the individual values and of the mean?

$$\mu = 9.30\%$$

x_i	Absolute error, % sulfate
9.35	$9.35 - 9.30 = 0.05$
9.29	$9.29 - 9.30 = -0.01$
9.38	$9.38 - 9.30 = 0.08$
9.27	$9.27 - 9.30 = -0.03$

Absolute error of mean $= 9.32 - 9.30 = 0.02\%$

Relative Error

Often it is more meaningful to put error into relative terms as a fraction, percentage, or in parts per thousand. Error thus expressed is **relative error**. The equations below apply:

$$\text{Relative error of an individual value as a fraction} = \frac{\text{absolute error}}{\text{true value}} = \frac{x_i - \mu}{\mu} \tag{3.8}$$

$$\text{Relative error of an individual value as percent} = \frac{x_i - \mu}{\mu} \times 100 \tag{3.9}$$

$$\text{Relative error of an individual value as parts per thousand} = \frac{x_i - \mu}{\mu} \times 1000 \tag{3.10}$$

$$\text{Relative error of the mean as a fraction} = \frac{\bar{x} - \mu}{\mu} \tag{3.11}$$

The relative errors of the mean in percentages and parts per thousand (ppt) are calculated simply by multiplying the expression in Equation 3.11 by 100 and 1000, respectively.

EXAMPLE

For the sulfate analysis example, calculate the relative error of the mean in percentage and parts per thousand.

$$\text{Relative error, } \% = \frac{9.32 - 9.30}{9.30} \times 100 = 0.22$$

$$\text{Relative error, ppt} = \frac{9.32 - 9.30}{9.30} \times 1000 = 2.2$$

3.5   Mathematical Expressions of Precision

In most cases the true value is not known; if it were, there would be no need to perform a chemical analysis. Whenever several analyses are performed on the same sample, however, it is always possible to express the precision mathematically. This is done by several means.

The simplest means of expressing precision is the **range**, or **spread**, which is simply the difference between the highest and lowest values of a set of results.

$$\text{Range} = x_{\text{highest}} - x_{\text{lowest}} \tag{3.12}$$

EXAMPLE

What is the range for the set of sulfate analysis results discussed in the preceding section?

$$\text{Range} = 9.38 - 9.27 = 0.11\%$$

Range is commonly given the symbol w. The relative range of a set of results is expressed as

$$\text{Relative } w = \frac{w}{\bar{x}} \tag{3.13}$$

which is simply the range divided by the mean value of the results. Multiplying by 100 gives the relative range in percent, and multiplying by 1000 gives the relative range in parts per thousand.

EXAMPLE

Calculate the relative range in parts per thousand for the sulfate analysis example in the preceding section.

$$\text{Relative range, ppt} = \frac{w}{\bar{x}} \times 1000 = \frac{0.11}{9.32} \times 1000 = 12$$

The **deviation** d_i from the mean of a particular result is given by

$$d_i = x_i - \bar{x} \tag{3.14}$$

The **average deviation** $\bar{d}$ from the mean is calculated by summing the *absolute values* of all the individual deviations and dividing by the number of results as given by the equation

$$\bar{d} = \frac{\sum |d_i|}{n} \tag{3.15}$$

The relative average deviation is, of course,

$$\text{Relative average deviation} = \frac{\bar{d}}{\bar{x}} \tag{3.16}$$

and multiplying this value by 100 and 1000 gives the relative average deviation in percent and parts per thousand, respectively.

EXAMPLE

Four determinations of the molar concentration of an acid solution gave values of 0.1113, 0.1092, 0.1120, and 0.1105 M. Calculate the mean, the deviation from the mean of each result, and the relative average deviation in parts per thousand.

$$\bar{x} = \frac{0.1113 + 0.1092 + 0.1120 + 0.1105}{4} = 0.1108$$

x_i	d_i	$|d_i|$
0.1113	0.0005	0.0005
0.1092	−0.0016	0.0016
0.1120	0.0012	0.0012
0.1105	−0.0003	0.0003
	$\sum |d_i| = 0.0036$	

$$\text{Relative average deviation, ppt} = \frac{0.0036/4}{0.1108} \times 1000 = 8.1$$

3.6   Standard Deviation

Calculation of Standard Deviation

Standard deviation is a statistically significant measure of precision for a large set of values. When the true value μ is known, the standard deviation is given the symbol σ and is expressed by the formula

$$\sigma = \sqrt{\frac{\sum (x_i - \mu)^2}{n - 1}} \tag{3.17}$$

where n and x_i are as defined earlier in this chapter. When the true value is not known, the mean value $\bar{x}$ is substituted for μ, giving s, the **standard deviation from the mean**, usually simply called standard deviation.

$$s = \sqrt{\frac{\sum (x_i - \bar{x})^2}{n - 1}} = \sqrt{\frac{\sum d_i^2}{n - 1}} \tag{3.18}$$

An exactly equivalent form of the above equation that makes calculation easier is

$$s = \sqrt{\frac{\sum x_i^2 - (\sum x_i)^2/n}{n - 1}} \tag{3.19}$$

The formula for *relative standard deviation* is

$$\text{Relative standard deviation} = \frac{s}{\bar{x}} \qquad (3.20)$$

This fraction multiplied by 100 or 1000 gives the relative standard deviation in percent and parts per thousand, respectively.

EXAMPLE

From the four molar concentrations cited at the end of Section 3.5, calculate s and the relative standard deviation in percent.

The value of s can be calculated by substitution into Equation 3.18. The values of d_i needed for this calculation are cited in the example at the end of Section 3.5.

$$s = \sqrt{\frac{d_1^2 + d_2^2 + d_3^2 + d_4^2}{4 - 1}}$$

$$= \sqrt{\frac{(0.0005)^2 + (0.0016)^2 + (0.0012)^2 + (0.0003)^2}{3}}$$

$$= 1.2 \times 10^{-3}$$

$$\text{Relative standard deviation} = \frac{1.2 \times 10^{-3}}{0.1108} \times 100 = 1.1\%$$

In using the alternative formula for standard deviation, Equation 3.19, the values of x_i^2 should be calculated to twice as many significant figures as the values of x_i, and rounding off should be delayed until the end of the calculation; otherwise an erroneous value of standard deviation is obtained. For the example just cited above, the first value of x_i is 0.1113. Squaring this should give 0.01238769. Some hand calculators refuse to carry this many digits if used in the conventional manner; for example, an attempt to square 0.1113 will cause the calculator to display 1.2387 -02 (1.2387×10^{-2}), which has only five significant figures. This problem can be overcome by squaring 1.113 and keeping track of the exponent—that is $(10^{-1})^2$. Thus, squaring 1.113 gives 1.238769, and multiplying by 10^{-2} (moving the decimal two places to the left) gives the correct answer.

Standard Deviation of the Mean

Whereas s refers to the standard deviation of individual measurements, it is sometimes desirable to express the standard deviation of the mean value of a set of experimental measurements. This parameter is called the **standard deviation of the mean**, designated s_m, and given by the formula

$$s_m = \frac{s}{\sqrt{n}} \qquad (3.21)$$

For this particular example of the standard deviation of four concentrations in molar units, the standard deviation of the mean is

$$s_m = \frac{s}{\sqrt{n}} = \frac{1.2 \times 10^{-3}}{\sqrt{4}} = 6.0 \times 10^{-4} \tag{3.22}$$

3.7   Distribution of Random Errors

Data with Indeterminate Error

As mentioned in Section 3.2, random or indeterminate errors are those that vary in magnitude and direction and cannot be assigned a source. If a large number of analyses of the same sample are run by an individual using procedures in which only random errors are present, the results will show a spread of values, most of which are relatively close to a central value, whereas a few are relatively very high and a few are very low. Consider, for example, the data in Table 3.2 for percent Fe_2O_3 determined in an iron ore sample.

TABLE 3.2   Percent Fe_2O_3 found in many analyses of the same iron ore sample

30.12	30.17	30.18	30.23	30.26
30.27	30.29	30.32	30.33	30.36
30.36	30.37	30.38	30.39	30.41
30.41	30.41	30.42	30.43	30.43
30.43	30.44	30.44	30.46	30.47
30.48	30.48	30.48	30.48	30.51
30.53	30.53	30.53	30.57	30.57
30.59	30.63	30.67	30.68	30.78

These data can be plotted on a **histogram**, or bar graph, as shown in Figure 3.2. Such a plot groups data into *cells*, each of which spans a range of percent Fe_2O_3. For example, examination of the data in Table 3.2 reveals one value, 30.12, that lies between 30.10 and 30.15. This is plotted as a bar corresponding to a frequency of 1. However, there are nine values lying between 30.40 and 30.45, so the cell encompassing these values is plotted as a bar nine units high. The midpoints at the top of each bar may be connected by lines composing a **frequency polygon**, which gives the rough shape of the plot.

Gaussian Distribution of Data

The shape of the frequency polygon approaches that of a bell–shaped curve. In fact, if a vast number of points were taken, a smooth curve known as a **Gaussian distribution** would be obtained.

Figure 3.3 is a Gaussian distribution or **normal error curve**. It is the kind of curve that would result from a frequency polygon involving an enormous

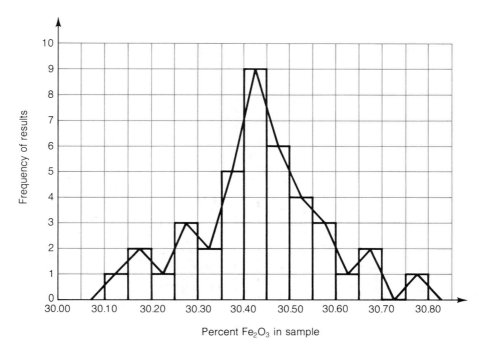

FIGURE 3.2 Histogram plotting frequency of results versus range in percent Fe_2O_3 for the data given in Table 3.2

number of results for the analysis of the same sample in which only indeterminate errors were involved. The properties of this curve include the following:

1. The curve is symmetrical, showing an equal probability of positive and negative errors.
2. There is a maximum frequency of occurrence of results at the central value, which is the value of zero indeterminate error.
3. The exponential decrease in frequency on either side of the central value shows that a large indeterminate error is much less likely to occur than a small one.

If the Gaussian distribution is plotted with the value of the result as the abscissa, the central value is the mean of the results, $\bar{x}$. The abscissa may also be plotted as the deviation from the mean or as the number of standard deviations; in these two cases both positive and negative values are plotted on the abscissa, and the central value is zero. Where y is the frequency, the shape of a Gaussian curve is given by the equation

$$y = \frac{1}{s\sqrt{2\pi}} \exp \frac{-(x - \bar{x})^2}{2s^2} \tag{3.23}$$

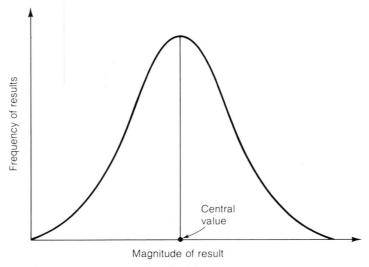

FIGURE 3.3   Gaussian distribution

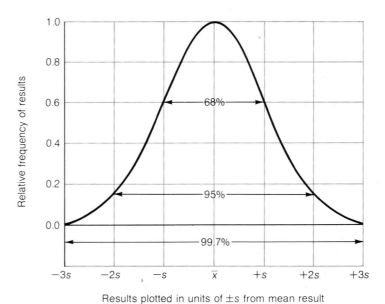

FIGURE 3.4   Gaussian plot of relative frequency of result versus magnitude of the result plotted in units of $\pm s$ from the mean. Percentages denote percent of all values falling under the area of the curve indicated by the arrows.

in which s, x, $\bar{x}$ have been defined previously. The value of π is 3.14159 ... and "exp" stands for e (the base of natural logarithms) taken to the power designated, in this case, $-(x - \bar{x})^2 / 2s^2$.

The precision of a large set of measurements is indicated by the shape of a Gaussian curve. A set of highly precise results gives a narrow, high curve; a set of imprecise results with a high standard deviation gives a broad, low curve.

The relationship of standard deviation to the Gaussian error curve is shown in Figure 3.4. It is significant to note that 68% of all results lie within $\pm 1s$, 95% lie within $\pm 2s$, and 99.7% are within $\pm 3s$. Only 0.3% of the results are outside $\pm 3s$.

As noted previously, a Gaussian distribution of experimental results is always centered around $\bar{x}$. Frequently, because of determinate error, $\bar{x}$ differs from the true value μ. In such a case the curve is offset to one or the other side of μ.

3.8   Confidence Limits

If a huge number of analyses—1000 or 10,000—were run on the same sample, the results would give a Gaussian distribution like that in Figure 3.4, and an excellent value of $\bar{x}$ would be obtained. In practice it is usually impractical to do more than three or four analyses of the same sample. It is thus desirable to have some measure of confidence that the mean of these few samples lies within a certain distance of the exact mean that would be obtained by running an enormous number of samples. The **confidence limit** expresses the range within which the true mean is likely to lie, within a certain percentage of probability. For example, expressing the percent Fe_2O_3 determined in a sample as,

$$\text{Percent } Fe_2O_3 = 27.25 \pm 0.17\% \qquad (95\% \text{ confidence level}) \qquad (3.24)$$

states that there is a 95% probability that the true mean in the determination of Fe_2O_3 lies within the range of $27.25 \pm 0.17\%$ assuming no determinate error.

Use is made of a parameter t to estimate the precision of a mean value when only a limited number of replicate determinations must be used for calculating both the mean and a realistic estimate of the precision. Values of t are tabulated as shown in Table 3.3. In this tabulation the values of degrees of freedom are equal to $n - 1$, where n is the number of replicate determinations made. In the absence of determinate error, the value of a result based upon a limited number of replicate measurements can be given as the following:

$$\mu = \bar{x} \pm \frac{ts}{\sqrt{n}} \qquad (3.25)$$

TABLE 3.3   Values of t for estimating confidence limits

Degrees of freedom	Factor for confidence interval, %				
	80	90	95	99	99.9
1	3.08	6.31	12.7	63.7	637
2	1.89	2.92	4.30	9.92	31.6
3	1.64	2.35	3.18	5.84	12.9
4	1.53	2.13	2.78	4.60	8.60
5	1.48	2.02	2.57	4.03	6.86
6	1.44	1.94	2.45	3.71	5.96
7	1.42	1.90	2.36	3.50	5.40
8	1.40	1.86	2.31	3.36	5.04
9	1.38	1.83	2.26	3.25	4.78
10	1.37	1.81	2.23	3.17	4.59
11	1.36	1.80	2.20	3.11	4.44
12	1.36	1.78	2.18	3.06	4.32
13	1.35	1.77	2.16	3.01	4.22
14	1.34	1.76	2.14	2.98	4.14
∞	1.29	1.64	1.96	2.58	3.29

In the preceding equation t is a number obtained from a table of t values and corresponding to a designated confidence interval and the number of degrees of freedom, $n - 1$.

When the correct value is not known, the parameter $ts/\sqrt{n}$ can still be used to calculate the confidence limits of the mean of several replicate measurements as shown in the following example.

EXAMPLE

Four replicate determinations of dissolved oxygen in water gave values of 6.34, 6.67, 5.99, and 6.08 mg/L dissolved O_2. What are the confidence limits for the mean value at the 95% probability level?

Degrees of freedom $= n - 1 = 4 - 1 = 3$

From Table 3.3, at 3 degrees of freedom and 95% probability level, $t = 3.18$.

$$\bar{x} = \frac{6.34 + 6.67 + 5.99 + 6.08}{4} = 6.27 \text{ mg/L}$$

$$s = \sqrt{\frac{\sum x_i^2 - (\sum x_i)^2/n}{n - 1}} = \sqrt{\frac{157.531 - 629.0064/4}{3}} = 0.305$$

$$\text{Confidence limit} = \pm\frac{ts}{\sqrt{n}} = \pm\frac{3.18 \times 0.305}{\sqrt{4}} = \pm 0.48$$

The concentration of dissolved oxygen is 6.27 ± 0.48 mg/L at the 95% probability level.

It should be noted that the higher the standard deviation, the wider the range of confidence limits. Also, confidence limits are wider for higher probability levels. Thus the confidence limits are considerably wider at the 99% probability level than at the 80% probability level.

3.9   ## Student's *t* Test

Student's *t* test is a way of comparing two means from two different sets of data to determine the probability that they are the same. For example, suppose that a federal laboratory and a state environmental protection laboratory both analyze the same soil sample suspected of being contaminated with toxic 2, 3, 7, 8-tetrachloro-*p*-dioxin, the infamous "dioxin" associated with Agent Orange herbicide. Each agency analyzes several samples. One set of m analyses of the sample gives a mean value of $\bar{x}$, and the other set of n analyses gives a mean value of $\bar{y}$. The symbol s is given to the **pooled standard deviation** of both sets of data and is calculated by the following formula:

$$s = \left[\frac{\sum(x_i - \bar{x})^2 + \sum(y_i - \bar{y})^2}{(m-1)+(n-1)}\right]^{1/2} = \left[\frac{\sum x_i^2 - (\sum x_i)^2/m + \sum y_i^2 - (\sum y_i)^2/n}{(m-1)+(n-1)}\right]^{1/2} \tag{3.26}$$

In the preceding equation, x_i and y_i represent the ith value for each of two sets of data, respectively. With the above equations it is now possible to compare the means of the two following sets of data:

Results, parts per billion (ppb) dioxin in soil, federal agency laboratory: 91, 87, 101, 99, 95   ($m = 5$)

Results, ppb dioxin in soil, state agency laboratory: 95, 92, 81, 89, 94, 93   ($n = 6$)

If x and m are used to represent federal results and y and n are employed to represent state results, the pooled standard deviation is calculated as follows:

$$\bar{x} = \frac{91 + 87 + 101 + 99 + 95}{5} = 95 \tag{3.27}$$

$$\bar{y} = \frac{95 + 92 + 81 + 89 + 94 + 93}{6} = 91 \tag{3.28}$$

$$\sum x_i^2 = 91^2 + 87^2 + 101^2 + 99^2 + 95^2$$
$$= 8281 + 7569 + 10{,}201 + 9801 + 9025 = 44{,}877 \tag{3.29}$$

$$(\sum x_i)^2 = (91 + 87 + 101 + 99 + 95)^2 = 473^2 = 223{,}729 \tag{3.30}$$

$$\sum y_i^2 = 95^2 + 92^2 + 81^2 + 89^2 + 94^2 + 93^2$$
$$= 9025 + 8464 + 6561 + 7921 + 8836 + 8649 = 49{,}456 \tag{3.31}$$

$$(\sum y_i)^2 = (95 + 92 + 81 + 89 + 94 + 93)^2 = 544^2 = 295{,}936 \tag{3.32}$$

Substituting these values into Equation 3.26 gives,

$$s = \left[\frac{44{,}877 - 223{,}729/5 + 49{,}456 - 295{,}936/6}{(5-1) + (6-1)} \right]^{1/2} = 5 \tag{3.33}$$

Student's t, t_{exp}, may now be calculated from the following equation

$$t_{exp} = \frac{(\bar{x} - \bar{y})}{s} \sqrt{\frac{mn}{(m+n)}} = \frac{95 - 91}{5} \sqrt{\frac{5 \times 6}{5 + 6}} = 1 \tag{3.34}$$

$$\text{Degrees of freedom} = m - 1 + n - 1 = 5 - 1 + 6 - 1 = 9 \tag{3.35}$$

Reference to Table 3.3 reveals that for 9 degrees of freedom at the 95% confidence level, $t = 2.26$. Therefore,

$$t_{exp} < t \tag{3.36}$$

$$1 < 2.26$$

Since t_{exp} is less than t, there is at least a 95% probability that the values for the two portions of the sample are the same.

3.10   Rejection of an Experimental Result

One of the greater concerns for students in the quantitative analysis laboratory is whether or not to reject a result that appears out of place. The quantitative method for rejecting a result is based upon the **Q test**. To see how this is done, consider the following values for percent sulfur in a bituminous coal sample: 4.01, 3.99, 3.96, 4.13. The last value may be sufficiently higher than the other three to justify discarding it; such a suspicious value is called a *discordant result*. In deciding whether or not to reject a discordant value, the experimental aspects of the analysis should be reviewed first. This may reveal that an experimental error was made in obtaining the suspicious value, providing grounds for its rejection. Otherwise, an *experimental Q value* called Q_{exp} should be calculated from the formula

$$Q_{exp} = \frac{\Delta_{dev}}{w} \tag{3.37}$$

where w has previously been defined as the range and Δ_{dev} is the difference between the discordant result and the value closest to it in the remaining results. In this case,

$$w = 4.13 - 3.96 = 0.17 \tag{3.38}$$

$$\Delta_{dev} = 4.13 - 4.01 = 0.12 \tag{3.39}$$

$$Q_{exp} = \frac{0.12}{0.17} = 0.71 \tag{3.40}$$

TABLE 3.4 Critical Q values at three probability levels

Number of observations	Q_{crit} (reject if $Q_{exp} > Q_{crit}$)		
	90% confidence	96% confidence	99% confidence
3	0.94	0.98	0.99
4	0.76	0.85	0.93
5	0.64	0.73	0.82
6	0.56	0.64	0.74
7	0.51	0.59	0.68
8	0.47	0.54	0.63
9	0.44	0.51	0.60
10	0.41	0.48	0.57

Source: Reproduced from W. J. Dixon, *Ann. Math. Stat.*, 22, 68 (1951).

The value of Q_{exp} calculated above is to be compared to a table of critical values of Q, Q_{crit}, as shown in Table 3.4. At the 90% probability level for four replicates, the value of Q_{crit} is 0.76. Therefore, it is seen that

$$Q_{exp} < Q_{crit} \tag{3.41}$$

$$0.71 < 0.76$$

Since Q_{exp} is less than Q_{crit} at the 90% probability level, the discordant value *should not* be discarded; conversely, if it is greater, it may be discarded.

3.11 Plotting Experimental Data

The Nature of Plots of Data

Simple instruments are commonly used in quantitative analysis laboratories. *Plotting* or *graphing* of the data from these instruments is often required. Such data include plots of ion–selective electrode potential versus log of ion concentration for potentiometric analysis (Chapter 13), plots of absorbance of light versus concentration for spectrophotometric analysis (Chapter 21), and plots of chromatographic peak height versus quantity of analyte injected into the chromatograph (Chapter 19).

For most uses in the quantitative analysis laboratory, rectilinear coordinate paper ruled 10 squares to the centimeter with darker lines every 5 or 10 cm can be employed for graphing. Other types, such as semilogarithmic graph paper, are available as well. The first step in graphing is to define the parameters and scale of each of the two axes. Conventionally, the *independent variable*, such as concentration of standard or volume of titrant added in a titration, is plotted on the *x* axis, or *abscissa*. The *dependent variable*, such as absorbance or electrode potential, is plotted on the *y* axis or *ordinate*. These axes should be labeled carefully, both as to their identity and scale (units per division).

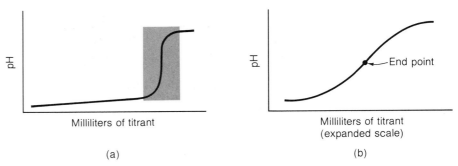

FIGURE 3.5   Titration curve for the titration of an acid with a base. Entire titration curve is shown in (a); the shaded portion encompassing the end point is shown in (b), where an expanded abscissa scale enables much more accurate reading of the volume of titrant at the end point.

Choosing the scale for each axis of graph paper is an important operation, now often simplified by programming computers to plot the data on the optimum scales with a printer. Generally the scales should be chosen such that the entire plot fills the sheet of graph paper as nearly as possible. This allows reading the graph to the maximum number of significant figures. In some cases it may be desirable to plot only part of the data by expanding at least one of the scales. This is true of a titration curve (Chapter 5), typically a plot of measured pH versus volume of titrant added. Most of the plot of an entire titration curve [Figure 3.5(a)] is not needed for measurement of the end point volume, which occurs midway (at the inflection point) of the S-shaped break in the titration curve. The volume of titrant can be read much more closely by expanding the volume scale over the area encompassed by the titration curve break, as shown in Figure 3.5(b). This plot may not be so esthetically pleasing, because of the emphasized scatter of fewer data points, but it does enable reading the end point volume to the required degree of accuracy.

Fitting Lines to Data

After the units for the axes have been chosen, it is necessary to draw a curve which is the *best fit*. Except in the case where the plot is known to be a straight line (to be considered later), such a plot can be prepared simply with the use of a French curve (curved plastic template). For example, consider the data in Table 3.5, giving pH as a function of volume of NaOH titrant solution for the titration of an acid with a base. The data are given only over a 2-mL range encompassing the S-shaped end point break in the titration curve. Examination of the curve in Figure 3.6 shows that there is some scatter in the data. Plotting on an expanded scale tends to make this scatter appear relatively larger, but in reality the plot is much better on the expanded scale because it enables reading the volume much more accurately. As will be explained in Chapter 5,

TABLE 3.5 Values of pH as a function of volume of added base for the titration of acid with NaOH

Volume of NaOH, mL	pH
46.31	2.6
46.60	3.0
46.78	4.2
46.95	5.4
47.12	7.1
47.30	8.5
47.43	9.9
47.66	10.4
47.93	12.0

there are simple graphical means for determining the midpoint of a titration curve break. For the curve shown in Figure 3.6, a vertical dashed line has been drawn downward from the midpoint to the volume axis corresponding to a volume of 47.15 mL. Plotted on this scale, the end point volume can be read as closely as the volume on a 50-mL laboratory buret.

Whenever possible, mathematical relationships are developed so that plots of data are straight lines. After choosing the scales for the axes of such plots and placing the data points in the right location, the first step is to draw the best straight line through the points. Sometimes the plot appears like the one

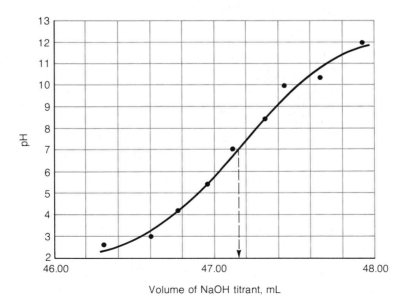

FIGURE 3.6 Plot of end point break for the titration of acid with base; data from Table 3.5.

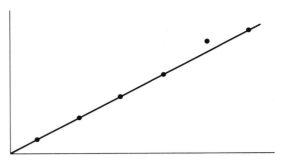

FIGURE 3.7   Linear plot for six data points, one of which is off the line

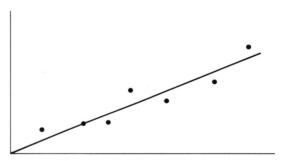

FIGURE 3.8   Straight-line plot of laboratory data showing scatter from indeterminate error

in Figure 3.7 where all but one of the points fall exactly on the straight line. The one point that does not fall on the line with the rest is clearly irregular and should be discarded. Mathematical techniques for calculating the straight line best fitting the data (described below) should not be employed on all six of these data points, because the weight given to the aberrant point will give the line an incorrect slope.

Laboratory data are often scattered as shown in Figure 3.8. Here there is no point that is clearly out of place; the data are simply scattered due to high indeterminate error, and all points should be considered in making the plot.

Least-Squares Method

Data with significant scatter, like that plotted in Figure 3.8, can be handled mathematically to obtain a straight line that best fits the data points. The method used is that of **least squares**, also called **linear regression analysis**. We will discuss this method in terms of the equation for a linear plot

$$y = mx + b \tag{3.42}$$

where m is the slope of the linear plot and b is the y intercept. The method in its simplest form is applicable when there is essentially no error in x, whereas the errors in y are random and presumed to fit a Gaussian distribution. The method minimizes the squares of the deviations of the data points from the straight line, thus giving the best values of y and m.

Consider as an example that an analyst was given the job of determining the total amount of sulfur in coal by analyzing portions of the same coal sample ranging in mass from exactly 1 g to exactly 5 g. These samples are weighed on an analytical balance so that the independent variable, mass of sample, is known to within ±0.0001 g (0.1 ppt for a 1-g sample). The sulfur is determined by burning the coal in a horizontal column of silica tubing that is heated in a tube furnace and through which a stream of oxygen flows

$$S(\text{in coal}) + O_2 \longrightarrow SO_2(g) \tag{3.43}$$

allowing the stream of gas with combustion products to flow into a trap containing basic hydrogen peroxide

$$SO_2 + H_2O_2 + 2OH^- \longrightarrow SO_4^{2-} + 2H_2O \tag{3.44}$$

eliminating the hydrogen peroxide by boiling, adjusting the pH appropriately, and precipitating $BaSO_4(s)$ by adding $BaCl_2$ solution

$$Ba^{2+} + SO_4^{2-} \longrightarrow BaSO_4(s) \tag{3.45}$$

The $BaSO_4$ is ignited (see Section 2.6) and weighed, and the mass of sulfur in the original sample is calculated. The analytical data are shown in Table 3.6.

TABLE 3.6 Data for least-squares plot of mass of sulfur in coal versus sample mass

Mass of coal, g (x_i)	x_i^2	Mass of S in coal, mg (y_i)	$x_i y_i$
1	1	6.5	6.5
2	4	26.0	52.0
3	9	26.0	78.0
4	16	42.8	171
5	25	47.8	239

The least-squares method gives the values of m and b for Equation 3.42 from the two following equations.

$$m = \frac{n\sum x_i y_i - \sum x_i \sum y_i}{n\sum x_i^2 - (\sum x_i)^2} \tag{3.46}$$

$$b = \frac{\sum x_i^2 \sum y_i - \sum x_i \sum x_i y_i}{n\sum x_i^2 - (\sum x_i)^2} \tag{3.47}$$

For the data given in Table 3.6,

$$\sum x_i = 1 + 2 + 3 + 4 + 5 = 15 \tag{3.48}$$

$$\sum y_i = 6.5 + 26.0 + 26.0 + 42.8 + 47.8 = 149 \tag{3.49}$$

$$\sum x_i^2 = 1 + 4 + 9 + 16 + 25 = 55 \tag{3.50}$$

$$\sum x_i y_i = 6.5 + 52.0 + 78.0 + 171 + 239 = 547 \tag{3.51}$$

The calculation of m is

$$m = \frac{5 \times 547 - 15 \times 149}{5 \times 55 - 15^2} = 10.0 \text{ mg S/g coal} \tag{3.52}$$

The calculation of b is

$$b = \frac{55 \times 149 - 15 \times 547}{5 \times 55 - 15^2} = -0.2 \text{ mg S} \tag{3.53}$$

Thus, according to the least-squares calculation of the data, the y intercept of the line is at -0.2 mg S, and the line has a positive slope of 10.0 mg S/g coal. These factors are used in Figure 3.9 to draw the best straight-line fit to the data.

The mathematical tedium of the least-squares method is relieved considerably by modern pocket calculators or personal computers programmed to do it. There are also mathematical processes for fitting data that produce curved plots. These are generally too involved to do with a calculator, but they are readily done with a properly programmed computer.

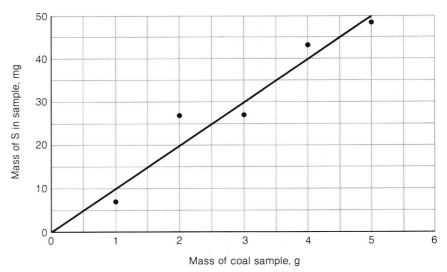

FIGURE 3.9   Least-squares plot of mass of sulfur in coal versus sample mass

Programmed Summary of Chapter 3

This summary contains in a programmed format the major terms and concepts introduced in this chapter. To get the most out of it, you should fill in the blanks for each question and then check the answers at the end of the book to see if you are correct.

Indeterminate or random errors are manifested by a lack of (1) _____ in laboratory data. An error with a definite cause that is always either positive or negative is a (2) _____ error. The size of the uncertainty in a number is called the (3) _____, whereas the uncertainty divided by the number itself is the (4) _____. The term *accuracy* describes how close data values or their mean are to the (5) _____. In expressing a number, retain no digits beyond the (6) _____. The following numbers properly rounded to three significant digits are $4.937 = $ (7) _____, $4.932 = $ (8) _____, $4.935 = $ (9) _____, and $4.965 = $ (10) _____. The sum $31.3 + 0.9625 + 337.98 = 370.2425$ properly rounded is (11) _____. The product $3.433 \times 9.8 \times 3.38041 = 113.72849$ properly rounded is (12) _____. The logarithm of a number should contain as many significant digits in its (13) _____ as there are in the number. The number of significant digits in 0.0291 is (14) _____, and the number in 0.023700 is (15) _____. Confusion in the number of significant digits in a number such as 42,000 can be eliminated by using (16) _____. The mean of a set of measurements with values of x_1, x_2, x_3, and x_4 is denoted by the symbol (17) _____ and is given by the formula $\bar{x} = $ (18) _____. The true value of a parameter is designated by (19) _____ and the absolute error of the mean of a set of measurements is (20) _____. The relative error of a mean of a set of measurements in parts per thousand is (21) _____. A set of measurements ranks $x_3 > x_1 > x_4 > x_2$ in magnitude; the value of w is (22) _____. The symbol and name of the quantity $x_i - \bar{x}$ is (23) _____. The average deviation $\bar{d}$ of a set of n measurements is $\bar{d} = $ (24) _____. The standard deviation s is given by the two equivalent formulas, $s = $ (25) _____. For a very large number of results, a plot of frequency of result versus magnitude of result has the shape of a (26) _____. For a Gaussian distribution, the percentages of results that lie within ± 1, ± 2, and ± 3 standard deviations are, respectively, (27) _____. When only several values of an analysis are available, the range, within a certain percentage of probability, in which the true mean is likely to lie is called the (28) _____. In terms of t values, the confidence limit for n results is (29) _____. For two sets of data on the analysis of the same sample, where s is the pooled standard deviation, $\bar{x}$ is the mean of one set involving m analyses, and $\bar{y}$ is the mean of the other set involving n analyses, the formula for Student's t is $t_{exp} = $ (30) _____. When $t_{exp} > t$ at the $X\%$ confidence level, there is an $X\%$ probability that the means of the two results are (31) _____. The test to see whether one apparently aberrant datum should be discarded from a set of values is the (32) _____. Where x_o is the "outlying" value of suspicion in a set of data, x_l is the lowest value, x_h is the highest value (one of these will also be x_o), and x_n is the value nearest x_o, the formula for the experimental Q value, $Q_{exp} =$ (33) _____. At a particular percentage confidence level a suspicious outlying value should be discarded when Q_{exp} (34) _____. In graphing data the value that is conventionally plotted on the abscissa is the (35) _____, whereas that plotted on the ordinate is the (36) _____. In graphing data, generally a scale should be chosen such that (37) _____, whereas in some cases, such as titrations, it may be desirable to (38) _____. In graphing, a line is drawn that gives the (39) _____ for the data. The mathematical technique for drawing the best straight line through scattered data is the (40) _____. The fundamental equation for a straight line is (41) _____.

Questions

1. A gold nugget weighing 1.3072 g was weighed by four different students on an analytical balance accurate to 0.1 mg. The values reported by the four students were 1.3055, 1.3091, 1.3097, and 1.3041 g. What may be said about the accuracy and precision of these results?

2. What is wrong with expressing the volume delivered from a 50-mL laboratory buret as 37.3 mL?

3. Round off the following numbers to one less digit: (a) 57.32, (b) 4.837, (c) 811.35, (d) 0.1185.

4. Where X is the relative uncertainty of the factor with the greatest uncertainty in a product or quotient, what should be the range of the relative uncertainty of the result?

5. Express each of the following to the correct number of significant digits:
(a) log 3185, (b) log 0.00076,
(c) log 2.97×10^5, (d) log 6.0400×10^{-3}.

6. Write the following numbers in a manner that correctly denotes the number of significant digits in each: (a) 37,800 to three

significant digits, (b) 72,000 to four significant digits, (c) 90,083 to three significant digits, (d) 129,000 to six significant digits.

7. Define the following symbols pertaining to the expression of quantitative results: (a) x_i, (b) μ, (c) $\bar{x}$, (d) $x_i - \mu$, (e) $\bar{x} - \mu$.

8. Using the appropriate symbols, define: (a) range of a set of results, (b) relative range, (c) deviation from the mean of an individual result, (d) average deviation from the mean.

9. What is the formula for the relative average deviation from the mean in (a) percent and (b) parts per thousand?

10. What is the distinction between σ and s in respect to standard deviation?

11. Into what form does a frequency polygon representing a random distribution of errors evolve if a very large number of analyses of the same sample are performed?

12. Regarding the Gaussian distribution, what is indicated by (a) symmetry of the curve, (b) maximum frequency of occurrence of results at the central value, and

(c) exponential decrease in frequency on either side of the central value.

13. What are the effects of (a) relatively higher standard deviation and (b) relatively fewer analyses upon confidence limits.

14. What is the main application of the Q test?

15. The data in the table below were obtained for a plot of the concentration of a light-absorbing species C in units of moles per liter versus absorbance A. The results are to be plotted on a graph within a 10×10 cm square of graph paper ruled at 1-mm intervals with heavier rulings at each cm units. Describe how the data should be graphed, including choice of axes, scale, whether or not to make the plot a straight line, and any other significant aspects of the process.

C, $\times 10^{-4}$	A
0	0
1.0	0.43
2.0	0.63
3.0	0.84
4.0	0.87
5.0	1.00

16. What is the basic principle of the method of least squares as applied to a linear plot?

17. Of the following, the one that is most indicative of low (absence of) indeterminate (random) error is (a) high precision, (b) agreement with the true value of the mean of a large number of analyses of a sample of known concentration, (c) a high value of the standard deviation, (d) meticulous calibration of the glassware used, (e) expression of all results to at least four digits.

18. Of the following, the one that is most indicative of the absence of determinate error is (a) high precision, (b) agreement with the true value of the mean of several analyses of a sample of known concentration, (c) a high value of the standard deviation, (d) meticulous calibration of glassware used, (e) expression of all results to at least four digits.

19. In the following figure, use a number, letter, or name to designate the following: X, Y, Z, F.

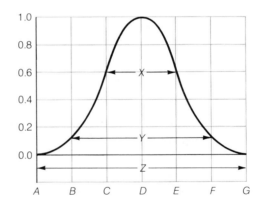

Problems

1. Express the answer to the following to the correct number of significant digits: $0.01 + 323.419 + 1.70031 - 21.228$.

2. What is the sum of $81.5401 - 57.902 + 0.0009 + 1.4611$, rounded properly?

3. What is the sum of $0.9318 + 32.5 - 27.099 + 0.059$, rounded properly?

4. Give the correctly rounded product, $3.25 \times 10^4 \times 0.07718 \times 0.40106$.

5. Give the properly rounded answer to $0.4701 \times 7.7328 \times 10^3 \times 2.238/981.733$.

6. Give the correctly rounded answer to $6317 \times 0.00340822 \times 99/(20.337 \times 0.41709)$.

7. Express the number whose log is 1.3317 to the correct number of significant digits.

8. Express the number whose log is −0.80173 to the correct number of significant figures.

9. Express the number whose log is −13.42 to the correct number of significant figures.

10. A sample of asphalt-based road surfacing material was analyzed for asphalt content by freezing in liquid nitrogen, grinding the cold sample to a powder, extracting asphalt from weighed portions of the ground sample with dichloromethane, and weighing the residues left from the evaporation of the solvent. The values of five different analyses were 5.18, 5.03, 5.11, 5.07, and 5.15%. The true value from formulation of the material was 5.20%. Please calculate the following: (a) $\bar{x}$ for the set of analyses, (b) the absolute error of $\bar{x}$, (c) the relative error of $\bar{x}$ in parts per thousand, (d) w, (e) d_i for each result, and (f) $\bar{d}$.

11. Calculate the same parameters as for Problem 10 with the following numbers determined for the molar concentration of a hydrochloric acid solution that was actually 0.1237 M in HCl: 0.1245, 0.1234, 0.1239, and 0.1246 M.

12. Calculate the same parameters calculated in the two preceding problems for the percentage of N in a commercial fertilizer for which the actual value was 32.18% N: 31.98, 32.34, 32.21, 32.12, and 32.25%.

13. For the data given in Problem 10, calculate the standard deviation from the mean.

14. Calculate the standard deviation from the mean for the data given in Problem 11.

15. Calculate the standard deviation from the mean for the data given in Problem 12.

16. Four different analyses for the percentage of water in lignite, a low-rank coal, gave values of 18.3, 18.9, 17.7, and 18.6%. What is the confidence limit at the 95% probability level?

17. What is the confidence limit at the 90% probability level for the analytical values of 87.3, 86.2, 88.6, 86.5, and 88.4 mg/L chloride determined in a sample of wastewater?

18. What is the confidence level at the 95% probability level for the following values of cadmium content in sewage sludge: 2.07, 1.93, 2.14, 1.98 mg Cd/kg sludge.

19. Two analysts each determined the molar concentration of HCl in a sample of hydrochloric acid. Analyst A got values of 0.1382, 0.1360, 0.1375, and 0.1366 M. Analyst B obtained values of 0.1364, 0.1358, 0.1354, and 0.1360 M. What may be said about the probability that the two means are the same?

20. Five different determinations were made of the calcium content of a calcium silicate furnace slag used as a substitute for lime to neutralize acid in soil. The calcium was isolated as a calcium oxalate precipitate, CaC_2O_4, and heated (ignited) to CaO, which was weighed. The values as % CaO obtained were 35.61, 35.32, 35.78, 35.22, and 34.52. Use the Q test to show if the last, possibly discordant value should be discarded.

21. Apply the Q test to the four acid-titration replicates, 0.0953, 0.0959, 0.0950, and 0.0969 M.

22. Apply the Q test to the three acid-titration replicates, 0.1122, 0.1125, and 0.1301 M.

23. Using the least-squares method on the data in the following table, calculate the parameters for a straight-line plot of A (measure of light absorbed) versus concentration of light-absorbing substance in solution.

C, $M \times 10^4$	A
1.00	0.150
2.00	0.279
3.00	0.505
4.00	0.679
5.00	0.758

24. The data in the table below give the potential of a fluoride ion–selective electrode (versus a reference electrode) as a function of log [F$^-$]. Such a plot should be linear. Fit a straight line to the data using the least–squares method.

log [F$^-$]	E, mV
−5.00	−100.0
−4.50	−120.2
−4.00	−162.5
−3.50	−194.5
−3.00	−211.5

4 Acid-Base Equilibria

LEARNING GOALS

1. Definitions of acids and bases as applied to quantitative analysis.
2. Meaning of conjugate acids and bases.
3. Role of solvents in acid-base phenomena, including amphiprotic solvents, leveling solvents, and differentiating solvents.
4. Meaning of acid-base equilibrium and equilibrium constants, including K_a, K_b, and K_w, and the relationships among these constants.
5. Principles of buffers and what is meant by buffer capacity.
6. Polyprotic acids and their equilibria.
7. Distribution of species diagram as it applies to polyprotic acids.

4.1   Introduction to Acids and Bases

Acids and bases are widely used reagents in quantitative analysis. In many cases the control of acidity or basicity of an analyte solution is crucial in performing a quantitative determination. Acids and bases themselves are common analytes in quantitative analysis. It is important for the analytical chemistry student to have a fundamental understanding of acids and bases and their equilibria in water.

Definitions of Acids and Bases

For the analytical chemist the most useful definition of an **acid** is *a substance capable of donating a proton*, H^+, whereas a **base** *is a substance capable of accepting a proton*. These definitions fit the Brønsted-Lowry theory of acids and bases formulated independently by these two scientists in 1923. The term **proton** is used to designate the H^+ ion nucleus. According to this definition, acetic acid dissolved in water functions as an acid by donating a proton to a water molecule

$$\underset{\text{acid}_1}{\text{HO}-\overset{\overset{\text{O}}{\|}}{\text{C}}-\overset{\overset{\text{H}}{|}}{\underset{\underset{\text{H}}{|}}{\text{C}}}-\text{H}} + \underset{\text{base}_2}{\text{H}_2\text{O}} \rightleftharpoons \underset{\text{base}_1}{{}^{(-)}\text{O}-\overset{\overset{\text{O}}{\|}}{\text{C}}-\overset{\overset{\text{H}}{|}}{\underset{\underset{\text{H}}{|}}{\text{C}}}-\text{H}} + \underset{\text{acid}_2}{\text{H}_3\text{O}^+} \tag{4.1}$$

to produce an acetate ion and a hydronium ion, H_3O^+. Ammonia dissolved in water accepts a proton from a water molecule

$$NH_3 + H_2O \rightleftarrows NH_4^+ + OH^- \qquad (4.2)$$

$\text{base}_1 \quad \text{acid}_2 \qquad \text{acid}_1 \quad \text{base}_2$

to produce an ammonium ion and a hydroxide ion, OH^-. As seen from the above reactions, an acid can donate a proton only if there is a base present to accept it, and a base can accept a proton only in the presence of a proton-donor acid. It is seen here that the solvent, water, can act as both an acid and a base depending upon the nature of the solute; such a solvent is said to be **amphiprotic**.

The double arrows in the reactions above indicate that these reactions are *reversible*. Acetic acid molecules are continually losing protons to form acetate ions, which in turn are continually accepting protons to form acetic acid molecules. However, as long as the total solute concentrations and temperature of the solution remain unchanged, the proportions of unionized acetic acid and acetate ions remain the same. Similarly, under constant conditions, a solution with a particular concentration of NH_3 retains the same proportions of NH_3 and NH_4^+, although individual NH_3 molecules may become NH_4^+ ions and vice versa.

Conjugate Acids and Bases

The loss of a proton from a Brønsted-Lowry acid results in the formation of the **conjugate base** of that acid, whereas the acceptance of a proton by a Brønsted-Lowry base yields the **conjugate acid** of that base. Every Brønsted-Lowry acid has a conjugate base and vice versa. In Reaction 4.1, acetic acid and acetate ion are the conjugate acid and conjugate base of each other. In that same reaction the hydronium ion H_3O^+ is paired with its conjugate base H_2O. In Reaction 4.2, NH_4^+ and NH_3 make up a conjugate acid-base pair, as do OH^- and H_2O.

Autoionization of Amphiprotic Solvents

Amphiprotic solvents undergo a process known as **autoionization**, in which a proton is transferred from one solvent molecule to another. In the case of water this reaction is

$$H_2O + H_2O \rightleftarrows H_3O^+ + OH^- \qquad (4.3)$$

$\text{base}_1 \quad \text{acid}_2 \qquad \text{acid}_1 \quad \text{base}_2$

Other solvents with ionizable protons act similarly. A typical example is liquid ammonia:

$$NH_3 + NH_3 \rightleftarrows NH_4^+ + NH_2^- \qquad (4.4)$$

$\text{base}_1 \quad \text{acid}_2 \qquad \text{acid}_1 \quad \text{base}_2$

In this discussion so far, the conjugate acid of the base, water, has been designated as H_3O^+ and called **hydronium ion**. This species is a **hydrate** of the proton H^+. Higher hydrates such as $H_5O_2^+$ and $H_9O_4^+$ also exist. Although unhydrated hydrogen ions do not exist in water, for the most part we will designate hydrated protons simply as H^+ as a matter of convenience. The H_3O^+ designation will be used when it is important to show H_3O^+ as a participant in an acid–base reaction. It should be kept in mind, however, that this is simply an abbreviation for the hydrated proton.

4.2   **Acid and Base Strength**

Strong and Weak Electrolytes

Ionizable compounds dissolved in water can be classified as strong and weak electrolytes. **Strong electrolytes** are those compounds that are completely ionized in water. For example, when HCl gas is bubbled into water it completely ionizes to produce $H^+(H_3O^+)$ ions

$$HCl(g) \xrightarrow[\text{water}]{} H^+(aq) + Cl^-(aq) \tag{4.5}$$

The strong electrolytes belong to the three major classes of (1) inorganic acids, such as HCl, HI, HBr, HNO_3, H_2SO_4, and $HClO_4$; (2) alkali and alkaline earth hydroxides; and (3) most salts. **Weak electrolytes** belong to the following classes: (1) some inorganic acids, such as HCN, $CO_2(aq)$, H_3PO_3, and H_2S; (2) most organic acids; (3) most organic bases and NH_3.

Strength of Acids and Bases

Because of the complete ionization of HCl in water as shown by Reaction 4.5, HCl is a strong electrolyte in water. Since all of the ionizable hydrogen is ionized in water, this acid is called a **strong acid**. Other strong acids include HNO_3, $HClO_4$, and H_2SO_4 (for its first ionizable hydrogen). The alkali and alkaline earth hydroxides are strong electrolytes and are therefore strong bases. For example, when sodium hydroxide is dissolved in water,

$$NaOH(s) \xrightarrow[\text{water}]{} Na^+(aq) + OH^-(aq) \tag{4.6}$$

it completely ionizes to Na^+ and OH^- ions, making it a **strong base**. Except at very low concentrations, all but a very small percentage of acetic acid molecules in water remain unionized, and the equilibrium of Reaction 4.1 lies far to the left. Therefore, acetic acid is a **weak acid**. Similarly, only a small percentage of ammonia molecules react with water to produce NH_4^+ and OH^- ions as shown in Reaction 4.2; therefore, NH_3 is a **weak base**.

Role of Solvent in Acid Strength

The strengths of acids and bases are strongly dependent upon the solvent in which they are dissolved. Water is such a good proton acceptor that several commonly used acids, including $HClO_4$, HCl, and HBr, are completely ionized to H_3O^+ and the conjugate bases of these acids in water. Because of its ability to ionize a number of acids completely to strong acids of equal acidity, water is said to be a good **leveling solvent**.

In contrast to water, anhydrous acetic acid is a relatively poor proton acceptor. Perchloric acid, $HClO_4$, is the strongest common acid in acetic acid solvent, undergoing the following ionization:

$$HClO_4 + HC_2H_3O_2 \rightleftarrows ClO_4^- + H_2C_2H_3O_2^+$$
$$\text{acid}_1 \qquad \text{base}_2 \qquad \text{base}_1 \qquad \text{acid}_2$$

Some other acids that are strong in water are weak in acetic acid solvent. For example, HCl is about three orders of magnitude weaker in acetic acid solvent than in water. A solvent such as acetic acid that brings out differences in acid strength is called a good **differentiating solvent**. These concepts are discussed further in Chapter 6.

4.3   Acid-Base Equilibrium Constant Expressions

Calculations with the Ion Product of Water

The degree of ionization of an acid or base can be expressed in terms of an **equilibrium constant**. One of the simplest of these is that of the autoionization of water:

$$H_2O \rightleftarrows H^+ + OH^- \tag{4.7}$$

$$K_w = [H^+][OH^-] \tag{4.8}$$

K_w is a constant and is called the **ion product** of water. It does not include a term for H_2O because in dilute solutions the concentration of water (55 mol/L in pure water) does not change appreciably with changes in solute concentration and with solute reactions. The value of K_w does change with temperature. At 25° C, K_w is 1.01×10^{-14}, usually rounded off to 1.00×10^{-14}, for convenience in calculations. Other values of K_w at different temperatures are 1.114×10^{-15} at 0° C, 5.47×10^{-14} at 50° C, and 4.9×10^{-13} at 100° C.

With a knowledge of K_w, $[H^+]$ can be calculated if $[OH^-]$ is known, and $[OH^-]$ can be calculated if $[H^+]$ is known. Consider the following examples.

EXAMPLE

What is $[H^+]$ in a 0.0271 M solution of NaOH?

NaOH is a strong electrolyte and is completely ionized in a 0.0271 M solution. Therefore,

$$[OH^-] = 2.71 \times 10^{-2} \; M$$

Solving K_w for $[H^+]$ yields

$$[H^+] = \frac{K_w}{[OH^-]} = \frac{1.00 \times 10^{-14}}{2.71 \times 10^{-2}} = 3.69 \times 10^{-13}$$

EXAMPLE

What is $[OH^-]$ in a $7.12 \times 10^{-3} \; M$ solution of HCl?

HCl is a strong electrolyte and is completely ionized in a $7.12 \times 10^{-3} \, M$ solution. Therefore,

$$[H^+] = 7.12 \times 10^{-3} \, M$$

Solving the K_w expression for $[OH^-]$ yields

$$[OH^-] = \frac{K_w}{[H^+]} = \frac{1.00 \times 10^{-14}}{7.12 \times 10^{-3}} = 1.40 \times 10^{-12} \, M$$

EXAMPLE

What is $[H^+]$ in pure water?

In pure water the only source of H^+ and OH^- is the autoionization of water shown by the equation

$$H_2O \rightleftharpoons H^+ + OH^-$$

For every H^+ ion produced, an OH^- ion is also produced, leading to the relationship

$$[H^+] = [OH^-]$$

Substitution into the K_w expression yields

$$[H^+][OH^-] = [H^+]^2 = K_w = 1.00 \times 10^{-14}$$

$$[H^+] = (K_w)^{1/2} = (1.00 \times 10^{-14})^{1/2} = 1.00 \times 10^{-7}$$

$$pH = -\log[H^+] = -\log 1.00 \times 10^{-7} = 7.00$$

$$[OH^-] = [H^+] = 1.00 \times 10^{-7}$$

The preceding example shows why pure CO_2-free water has a pH of 7.00.

EXAMPLE

What is the pH of a $1.00 \times 10^{-7}\,M$ solution of HCl? The pH is not 7.00 because that is the pH of a neutral solution, and this solution contains some added strong acid. The solution to this problem is the following:

In $1.00 \times 10^{-7}\,M$ HCl there are two sources of H^+, one from the HCl and one from the autoionization of water.

$$[H^+]_{HCl} = 1.00 \times 10^{-7}\,M$$

$$H_2O \rightleftharpoons H^+ + OH^-$$

$$[H^+]_{autoionization} = [OH^-]$$

In terms of total hydrogen ion concentration and K_w,

$$[OH^-] = \frac{K_w}{[H^+]}$$

$$[H^+] = [H^+]_{HCl} + [H^+]_{autoionization}$$

$$[H^+] = 1.00 \times 10^{-7} + \frac{K_w}{[H^+]} = 1.00 \times 10^{-7} + \frac{1.00 \times 10^{-14}}{[H^+]}$$

$$[H^+]^2 - 1.00 \times 10^{-7}[H^+] - 1.00 \times 10^{-14} = 0$$

Solving the quadratic yields

$$[H^+] = 1.62 \times 10^{-7}\,M$$

$$pH = 6.79$$

Weak Acid Dissociation Constants

The use of a **weak acid dissociation constant** may be illustrated by the case of acetic acid, $HC_2H_3O_2$. This acid ionizes according to the equilibrium reaction

$$HC_2H_3O_2 \rightleftharpoons H^+ + C_2H_3O_2^- \tag{4.9}$$

The equilibrium of this reaction is expressed by the acid dissociation constant, K_a, given by the following:

$$K_a = \frac{[H^+][C_2H_3O_2^-]}{[HC_2H_3O_2]} = 1.75 \times 10^{-5} \tag{4.10}$$

Of course, the higher the value of K_a for a weak acid, the greater its dissociation in solution at a given concentration. The following examples illustrate the use of K_a for the calculation of $[H^+]$.

EXAMPLE

What is $[H^+]$ in 0.100 M acetic acid?

In 0.100 M $HC_2H_3O_2$ the stoichiometry of the reaction

$$HC_2H_3O_2 \rightleftharpoons H^+ + C_2H_3O_2^-$$

shows that $[H^+] = [C_2H_3O_2^-]$. Since this is a weak acid at a relatively high concentration, it may be assumed that only a small percentage of the acid is dissociated, so that

$$[HC_2H_3O_2] = 0.100 \ M$$

$$K_a = \frac{[H^+][C_2H_3O_2^-]}{[HC_2H_3O_2]} = \frac{[H^+]^2}{0.100} = 1.75 \times 10^{-5}$$

$$[H^+] = (0.100 \times 1.75 \times 10^{-5})^{1/2} = 1.32 \times 10^{-3} \ M$$

Subtraction of acetate ion concentration from that of undissociated acetic acid shows that the latter is very close to 0.100, as was assumed.

$$[HC_2H_3O_2] = 0.100 - 0.001 = 0.099 \ M$$

EXAMPLE

What is $[H^+]$ in a solution formed by dissolving 1.00×10^{-4} mol of $HC_2H_3O_2$ in water and diluting to 1.00 L? Because this solution is relatively dilute, it cannot be assumed that $[HC_2H_3O_2] = 1.00 \times 10^{-4} \ M$. Such a solution is said to be 1.00×10^{-4} formal (F) in acetic acid (see Section 1.8).

In $1.00 \times 10^{-4} \ F$ acetic acid,

$$[H^+] = [C_2H_3O_2^-]$$

$$[HC_2H_3O_2] = 1.00 \times 10^{-4} - [C_2H_3O_2^-] = 1.00 \times 10^{-4} - [H^+]$$

$$K_a = \frac{[H^+][C_2H_3O_2^-]}{[HC_2H_3O_2]} = \frac{[H^+]^2}{1.00 \times 10^{-4} - [H^+]} = 1.75 \times 10^{-5}$$

$$[H^+]^2 + 1.75 \times 10^{-5}[H^+] - 1.75 \times 10^{-9} = 0$$

$$[H^+] = 3.40 \times 10^{-5} \ M = [C_2H_3O_2^-]$$

$$[HC_2H_3O_2] = 1.00 \times 10^{-4} - [C_2H_3O_2^-] = 1.00 \times 10^{-4} - 3.40 \times 10^{-5}$$

$$[HC_2H_3O_2] = 6.60 \times 10^{-5}$$

Check by substituting into the K_a expression.

$$K_a = \frac{[H^+][C_2H_3O_2^-]}{[HC_2H_3O_2]} = \frac{(3.40 \times 10^{-5})^2}{6.60 \times 10^{-5}} = 1.75 \times 10^{-5}$$

In the last two examples it has not been necessary to consider the autoionization of water as a contribution to $[H^+]$. It is necessary to consider this contribution, however, when a solution of a weak acid is present at a concentration around 1×10^{-7} M. Such a case is shown by the following example.

EXAMPLE

Hypochlorous acid, HOCl, is formed by the reaction of disinfectant Cl_2 with water. It has a K_a of 3.0×10^{-8}. What is $[H^+]$ in 5.00×10^{-7} F HOCl? In this case a significant amount of HOCl is dissociated so that [HOCl] cannot be assumed to be 5.00×10^{-7} M. Furthermore, the pH is around 7, so that the autoionization of water must be considered. These conditions lead to a cubic equation, which may be solved using Newton's approximation (see Appendix).

In 5.00×10^{-7} F HOCl the two sources of H^+ are

$$HOCl \rightleftharpoons H^+ + OCl^- \qquad [H^+]_{HOCl} = [OCl^-]$$

$$[HOCl] = 5.00 \times 10^{-7} - [OCl^-]$$

$$H_2O \rightleftharpoons H^+ + OH^- \qquad [H^+]_{H_2O} = [OH^-]$$

$$[H^+] = [H^+]_{HOCl} + [H^+]_{H_2O}$$

$$[H^+] = [OCl^-] + [OH^-]$$

$$K_a = \frac{[H^+][OCl^-]}{[HOCl]} = \frac{[H^+][OCl^-]}{5.00 \times 10^{-7} - [OCl^-]}$$

$$[OCl^-] = \frac{5.00 \times 10^{-7} K_a}{K_a + [H^+]} \qquad [H^+] = \frac{5.00 \times 10^{-7} K_a}{K_a + [H^+]} + \frac{K_w}{[H^+]}$$

$$[H^+]^3 + K_a[H^+]^2 - (5.00 \times 10^{-7} K_a + K_w)[H^+] - K_a K_w = 0$$

$$[H^+] = 1.50 \times 10^{-7} \ M$$

Weak Base Dissociation Constants

The degree of ionization of a **weak base dissociation constant** in water is expressed in terms of K_b, which relates the concentrations of the base, its conjugate acid, and OH^-. This can be illustrated by the following reaction of NH_3 in water:

$$NH_3 + H_2O \rightleftharpoons NH_4^+ + OH^- \tag{4.11}$$

$$\text{base}_1 \quad \text{acid}_2 \quad \quad \text{acid}_1 \quad \text{base}_2$$

$$K_b = \frac{[NH_4^+][OH^-]}{[NH_3]} = 1.76 \times 10^{-5} \tag{4.12}$$

The concentration of H_2O is not contained in the K_b expression because water is normally present in a large excess so that its concentration is constant. To calculate $[H^+]$ in a solution of a weak base, it is first necessary to calculate $[OH^-]$, then substitute into the expression for K_w (Equation 4.8) to obtain $[H^+]$. This kind of calculation can be shown by an example.

EXAMPLE

What is $[H^+]$ in 0.125 M NH_3? As shown by its K_b value, NH_3 is a weak enough base that it is reasonable to assume that only a small fraction of it goes to form the conjugate acid, NH_4^+, and $[NH_3] = 0.125$ M at equilibrium.

In 0.125 M NH_3

$$NH_3 + H_2O \rightleftharpoons NH_4^+ + OH^-$$

$$[NH_4^+] = [OH^-]$$

$$[NH_3] = 0.125 \ M$$

$$K_b = \frac{[NH_4^+][OH^-]}{[NH_3]} = \frac{[OH^-]^2}{0.125} = 1.76 \times 10^{-5}$$

$$[OH^-] = (0.125 \times 1.76 \times 10^{-5})^{1/2} = 1.48 \times 10^{-3} \ M$$

$$[H^+] = \frac{K_w}{[OH^-]} = \frac{1.00 \times 10^{-14}}{1.48 \times 10^{-3}} = 6.76 \times 10^{-12} \ M$$

$$pH = 11.17$$

4.4   **Relationships among K_a, K_b, and K_w**

The degree of dissociation of a weak acid is expressed by its K_a value, whereas the degree to which a solution of its conjugate base forms OH^- is calculated by means of a K_b expression. If the value of K_a of a weak acid is known, the K_b value of its conjugate base may be calculated and vice versa. Consider sodium acetate, $NaC_2H_3O_2$, the salt of acetic acid, and the strong base $NaOH$. When this salt is dissolved in water, the acetate anion reacts with water to a slight extent producing OH^- and acetic acid. Such a reaction is called a **hydrolysis reaction**.

$$C_2H_3O_2^- + H_2O \rightleftharpoons HC_2H_3O_2 + OH^- \tag{4.13}$$

$$K_b = \frac{[HC_2H_3O_2][OH^-]}{[C_2H_3O_2^-]} \tag{4.14}$$

The value of K_b may be calculated from two reactions with known equilibrium constants. To do this, it is necessary to know that two or more chemical reactions (that is, the chemical equations expressing these reactions) may be added together to give another chemical reaction. The equilibrium constants of the reactions that were added can then be multiplied together to give the equilibrium constant of their sum reaction. This is an important general approach to multiple equilibrium problems. In this case the reaction for the autoionization of water will be added to the reverse of the reaction for the dissociation of acetic acid. The equilibrium constant for the latter reaction is simply the inverse of the K_a expression for acetic acid.

$$H_2O \rightleftharpoons H^+ + OH^- \tag{4.15}$$

$$K_w = [H^+][OH^-] = 1.00 \times 10^{-14} \tag{4.16}$$

$$C_2H_3O_2^- + H^+ \rightleftharpoons HC_2H_3O_2$$

$$\frac{1}{K_a} = \frac{[HC_2H_3O_2]}{[H^+][C_2H_3O_2^-]} = \frac{1}{1.75 \times 10^{-5}} \tag{4.17}$$

$$C_2H_3O_2^- + H_2O \rightleftharpoons HC_2H_3O_2 + OH^- \tag{4.18}$$

$$K_b = K_w \frac{1}{K_a} = [H^+][OH^-] \frac{[HC_2H_3O_2]}{[H^+][C_2H_3O_2^-]} = \frac{[HC_2H_3O_2][OH^-]}{[C_2H_3O_2^-]} \tag{4.19}$$

$$= \frac{K_w}{K_a} = \frac{1.00 \times 10^{-14}}{1.75 \times 10^{-5}} = 5.71 \times 10^{-10}$$

As seen in the preceding example, the value of K_b for a weak base may be calculated from K_a of its conjugate acid by the relationship

$$K_b = \frac{K_w}{K_a} \tag{4.20}$$

Consider next the case in which K_b of a weak base is known, and it is desired to calculate K_a of its conjugate acid. Ammonium ion, NH_4^+, behaves as a weak acid, dissociating as follows:

$$NH_4^+ \rightleftharpoons H^+ + NH_3 \tag{4.21}$$

The value of K_a for this reaction may be calculated from K_b of NH_3 and K_w.

$$H_2O \rightleftharpoons H^+ + OH^- \tag{4.22}$$

$$K_w = [H^+][OH^-] = 1.00 \times 10^{-14} \tag{4.23}$$

$$NH_4^+ + OH^- \rightleftharpoons NH_3 + H_2O \tag{4.24}$$

$$\frac{1}{K_b} = \frac{[NH_3]}{[NH_4^+][OH^-]} = \frac{1}{1.76 \times 10^{-5}} \tag{4.25}$$

$$NH_4^+ \rightleftharpoons H^+ + NH_3 \tag{4.26}$$

$$K_a = K_w \frac{1}{K_b} = [H^+][OH^-] \frac{[NH_3]}{[NH_4^+][OH^-]} = \frac{[H^+][NH_3]}{[NH_4^+]} \tag{4.27}$$

$$= \frac{K_w}{K_b} = \frac{1.00 \times 10^{-14}}{1.76 \times 10^{-5}} = 5.68 \times 10^{-10}$$

As shown by this calculation, the value of K_a for a weak acid may be determined from K_b of its conjugate base by the relationship

$$K_a = \frac{K_w}{K_b} \tag{4.28}$$

From above, it can be seen that the relationship between K_a for a weak acid and K_b for its conjugate base, or vice versa, is simply the following:

$$K_a K_b = K_w \tag{4.29}$$

4.5   Solutions of Conjugate Acid-Base Pairs

In a solution containing a weak acid, HA, and its conjugate base, A^-, the H^+ ion concentration is determined by the relative strengths and concentrations of HA and A^-, which react according to the following two reactions:

$$HA \rightleftarrows H^+ + A^- \tag{4.30}$$

$$K_a = \frac{[H^+][A^-]}{[HA]} \tag{4.31}$$

$$A^- + H_2O \rightleftarrows HA + OH^- \tag{4.32}$$

$$K_b = \frac{[HA][OH^-]}{[A^-]} = \frac{K_w}{K_a} \tag{4.33}$$

For cases in which $[HA] = [A^-]$, the solution will be acidic (pH less than 7) if the first reaction predominates, and basic (pH greater than 7) if the second reaction is dominant. This, in turn, depends upon the value of K_a and the related value of K_b. The K_a expression may be rearranged and [HA] and $[A^-]$ cancelled, if they are equal, to give

$$[H^+] = \frac{[HA]}{[A^-]} K_a = K_a \tag{4.34}$$

$$pH = pK_a \qquad \text{where } pK_a = -\log K_a \tag{4.35}$$

Thus $[H^+]$ will be greater than 1.00×10^{-7} (pH less than 7) if K_a is greater than 1.00×10^{-7} (pK_a less than 7). Furthermore, $[H^+]$ will be less than 1.00×10^{-7} (pH greater than 7) if K_a is less than 1.00×10^{-7} (pK_a greater than 7).

To take account of different concentrations of HA and A^-, the K_a expression in Equation 4.34 may be used (see the example on page 97).

For very dilute solutions, or when the values of K_a and K_b are especially high, it cannot be assumed that $[HA] = C_{HA}$ and $[A^-] = C_{NaA}$, where the C's are the analytical concentrations (see Section 1.8) of the species denoted. From the reaction

$$HA \rightleftarrows H^+ + A^- \tag{4.36}$$

EXAMPLE

A solution is 0.0764 M in HA, $K_a = 4.42 \times 10^{-6}$, and 0.0237 M in NaA. What is the pH of the solution?

In the case where [HA] = 0.0764 M, $K_a = 4.42 \times 10^{-6}$, and [A$^-$] = 0.0237,

$$[\text{H}^+] = \frac{[\text{HA}]}{[\text{A}^-]} K_a = \frac{0.0764}{0.0237} 4.42 \times 10^{-6} = 1.42 \times 10^{-5} \ M$$

$$\text{pH} = 4.85$$

Alternatively,

$$\log [\text{H}^+] = \log [\text{HA}] - \log [\text{A}^-] + \log K_a$$

$$\text{pH} = -\log [\text{HA}] + \log [\text{A}^-] + pK_a$$

$$= 1.117 - 1.625 + 5.355 = 4.85$$

it is seen that formation of an H$^+$ ion from HA results in the loss of an HA molecule and production of an A$^-$ ion. Furthermore, from the reaction

$$\text{A}^- + \text{H}_2\text{O} \rightleftharpoons \text{HA} + \text{OH}^- \qquad (4.37)$$

it is obvious that hydrolysis of an A$^-$ ion results in the formation of HA and production of an OH$^-$ ion. Therefore, the true concentration of HA is given by

$$[\text{HA}] = C_{\text{HA}} - [\text{H}^+] + [\text{OH}^-] \qquad (4.38)$$

and the true concentration of A$^-$ is given by the following:

$$[\text{A}^-] = C_{\text{NaA}} + [\text{H}^+] - [\text{OH}^-] \qquad (4.39)$$

The two preceding equations can be written in a more general form for solutions of conjugate acid-base pairs, where a represents the acid species and b the base species. For example, a might be acetic acid, HC$_2$H$_3$O$_2$, and b its conjugate base, acetate ion, C$_2$H$_3$O$_2^-$; b might be ammonia, NH$_3$, and a its conjugate acid, ammonium ion, NH$_4^+$. The equations applicable to mixtures of a and b are

$$[a] = C_a - [\text{H}^+] + [\text{OH}^-] \qquad (4.40)$$

$$[b] = C_b + [\text{H}^+] - [\text{OH}^-] \qquad (4.41)$$

where $[a]$ and $[b]$ are the actual concentrations of a and b and C_a and C_b are their analytical concentrations.

The preceding equations for $[a]$ and $[b]$ simplify because at any pH, except a pH close to 7, [OH$^-$] is negligible compared to [H$^+$] or vice versa. In most cases of interest, C_a and C_b are of relatively comparable magnitudes; therefore, if pK_a of the weak acid a is less than 7, [OH$^-$] is negligible compared to [H$^+$], and if pK_a is greater than 7, $[H^+]$ is negligible compared to $[OH^-]$.

The two following examples illustrate applications of the equations just discussed.

EXAMPLE

For chloroacetic acid, $HC_2H_2ClO_2$, the value of K_a is 1.36×10^{-3}. What is $[H^+]$ in a solution for which $C_{HC_2H_2ClO_2} = 0.0850$ and $C_{NaC_2H_2ClO_2} = 0.0775$?

A solution in which $C_{HC_2H_2ClO_2} = 0.0850$ and $C_{NaC_2H_2ClO_2} = 0.0775$ will have a low pH because of the high value of K_a, 1.36×10^{-3}, so that $[OH^-]$ is negligible relative to $[H^+]$. Therefore,

$$[HC_2H_2ClO_2] = 0.0850 - [H^+] + [OH^-]$$

$$[C_2H_2ClO_2] = 0.0775 + [H^+] - [OH^-]$$

$$K_a = \frac{[H^+](0.0775 + [H^+])}{0.0850 - [H^+]} = 1.36 \times 10^{-3}$$

$$[H^+]^2 + 0.07886[H^+] - 1.156 \times 10^{-4} = 0$$

$$[H^+] = 1.44 \times 10^{-3}\ M \qquad pH = 2.84$$

EXAMPLE

What is $[H^+]$ of a solution containing acetic acid and sodium acetate at the following analytical concentrations:

$$C_{HC_2H_3O_2} = 2.50 \times 10^{-5} \qquad C_{NaC_2H_3O_2} = 4.00 \times 10^{-5}$$

For acetic acid, $K_a = 1.75 \times 10^{-5}$

Since both acetic acid and sodium acetate are present at very low concentrations, neither $HC_2H_3O_2$ nor $NaC_2H_3O_2$ may be assumed to be present at their analytical concentrations. Because both these species are present at similar concentrations, it is likely that $[H^+]$ is close to K_a, so that the solution is slightly acidic and $[OH^-]$ is negligible relative to $[H^+]$. Therefore,

$$[HC_2H_3O_2] = 2.50 \times 10^{-5} - [H^+] + [OH^-]$$

$$[C_2H_3O_2^-] = 4.00 \times 10^{-5} + [H^+] - [OH^-]$$

$$K_a = \frac{[H^+](4.00 \times 10^{-5} + [H^+])}{2.50 \times 10^{-5} - [H^+]} = 1.75 \times 10^{-5}$$

$$[H^+]^2 + 5.75 \times 10^{-5}[H^+] - 4.38 \times 10^{-10} = 0$$

$$[H^+] = 6.80 \times 10^{-6} \qquad [C_2H_3O_2^-] = 4.68 \times 10^{-5}$$

$$[HC_2H_3O_2] = 1.82 \times 10^{-5}$$

4.6   Buffer Solutions

Definition of Buffers

A **buffer solution** *is a solution that resists changes in pH upon addition of strong acid or strong base or when diluted by water.* A buffer solution consists of a conjugate acid–base pair. Using a and b to represent the acid and base species of such a pair, respectively, gives the following general equation for $[H^+]$ of a buffer:

$$[H^+] = K_a \frac{[a]}{[b]} \tag{4.42}$$

where K_a is the acid dissociation constant of the acid species. For example, in a mixture of acetic acid and sodium acetate, where acetate ion is the conjugate base of acetic acid, the value of $[H^+]$ is given in terms of K_a of acetic acid and the concentrations of acetic acid and acetate ion as follows:

$$K_a = \frac{[H^+][C_2H_3O_2^-]}{[HC_2H_3O_2]} = 1.75 \times 10^{-5} \tag{4.43}$$

$$[H^+] = K_a \frac{[HC_2H_3O_2]}{[C_2H_3O_2^-]} = 1.75 \times 10^{-5} \frac{[HC_2H_3O_2]}{[C_2H_3O_2^-]} \tag{4.44}$$

In the case where the base in a conjugate acid–base pair is uncharged, the value of K_b for the base is commonly given. In such a case it is more convenient to first calculate K_a of the conjugate acid from the following equation (see Section 4.4):

$$K_a = \frac{K_w}{K_b} = \frac{1.00 \times 10^{-14}}{K_b} \tag{4.45}$$

Once that is done, Equation 4.42, can be used to express $[H^+]$.

The action of a buffer solution is most effective when the conjugate acid and base are present at nearly equal concentrations. Examination of Equation 4.42 shows that under those circumstances—that is $[a] \cong [b]$—the value of $[H^+]$ is close to that of K_a and pH is very close to pK_a. Thus, preparation of a buffer of a particular pH is best accomplished by selecting an acid–base pair in which pK_a of the acid is close to the value of the desired pH.

Calculation of Buffer pH

The pH of a buffer solution is readily calculated with a knowledge of the species concentrations and K_a using Equation 4.42. This is shown by the following examples.

EXAMPLE

A solution is 0.0775 M in formic acid, HCO_2H, and 0.0460 M in sodium formate, $NaCO_2H$. The value of K_a of formic acid is 1.77×10^{-4}. What is $[H^+]$ and pH of the buffer solution?

Substitution into Equation 4.42 gives the following:

$$[H^+] = K_a \frac{[HCO_2H]}{[CO_2H^-]}$$

$$[H^+] = 1.77 \times 10^{-4} \frac{0.0775}{0.0460} = 2.98 \times 10^{-4} \ M$$

$$pH = 3.53$$

EXAMPLE

A solution is 0.0235 M in ammonia, NH_3, and 0.0460 M in ammonium chloride, NH_4Cl. What are $[H^+]$ and the pH given K_b of $NH_3 = 1.76 \times 10^{-5}$?

The first step is to calculate K_a of NH_4^+.

$$K_a = \frac{K_w}{K_b} = \frac{1.00 \times 10^{-14}}{1.76 \times 10^{-5}} = 5.68 \times 10^{-10}$$

The value of $[H^+]$ is calculated as follows:

$$[H^+] = K_a \frac{[NH_4^+]}{[NH_3]} = 5.68 \times 10^{-10} \frac{0.0460}{0.0235} = 1.11 \times 10^{-9} \ M$$

$$pH = 8.95$$

Within reasonable limits, the pH of a buffer solution does not change with dilution. Suppose that the buffer solution described in the preceding example were diluted tenfold with water. The pH of the diluted solution is calculated as follows:

$$[NH_4^+] = 0.0460 \times 0.1 = 0.00460 \ M \tag{4.46}$$

$$[NH_3] = 0.0235 \times 0.1 = 0.00235 \ M \tag{4.47}$$

$$[H^+] = K_a \frac{[NH_4^+]}{[NH_3]} = 5.68 \cdot \times 10^{-10} \frac{0.00460}{0.00235}$$

$$= 1.11 \times 10^{-9} \ M \tag{4.48}$$

$$pH = 8.95$$

It is seen that dilution did not change the pH.

Calculation of Buffer Composition

The most common buffer calculation is that of the composition of a buffer solution that gives a specified pH. Such a calculation is shown in the following example.

EXAMPLE

A buffer solution is to be prepared with a pH of 9.50 and having $[NH_3] = 0.100\ M$. What should be the concentration of NH_4Cl ($[NH_4^+]$)?

At a pH of 9.50,

$$[H^+] = 10^{-9.50} = 3.16 \times 10^{-10}\ M$$

The value of $[NH_4^+]$ is calculated from the expression for K_a of NH_4^+:

$$[NH_4^+] = \frac{[H^+][NH_3]}{K_a} = \frac{3.16 \times 10^{-10} \times 0.100}{5.68 \times 10^{-10}} = 0.0556\ M$$

For the buffer specified, the analytical concentration of NH_4Cl should be $0.0553\ M$.

Polyprotic acids (those with several ionizable hydrogens) and their anions can be used to prepare buffer solutions. The most common such buffers are those prepared from salts of phosphoric acid, H_3PO_4. Suppose that it is desired to prepare a neutral buffer, pH 7.00, by an appropriate mixture of sodium salts of H_3PO_4. For the dissociation of the second hydrogen from H_3PO_4, the following apply:

$$H_2PO_4^- \rightleftarrows H^+ + HPO_4^{2-} \tag{4.49}$$

$$K_{a2} = \frac{[H^+][HPO_4^{2-}]}{[H_2PO_4^-]} = 6.17 \times 10^{-8} \qquad pK_{a2} = 7.21 \tag{4.50}$$

The value of pK_{a2} is close to 7, as is required to make a pH 7 buffer. For a buffer of pH 7.00, the concentration ratio, $[HPO_4^{2-}]/[H_2PO_4^-]$, is calculated as follows:

$$\frac{[HPO_4^{2-}]}{[H_2PO_4^-]} = \frac{K_{a2}}{[H^+]} = \frac{6.17 \times 10^{-8}}{1.00 \times 10^{-7}} = 0.617 \tag{4.51}$$

Therefore, a pH 7.00 solution may be prepared by mixing Na_2HPO_4 and NaH_2PO_4 in a molar ratio of 0.617:1.

Effect of the Addition of Strong Acid or Base upon Buffer pH

To see the effects of strong acid or strong base upon a buffer solution, again consider a pH 9.50 buffer in which $[NH_3] = 0.100\ M$ and $[NH_4^+] = 0.0556\ M$.

This can be compared to a solution of the same pH made with $3.16 \times 10^{-5}\ M$ NaOH. Suppose that 0.0100 mol of solid NaOH were mixed with a liter of the NH_3–NH_4^+ buffer and a liter of the NaOH solution. (Solid NaOH is used so that there is negligible dilution.) What is the new pH of each of the solutions? The conjugate acid in the buffer reacts as follows:

$$\underset{0.0556\ \text{mol}}{NH_4^+} \quad + \quad \underset{0.0100\ \text{mol}}{OH^-} \quad \rightarrow \quad NH_3 + H_2O \tag{4.52}$$

Stoichiometric calculations readily show that in the buffer solution to which OH^- has been added, the new concentrations of NH_4^+ and NH_3 are

$$[NH_4^+] = 0.0556\ M - 0.0100\ M = 0.0456\ M \tag{4.53}$$

$$[NH_3] = 0.100\ M + 0.0100\ M = 0.110\ M \tag{4.54}$$

$$[H^+] = \frac{[NH_4^+]}{[NH_3]}\ K_a = \frac{0.0456}{0.110} \times 5.68 \times 10^{-10} = 2.35 \times 10^{-10} \tag{4.55}$$

$$pH = 9.63$$

It is seen that the pH changes by only 0.13 units, from 9.50 to 9.63. Now consider what happens when 0.0100 mol of NaOH is added to 1 L of $3.16 \times 10^{-5}\ M$ NaOH, pH 9.50. The new OH^- concentration is given by

$$[OH^-] = [OH^-]_{original} + [OH^-]_{added} \tag{4.56}$$

$$[OH^-] = 3.16 \times 10^{-5} + 0.010 = 0.01003 \tag{4.57}$$

$$[H^+] = \frac{1.00 \times 10^{-14}}{0.01003} = 9.97 \times 10^{-13}$$

$$pH = 12.00$$

There is a change of 2.50 pH units from pH 9.50 to 12.00. Thus, the dilute solution of NaOH is a very poor buffer.

Buffer Capacity

The **buffer capacity** of a solution is its ability to resist changes in pH. Conceptually, the buffer capacity may be viewed as the number of moles of H^+ from a strong acid or the number of moles of OH^- from a strong base that, when added to 1 L of buffer solution, will change the pH of the buffer solution by 1 unit. To avoid changing the buffer capacity itself by the addition of acid or base, only a small increment of either should be considered in measuring or calculating the buffer capacity of a solution. Therefore, the buffer capacity is given by the equation,

$$\text{Buffer capacity} = \frac{\Delta\ \text{equivalents } H^+ \text{ or } OH^- \text{ added per liter}}{|\Delta pH|} \tag{4.58}$$

where the absolute value of ΔpH is taken because it decreases with added acid, and buffer capacity is expressed as a positive value. (Note that one equivalent of H^+ or OH^- is the same as one mole of H^+ or OH^-.) The buffer capacity increases with the total concentration of the acid–base conjugate pair and has a maximum value when their ratio is exactly 1.

Standard Buffer Solutions

Often the concentrations of buffer constituents are so high that the concentration equilibrium constants are not known accurately, thus making the calculation of exact buffer pH values from equilibrium constant expressions impossible. The preparation of standard buffer solutions with relatively high constituent concentrations has been the subject of a great deal of study; directions for preparing such buffer solutions are given in handbooks and references.[1] Clark and Lubs buffers make use of three systems—phthalic acid–potassium hydrogen phthalate, potassium dihydrogen phosphate–dipotassium hydrogen phosphate, and boric acid–sodium borate—to span the pH range 2 to 10. Buffers ranging in pH from approximately 2 to 8 can be prepared from mixtures of citric acid and disodium hydrogen phosphate according to instructions given by McIlvaine. Standard buffer solutions and powdered mixtures for preparing them are readily available from commercial sources.

4.7   Ionization of Polyprotic Acids

Stepwise Ionization of Acids

A number of acids have more than one ionizable hydrogen and dissociate in a stepwise fashion as represented below.

$$H_nA \rightleftarrows H^+ + H_{n-1}A^- \tag{4.59}$$

$$K_{a1} = \frac{[H^+][H_{n-1}A^-]}{[H_nA]} \tag{4.60}$$

$$H_{n-1}A^- \rightleftarrows H^+ + H_{n-2}A^{2-} \tag{4.61}$$

$$K_{a2} = \frac{[H^+][H_{n-2}A^{2-}]}{[H_{n-1}A^-]} \tag{4.62}$$

$$\vdots$$

$$HA^{(n-1)-} \rightleftarrows H^+ + A^{n-} \tag{4.63}$$

$$K_{an} = \frac{[H^+][A^{n-}]}{[HA^{(n-1)-}]} \tag{4.64}$$

Normally we will consider only diprotic acids, which have two ionizable hydrogens; some triprotic acids, such as H_3PO_4, and some tetraprotic acids, such as EDTA (see Section 10.7) are also important in analytical chemistry.

---

[1] L. Meites (ed.): *Handbook of Analytical Chemistry*. McGraw-Hill, New York, 1963, pp. 5–112 and 11–5 to 11–7.

Acidic Properties of Carbon Dioxide

One of the most important diprotic acids in chemical analysis, natural waters, wastewaters, and physiological systems is carbon dioxide, CO_2. Although CO_2 in water is often represented as H_2CO_3, the equilibrium constant for the reaction

$$CO_2(aq) + H_2O \rightleftharpoons H_2CO_3(aq) \tag{4.65}$$

is only around 2×10^{-3} at 25° C, so just a small fraction of the dissolved carbon dioxide is actually present as the species H_2CO_3. In this text, non-ionized carbon dioxide in water will be designated simply as CO_2, standing for the total of dissolved molecular CO_2 and undissociated H_2CO_3.

The acid–base character of CO_2 in water may be described by the following reactions and equilibrium constants:

$$CO_2(aq) + H_2O \rightleftharpoons H^+ + HCO_3^- \tag{4.66}$$

$$K_{a1} = \frac{[H^+][HCO_3^-]}{[CO_2]} = 4.45 \times 10^{-7} \qquad pK_{a1} = 6.35 \tag{4.67}$$

$$HCO_3^- \rightleftharpoons H^+ + CO_3^{2-} \tag{4.68}$$

$$K_{a2} = \frac{[H^+][CO_3^{2-}]}{[HCO_3^-]} = 4.69 \times 10^{-11} \qquad pK_{a2} = 10.33 \tag{4.69}$$

The products of the ionization of CO_2 in water are bicarbonate ion, HCO_3^-, and carbonate ion, CO_3^{2-}. With a knowledge of the equilibrium constants involved with the CO_2–HCO_3^-–CO_3^{2-} system, we can now work some problems involving this system.

EXAMPLE

The molar concentration of CO_2 in water saturated with this gas at a pressure of 1.00 atm at 25° C is 3.27×10^{-2} M. What is the pH of such a solution?

In 3.27×10^{-2} M CO_2,

$$CO_2 + H_2O \rightleftharpoons H^+ + HCO_3^-$$

$$[H^+] = [HCO_3^-]$$

$$K_{a1} = \frac{[H^+][HCO_3^-]}{[CO_2]} = \frac{[H^+]^2}{3.27 \times 10^{-2}} = 4.45 \times 10^{-7}$$

$$[H^+] = (3.27 \times 10^{-2} \times 4.45 \times 10^{-7})^{1/2} = 1.21 \times 10^{-4}$$

$$pH = 3.92$$

Is HCO_3^- appreciably dissociated to CO_3^{2-} at this pH? From K_{a2}

$$\frac{[CO_3^{2-}]}{[HCO_3^-]} = \frac{K_{a1}}{[H^+]} = \frac{4.69 \times 10^{-11}}{1.21 \times 10^{-4}} = 3.88 \times 10^{-7}$$

There is no significant ionization of HCO_3^- to CO_3^{2-}.

pH of CO_2–HCO_3^- Mixtures

Exactly 1 L of water was saturated with CO_2 at 25°C, then covered to prevent loss of the gas. Next, 0.0100 mol of NaOH was added to the CO_2 solution, and the solution allowed to come to equilibrium. What was the equilibrium pH?

The OH^- from NaOH reacts with CO_2 as follows:

$$CO_2 + OH^- \rightarrow HCO_3^-$$

$$[CO_2] = 3.27 \times 10^{-2} \ M - 0.0100 \ M = 2.27 \times 10^{-2} \ M$$

$$[HCO_3^-] = 1.00 \times 10^{-2} \ M$$

From K_{a1},

$$[H^+] = \frac{[CO_2]}{[HCO_3^-]} \times K_{a1} = \frac{2.27 \times 10^{-2}}{1.00 \times 10^{-2}} \times 4.45 \times 10^{-7} = 1.01 \times 10^{-6}$$

$$pH = 6.00$$

pH of HCO_3^- Solutions

Suppose that exactly enough solid NaOH was added to a $3.27 \times 10^{-2} \ M$ solution of CO_2 to convert it all to HCO_3^-. What is the pH of the resulting $3.27 \times 10^{-2} \ M$ HCO_3^- solution? For the most general case, the equation leading to the pH of this solution can be calculated as follows:

$$HCO_3^- \rightleftarrows H^+ + CO_3^{2-} \tag{4.70}$$

$$HCO_3^- + H^+ \rightleftarrows H_2O + CO_2 \tag{4.71}$$

$$H_2O \rightleftarrows H^+ + OH^- \tag{4.72}$$

$$[H^+] = [CO_3^{2-}] + [OH^-] - [CO_2]$$

$$[H^+] = \frac{[HCO_3^-]}{[H^+]} \times K_{a2} + \frac{[K_w]}{[H^+]} - \frac{[H^+][HCO_3^-]}{K_{a1}} \tag{4.73}$$

$$[H^+]^2 \left(1 + \frac{[HCO_3^-]}{K_{a1}} \right) = [HCO_3^-]K_{a2} + K_w \tag{4.74}$$

$$[H^+] = \left(\frac{[HCO_3^-]K_{a1}K_{a2} + K_{a1}K_w}{K_{a1} + [HCO_3^-]} \right)^{1/2} \tag{4.75}$$

This equation is an important one in that it can be used to calculate $[H^+]$ in solutions of the conjugate base HA^- of a diprotic acid H_2A. Such a solution is formed, for example, when just enough strong base has been added to a solution of diprotic acid to convert the acid to HA^-. It can also be formed by making a solution of an acid salt, such as $NaHCO_3$. Normally, Equation 4.75 simplifies greatly as seen by finishing the calculation of the pH of $3.27 \times 10^{-2} \, M \, HCO_3^-$. First the values of each of the terms in the equation can be calculated to see if any are negligible when added to the others:

$$[HCO_3^-]K_{a1}K_{a2} = 3.27 \times 10^{-2} \times 4.45 \times 10^{-7} \times 4.69 \times 10^{-11}$$
$$= 6.82 \times 10^{-19} \tag{4.76}$$

$$K_{a1}K_w = 4.45 \times 10^{-7} \times 1.00 \times 10^{-14} = 4.45 \times 10^{-21} \tag{4.77}$$

$$K_{a1} = 4.45 \times 10^{-7} \tag{4.78}$$

$$[HCO_3^-] = 3.27 \times 10^{-2} \tag{4.79}$$

It is seen here, as is commonly the case, that

$$K_{a1}K_w \ll [HCO_3^-]K_{a1}K_{a2} \tag{4.80}$$

$$K_{a1} \ll [HCO_3^-] \tag{4.81}$$

so that Equation 4.75 becomes

$$[H^+] = \left(\frac{[HCO_3^-]K_{a1}K_{a2}}{[HCO_3^-]}\right)^{1/2} = (K_{a1}K_{a2})^{1/2} \tag{4.82}$$

$$= (4.45 \times 10^{-7} \times 4.69 \times 10^{-11})^{1/2} = (2.09 \times 10^{-17})^{1/2} \tag{4.83}$$

$$= 4.57 \times 10^{-9}, \; pH = 8.34$$

Alternatively, from Equation 4.82:

$$pH = \frac{1}{2}(pK_{a1} + pK_{a2}) = 8.34 \tag{4.84}$$

EXAMPLE

What is $[H^+]$ in a solution formed by adding 4.00×10^{-2} mol solid NaOH (no dilution) to 1 L of $3.27 \times 10^{-2} \, M \, CO_2$?

When 4.00×10^{-2} mol of NaOH are added to a 1 L solution of $3.27 \times 10^{-2} \, M \; CO_2$, the first 3.27×10^{-2} mol of OH^- is consumed in converting CO_2 to HCO_3^-, leaving $4.00 \times 10^{-2} - 3.27 \times 10^{-2} = 0.73 \times 10^{-2}$ mol of NaOH to undergo the reaction

$$HCO_3^- + OH^- \rightarrow H_2O + CO_3^{2-}$$

Moles HCO_3^- remaining $= 3.27 \times 10^{-2} - 0.73 \times 10^{-2} = 2.54 \times 10^{-2}$.

$$[HCO_3^-] = 2.54 \times 10^{-2} \, M$$

Moles CO_3^{2-} produced $= 0.73 \times 10^{-2}$.

$$[CO_3^{2-}] = 0.73 \times 10^{-2} \, M$$

$$[H^+] = \frac{[HCO_3^-]}{[CO_3^{2-}]} K_{a2} = \frac{2.54 \times 10^{-2}}{0.73 \times 10^{-2}} \times 4.69 \times 10^{-11} = 1.63 \times 10^{-10}$$

$$pH = 9.79$$

pH of $HCO_3^- - CO_3^{2-}$ Mixture

When more than enough NaOH has been added to convert all the CO_2 in a solution to HCO_3^-, the latter reacts to yield CO_3^{2-} according to the reaction

$$HCO_3^- + OH^- \rightleftharpoons H_2O + CO_3^{2-} \tag{4.85}$$

Consider the example on page 106.

pH of a CO_3^{2-} Solution

As a final example, consider adding enough NaOH to convert all the CO_2 to CO_3^{2-}.

EXAMPLE

What is $[H^+]$ after the addition of $2 \times 3.27 \times 10^{-2} = 6.54 \times 10^{-2}$ mol of NaOH to 1 L of 3.27×10^{-2} M CO_2?

Addition of 6.54×10^{-2} mol of NaOH to 1 L of 3.27×10^{-2} M CO_2 is exactly the right amount to complete the reaction

$$CO_2 + 2OH^- \rightarrow H_2O + CO_3^{2-}$$

This leaves a 3.27×10^{-2} M solution of CO_3^{2-}, the $[H^+]$ of which is calculated as follows:

$$CO_3^{2-} + H_2O \rightarrow HCO_3^- + OH^- \qquad [HCO_3^-] = [OH^-]$$

$$K_b = \frac{[HCO_3^-][OH^-]}{[CO_3^{2-}]} = \frac{K_w}{K_{a2}} = \frac{1.00 \times 10^{-14}}{4.69 \times 10^{-11}} = 2.13 \times 10^{-4}$$

Because this equilibrium constant is relatively high and $[CO_3^{2-}]$ is relatively low, $[CO_3^{2-}] = 3.27 \times 10^{-2} - [HCO_3^-]$ or $[CO_3^{2-}] = 3.27 \times 10^{-2} - [OH^-]$.

$$\frac{[OH^-]^2}{3.27 \times 10^{-2} - [OH^-]} = 2.13 \times 10^{-4} \qquad [OH^-] = 2.54 \times 10^{-3}$$

$$[H^+] = 3.94 \times 10^{-12}$$

$$pH = 11.40$$

Generalized pH Calculations for Diprotic Acid Species

The pH calculations just discussed for the $CO_2 - HCO_3^- - CO_3^{2-}$ system apply in general to a diprotic acid H_2A and the HA^- and A^{2-} species produced from its deprotonation. Consider solutions composed of various combinations of H_2A and salts of its deprotonated forms with Group IA metal ions (Na^+, K^+). Such solutions could be prepared from pure H_2A, H_2A plus NaHA, pure NaHA, NaHA plus Na_2A, and pure Na_2A. In order to calculate the pH's of

TABLE 4.1   Relationships used to calculate $[H^+]$ of solutions derived from a diprotic acid, H_2A, and its salts

Solution	Calculation of $[H^+]$	
Pure H_2A	$[H^+] = [HA^-]$, $\dfrac{[H^+]^2}{C_{H_2A} - [H^+]} = K_{a1}$	(Often $[H^+]$ is negligible compared to $[H_2A]$)
H_2A–HA^- mixture	$[H^+] = \dfrac{[H_2A]}{[HA^-]} K_{a1}$	
Pure HA^-	$[H^+] = \left(\dfrac{[HA^-]K_{a1}K_{a2} + K_{a1}K_w}{K_{a1} + [HA^-]} \right)^{1/2}$	(Often simplifies to $[H^+] = \sqrt{K_{a1}K_{a2}}$)
HA^-–A^{2-} mixture	$[H^+] = \dfrac{[HA^-]}{[A^{2-}]} K_{a2}$	
Pure A^{2-}	$[OH^-] = [HA^-]$, $\dfrac{[OH^-]^2}{C_{A^{2-}} - [OH^-]} = K_b$, $K_b = \dfrac{K_w}{K_{a2}}$,	
	$[H^+] = \dfrac{K_w}{[OH^-]}$	(Often $[OH^-]$ is negligible compared to $C_{A^{2-}}$)

these solutions, it is necessary to know the analytical concentrations of the constituents and the K_a's of H_2A:

$$K_{a1} = \frac{[H^+][HA^-]}{[H_2A]} \tag{4.86}$$

$$K_{a2} = \frac{[H^+][A^{2-}]}{[HA^-]} \tag{4.87}$$

The solutions that could be prepared from H_2A and its salts and the relationships used to calculate the values of $[H^+]$ (and pH) of these solutions are summarized in Table 4.1.

4.8   Graphical Representation of Equilibrium Relationships

Fraction of Species

From the preceding examples it has been seen that as more NaOH is added to a CO_2 solution, there is a progression of predominance of species from CO_2, through HCO_3^-, to CO_3^{2-} with an accompanying increase in pH. This can be shown more systematically with a *distribution of species diagram*, consisting of the fractions of species present as CO_2, HCO_3^-, and CO_3^{2-} as a function of pH. The fraction that each of these species is of the total may be designated by an α with a subscript to show the species represented, and calculated from the expressions

$$\alpha_{CO_2} = \frac{[CO_2]}{[CO_2] + [HCO_3^-] + [CO_3^{2-}]} \tag{4.88}$$

$$\alpha_{HCO_3^-} = \frac{[HCO_3^-]}{[CO_2] + [HCO_3^-] + [CO_3^{2-}]} \tag{4.89}$$

$$\alpha_{CO_3^{2-}} = \frac{[CO_3^{2-}]}{[CO_2] + [HCO_3^-] + [CO_3^{2-}]} \tag{4.90}$$

In order to prepare a distribution of species diagram, it is necessary to express these fractions in terms of K_a's and $[H^+]$ values. This can be done for α_{CO_2}, for example, by expressing $[HCO_3^-]$ and $[CO_3^{2-}]$ in terms of K_{a1}, K_{a2}, $[H^+]$, and $[CO_2]$, giving

$$[HCO_3^-] = \frac{K_{a1}[CO_2]}{[H^+]} \tag{4.91}$$

$$[CO_3^{2-}] = \frac{K_{a1}K_{a2}[CO_2]}{[H^+]^2} \tag{4.92}$$

$$\alpha_{CO_2} = \frac{[CO_2]}{[CO_2] + \dfrac{K_{a1}[CO_2]}{[H^+]} + \dfrac{K_{a1}K_{a2}[CO_2]}{[H^+]^2}} \tag{4.93}$$

Cancelling $[CO_2]$ and multiplying numerator and denominator by $[H^+]^2$ yields

$$\alpha_{CO_2} = \frac{[H^+]^2}{[H^+]^2 + K_{a1}[H^+] + K_{a1}K_{a2}} \tag{4.94}$$

Similar derivations give

$$\alpha_{HCO_3^-} = \frac{K_{a1}[H^+]}{[H^+]^2 + K_{a1}[H^+] + K_{a1}K_{a2}} \tag{4.95}$$

$$\alpha_{CO_3^{2-}} = \frac{K_{a1}K_{a2}}{[H^+]^2 + K_{a1}[H^+] + K_{a1}K_{a2}} \tag{4.96}$$

Distribution of Species Plot

From these expressions it is possible to plot the fraction of each of these species versus pH. Such a plot is shown in Figure 4.1. Examination of Figure 4.1 shows that at very low pH values, α_{CO_2} approaches 1, and at very high pH values, $\alpha_{CO_3^{2-}}$ approaches 1. The value of $\alpha_{HCO_3^-}$ reaches a maximum value at an intermediate pH shown by the following calculation.

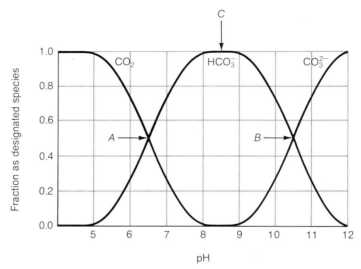

FIGURE 4.1 Distribution of species for the CO_2–HCO_3^-–CO_3^{2-} system as a function of pH

EXAMPLE

At what pH is $\alpha_{HCO_3^-}$ maximum?

Represent the denominator in the expression for $\alpha_{HCO_3^-}$ by G, so that

$$\alpha_{HCO_3^-} = \frac{K_{a1}[H^+]}{[H^+]^2 + K_{a1}[H^+] + K_{a1}K_{a2}} = \frac{K_{a1}[H^+]}{G}$$

For the maximum $\alpha_{HCO_3^-}$

$$\frac{d\alpha_{HCO_3^-}}{d[H^+]} = 0$$

From calculus,

$$\frac{d}{dx}\left(\frac{u}{v}\right) = \frac{v\,\dfrac{du}{dx} - u\,\dfrac{dv}{dx}}{v^2}$$

therefore,

$$\frac{d\alpha_{HCO_3^-}}{d[H^+]} = \frac{([H^+]^2 + K_{a1}[H^+] + K_{a1}K_{a2})K_{a1} - K_{a1}[H^+](2[H^+] + K_{a1})}{G^2} = 0$$

$$\cancel{K_{a1}[H^+]^2} + \cancel{K_{a1}^2[H^+]} + K_{a1}^2 K_{a2} - 2K_{a1}[H^+]^2 - \cancel{K_{a1}^2[H^+]} = 0$$

$$[H^+]^2 = K_{a1}K_{a2}$$

$$[H^+] = \sqrt{K_{a1}K_{a2}}$$

$$pH = \tfrac{1}{2}(pK_{a1} + pK_{a2})$$

From the preceding example it is seen that $\alpha_{HCO_3^-}$ has its maximum value at

$$pH = \frac{1}{2}(pK_{a1} + pK_{a2}) = \frac{1}{2}(6.35 + 10.33) = 8.34 \tag{4.97}$$

This occurs at point C in Figure 4.1. At this point $[H^+] = 4.57 \times 10^{-9}\ M$, and substitution of this value of $[H^+]$ into Equation 4.95 gives

$$\alpha_{HCO_3^-} = 0.989 \tag{4.98}$$

Point A from Figure 4.1, where the fraction of CO_2 is equal to the fraction of HCO_3^-, is readily obtained by setting the expressions in Equations 4.94 and 4.95 equal to each other so that

$$\alpha_{CO_2} = \alpha_{HCO_3^-} \qquad \text{when } [H^+] = K_{a1}, \ pH = 6.35 \tag{4.99}$$

At this point the two fractions are equal to 0.50. Similarly, at point B in Figure 4.1,

$$\alpha_{HCO_3^-} = \alpha_{CO_3^{2-}} = 0.50 \qquad \text{when } [H^+] = K_{a2}, \ pH = 10.33 \tag{4.100}$$

Importance of $CO_2(aq)$–HCO_3^-–CO_3^{2-} Species in Water

The distribution of $CO_2\ (aq)$–HCO_3^-–CO_3^{2-} species is particularly important to the quality of natural water (such as that in lakes, streams, and underground aquifers), in the treatment of water for municipal water systems, and in the treatment of wastewater. Most such water falls in a pH range of roughly 7 to 9 where, as shown by examination of Figure 4.1, HCO_3^- is the predominant species. This species is the major contributor to water **alkalinity**, which is defined as the capacity of solutes in water to neutralize added strong acid. Bicarbonate ion neutralizes acid by the reaction

$$HCO_3^- + H^+ \rightarrow H_2O + CO_2(g) \tag{4.101}$$

Other common contributors to alkalinity are carbonate ions each of which accepts 2 H^+ ions according to the two reactions

$$CO_3^{2-} + H^+ \rightarrow HCO_3^- \tag{4.102}$$

and

$$HCO_3^- + H^+ \rightarrow H_2O + CO_2(g) \tag{4.103}$$

and hydroxide ion

$$OH^- + H^+ \rightarrow H_2O \tag{4.104}$$

Distribution of Species for Polyprotic Acids

The distribution of species as a function of pH may be used for any polyprotic acid using the approach just shown for the $CO_2(aq)$–HCO_3^-–CO_3^{2-} system. For a polyprotic acid with n ionizable hydrogens, H_nA, the fraction present as the species $H_{n-x}A^{x-}$, where x ranges from 0 to n, is the following:

$$\alpha_{H_{n-x}A^{x-}} = \frac{K_{a1}\cdots K_{ax}[H^+]^{n-x}}{[H^+]^n + K_{a1}[H^+]^{n-1} + \cdots + K_{a1}K_{a2}\cdots K_{an}} \tag{4.105}$$

This equation may be applied to nitrilotriacetic acid (NTA), also abbreviated H_3T. The pK_a values for this acid are pK_{a1}, 1.66; pK_{a2}, 2.95; and pK_{a3}, 10.28. As an example of the application of Equation 4.105, consider the expression for $\alpha_{HT^{2-}}$. In this case n is 3 and x is 2, leading to the following:

$$\alpha_{HT^{2-}} = \frac{K_{a1}K_{a2}[H^+]}{[H^+]^3 + K_{a1}[H^+]^2 + K_{a1}K_{a2}[H^+] + K_{a1}K_{a2}K_{a3}} \tag{4.106}$$

Programmed Summary of Chapter 4

This summary contains in a programmed format the major terms and concepts introduced in this chapter. To get the most out of it, you should fill in the blanks for each question, and then check the answers at the end of the book to see if you are correct.

According to the Brønsted-Lowry theory an acid is (1) _____ and a base is (2) _____. A solvent that may act as either an acid or a base is called an (3) _____ solvent. Where HAc represents acetic acid in the reaction $HAc + H_2O \rightleftarrows H_3O^+ + Ac^-$, the acetate ion Ac^- is called the (4) _____ of acetic acid, and H_3O^+ is the (5) _____ of the base, H_2O. An electrolyte that is completely ionized in water is called a (6) _____ electrolyte. Because of its ability to ionize a number of acids completely to strong acids of equal acidity, water is called a (7) _____, whereas acetic acid, which brings out differences in acid strength is called a (8) _____. The expression $[H^+][OH^-]$ is called the (9) _____ of water, is given the symbol (10) _____, and describes the equilibrium of the reaction (11) _____. A base B reacts with water according to the reaction $B + H_2O \rightleftarrows HB^+ + OH$. The equilibrium constant expression for this reaction is (12) _____ and is given the symbol (13) _____. In the preceding example, the value of K_a of HB^+ is given by the expression $K_a = (14)$ _____. The reaction of acetate ion Ac^- with water produces (15) _____ ions in water and is called a (16) _____ reaction. A solution containing approximately equal concentrations of NH_4^+ and NH_3 is called a (17) _____ solution, and it resists changes in (18) _____. The expression (Δ moles H^+ or OH^- added per liter)/$|\Delta pH|$ defines the (19) _____ of a solution and has its maximum value when the ratio of the concentrations of conjugate acid-base pair species is (20) _____.

Questions

1. Show how the reaction of carbonate ion with water (hydrolysis of CO_3^{2-}) illustrates the Brønsted-Lowry definition of acids and bases.

2. What is a solvent called that can both donate and accept a proton?

3. The ammonium ion NH_4^+ is called the _____ of the base NH_3.

4. Illustrate with a chemical reaction how the hypothetical solvent HX might undergo autoionization; give the form of its autoionization constant, or ion product, K.

5. From examples cited in this chapter, classify each of the following as a strong (s) or weak (w) electrolyte: (a) NaCl, (b) NH_3, (c) HCl, (d) acetic acid, (e) $CO_2(aq)$, (f) NaOH.

6. Classify (a) acetic acid and (b) water in terms of their abilities to bring out differences in strengths of acidic solutes.

7. What is the mathematical relationship between the ion product of water, K_a of a weak acid, and K_b of its conjugate weak base?

8. Consider the weak acid HA and its conjugate base A^-. What does the pH value of a solution in which $[HA] = [A^-]$ reveal about the value of pK_a?

9. Give the definition and general composition of a buffer solution.

10. What is the quantitative definition of buffer capacity?

11. The distribution of species diagram for the diprotic acid H_2A is shown opposite. Estimate K_{a1} and K_{a2} of this acid.

12. Of the following, the true statement is (a) H^+ ion is not strongly associated with H_2O molecules in aqueous solution; (b) NH_4^+ ion acts as a weak acid; (c) acetate ion is the conjugate acid of acetic acid; (d) H_2O cannot act as both an acid and a base.

13. Consider a solution made by adding a pure acid to pure water. Of the following, the true statement is (a) the percentage of a weak acid that is ionized is greater for lower concentrations of the acid, (b) the terms *weak* and *strong* refer to the acid's molar concentration, (c) the percentage of ionization of a strong acid changes with its concentration, (d) preparation of an acid solution containing 1.26×10^{-7} mol of strong acid (such as HCl) per liter of solution gives a solution with pH 6.90, (e) a 0.0100 M solution of a strong acid has the same pH as a 0.0100 M solution of a weak acid.

14. Starting with 25 mL of 0.10 M acetic acid, which of the following is *not* true regarding the preparation of a buffer from this solution: (a) a buffer could be prepared by adding 2.5 mmol of solid sodium acetate, (b) the acetic acid itself is a buffer, (c) a buffer could be prepared by adding 5 mL of 0.10 M HCl and 30 mL of 0.100 M sodium acetate to the acetic acid, (d) a buffer could be prepared by adding 25 mL of 0.100 M sodium acetate, (e) a buffer could be prepared by adding 1.25 mmol of solid NaOH, (f) a buffer could be prepared by adding 12.5 mL of 0.10 M NaOH.

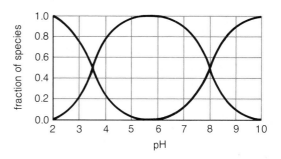

Problems

1. What is the pH of a solution prepared by dissolving 7.50×10^{-8} mol of NaOH in a liter of pure water?
2. What is the pH of a solution prepared by dissolving 2.00×10^{-7} mol of HCl and 4.00×10^{-8} mol of NaOH in a liter of pure water?
3. What is the pH of a solution prepared by dissolving 6.75×10^{-8} mol of HCl and 3.00×10^{-7} mol of NaOH in a liter of pure water?
4. What is the pH of a solution produced by dissolving 2.00×10^{-5} mol of acetic acid, $K_a = 1.75 \times 10^{-5}$, and diluting to 1 L?
5. What is the pH of a solution prepared by dissolving 2.50×10^{-5} mol of acetic acid, $K_a = 1.75 \times 10^{-5}$, and diluting to 1 L?
6. What is the pH of a solution prepared by dissolving 1.80×10^{-5} mol of NH_3, $K_b = 1.76 \times 10^{-5}$, in pure water and diluting to 1 L?
7. What is the pH of a solution $3.50 \times 10^{-7} F$ in HOCl, $K_a = 3.0 \times 10^{-8}$?
8. For chloroacetic acid, HClA, K_a is 1.36×10^{-3}. What is the H^+ concentration in a solution $2.00 \times 10^{-3} M$ in HCl and $0.050 F$ in HClA?
9. Benzoic acid

abbreviated HBz, has a K_a of 6.14×10^{-5}. What are $[H^+]$ and pH of a solution $0.0100 F$ in sodium benzoate, NaBz?
10. The value of K_b of piperidine, $C_5H_{11}N$, abbreviated Pn, is 1.30×10^{-3}. Calculate $[H^+]$ and pH in a solution $0.0222 F$ in Pn and $0.0275 F$ in its conjugate acid, piperidine hydrochloride, HPn^+Cl^-.
11. Given the reaction below and its equilibrium constant,

$$HCO_3^- \rightleftharpoons H^+ + CO_3^{2-}$$

$$K_{a2} = \frac{[H^+][CO_3^{2-}]}{[HCO_3^-]} = 4.69 \times 10^{-11}$$

$$pK_{a2} = 10.33$$

how many moles of CO_2 must be absorbed by 1.00 L of $0.0500 M$ OH^- to give a pH 10.00 buffer?
12. Describe the composition of 1 L of a pH 10.00 ammonia–ammonium chloride buffer in which $[NH_3] + [NH_4^+] = 0.0385 M$.
13. The buffer capacity of a buffer solution may be calculated from Equation 4.58 from the very small difference in pH resulting from a very small addition of acid or base. Consider a buffer solution exactly 0.05 M in acetic acid and 0.05 M in sodium acetate. The value of H^+ ion concentration is given by

$$[H^+] = \frac{[HC_2H_3O_2]}{[C_2H_3O_2^-]} K_a$$

$$= \frac{0.05}{0.05} \times 1.75 \times 10^{-5} = 1.75 \times 10^{-5}$$

Assuming that this result is the exact H^+ ion concentration to as many figures as an electronic calculator can handle, the pH is given by

$$pH = -\log[H^+] = -(\log 1.75 - 5)$$
$$= -(0.243038 - 5) = 4.756962$$

Calculate the pH change resulting from the addition of exactly 1×10^{-5} mol of NaOH to a liter of this buffer solution, and from this the buffer capacity.
14. Repeat Problem 13 for a buffer solution $0.09 M$ in acetic acid and $0.01 M$ in sodium acetate.
15. What is the pH of a $1.00 \times 10^{-5} M$ solution of $NaHCO_3$?

16. Prepare a distribution of species diagram for diprotic oxalic acid, HOOCCOOH, abbreviated H_2Ox. For this acid, $K_{a1} = 5.36 \times 10^{-2}$ and $K_{a2} = 5.42 \times 10^{-5}$. Plot points at pH 0; where pH is equal to, one unit less, and one unit more than the pH at which $\alpha_{H_2Ox} = \alpha_{HOx^-}$; pH at which α_{HOx^-} is at its maximum value; where pH is equal to, one unit less, and one unit more than the pH at $\alpha_{HOx^-} = \alpha_{Ox^{2-}}$; and at pH 7.

17. The weak monoprotic acid HB was 10.3% ionized in a solution for which $C_{HB} = 0.0286 \ M$. The value of K_a for the acid is (a) 3.38×10^{-4}, (b) 2.77×10^{-4}, (c) 3.01×10^{-4}, (d) 2.49×10^{-4}, (e) 3.32×10^{-4}?

5

Acid-Base Titrations

5.1   Introduction to Acid-Base Titrations

Nature of Acid-Base Titrations

An **acid-base titration** is an analytical procedure in which the quantity of a standard base solution of known concentration necessary to neutralize a sample of acid is measured, or the quantity of standard acid required to neutralize a sample of base is measured. The standard solution is called the **titrant**. Normally the volume of titrant is measured with a buret (see Section 2.7). The experimental objective of an acid-base titration is to find the **equivalence point** at which the exact quantity of standard base (or acid) has been added to neutralize the acid (or base) in the sample. The experimental estimate of the equivalence point is the **end point**, which is manifested by the change of color of a pH-sensitive dye (indicator) dissolved in the solution being titrated (Figure 5.1), or by other means discussed later in this chapter.

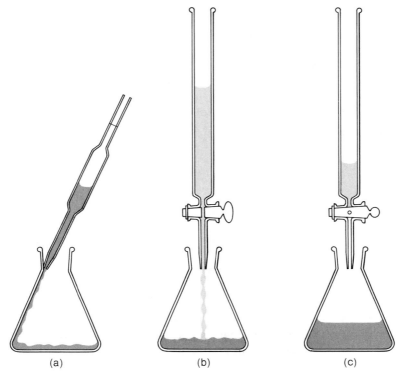

FIGURE 5.1 (a) The titration of an acid of unknown concentration begins with the addition of a measured volume or mass of the acid to a conical flask or other suitable container. (b) Water may be added as a diluent. Standard base is added from a buret and mixed with the acid solution. (c) The end point is perceived by the change in color of an indicator dye in solution or by other means.

Acid-base titrations in water are **neutralization reactions** in which the net ionic reaction is

$$H^+ + OH^- \rightarrow H_2O \tag{5.1}$$

The $H^+(H_3O^+)$ may be from a completely dissociated strong acid (see Section 4.2) in water or it may be from a weak acid, of which only a small fraction is dissociated in water. Similarly, the OH^- ion may come from a completely dissociated strong base, such as NaOH, or from the reaction of a weak base, such as NH_3, with water. Strong acids and strong bases are almost always employed as standard titrants because they produce the most complete reactions.

5.2   **The Nature of Acid-Base Titration Curves**

Titration of Strong Acid with Strong Base

A titration is best visualized as a **titration curve**. Such a curve is "S-shaped" and consists of a plot of some calculated or measured parameter, pH in the case of an acid–base titration, versus the volume of added titrant. The equivalence point in an acid–base titration curve is manifested by an abrupt change in pH appearing as an **S**-shaped *break* in the titration curve. Two or even three such breaks may be observed in the titrations of weak acids with two or three ionizable hydrogens, or bases that may accept two or three H^+'s.

To understand the concept of a titration curve, consider first the titration of a strong acid with a strong base, specifically the titration of 50.00 mL of 0.0500 M HCl with 0.1000 M standard NaOH titrant. The titration curve may be divided into four distinct points or regions. The first of these is the *initial* point before the addition of any base. The values of $[H^+]$ and pH at this point are given by the following:

$$[H^+] = C_{HCl} = 0.0500 \ M$$

$$pH = -\log[H^+] = -\log 0.0500 = 1.30 \tag{5.2}$$

Next comes the region of the titration curve in which base has been added prior to the equivalence point. In this region there is still an excess of H^+ in solution. A typical point is after the addition of 5.00 mL of standard NaOH, for which the pH is given by the following calculation:

$$\text{mmol } H^+ = \text{mmol } H^+ \text{ initially present} - \text{mmol } OH^- \text{ added} \tag{5.3}$$

$$= 0.0500 \ \text{mmol/mL} \times 50.00 \ \text{mL} - 0.1000 \ \text{mmol/mL} \times 5.00 \ \text{mL}$$
$$= 2.00 \ \text{mmol} \tag{5.4}$$

$$[H^+] = \frac{\text{mmol } H^+}{\text{total mL of solution after addition of titrant}} \tag{5.5}$$

$$= \frac{2.00 \ \text{mmol}}{50.00 \ \text{mL} + 5.00 \ \text{mL}} = 3.64 \times 10^{-2} M$$

$$pH = -\log[H^+] = -\log 3.64 \times 10^{-2} = 1.44 \tag{5.6}$$

This value of pH, along with other values calculated prior to the equivalence point by exactly the same approach, are listed in Table 5.1. Note that several values of pH are calculated at titrant volumes just less than the 25.00 mL required to reach the equivalence point in order to show the sharpness in the break of a strong acid–strong base titration curve. At the equivalence point where exactly 25.00 mL of base have been added, there is neither excess acid nor base, the solution is neutral, and $[H^+]$ and $[OH^-]$ are given by

$$[H^+] = [OH^-] = \sqrt{K_w} = \sqrt{1.00 \times 10^{-14}} = 1.00 \times 10^{-7} M \tag{5.7}$$

$$pH = -\log[H^+] = -\log 1.00 \times 10^{-7} = 7.00 \tag{5.8}$$

TABLE 5.1   Data for construction of the titration curve for the titration of 50.00 mL of 0.0500 M HCl with 0.1000 M NaOH

NaOH added, mL	Total volume, mL	H^+ excess, mmol	OH^- excess, mmol	$[H^+]$, M	pH
0.00	50.00	2.50	—	5.00×10^{-2}	1.30
5.00	55.00	2.00	—	3.64×10^{-2}	1.44
12.50	62.50	1.25	—	2.00×10^{-2}	1.70
20.00	70.00	0.50	—	7.14×10^{-3}	2.15
24.00	74.00	0.10	—	1.35×10^{-3}	2.87
24.90	74.90	0.010	—	1.34×10^{-4}	3.87
24.99	74.99	0.0010	—	1.33×10^{-5}	4.88
25.00	75.00	—	—	1.00×10^{-7}	7.00
25.01	75.01	—	0.001	7.50×10^{-10}	9.12
25.10	75.10	—	0.010	7.51×10^{-11}	10.12
26.00	76.00	—	0.100	7.60×10^{-12}	11.12
30.00	80.00	—	0.500	1.60×10^{-12}	11.80
35.00	85.00	—	1.00	8.50×10^{-13}	12.07

Beyond the equivalence point excess base is present. For example, after the addition of only 25.01 mL of standard base, there is an excess of 0.01 mL (0.001 mmol) of OH^- in a total volume of 75.01 mL of solution, yielding the following:

$$[OH^-] = \frac{1.00 \times 10^{-3} \text{ mmol}}{75.01 \text{ mL}} = 1.33 \times 10^{-5} M \tag{5.9}$$

$$[H^+] = \frac{K_w}{[OH^-]} = \frac{1.00 \times 10^{-14}}{1.33 \times 10^{-5}} = 7.52 \times 10^{-10} M \tag{5.10}$$

$$pH = -\log[H^+] = -\log 7.52 \times 10^{-10} = 9.12 \tag{5.11}$$

Note the change in pH of more than 2 units from the addition of only 0.01 mL of excess base. Additional values of pH in the presence of increasing volumes of excess base are given in Table 5.1.

Plot of Curve for the Titration of Strong Acid with Strong Base

Examination of Figure 5.2 shows that the break in the titration curve is very sharp, a characteristic of strong acid–strong base titration curves. The equivalence point in the titration curve comes exactly halfway up the steeply rising portion of the curve.

Figure 5.3 shows the curve for the titration of a more dilute acid, 50.00 mL of 5.00×10^{-4} M HCl with a more dilute base, 1.00×10^{-3} M NaOH. Here it is seen that the titration curve is very similar in shape to the one for the more concentrated acid and base shown in Figure 5.2. The break is still very sharp for the curve in Figure 5.3, except that it encompasses much less of a vertical distance than the one in Figure 5.2. Of course, with increasing

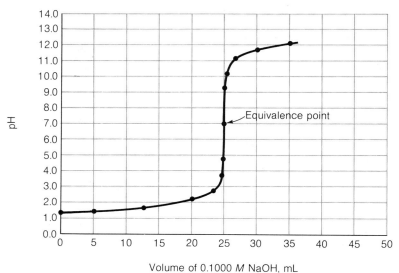

FIGURE 5.2   Curve for the titration of 50.00 mL of 0.0500 M HCl with 0.100 M NaOH

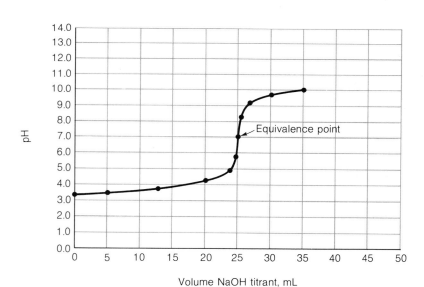

FIGURE 5.3   Curve for the titration of 50.00 mL of 5.00×10^{-4} M HCl with 1.00×10^{-3} M NaOH

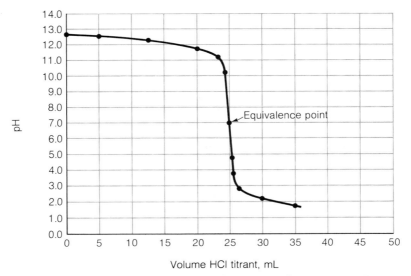

FIGURE 5.4   Curve for the titration of 50.00 mL of 0.0500 M NaOH with 0.1000 M HCl

dilution of acid and base the curves become less sharp and a point will be reached where there is no visible break at all.

Titration of Strong Base with Strong Acid

Figure 5.4 shows the curve for the titration of 50.00 mL of 0.0500 M NaOH with 0.1000 M HCl. The procedure for calculating the points on this curve is the reverse of that used for the points on the curve in Figure 5.2. Initially, $[OH^-] = 0.0500$, $[H^+] = K_w/[OH^-] = 2.00 \times 10^{-13}\ M$, pH $= 12.70$. From the beginning of the titration to the equivalence point, the pH is calculated from the concentration of OH^- remaining divided into K_w, as shown below after the addition of 12.50 mL of standard HCl:

$$\text{mmol } OH^- = 2.50 \text{ mmol initially present} - 1.25 \text{ mmol neutralized by HCl} \tag{5.12}$$

$$= 1.25 \text{ mmol} \tag{5.13}$$

$$[OH^-] = \frac{1.25 \text{ mmol}}{\text{total mL of solution}} = \frac{1.25 \text{ mmol}}{50.00 \text{ mL} + 12.50 \text{ mL}} = 2.00 \times 10^{-2}\ M \tag{5.14}$$

$$[H^+] = \frac{K_w}{[OH^-]} = \frac{1.00 \times 10^{-14}}{2.00 \times 10^{-2}} = 5.00 \times 10^{-13}\ M \tag{5.15}$$

$$pH = 12.30$$

As in the case of the titration of a strong acid with a strong base, at the equivalence point, 25.00 mL of added acid, the solution is neutral and pH = 7.00. Beyond the equivalence point, H^+ is in excess and is calculated from the relationship

$$[H^+] = \frac{0.1000 \text{ mmol HCl/mL titrant added} \times \text{mL titrant added} - 2.50 \text{ mmol HCl}}{\text{total mL of solution}} \qquad (5.16)$$

Note that 2.50 mmol of HCl are subtracted from the numerator because that quantity of HCl was neutralized by the NaOH originally present.

5.3   Experimental Measurement of Titration Curves—The pH Meter

Apparatus for pH Measurement

In the preceding pages, titration curves have been plotted from points obtained by calculating pH as a function of volume of added titrant. An acid-base titration curve can be obtained in the laboratory by measuring the pH of a solution of acid or base at increasing volumes of added titrant. The pH of a solution is sensed by the electrical potential at a glass electrode as measured by a potentiometer calibrated in pH units called a **pH meter**. A combination glass and reference electrode hooked to a pH meter is shown in Figure 5.5. The sensing element of this system is a thin glass membrane across which an electrical potential is developed that changes in magnitude with changes in the pH of the medium in which it is immersed. This potential is measured by the pH meter versus the potential of an external reference electrode, which, in a combination electrode, surrounds the glass electrode contained in a concentric cylinder of glass tubing. Additional details of the construction of glass electrodes are given in Section 13.3.

Glass Electrode Response

A glass electrode responds to $[H^+]$ according to an equation of the Nernstian type (see Section 13.1) which, at 25° C, is the following:

$$E = E_a + 0.0591 \log [H^+] \qquad (5.17)$$

In this equation E is the potential in volts of the glass electrode measured against the external reference electrode, E_a is a constant for the electrode system, and 0.0591 is a constant for 25° C; it has different values at other temperatures, compensation for which is made with the "temperature-adjust" control on the pH meter. In terms of pH, the preceding equation is

$$E = E_a - 0.0591 \times \text{pH} \qquad (5.18)$$

pH Meter Calibration

In practice a pH meter with its glass and reference electrode system is calibrated directly in pH units. The meter temperature adjust is set to the

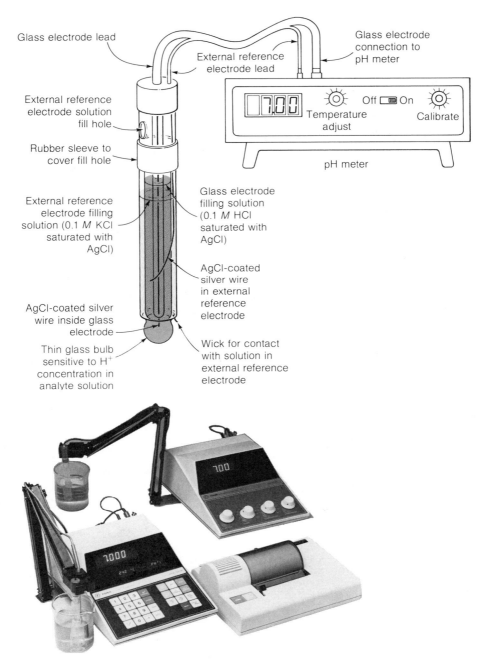

Glass electrode lead

External reference electrode lead

Glass electrode connection to pH meter

External reference electrode solution fill hole

Rubber sleeve to cover fill hole

External reference electrode filling solution (0.1 M KCl saturated with AgCl)

Glass electrode filling solution (0.1 M HCl saturated with AgCl)

AgCl-coated silver wire in external reference electrode

AgCl-coated silver wire inside glass electrode

Thin glass bulb sensitive to H^+ concentration in analyte solution

Wick for contact with solution in external reference electrode

Temperature adjust   Off   On   Calibrate

pH meter

FIGURE 5.5   *Above*: Combination glass and reference electrode attached to a pH meter for measuring pH. The external reference electrode is constructed as a glass cylinder surrounding the glass body of the pH-responsive glass electrode (named for the special glass membrane at its end that gives different electrical potentials with differences in H^+ concentration in the solution in which it is immersed). Properly calibrated with a buffer of known pH, the entire electrode assembly can be immersed in a solution of unknown pH, and the pH read directly from the pH meter. *Below*: pH meters each with electrode holder and combination glass/reference electrode.

temperature of the solution being measured. The electrode system is then immersed in a buffer of known pH and the calibrate control on the pH meter is used to set the meter reading to the standard buffer pH. Subsequently the electrode system is placed in solutions for which the pH is to be measured and the value of pH read directly from the meter.

The standard pH at which a pH meter is calibrated should be as close to that to be measured as possible. For best results when a relatively large range of pH values is to be measured, the meter should be calibrated using the buffer solutions with the highest and lowest pH values. The electrode system is dipped into one buffer and the meter adjusted with the calibrate control to read the pH of that buffer. The electrode system is next placed in the second buffer, and the meter adjusted to read its value with the temperature-adjust control. These operations are repeated until the meter reads the correct pH in each buffer without further calibration adjustment.

Care of Glass Electrodes

The glass electrode is a fragile device and must be treated carefully to avoid breaking or scratching the pH-sensitive glass membrane. The electrode system should be rinsed thoroughly with deionized water, and water droplets hanging from the ends of the electrodes should be removed by blotting with laboratory tissue paper each time the electrode system is transferred to a different solution. The glass electrode should be stored in deionized water; if the pH-sensitive glass membrane becomes dry, it should be soaked in water for several hours before use.

More detailed theory of glass electrode operation is discussed in Section 13.3. The measurement of pH with a pH meter is done at many stages of quantitative chemical analysis procedures and excellent results can be obtained by carefully following directions even though you may be lacking details of the theory of glass electrodes.

5.4   Determination of End Point

End Point from Plot of pH versus Volume of Titrant

Often the most challenging aspect of a titration is the determination of the equivalence point which, as mentioned in Section 5.1, is approximated by the end point. In the case of acid-base titrations, the end point is readily determined from a plot of pH versus volume of titrant. Since in most titrations volumes should be estimated to the nearest 0.01 mL, the break in the titration curve over which the end point is to be determined should be spread over a relatively small volume so that the end point volume can be estimated very accurately. This is shown for the titration of 50.00 mL of 0.0500 M HCl with

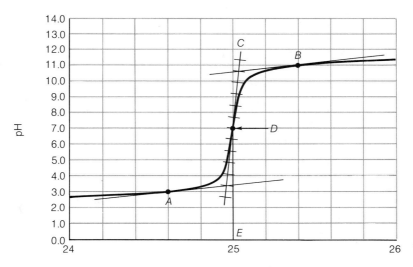

Volume of 0.1000 *M* NaOH, mL

FIGURE 5.6 Portion of the titration curve at the end point for the titration of 50.00 mL of 0.0500 *M* HCl with 0.1000 *M* NaOH. This kind of plot is readily obtained experimentally with a pH meter and glass electrode as discussed in the preceding section.

0.1000 *M* NaOH in Figure 5.6. Here the titration curve break is plotted over added volumes of titrant from 24.00 to 26.00 mL, whereas in Figure 5.2, the whole titration curve was shown.

The procedure for graphically measuring the end point in a titration curve such as that shown in Figure 5.6 is very simple. First two sloping straight lines are drawn tangent to the titration curve intersecting it at points (*A* and *B* in Figure 5.6) estimated to be equal distances on either side of the end point. Next a steeply sloping straight line is drawn tangent to the titration curve at its most steeply rising portion and the point (point *D*) is found half way between the intersections of this straight line with the two drawn previously. Finally, from this halfway point a vertical line (*DE*) is dropped down to the volume axis and the end point volume (point *E*) is estimated. Here it is seen to be 25.00 mL, which is exactly the equivalence-point volume.

Modifications of the simple method for determining end point from a titration curve can be used on curves for precipitation titrations (Chapter 9), complexation titrations (Chapter 10), and oxidation-reduction titrations (Chapter 12), all of which involve a plot of the potential at an electrode versus volume of titrant.

End Point from a Differential Titration Plot

The end point can also be determined from a **differential titration** curve in which the change in pH per unit change in volume of titrant (ΔpH/ΔpH) is plotted. For such a plot, data are taken for short distances either side of the end point. For example, suppose that the volume increment of added standard base is 0.10 mL as a result of addition of base from an initial reading of 24.70 to 24.80 mL, and that the change in pH resulting from this addition was 0.24 pH units. This gives

$$\frac{\Delta pH}{\Delta V} = \frac{0.24 \text{ pH units}}{0.10 \text{ mL}} = 2.4 \text{ pH units/mL} \tag{5.19}$$

over the *average* added volume for the increment of

$$\frac{24.70 \text{ mL} + 24.80 \text{ mL}}{2} = 24.75 \text{ mL} \tag{5.20}$$

This data point would be plotted as 2.4 pH units/mL at 24.75 mL. Laboratory data for such a plot taken for the titration of 50.00 mL of 0.0500 M HCl with 0.1000 M NaOH are given in Table 5.2, and the differential titration plot is

TABLE 5.2   Differential titration data for the titration of 50.00 mL of 0.0500 M HCl with 0.1000 M NaOH

Volume NaOH, mL	pH	Average volume NaOH for first-derivative point, mL	$\Delta pH/\Delta V$, pH units/mL	Average volume NaOH for second-derivative point, mL	$\Delta^2 pH/\Delta V^2$, pH units/mL^2
24.00	2.82	—	—	—	—
		24.20	0.70	—	—
24.40	3.10	—	—	—	—
		24.50	0.65	—	—
				24.60	+6.0
24.60	3.23	—	—	—	—
		24.70	1.85	—	—
				24.775	+7.7
24.80	3.60	—	—	—	—
		24.85	3.00	—	—
				24.90	+280
24.90	3.90	—	—	—	—
		24.95	31	—	—
				25.00	0.00
25.00	7.00	—	—	—	—
		25.05	31	—	—
				25.10	−285
25.10	10.10	—	—	—	—
		25.15	2.50	—	—
				25.225	−6.7
25.20	10.35	—	—	—	—
		25.30	1.50	—	—
				25.40	−3.8
25.40	10.65	—	—	—	—
		25.50	0.75	—	—
25.60	10.80	—	—	—	—
		25.80	1.0	—	—
26.00	11.20	—	—	—	—

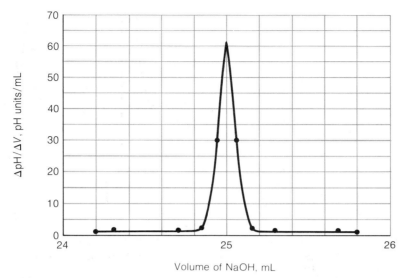

FIGURE 5.7   First-derivative plot for the titration of 50.00 mL of 0.0500 M HCl with 0.1000 M NaOH

shown in Figure 5.7. The end point of 25.00 mL is taken as the volume of titrant corresponding to the peak in the plot; it corresponds to the inflection point (point D) of the break in the titration curve plotted in Figure 5.6.

End Point from a Second-Derivative Plot

The end point can also be determined as a *second-derivative plot*, as shown by the calculations in the last two columns in Table 5.2 and the graph in Figure 5.8. With such a plot, there is an abrupt reversal of sign at the inflection point in the titration break that corresponds to the peak of the first-derivative curve. A second-derivative plot is especially amenable to the determination of the end point by computer.

5.5   **Titration of a Weak Acid**

Calculation of pH at Various Titrant Volumes for Acetic Acid Titrated with NaOH

Consider next the titration of a weak acid with a strong base, specifically the titration of 50.00 mL of 0.0500 M acetic acid, $HC_2H_3O_2$, with 0.1000 M standard NaOH. The pH will be calculated initially and after the addition of 1.00, 12.50, 24.00, 25.00, 26.00, and 30.00 mL of standard NaOH. The initial

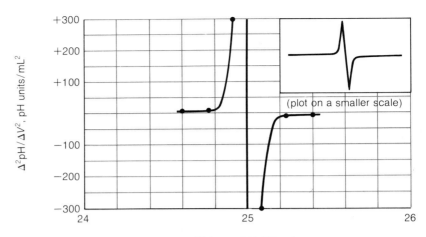

FIGURE 5.8   Second-derivative plot for the titration of 50.00 mL of 0.0500 M HCl with 0.1000 M NaOH

pH is simply that of 0.0500 M $HC_2H_3O_2$ calculated as follows:

$$HC_2H_3O_2 \rightleftarrows H^+ + C_2H_3O_2^- \tag{5.21}$$

$$[H^+] = [C_2H_3O_2^-] \tag{5.22}$$

$$[HC_2H_3O_2] = 0.0500 \; M \tag{5.23}$$

$$K_a = \frac{[H^+][C_2H_3O_2^-]}{[HC_2H_3O_2]} = \frac{[H^+]^2}{0.0500} = 1.75 \times 10^{-5} \tag{5.24}$$

$$[H^+] = (0.0500 \times 1.75 \times 10^{-5})^{1/2} = 9.35 \times 10^{-4} \; M \tag{5.25}$$

$$pH = 3.03$$

The addition of 1.00 mL of 0.1000 M NaOH converts 0.100 mmol of the original 2.500 mmol of acetic acid to acetate ion in a total volume of 51.00 mL, so that

$$[C_2H_3O_2^-] = \frac{0.100 \; \text{mmol}}{51.00 \; \text{mL}} = 1.96 \times 10^{-3} \; M \tag{5.26}$$

$$[HC_2H_3O_2] = \frac{2.500 - 0.100 \; \text{mmol}}{51.00 \; \text{mL}} = \frac{2.400 \; \text{mmol}}{51.00 \; \text{mL}} = 4.71 \times 10^{-2} \; M \tag{5.27}$$

$$[H^+] = \frac{[HC_2H_3O_2]}{[C_2H_3O_2^-]} K_a = \frac{4.71 \times 10^{-2}}{1.96 \times 10^{-3}} \times 1.75 \times 10^{-5} = 4.20 \times 10^{-4} \; M \tag{5.28}$$

$$pH = 3.38$$

After the addition of 12.50 mL of NaOH there are 1.25 mmol of acetic acid and 1.25 mmol of acetate ion in 62.50 mL of solution, yielding

$$[HC_2H_3O_2] = [C_2H_3O_2^-] = \frac{1.25 \text{ mmol}}{62.50 \text{ mL}} = 2.00 \times 10^{-2} \; M \tag{5.29}$$

$$[H^+] = \frac{[\cancel{HC_2H_3O_2}]}{[\cancel{C_2H_3O_2^-}]} \; K_a = K_a = 1.75 \times 10^{-5} \; M \tag{5.30}$$

$$pH = 4.76$$

Note that this is the pH half way to the equivalence point, and $pH = pK_a$. After the addition of 24.00 mL of NaOH, there are 0.100 mmol of $HC_2H_3O_2$ and 2.400 mmol of $C_2H_3O_2^-$ in 74.00 mL of solution, giving

$$[C_2H_3O_2^-] = \frac{2.400 \text{ mmol}}{74.00 \text{ mL}} = 3.24 \times 10^{-2} \; M \tag{5.31}$$

$$[HC_2H_3O_2] = \frac{2.500 - 2.400 \text{ mmol}}{74.00 \text{ mL}} = \frac{0.100 \text{ mmol}}{74.00 \text{ mL}} = 1.35 \times 10^{-3} \; M \tag{5.32}$$

$$[H^+] = \frac{[HC_2H_3O_2]}{[C_2H_3O_2^-]} \; K_a = \frac{1.35 \times 10^{-3}}{3.24 \times 10^{-2}} \times 1.75 \times 10^{-5} = 7.29 \times 10^{-7} \; M \tag{5.33}$$

$$pH = 6.14$$

After the addition of exactly 25.00 mL of NaOH, the solution is of pure sodium acetate, with 2.500 mmol of $C_2H_3O_2^-$ ion in 75.00 ml of solution. The pH of this solution is calculated from the reaction of acetate anion as a base as follows:

$$C_2H_3O_2^- + H_2O \rightleftarrows HC_2H_3O_2 + OH^- \tag{5.34}$$

$$K_b = \frac{[HC_2H_3O_2][OH^-]}{[C_2H_3O_2^-]} = \frac{K_w}{K_a} = \frac{1.00 \times 10^{-14}}{1.75 \times 10^{-5}} = 5.71 \times 10^{-10} \tag{5.35}$$

$$[HC_2H_3O_2] = [OH^-] \tag{5.36}$$

$$[C_2H_3O_2^-] = \frac{2.500 \text{ mmol}}{75.00 \text{ mL}} = 3.33 \times 10^{-2} \; M \tag{5.37}$$

$$K_b = \frac{[HC_2H_3O_2][OH^-]}{[C_2H_3O_2^-]} = \frac{[OH^-]^2}{3.33 \times 10^{-2}} = 5.71 \times 10^{-10} \tag{5.38}$$

$$[OH^-] = (3.33 \times 10^{-2} \times 5.71 \times 10^{-10})^{1/2} = 4.36 \times 10^{-6} \; M \tag{5.39}$$

$$[H^+] = \frac{K_w}{[OH^-]} = \frac{1.00 \times 10^{-14}}{4.36 \times 10^{-6}} = 2.29 \times 10^{-9} \tag{5.40}$$

$$pH = 8.64$$

At 26.00 and 30.00 mL added NaOH, the pH is calculated from the excess NaOH present and is the same as the corresponding volumes added to the HCl solution described in Table 5.1—that is, at 26.00 mL added NaOH the pH is

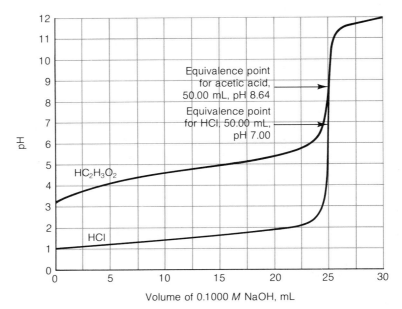

FIGURE 5.9   Titration of 50.00 mL of 0.0500 M HCl and 50.00 mL of 0.0500 M acetic acid with 0.1000 M NaOH

11.12, and at 30.00 mL it is 11.80. The curve for the titration of acetic acid with NaOH is plotted in Figure 5.9.

Titration Curve for a Weak Monoprotic Acid

Examination of Figure 5.9 shows the distinct differences in the curve for the titration of a weak monoprotic (one ionizable H) acid as compared to that for a strong acid. For the examples illustrated, the equivalence points for both the strong and weak acids occur at exactly 50.00 mL of added NaOH, but at a pH of 7.00 for HCl and 8.64 for acetic acid. As we will discuss in Section 5.6, special care must be exercised in choosing an indicator for the titration of a weak acid so that the indicator changes color near the pH of the equivalence point. The choice is especially critical when the break in the titration curve covers only a very narrow pH range, as is the case for an especially weak acid or a relatively dilute solution of a weak acid.

   The titration curve for the weak acid in Figure 5.9 shows a relatively steep rise at the beginning, from 0 to about 3 mL added base, that is not present in the curve for the titration of strong acid. From this initial sharp rise to shortly before the equivalence point in the weak acid titration curve is a relatively flat region in which the pH does not change greatly with added strong base. This is the **buffer region**. It is present because the titration reaction

$$HC_2H_3O_2 + OH^- \rightarrow C_2H_3O_2^- \tag{5.41}$$

results in a concentration of $C_2H_3O_2^-$ that is of the same order of magnitude as that of $HC_2H_3O_2$, thus producing a buffer solution consisting of a weak acid and its conjugate base.

Summary of the Titration Curve for a Weak Monoprotic Acid

Acetic acid is a prime example of a weak monoprotic (one ionizable H) acid. The approach just applied in calculating the curve for the titration of acetic acid with strong base can be applied in general to the titration of weak monoprotic acids. The calculation of the points required to construct the titration curve for a weak monoprotic acid HA requires consideration of four different parts of the titration curve. Initially the solution contains only the acid, and the pH is calculated in the same manner as for any weak monoprotic acid. The second part of the titration curve is one in which the solution contains significant quantities of both weak acid and its conjugate base—that is, the buffer region. The value of $[H^+]$ in the buffer region is calculated from the relationship

$$[H^+] = \frac{[HA]}{[A^-]} K_a \tag{5.42}$$

where $[HA]$ is the concentration of the weak acid and $[A^-]$ is the concentration of its conjugate base. It should be noted that exactly half way to the equivalence point $[HA]$ and $[A^-]$ are equal so that they cancel in Equation 5.42 and $[H^+] = K_a$ (pH = pK_a). The third distinct part of the titration curve is at the equivalence point, where enough strong base has been added to convert all the HA to A^-. Here the reaction that determines the pH is

$$A^- + H_2O \rightleftarrows HA + OH^- \tag{5.43}$$

such that

$$[HA] = [A^-] \tag{5.44}$$

and $[OH^-]$ is calculated from K_b of A^-.

$$K_b = \frac{K_w}{K_a} = \frac{[HA][OH^-]}{[A^-]} = \frac{[OH^-]^2}{[A^-]} \tag{5.45}$$

The value of $[H^+]$ is then simply calculated by substituting the OH^- concentration into the K_w expression. The fourth distinct region of the titration curve is beyond the equivalence point where $[OH^-]$ is calculated from

$$[OH^-] = \frac{\text{millimoles excess } OH^-}{\text{total volume of solution}} \tag{5.46}$$

and $[H^+]$ is calculated from the expression for K_w.

5.6   **Acid-Base Indicators**

What an Acid-Base Indicator Does

An **acid–base indicator** is a dye that changes color in dilute solution when the solution pH changes. The acid form of the indicator may be neutral and represented by HIn; its conjugate base may be negatively charged and represented by In^-. In some cases the acid form has a positive charge and is represented as HIn^+, whereas the base form is neutral, represented by In. Indicators behave as weak acids and weak bases in solution.

The equilibrium of acid-base forms of indicators in solution may be represented by the following:

$$HIn \rightleftharpoons H^+ + In^-$$

acid color   base color

(5.47)

Obviously, HIn and its color will predominate in acidic solution, whereas In^- and its color will predominate in basic solution. More specifically, the predominant species are determined by the pH in relationship to the pK_a of the indicator. To see how this is so, consider the following K_a expression for HIn and its logarithmic form:

$$K_a = \frac{[H^+][In^-]}{[HIn]}$$

$$[H^+] = K_a \frac{[HIn]}{[In^-]}$$

$$pH = pK_a - \log \frac{[HIn]}{[In^-]} = pK_a + \log \frac{[In^-]}{[HIn]}$$

It is seen from this expression that when $[In^-] = [HIn]$, $\log [In^-]/[HIn] = \log 1 = 0$, so that $pH = pK_a$. If In^- is present in a tenfold excess over HIn— that is, $[In^-]/[HIn] = 10$, $pH = pK_a + 1$. When In^- is present in a tenfold excess over HIn, the color of the solution is generally indistinguishable from that of a solution of pure In^- ($pH \gg pK_a$). Similarly, when HIn is present in a tenfold excess over In^-, so that $[In^-]/[HIn] = 1/10$, $pH = pK_a - 1$. Under these conditions the color of the solution appears the same as that of pure HIn solution ($pH \ll pK_a$). In between these two extremes of pH there is a range of indicator color. As a general approximation, the indicator changes color over the range $pH = pK_a \pm 1$, but this varies depending upon the relative intensities of indicator colors, the ability of the human eye to perceive these colors (a problem for the color-blind), and solution conditions. The range of pH over which the color change is perceived by the human eye is commonly called the *indicator pH range*; as shown in Table 5.3, it is usually less than 2 pH units. Furthermore, the midpoint of the indicator pH range does not always coincide with pK_a of the indicator.

EXAMPLE

The K_a of an indicator is 7.31×10^{-9}. What is the approximate indicator pH range?

If K_a of an indicator is 7.31×10^{-9}, $pK_a = 8.14$. The indicator pH range is

$$pK_a \pm 1 = 8.14 \pm 1$$

Therefore, the indicator changes color over an approximate pH range from 7.14 to 9.14.

5.7   Chemical Nature of Acid-Base Indicators

List of Common Indicators

A very large number of organic compounds have been used as acid–base indicators. Table 5.3 lists some of these, spanning very low to high pH values.

Phthalein Indicators

A number of indicators fall into one of several major chemical classifications. One of the most common of these is the *phthalein indicators*, of which phenolphthalein is the best known. In the pH range of 8.3 to 10.0 over which phenolphthalein changes color, it undergoes the reaction

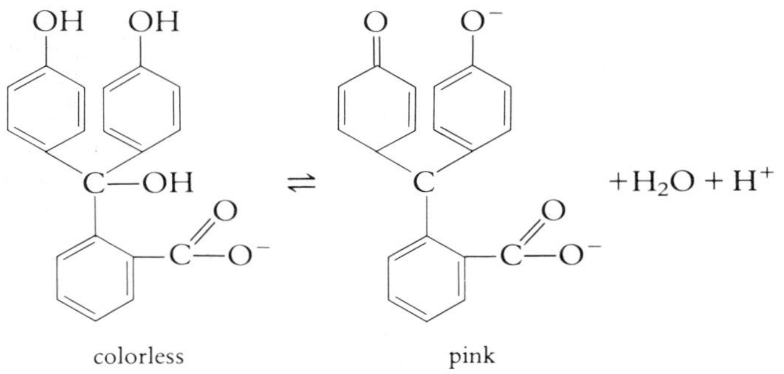

           colorless   pink

The pH at which the pink color of the phenolphthalein appears depends upon the concentration of the indicator and the ability of the individual to perceive the color. The colored form of the indicator has a quinoid ring

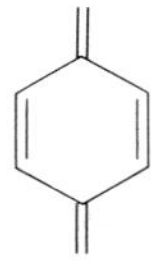

which is a structure often associated with colored organic compounds.

TABLE 5.3   Some common acid–base indicators

Name	Acid color	pH range*	Base color	pK_a
Methyl violet	Yellow	0–1.5	Blue	
Cresol red	Red	1.0–2.0	Yellow	
Thymol blue[†]	Red	1.2–2.8	Yellow	1.65
2,4-Dinitrophenol	Colorless	2.8–4.0	Yellow	
Methyl orange	Red	3.1–4.4	Orange	3.46
Bromocresol green	Yellow	3.8–5.4	Blue	4.66
Methyl red	Red	4.2–6.3	Yellow	5.00
p-Nitrophenol	Colorless	5.5–6.5	Yellow	
Bromothymol blue	Yellow	6.2–7.6	Blue	7.10
Phenol red[†]	Yellow	6.8–8.4	Red	7.81
Thymol blue	Yellow	8.0–9.6	Blue	8.90
Phenolphthalein	Colorless	8.3–10.0	Pink	
Thymolphthalein	Colorless	9.3–10.5	Blue	
Alizarin yellow GG	Colorless	10–12	Yellow	

* pH range over which indicator is perceived to change color
† Changes color over two pH ranges

   Another common indicator of this class, thymolphthalein, has two alkyl groups on each of the two phenolic rings associated with the phthalein structure. Other phthalein indicators also differ in the substituents on the phenolic rings. Most are colorless on the acidic side of the transition range and exhibit one of several colors on the basic side. These colors may fade in high-pH media. Most phthalein indicator solutions are made up in ethanol because of the low solubility of the indicators in water.

Sulfonphthalein Indicators

The *sulfonphthalein* indicators are similar to the phthalein indicators in having three 6-carbon rings attached to a central carbon atom. The nonphenolic ring (the bottom one in the structure below) has an —SO_3^- group attached to it. Different indicator compounds are formed by the attachment of alkyl groups or halogens to the phenolic rings. Phenolsulfonphthalein (phenol red) undergoes a color change that is useful for indicator purposes in the pH range 6.4 to 8.0, as shown by the reaction

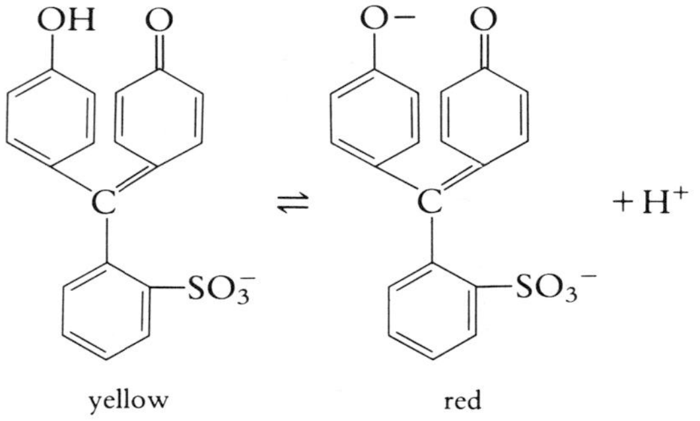

yellow   red   (5.49)

In general, the base color of sulfonphthalein indicators does not fade at high pH. Another advantage of these indicators is the adequate water solubility of their sodium salts, so that stock indicator may be made up in aqueous solution.

Azo Indicators

Methyl orange and methyl red are typical of *azo indicators*. These generally have transition ranges below pH 7. One of the most common azo indicators is methyl orange, which behaves as follows:

$$+ H^+ \qquad (5.50)$$

Methyl red differs from methyl orange in having a carboxylic acid group, $-CO_2H$, in place of the sulfonate group, $-SO_3^-$. Different substituents on the rings and on the amino nitrogen give indicators with somewhat different properties.

5.8   Indicator Errors and Choice of Indicator

The indicator pH range may vary with solution conditions including the ionic strength and temperature of the medium. Organic solvents and colloidal particles can also affect indicator performance. Some of these effects can change the indicator pH ranges by 1 or even more pH units.

A major determinate error (see Section 3.1 for a discussion of determinate error) associated with indicators is a change in color at a pH different from that of the equivalence point in the titration. The magnitude of this effect depends upon the pH range covered by the steeply rising S-shaped portions of the titration curve. Consider the two titration curves shown in Figure 5.10. Suppose that methyl orange, pH transition range 3.1 to 4.4, was used as an indicator for these titrations. For the titration of acetic acid, it is seen in

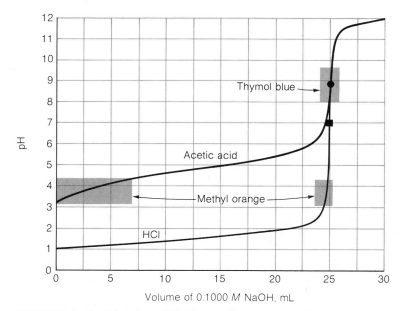

FIGURE 5.10   Titration of 50.00 mL of 0.0500 M HCl and 50.00 mL of 0.0500 M acetic acid with 0.1000 M NaOH. ■ = HCl equivalence point, ● = acetic acid equivalence point.

Figure 5.10 that the color transition begins as soon as the titration starts and is complete when the acid is only about one–fourth titrated. Obviously, methyl orange would not be a suitable indicator for this titration. The break in the titration curve for HCl, however, is so large and extends to such a low pH that methyl orange could be used as an indicator with only a minute negative error. Thymol blue, with a second pH transition range of 8.0 to 9.6, changes color right at the equivalence point for acetic acid and could be used very well for this titration. It is also close enough to the equivalence point for HCl so that it could be used for this acid without error. In some cases where an indicator exactly suited to a particular titration is not available, a blank (solution containing the indicator and all reagents except the acid or base analyte) can be used to reduce error.

A major indeterminate error that occurs with the use of indicators is the variable ability of the eye to determine color changes. Such an error normally ranges between ±0.5 to 1 pH units. This uncertainty can be reduced to as little as ±0.1 pH unit by matching the indicator color to that of a solution containing the same concentration of the indicator at the pH of the equivalence point (for example, a sodium acetate solution in the titration of acetic acid with NaOH). Indeterminate errors in finding the equivalence point with an indicator depend upon the indicator concentration, sensitivity of the eye to the two indicator colors, and degree of pH change per unit volume of titrant added. Errors vary with different indicators and among different people.

5.9   **Limitations of K_a and Acid Concentration**

The greater the break at the equivalence point in a titration curve, the easier it is to determine the volume of added titrant at the equivalence point. For equal concentrations of acid, the magnitude of the break depends upon K_a and is greatest for a strong acid. (The magnitude of the break for any acid increases with increasing acid concentration.) The choice of an indicator pH range becomes more critical with decreasing acid strength. This is readily seen in Figure 5.11, which shows titration curves for acids of varying strengths.

Titration of the weaker acids using an indicator to determine the equivalence point may only be feasible by choosing an indicator with a pH transition range as close as possible to the equivalence point pH and titrating to a color that matches that of a solution containing the same indicator and a salt (conjugate base) of the acid, both at approximately the same concentrations as in the solutions being titrated at the equivalence point. The equivalence point pH of weak acids is always greater than 7, so indicators with pH transition ranges above 7 are chosen for titrations of weak acids.

As a practical limit, K_a of a weak monoprotic acid should be at least 1×10^{-8} to enable titration of a 0.1 M solution of the acid. If the concentration of acid is less than 0.1 M, the value of K_a must be correspondingly greater to permit a titration with a sufficiently large end point break. In some cases, particularly organic acids, the solubility of the acid is a limiting factor.

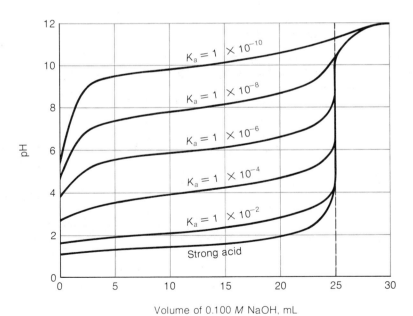

FIGURE 5.11 Curves for the titration of 25.00 mL of 0.100 M solutions of acids with different strengths

5.10   Calculation of End Point Error in Acid-Base Titrations

For the titration of an acid with a base, the equivalence point occurs when the exact quantity of base has been added that is required to neutralize the acid present. Indicators are used to estimate the volume of titrant at the equivalence point in the titration of an acid with a base, or vice versa, but the actual volume recorded at an indicator color change is the end point. The volume of added titrant at the end point may differ from that at the equivalence point, in which case there is a determinate *end point error*. If the end point volume is greater than that of the equivalence point, the end point error is positive; if the reverse is true, the end point error is negative. The equation for calculating end point error is

$$\text{End point error, \%} = \frac{\text{end point volume} - \text{equivalence-point volume}}{\text{equivalence-point volume}} \times 100 \qquad (5.51)$$

If a titration has an equivalence-point volume of 24.20 mL and an end point volume of 23.95 mL, the end point error is

$$\text{End point error} = \frac{23.95 \text{ mL} - 24.20 \text{ mL}}{23.95 \text{ mL}} \times 100 = -1.04\%$$

EXAMPLE

Consider the titration of 25.00 mL of 0.1000 M acetic acid with 0.1000 M NaOH similar to the titration discussed in Section 5.1. What is the end point error if the end point is taken at pH 6.50?

The equivalence point for the titration of 25.00 mL of 0.1000 M acetic acid with 0.1000 M NaOH occurs at 25.00 mL added NaOH at a pH of 8.72. If the end point is taken at pH 6.50, $[H^+] = 3.16 \times 10^{-7}$ M. Since the end point occurs at a pH below that of the equivalence point, the end point error is negative for the titration of an acid with a base. The end point error is given as the percentage of unreacted $HC_2H_3O_2$ out of the sum of $HC_2H_3O_2$ + $C_2H_3O_2^-$, which can be related to the pH through K_a.

$$\begin{aligned}
\text{End point error} &= \frac{[HC_2H_3O_2]}{[HC_2H_3O_2] + [C_2H_3O_2^-]} \times 100 \\[2mm]
&= \frac{[HC_2H_3O_2]}{[HC_2H_3O_2] + ([HC_2H_3O_2]/[H^+])K_a} \times 100 \\[2mm]
&= \frac{[H^+]}{[H^+] + K_a} \times 100 \\[2mm]
&= \frac{3.16 \times 10^{-7}}{3.16 \times 10^{-7} + 1.75 \times 10^{-5}} \times 100 = 1.77\% \, (-1.77\%)
\end{aligned}$$

5.11   **Titration of Weak Bases**

Titration of Ethanolamine

As an example of a titration of a weak base, consider the titration of 25.00 mL of 0.1000 M ethanolamine, $HOC_2H_4NH_2$, with 0.1000 M standard HCl. Ethanolamine acts as a base by reacting with water

$$HOC_2H_4NH_2 + H_2O \rightleftarrows HOC_2H_4NH_3^+ + OH^- \tag{5.52}$$

to produce hydroxide ion. For convenience, ethanolamine will be designated as En and its conjugate acid as EnH^+. The K_b expression for this weak base is

$$K_b = \frac{[EnH^+][OH^-]}{[En]} = 3.18 \times 10^{-5} \tag{5.53}$$

In a manner analogous to the titration curve of a weak acid, the titration curve of ethanolamine can be divided into the following four parts: (1) initial point consisting of a solution of pure ethanolamine; (2) buffer region extending from the initial point to the equivalence point where the pH is determined by the ratio of $[EnH^+]$ to $[En]$; (3) equivalence point where the pH is determined by the acid dissociation of EnH^+; and (4) region beyond the equivalence point where the pH is determined by excess H^+.

The initial pH is calculated as follows:

$$En + H_2O \rightleftarrows EnH^+ + OH^- \tag{5.54}$$

$$[EnH^+] = [OH^-] \tag{5.55}$$

$$[En] = 0.100 \tag{5.56}$$

$$K_b = \frac{[EnH^+][OH^-]}{[En]} = \frac{[OH^-]^2}{0.1000} = 3.18 \times 10^{-5} \tag{5.57}$$

$$[OH^-] = (0.1000 \times 3.18 \times 10^{-5})^{1/2} = 1.78 \times 10^{-3} \ M \tag{5.58}$$

$$[H^+] = \frac{K_w}{[OH^-]} = \frac{1.00 \times 10^{-14}}{1.78 \times 10^{-3}} = 5.62 \times 10^{-12} \tag{5.59}$$

$$pH = 11.25 \tag{5.60}$$

Two examples will be worked of the calculation of pH between the initial point and the equivalence point. Other values in this region are given in Table 5.4. After the addition of 5.00 mL of acid, 0.500 mmol of the 2.500 mmol of En initially present is converted to EnH^+ in a total volume of 30.00 mL. The hydroxide ion concentration is calculated as follows:

$$[En] = \frac{2.500 \ mmol - 0.500 \ mmol}{30.00 \ mL} = \frac{2.000 \ mmol}{30.00 \ mL} \tag{5.61}$$

$$[EnH^+] = \frac{0.500 \ mmol}{30.00 \ mL} \tag{5.62}$$

TABLE 5.4   Titration of 25.00 mL of 0.1000 M ethanolamine (En) with 0.1000 M HCl

Volume HCl added, mL	HCl added, mmol	En left, mmol	Total volume, mL	HCl excess, mmol	$[H^+]$, M	pH
0.00	0.000	2.500	25.00	—	5.62×10^{-12}	11.25
5.00	0.500	2.000	30.00	—	7.87×10^{-11}	10.10
12.50	1.250	1.250	37.50	—	3.14×10^{-10}	9.50
20.00	2.000	0.500	45.00	—	1.26×10^{-9}	8.90
25.00	2.500	—	50.00	—	3.96×10^{-6}	5.40
26.00	2.600	—	51.00	0.100	1.96×10^{-3}	2.71
30.00	3.000	—	55.00	0.500	9.09×10^{-3}	2.04

$$[OH^-] = \frac{[En]}{[EnH^+]} K_b = \frac{2.000 \text{ mmol}/30.00 \text{ mL}}{0.500 \text{ mmol}/30.00 \text{ mL}} \times 3.18 \times 10^{-5} = 1.27 \times 10^{-4} \ M \tag{5.63}$$

$$[H^+] = \frac{1.00 \times 10^{-14}}{1.27 \times 10^{-4}} = 7.87 \times 10^{-11} \ M \tag{5.64}$$

$$pH = 10.10 \tag{5.65}$$

The addition of 12.50 mL of HCl brings the titration exactly half way to the equivalence point. The pH is calculated by

$$[En] = [EnH^+] = \frac{1.250 \text{ mmol}}{37.50 \text{ mL}} \tag{5.66}$$

$$[OH^-] = \frac{[En]}{[EnH^+]} K_b = \frac{1.250 \text{ mmol}/37.50 \text{ mL}}{1.250 \text{ mmol}/37.50 \text{ mL}} \times 3.18 \times 10^{-5} = 3.18 \times 10^{-5} \ M \tag{5.67}$$

Note that at this point, half way to the equivalence point, $[OH^-] = K_b$ and

$$pOH = -\log [OH^-] = -\log 3.18 \times 10^{-5} = 4.50 = pK_b \tag{5.68}$$

The pH at this point is

$$[H^+] = \frac{1.00 \times 10^{-14}}{3.18 \times 10^{-5}} = 3.14 \times 10^{-10} \ M \tag{5.69}$$

$$pH = 9.50 \tag{5.70}$$

At the equivalence point there are 2.500 mmol of EnH^+ in 50.00 mL of solution. The concentration of EnH^+ is 0.0500. The pH is calculated from K_a of EnH^+ as follows:

$$EnH^+ \rightleftarrows H^+ + En \tag{5.71}$$

$$K_a = \frac{[H^+][En]}{[EnH^+]} = \frac{K_w}{K_a} = \frac{1.00 \times 10^{-14}}{3.18 \times 10^{-5}} = 3.14 \times 10^{-10} \tag{5.72}$$

$$[H^+] = [En] \tag{5.73}$$

$$\frac{[H^+]^2}{0.0500} = K_a = 3.14 \times 10^{-10} \tag{5.74}$$

$$[H^+] = (0.0500 \times 3.14 \times 10^{-10})^{1/2} = 3.96 \times 10^{-6} \ M \tag{5.75}$$

$$pH = 5.40 \tag{5.76}$$

The addition of a total of 26.00 mL of HCl results in an excess of 1.00 mL of standard acid beyond the equivalence point. This corresponds to 0.100 mmol of strong acid beyond the equivalence point in a total volume of 51.00 mL so that $[H^+]$ is given by

$$[H^+] = \frac{0.100 \ \text{mmol}}{51.0 \ \text{mL}} = 1.96 \times 10^{-3} \ M$$

$$pH = 2.71$$

Other values of pH beyond the equivalence point are calculated from the amount of excess HCl in the same way.

Titration Curve for a Weak Base

The pH values at different volumes of HCl needed to construct the titration curve are given in Table 5.4. The titration curve is shown in Figure 5.12.

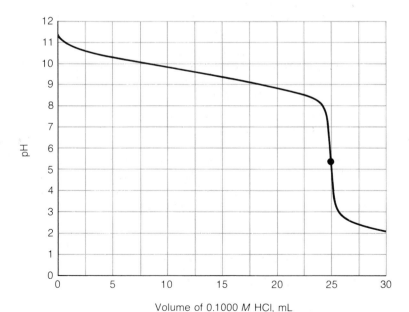

Volume of 0.1000 M HCl, mL

FIGURE 5.12   Titration of 25.00 mL of 0.1000 M ethanolamine (En) with 0.1000 M HCl; ● designates equivalence point.

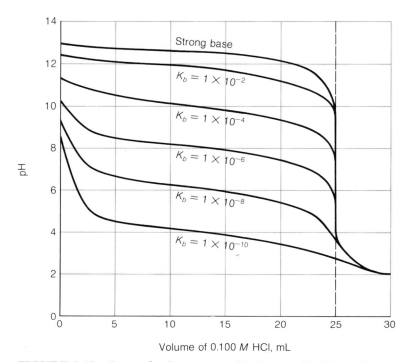

FIGURE 5.13   Curves for the titration of 25.00 mL of 0.100 M solutions of bases having different strengths

As is the case of all titrations of a weak base with a strong acid, the equivalence point pH for the titration of ethanolamine with HCl is in the acidic region (pH 5.40). Examination of Table 5.3 indicates that methyl red should be a suitable indicator for this titration.

The sharpness of the equivalence point break in the titration of bases with strong acids decreases with decreasing values of K_b. The variations in the shapes of the titration curves of bases as a function of K_b are shown in Figure 5.13. It is readily seen that indicators with pH transition ranges in the acidic region must be employed for the weaker bases. Analogous to weak acids (Section 5.9) a K_b of 1×10^{-8} is about the lower limit for the titration of 0.1 M solution of weak base; lower concentrations require higher values of K_b.

5.12   Titration of Acid Mixtures

Mixture of Acetic Acid and HCl

The titration of a mixture of a strong and a weak acid can give a titration curve with two breaks if the acid concentrations are high enough and K_a is sufficiently low. Consider the titration of 25.00 mL of a solution that is

0.1000 M in HCl, a strong acid, and 0.1000 M in acetic acid, $K_a = 1.75 \times 10^{-5}$, titrated with 0.100 M NaOH. It will take a total volume of 50.00 mL of NaOH to neutralize both the acids. Prior to the addition of that much NaOH, the value of $[H^+]$ is given by

$$[H^+] = C_{HCl} + [C_2H_3O_2^-] \qquad (5.77)$$

where C_{HCl} is the molar concentration of HCl that has not been neutralized and $[C_2H_3O_2^-]$ equals the contribution to $[H^+]$ from any dissociation of acetic acid, $HC_2H_3O_2$.

$$HC_2H_3O_2 \rightleftarrows H^+ + C_2H_3O_2^- \qquad (5.78)$$

Since acetic acid is a weak acid, its dissociation is suppressed by the presence of HCl. Therefore, initially it is reasonable to assume that

$$[H^+] = C_{HCl} = 0.1000 \qquad (5.79)$$

$$pH = 1.00$$

From the expression for K_a of acetic acid, it is possible to calculate the ratio

$$\frac{[C_2H_3O_2^-]}{[HC_2H_3O_2]} = \frac{K_a}{[H^+]} = \frac{1.75 \times 10^{-5}}{0.1000} = 1.75 \times 10^{-4} \qquad (5.80)$$

which shows that only a very small fraction of acetic acid is dissociated. After the addition of 24.00 mL of NaOH, the solution contains 0.100 mmol of HCl in a volume of 49.00 mL. If unreacted HCl is assumed to be the only contributor to H^+ in solution, the value of $[H^+]$ is given by

$$[H^+] = C_{HCl} = \frac{0.100 \text{ mmol}}{49.00 \text{ mL}} = 2.04 \times 10^{-3} \ M \qquad (5.81)$$

However, this concentration of H^+ is of comparable magnitude to that provided by a solution of 2.500 mmol of acetic acid in a volume of 49.00 mL, $[HC_2H_3O_2] = 0.0510 \ M$, so a more accurate value of $[H^+]$ is obtained by substituting into Equation 5.77.

$$[H^+] = C_{HCl} + [C_2H_3O_2^-] = 2.04 \times 10^{-3} + \frac{[HC_2H_3O_2]}{[H^+]} K_a \qquad (5.82)$$

$$= 2.04 \times 10^{-3} + \frac{0.0510}{[H^+]} \times 1.75 \times 10^{-5} \qquad (5.83)$$

$$[H^+]^2 - 2.04 \times 10^{-3}[H^+] - 8.93 \times 10^{-7} = 0 \qquad (5.84)$$

$$[H^+] = 2.41 \times 10^{-3} \ M \qquad (5.85)$$

$$pH = 2.62$$

This shows that near the first end point where virtually all the HCl has been neutralized, acetic acid begins to make a contribution to $[H^+]$. At exactly 25.00 mL of added NaOH, the solution is 0.0500 M in both acetic acid and NaCl; the latter does not affect pH (except for its contribution to the ionic

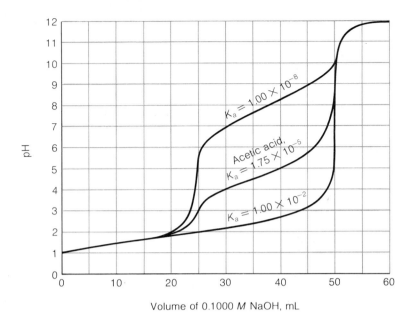

Volume of 0.1000 M NaOH, mL

FIGURE 5.14   Titration with 0.1000 M NaOH of 25.00 mL of three solutions, each 0.1000 M in HCl and 0.1000 M in weak acid with the K_a's shown

strength, which is not being considered here). Thus $[H^+]$ is given by

$$[H^+] = [C_2H_3O_2^-] = ([HC_2H_3O_2]K_a)^{1/2} \tag{5.86}$$

$$= (0.0500 \times 1.75 \times 10^{-5})^{1/2} = 9.35 \times 10^{-4} \, M \tag{5.87}$$

$$\text{pH} = 3.03$$

At this point and beyond, the calculation of $[H^+]$ is the same as for the titration of a solution of acetic acid as shown in Section 5.5. The shape of the titration curve is shown in Figure 5.14.

Mixtures of HCl with Very Weak and Slightly Weak Acids

In addition to the titration curve for the HCl–acetic acid mixture, Figure 5.14 also shows titration curves for two other 25-mL solutions each containing 0.1000 M HCl and 0.1000 M weak acid. In one case, K_a is 1.00×10^{-2}, such a high value that the acid can be classified as "slightly weak." This acid dissociates enough that a first break in the titration curve at 25.00 mL is not even observed because the "weak" acid is contributing appreciably to the H^+ ion concentration long before the HCl is exhausted. The total titration curve looks like that of a strong acid. The other extreme shown is the titration of a weak acid with a K_a of 1.00×10^{-8} in a solution with HCl. In this case the first break is large and the second break is relatively small.

Analogous titration curves can be plotted for mixtures of weak bases mixed with a strong base, such as NaOH, titrated with a strong acid. Here the pH starts out high and moves downward with added acid. The first break becomes progressively larger and the second progressively smaller with decreasing value of K_b of the weak base.

5.13 Titration of Diprotic Acids

Calculation of Points on Titration Curve

Diprotic acids are those with two ionizable hydrogens. In cases where K_{a1} and K_{a2} of the acid are separated by several orders of magnitude, and K_{a2} is greater than approximately 1×10^{-10}, diprotic acids give well-defined titration curves with two distinct breaks. Such an acid is phosphorous acid, H_3PO_3, for which $K_{a1} = 1.00 \times 10^{-2}$ and $K_{a2} = 2.6 \times 10^{-7}$. The third hydrogen on H_3PO_3 is bonded directly to P and is not ionizable.

Consider the titration of 50.00 mL of 0.0500 M H_3PO_3 (2.50 mmol H_3PO_3) with 0.1000 M standard NaOH. Since the value of K_{a1} is relatively large, it will be necessary to consider that H_3PO_3 is partially ionized prior to the first equivalence point. Consider the calculation of $[H^+]$ prior to the addition of any base. The equilibrium of the reaction

$$H_3PO_3 \rightleftharpoons H^+ + H_2PO_3^- \tag{5.88}$$

lies significantly to the right, so that

$$[H_3PO_3] = 0.0500 - [H_2PO_3^-] \tag{5.89}$$

Furthermore,

$$[H^+] = [H_2PO_3^-] \tag{5.90}$$

so that

$$\frac{[H^+][H_2PO_3^-]}{[H_3PO_3]} = K_{a1} = 1.00 \times 10^{-2} = \frac{[H^+]^2}{0.0500 - [H^+]} \tag{5.91}$$

The value of $[H^+]$ may be obtained by solving the quadratic equation

$$[H^+]^2 + 1.00 \times 10^{-2}[H^+] - 5.00 \times 10^{-4} = 0 \tag{5.92}$$

to give

$$[H^+] = 1.79 \times 10^{-2} \ M \tag{5.93}$$

$$pH = 1.75$$

After the addition of 5.00 mL of 0.1000 M NaOH titrant, the stoichiometry of the neutralization reaction

$$H_3PO_3 + OH^- \rightarrow H_2PO_4^- + H_2O \tag{5.94}$$

gives the following analytical concentrations:

$$C_{H_3PO_3} = \frac{2.50 \text{ mmol} - 0.50 \text{ mmol}}{50.00 \text{ mL} + 5.00 \text{ mL}} = 3.64 \times 10^{-2} \ M \tag{5.95}$$

$$C_{H_2PO_3^-} = \frac{0.50 \text{ mmol}}{50.00 \text{ mL} + 5.00 \text{ mL}} = 9.09 \times 10^{-3} \ M \tag{5.96}$$

However, because of the appreciable dissociation of the remaining H_3PO_3 (Reaction 5.88), the molar concentrations of these species are

$$[H_3PO_3] = 3.64 \times 10^{-2} \ M - [H^+] \tag{5.97}$$

$$[H_2PO_3^-] = 9.09 \times 10^{-3} + [H^+] \tag{5.98}$$

Substituted into the expression for K_{a1}, these concentrations give

$$\frac{[H^+](9.09 \times 10^{-3} + [H^+])}{(3.64 \times 10^{-2} - [H^+])} = 1.00 \times 10^{-2} \tag{5.99}$$

leading to the quadratic

$$[H^+]^2 + 1.91 \times 10^{-2}[H^+] - 3.64 \times 10^{-4} = 0 \tag{5.100}$$

the solution of which yields

$$[H^+] = 1.18 \times 10^{-2} \ M \tag{5.101}$$

$$pH = 1.93$$

The approach just used is employed to calculate $[H^+]$ and pH at any added volume of standard NaOH less than 25.00 mL—that is, prior to the first equivalence point. At 12.50 mL added NaOH (half way from the beginning of the titration to the first equivalence point), the value of $[H^+]$ is $5.62 \times 10^{-3} \ M$, pH = 2.25; at 20.00 mL added NaOH, $[H^+] = 1.77 \times 10^{-3} \ M$, pH = 2.75.

The addition of 25.00 mL of standard NaOH brings the titration to the first equivalence point at which there are 2.50 mmol of $H_2PO_3^-$ in 75.00 mL of solution, giving

$$[H_2PO_3^-] = \frac{2.50 \text{ mmol}}{75.00 \text{ mL}} = 3.33 \times 10^{-2} \ M \tag{5.102}$$

The formula for $[H^+]$ in a solution of an amphiprotic species such as $H_2PO_3^-$ was given in Section 4.7 as

$$[H^+] = \left(\frac{[H_2PO_3^-]K_{a2} + K_w}{1 + [H_2PO_3^-]/K_{a1}}\right)^{1/2} \tag{5.103}$$

Substitution of the values for the constants and for $[H_2PO_3^-]$ into this equation shows that

$$[H_2PO_3^-]K_{a2} \gg K_w \tag{5.104}$$

$$\frac{[H_2PO_3^-]}{K_{a1}} = \frac{3.33 \times 10^{-2}}{1.00 \times 10^{-2}} = 3.33 \tag{5.105}$$

leading to the following:

$$[H^+] = \left(\frac{[H_2PO_3^-]K_{a2}}{1 + [H_2PO_3^-]/K_{a1}} \right)^{1/2} = \left(\frac{3.33 \times 10^{-2} \times 2.6 \times 10^{-7}}{1 + 3.33 \times 10^{-2}/1.00 \times 10^{-2}} \right)^{1/2} = 4.47 \times 10^{-5}\ M \qquad (5.106)$$

$$pH = 4.35 \qquad (5.107)$$

(In most cases the value of $[H^+]$ at the first equivalence point in the titration of a weak diprotic acid with a strong base can be approximated by $[H^+] = \sqrt{K_{a1}K_{a2}}$, $pH = \frac{1}{2}(pK_{a1} + pK_{a2})$. If that approximation were made in this case, $[H^+] = \sqrt{1.00 \times 10^{-2} \times 2.6 \times 10^{-7}} = 5.10 \times 10^{-5}\ M$, $pH = 4.29$. The difference between this pH value and the exact value of 4.35 calculated above is about 12%, but is negligible in preparing the titration curve.

In the region between the first and second equivalence points (25.00 to 50.00 mL NaOH), the neutralization reaction is

$$H_2PO_3^- + OH^- \rightarrow HPO_3^{2-} \qquad (5.108)$$

After the addition of a total of 30.0 mL of NaOH, 25.00 mL of which were consumed in converting H_3PO_3 to $H_2PO_3^-$, stoichiometry gives the following concentrations:

$$[H_2PO_3^-] = \frac{2.00\ \text{mmol}}{80.00\ \text{mL}} = 2.50 \times 10^{-2}\ M \qquad (5.109)$$

$$[HPO_3^{2-}] = \frac{0.50\ \text{mmol}}{80.00\ \text{mL}} = 6.25 \times 10^{-3}\ M \qquad (5.110)$$

Substitution into the expression for K_{a2} yields the following values of $[H^+]$ and pH:

$$[H^+] = \frac{[H_2PO_3^-]}{[HPO_3^{2-}]}\ K_{a2} = \frac{2.00\ \text{mmol}/80.00\ \text{mL}}{0.50\ \text{mmol}/80.00\ \text{mL}} \times 2.6 \times 10^{-7} = 1.04 \times 10^{-6}\ M \qquad (5.111)$$

$$pH = 5.98 \qquad (5.112)$$

The addition of a total of 37.50 mmol of NaOH brings the titration midway between the first and second equivalence points, where

$$[H_2PO_3^-] = [HPO_3^{2-}] = \frac{1.25\ \text{mmol}}{87.50\ \text{mL}} \qquad (5.113)$$

$$[H^+] = \frac{[H_2PO_3^-]}{[HPO_3^{2-}]}\ K_{a2} = \frac{1.25\ \text{mmol}/87.50\ \text{mL}}{1.25\ \text{mmol}/87.50\ \text{mL}} \times 2.6 \times 10^{-7} = 2.6 \times 10^{-7}\ M \qquad (5.114)$$

$$pH = 6.59 \qquad (5.115)$$

Using the same approach applied to the two preceding examples, the value of $[H^+]$ after the addition of 45.00 mL of NaOH is $6.5 \times 10^{-8}\ M$ and $pH = 7.19$.

The addition of a total of exactly 50.00 mL of NaOH brings the titration to the second equivalence point, where 2.50 mmol of HPO_3^- are present in a total of 100.00 mL of solution, giving

$$[HPO_3^-] = \frac{2.50 \text{ mmol}}{100.00 \text{ mL}} = 2.50 \times 10^{-2} \, M \tag{5.116}$$

Here the pH is calculated considering the base (hydrolysis) reaction

$$HPO_3^{2-} + H_2O \rightleftharpoons H_2PO_3^- + OH^- \tag{5.117}$$

This reaction and the equilibrium constant for it are obtained by adding the following two reactions and multiplying their respective equilibrium constants:

$$HPO_3^{2-} + H^+ \rightleftharpoons H_2PO_3^- \qquad \frac{1}{K_{a2}} = \frac{[H_2PO_3^-]}{[HPO_3^{2-}][H^+]} \tag{5.118}$$

$$H_2O \rightleftharpoons H^+ + OH^- \qquad K_w = [H^+][OH^-] \tag{5.119}$$

$$\overline{HPO_3^{2-} + H_2O \rightleftharpoons H_2PO_3^- + OH^-} \qquad \frac{K_w}{K_{a2}} = \frac{[H_2PO_3^-]}{[HPO_3^{2-}][H^+]} [H^+][OH^-]$$

$$= \frac{[H_2PO_3^-][OH^-]}{[HPO_3^{2-}]} = K_b \tag{5.120}$$

$$K_b = \frac{K_w}{K_{a2}} = \frac{1.00 \times 10^{-14}}{2.6 \times 10^{-7}} = 3.8 \times 10^{-8} \tag{5.121}$$

From Reaction 5.117, the hydrolysis reaction for HPO_3^{2-}, it is seen that

$$[H_2PO_3^{2-}] = [OH^-] \tag{5.122}$$

Substitution into the expression for K_b yields the following value for $[OH^-]$:

$$\frac{[H_2PO_3^-][OH^-]}{[HPO_3^{2-}]} = K_b = \frac{[OH^-]^2}{[HPO_3^{2-}]} = \frac{[OH^-]^2}{2.50 \times 10^{-2}} = 3.8 \times 10^{-8} \, M \tag{5.123}$$

$$[OH^-] = \sqrt{2.50 \times 10^{-2} \times 3.8 \times 10^{-8}} = 3.08 \times 10^{-5} \, M \tag{5.124}$$

$$[H^+] = \frac{K_w}{[OH^-]} = \frac{1.00 \times 10^{-14}}{3.08 \times 10^{-5}} = 3.25 \times 10^{-10} \, M \tag{5.125}$$

$$pH = 9.49$$

The addition of totals of 55 and 60 mL of NaOH provides 0.50 and 1.00 mmol excess OH^-, respectively, in total volumes of 105 and 110 mL of solution, respectively. The values of $[H^+]$ and pH are given at these two points by the following calculations:

At 55 mL NaOH,

$$[H^+] = \frac{K_w}{[OH^-]} = \frac{1.00 \times 10^{-14}}{0.500 \text{ mmol}/105 \text{ mL}} = 2.10 \times 10^{-12} \, M \tag{5.126}$$

$$pH = 11.68$$

At 60 mL NaOH,

$$[H^+] = \frac{K_w}{[OH^-]} = \frac{1.00 \times 10^{-14}}{1.00 \text{ mmol}/110 \text{ mL}} = 1.10 \times 10^{-12} \ M \qquad (5.127)$$

$$pH = 11.96$$

Plot of Diprotic Acid Titration Curve

The titration curve may now be plotted from the volume of titrant versus pH data summarized in Table 5.5. The curve is plotted in Figure 5.15. It is seen to have two well-defined breaks characteristic of diprotic acids in which K_{a1} and K_{a2} are well separated and K_{a2} is not too small (in practice few acids meet these criteria).

TABLE 5.5   Data for the titration of 50.00 mL of 0.0500 M H_3PO_3 with 0.1000 M NaOH

Volume NaOH, mL	pH
0.00	1.75
5.00	1.93
12.50	2.25
20.00	2.75
25.00	4.29
30.00	5.98
37.50	6.59
45.00	7.19
50.00	9.49
55.00	11.68
60.00	11.96

Examination of Figure 5.15 shows that it ought to be possible to select indicators that would reveal titration to the first end point (first break in the curve), or to the second end point (second break in the curve). Table 5.3 shows that bromocresol green, changing color from pH 3.8 to 5.4, should change color at the first end point and phenolphthalein, with a pH color change range of 8.3 to 10.0, should serve for the second end point.

It is also seen from Figure 5.15 that the regions from the initial point to the first break and from the first to the second break in the titration curve are relatively flat. These are the two *buffer regions* of the titration curve. The first region is buffered by the presence of the acid H_3PO_3 and its conjugate base $H_2PO_3^-$. The second region is buffered by the acid $H_2PO_3^-$ and its conjugate base HPO_3^{2-}.

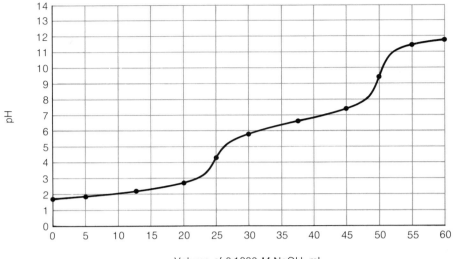

FIGURE 5.15   Curve for the titration of 50.00 mL of 0.0500 M H_3PO_4 with 0.1000 M NaOH

5.14   Titration of Na$_2$CO$_3$ with Acid

Calculation of Points on Titration Curve

The titration of a solution of Na_2CO_3 with HCl illustrates the titration of a base capable of accepting two H^+ ions stepwise. Consider 25.00 mL of a solution that is 0.1000 M in Na_2CO_3 titrated with 0.1000 M HCl. The equilibrium expressions and constants needed to construct the titration curve are

$$CO_2 + H_2O \rightleftharpoons H^+ + HCO_3^- \tag{5.128}$$

$$K_{a1} = \frac{[H^+][HCO_3^-]}{[CO_2]} = 4.45 \times 10^{-7} \tag{5.129}$$

$$pK_{a1} = 6.35$$

where CO_2 is carbon dioxide dissolved in water.

$$HCO_3^- \rightleftharpoons H^+ + CO_3^{2-} \tag{5.130}$$

$$K_{a2} = \frac{[H^+][CO_3^{2-}]}{[HCO_3^-]} = 4.69 \times 10^{-11} \tag{5.131}$$

$$pK_{a2} = 10.33$$

Initially the pH is determined by the base reaction of carbonate ion.

$$CO_3^{2-} + H_2O \rightleftharpoons HCO_3^- + OH^- \tag{5.132}$$

The equilibrium of this reaction is described by the K_b expression

$$K_b = \frac{[HCO_3^-][OH^-]}{[CO_3^{2-}]} = \frac{K_w}{K_{a2}} = \frac{1.00 \times 10^{-14}}{4.69 \times 10^{-11}} = 2.13 \times 10^{-4} \tag{5.133}$$

The remaining steps in the calculation of pH are

$$[HCO_3^-] = [OH^-] \tag{5.134}$$

$$[CO_3^{2-}] = 0.1000 \tag{5.135}$$

$$\frac{[HCO_3^-][OH^-]}{[CO_3^{2-}]} = K_b = \frac{[OH^-]^2}{0.1000} = 2.13 \times 10^{-4} \tag{5.136}$$

$$[OH^-] = (0.1000 \times 2.13 \times 10^{-4})^{1/2} = 4.62 \times 10^{-3} \tag{5.137}$$

$$pOH = 2.34$$

$$pH = 14.00 - pOH = 14.00 - 2.34 = 11.66 \tag{5.138}$$

After the addition of 5.00 mL (0.500 mmol) of HCl there are 2.00 mmol of CO_3^{2-} left from the original 2.50 mmol because of the reaction

$$CO_3^{2-} + H^+ \rightarrow HCO_3^- \tag{5.139}$$

This reaction also produces 0.500 mmol of HCO_3^- in 30.00 mL of solution. The pH is calculated as follows:

$$[H^+] = \frac{[HCO_3^-]}{[CO_3^{2-}]} K_{a2} = \frac{0.500 \text{ mmol}/30.00 \text{ mL}}{2.00 \text{ mmol}/30.00 \text{ mL}} \times 4.69 \times 10^{-11} = 1.17 \times 10^{-11} M \tag{5.140}$$

$$pH = 10.93$$

After the addition of 12.50 mL of HCl, the titration is half way to the first equivalence point, and $[HCO_3^-] = [CO_3^{2-}] = 1.25 \text{ mmol}/37.50 \text{ mL}$. The pH is simply pK_{a2} as given by

$$[H^+] = \frac{[HCO_3^-]}{[CO_3^{2-}]} K_{a2} = K_{a2} = 4.69 \times 10^{-11} M \tag{5.141}$$

$$pH = 10.33$$

The addition of 20.00 mL of HCl produces 2.00 mmol of HCO_3^- in 45.00 mL of solution and leaves 0.500 mmol of CO_3^{2-}, yielding the pH

$$[H^+] = \frac{[HCO_3^-]}{[CO_3^{2-}]} K_{a2} = \frac{2.00 \text{ mmol}/45.00 \text{ mL}}{0.500 \text{ mmol}/45.00 \text{ mL}} \times 4.69 \times 10^{-11} = 1.88 \times 10^{-10} M \tag{5.142}$$

$$pH = 9.73$$

The addition of 25.00 mL of HCl brings the titration to the first equivalence point where 2.500 mmol of HCO_3^- are contained in a volume of 50.00 mL. As noted previously, the formula for the value of $[H^+]$ in a solution of an amphiprotic species is, for the case of HCO_3^-,

$$[H^+] = \left(\frac{[HCO_3^-]K_{a2} + K_w}{1 + [HCO_3^-]/K_{a1}} \right)^{1/2} \tag{5.143}$$

At this point in the titration $[HCO_3^-] = 0.0500\ M$. Substitution of this value and the values of the constants into the expression above shows that

$$[HCO_3^-]K_{a2} \gg K_w \tag{5.144}$$

$$\frac{[HCO_3^-]}{K_{a1}} \gg 1 \tag{5.145}$$

leading to

$$[H^+] = \sqrt{K_{a1}K_{a2}} = \sqrt{4.45 \times 10^{-7} \times 4.69 \times 10^{-11}} = 4.57 \times 10^{-9}\ M \tag{5.146}$$

$$pH = \tfrac{1}{2}(pK_{a1} + pK_{a2}) = \tfrac{1}{2}(6.35 + 10.33) = 8.34 \tag{5.147}$$

A total of 30.00 mL of added HCl brings the titration 5.00 mL beyond the first equivalence point. This excess HCl reacts with HCO_3^-.

$$HCO_3^- + H^+ \rightarrow H_2O + CO_2(aq) \tag{5.148}$$

Assuming no loss of volatile CO_2, at this point there are 2.000 mmol of HCO_3^- and 0.500 mmol of CO_2 in 55.0 mL of solution, which gives

$$[H^+] = \frac{[CO_2]}{[HCO_3^-]} K_{a1} = \frac{0.500\ \text{mmol}/55.00\ \text{mL}}{2.00\ \text{mmol}/55.00\ \text{mL}} \times 4.45 \times 10^{-7} = 1.11 \times 10^{-7}\ M \tag{5.149}$$

$$pH = 6.95$$

The addition of 37.50 mL of NaOH brings the titration of half way between the first and second equivalence points, where

$$[HCO_3^-] = [CO_2] = \frac{1.25\ \text{mmol}}{62.50\ \text{mL}} \tag{5.150}$$

$$[H^+] = \frac{[CO_2]}{[HCO_3^-]} K_{a1} = K_{a1} = 4.45 \times 10^{-7}\ M \tag{5.151}$$

$$pH = pK_{a1} = 6.35 \tag{5.152}$$

After the addition of 45.00 mL of titrant, 2.00 mmol of CO_2 have been formed and 0.500 mmol of HCO_3^- is left in a volume of 70.00 mL. The value of $[H^+]$ is given by

$$[H^+] = \frac{[CO_2]}{[HCO_3^-]} K_{a2} = \frac{2.00\ \text{mmol}/70.0\ \text{mL}}{0.500\ \text{mmol}/70.0\ \text{mL}} \times 4.45 \times 10^{-7} = 1.78 \times 10^{-6}\ M \tag{5.153}$$

$$pH = 5.75$$

The addition of 50.00 mL of HCl brings the titration to the second equivalence point with 2.50 mmol of CO$_2$ in a volume of 75.0 mL. (The assumption is made that no gaseous CO$_2$ is lost from the solution.) The 0.0333 M CO$_2$ solution reacts slightly with water,

$$CO_2 + H_2O \rightleftharpoons H^+ + HCO_3^- \tag{5.154}$$

and the pH is calculated as follows:

$$[H^+] = [HCO_3^-] \tag{5.155}$$

$$\frac{[H^+][HCO_3^-]}{[CO_2]} = K_{a1} = \frac{[H^+]^2}{3.33 \times 10^{-2}} \times 4.45 \times 10^{-7} \tag{5.156}$$

$$[H^+] = (3.33 \times 10^{-2} \times 4.45 \times 10^{-7})^{1/2} = 1.22 \times 10^{-4} \, M \tag{5.157}$$

$$pH = 3.91$$

Finally, after the addition of 55.00 mL of titrant, there is an excess of 5.00 mL (0.500 mmol) of strong acid in 80.00 mL of solution. This gives a pH of

$$[H^+] = \frac{mmol \; H^+}{mL \; solution} = \frac{0.500 \; mmol}{80.0 \; mL} = 6.25 \times 10^{-3} \, M \tag{5.158}$$

$$pH = 2.20$$

The titration curve resulting from plotting the points calculated above is shown in Figure 5.16.

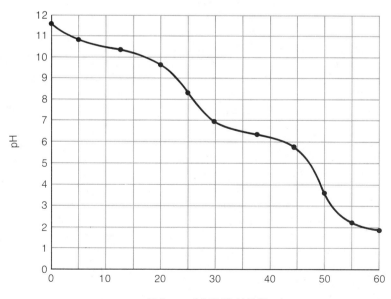

Volume of 0.1000 M HCl, mL

FIGURE 5.16   Titration of 25.00 mL of 0.1000 M Na$_2$CO$_3$ with 0.1000 M HCl

Programmed Summary of Chapter 5

This summary contains in a programmed format the major terms and concepts introduced in this chapter. To get the most out of it, you should fill in the blanks for each question and then check the answers at the end of the book to see if you are correct.

The procedure by which the quantity of a standard base required to neutralize a sample of acid is measured, or vice versa, is known as an (1) _____. The point at which the exact quantity of standard base (or acid) has been added to exactly neutralize the acid (or base) being titrated is known as the (2) _____, and its experimental estimate is known as the (3) _____. Because they produce the most complete reactions, (4) _____ acids and bases are almost always employed as titrants. When a parameter, such as pH, that changes abruptly at the equivalence point(s), is plotted versus volume of added titrant, the plot is known as a (5) _____. The four specific points or regions into which the curve for the titration of a monoprotic acid with a strong base may be broken are (6) _____, _____, _____, _____. A plot of the change in pH per unit change in volume versus volume is known as a (7) _____. Another way of plotting titration data such that there is an abrupt reversal of sign at the inflection point in the titration curve is the (8) _____. The pH at the equivalence point for the titration of a weak acid with strong base will occur at a (9) _____ value than the pH at the equivalence point for the titration of a strong acid with a strong base. A dye that changes color with pH changes in dilute solution is known as an (10) _____. The range of pH over which the indicator color change occurs is commonly called the (11) _____, and lacking other information, it is estimated to occur over the pH range of indicator pK_a (12) _____. Three major classes of indicators differentiated by their chemical structures are (13) _____, _____, and _____ indicators. The indeterminate error caused by the variable ability of the eye to determine color changes normally ranges over (14) _____ pH units. The formula for percent end point error is (15) _____. The four distinct regions of the curve for the titration of a weak acid with a strong base are (16) _____, _____, _____, _____. For a given concentration of acid titrated with a base, the magnitude of the break in the titration curve increases with increasing (17) _____. For titrations of a weak acid with a strong base the equivalence point pH is (18) _____ 7, and for the titration of a weak base with a strong acid, the equivalence point pH is (19) _____ 7. In order to see two well-defined breaks in the titration curve of a diprotic acid it is necessary that the K_a's (20) _____, and that K_{a2} (21) _____. The titration of Na_2CO_3 with strong acid gives a titration curve shaped such that it has (22) _____.

Questions

1. Define titrant.

2. Describe the general characteristics of a titration curve for an acid-base titration.

3. Describe where the equivalence point may be found in the titration curve of a strong acid titrated with a strong base.

4. Show by means of sketches how the end point in an acid-base titration may be determined graphically (a) from the titration curve, (b) from a first-derivative plot, and (c) from a second-derivative plot.

5. How do the equivalence point pH values compare between a weak monoprotic acid titrated with a strong base and a strong acid of the same concentration titrated with a strong base?

6. Comment on the statement that "K_a of an indicator to be used to show the end point in the titration of a weak acid with a strong base should be as close as possible to the K_a of the weak acid." Is this statement correct? If not, why not?

7. What factors cause the magnitude of the indicator pH range to vary for different indicators?

8. As applied to the titration of an acid with a base define (a) equivalence point, (b) end point, (c) positive end point error, (d) negative end point error.

9. From the curves in Figure 5.11 comment on the minimum value of K_a for a weak acid required for the titration of the acid with 0.100 M solutions of the acid with strong base.

10. Why does the titration curve for the titration of a mixture of HCl and an acid with a K_a of 1×10^{-2} show only one break?

11. What are the buffer regions of a titration curve and why are they so named?

12. In the titration of Na_2CO_3 with acid, what determines the initial pH?

13. Below is shown the curve for the titration of Na_2CO_3 with HCl. At point A CO_3^{2-} is the predominant anionic species in solution. Indicate the predominant species produced from the reaction of CO_3^{2-} with H^+ at points B through E.

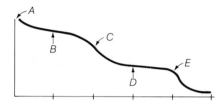

14. Addition of 1 mL of 0.100 M NaOH to 50 mL of water that may have something dissolved in it changed the pH from 8.9 to 10.3. From this information and what you know about titration curves, but without doing any detailed calculations, the statement of the following that is most likely true is (a) the liquid may have been pure water, (b) the liquid was an effective buffer solution, (c) the liquid may have been a solution of a weak acid, (d) the liquid may have been a solution of a sodium salt of a weak acid, (e) the liquid may have been a very dilute solution of a strong acid.

15. Regarding the pH range over which an acid-base indicator changes color, which of the following is true: (a) the range depends upon the relative intensities of indicator colors, (b) the change covers the range of $pK_a \pm 1$ for all indicators, (c) the change is independent of the human eye's perception of color, (d) the change is independent of solution conditions, (e) the observed change normally occurs over a range somewhat greater than 2 pH units.

16. Of the following, the true statement regarding choice of acid-base indicator is (a) an indicator with an indicator pH range below 7 (for example, one that changes color from pH 4.2 to 5.8) should be chosen for titration of a weak acid with a strong base, (b) an indicator with an indicator pH range above 7 should be chosen for titration of a weak base with a strong acid, (c) the choice of indicator is generally less critical for the titration of a strong acid with a strong base than when a weak acid or base is involved, (d) the same indicator can normally be used for the observation of both equivalence points in the titration of a diprotic acid, (e) the indicator chosen should change color in the buffer region of the titration curve.

17. Below is a graph to plot the distribution of species as the fractions present of H_2A, HA^-, and A^{2-} as a function of pH for a

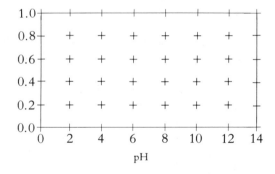

pH

diprotic acid with K_{a1} of 2.00×10^{-5} and K_{a2} of 3.00×10^{-9}. This was done in the text for CO_2, HCO_3^-, and CO_3^{2-}. Without doing any detailed calculations, designate three key points from which the plot may be sketched.

18. A 12.5-mL portion of a 0.100 M solution of a diprotic acid with a K_{a1} of 2.00×10^{-5} and K_{a2} of 3.00×10^{-9} was titrated with 0.100 M NaOH. From the text and lecture material you should know the shape of a diprotic acid titration curve and should be able to calculate at least three points on such a curve in just a few seconds. Without spending a lot of time doing calculations, sketch the titration curve below.

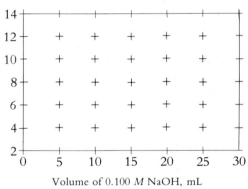

Volume of 0.100 M NaOH, mL

Problems

1. A 25.00-mL volume of strong acid that was actually 0.0857 M in H^+ was titrated with 0.0500 M standard NaOH. The end point was recorded at 42.57 mL NaOH. What was the percent end point error?

2. Calculate the end point error in the preceding problem if the end point were taken at pH 11.477.

3. For an acid-base indicator HIn, the color of a mixture of HIn and In^- is indistin-

guishable from pure HIn when $[HIn]/[In^-] = 5/1$ and is indistinguishable from pure In^- when $[In^-]/[HIn] = 3/1$. Estimate the indicator pH range in reference to pK_a of the indicator.

4. From the data in Table 5.3, calculate $[HIn]/[In^-]$ at the lower and upper limits of the indicator pH range of methyl orange.

5. For the titration of 20.00 mL of a solution 0.1000 M in HCl and 0.1000 M in HA, $pK_a = 6.80$, with 0.1000 M NaOH titrant, calculate pH at 0, 10.00, 19.00, 20.00, 21.00, 30.00, 39.00, 40.00, and 41.00 mL added titrant and sketch the titration curve.

6. For the titration of 25.00 mL of 0.0500 M weak acid HA with 0.1000 M standard NaOH, the pH half way to the equivalence point was 2.750. What was K_a of the acid?

7. For the titration of 50.00 mL of 0.0500 M HCl with 0.0800 M NaOH, construct a table in the form of Table 5.1, calculating pH initially; half way to the equivalence point; 1, 0.5, and 0.1 mL prior to the equivalence point; at the equivalence point; and 0.1, 0.5, 1, and 5 mL beyond the equivalence point. Plot the titration curve.

8. Repeat Problem 7 for the case in which the acid is a weak monoprotic acid with a K_a of 2.40×10^{-6}.

9. For the titration of an acid with a standard base the following values of volume of titrant in milliliters and pH (in parentheses) were obtained around the equivalence point: 24.6 (4.90), 24.65 (5.23), 24.80 (5.60), 24.84 (6.20), 24.86 (6.45), 24.88 (7.05), 25.00 (7.90), 25.02 (9.45), 25.04 (9.70), 25.06 (9.95), 25.08 (9.99). Plot the data as a standard titration curve, first-derivative plot, and second-derivative plot. Calculate the end point volume from each plot.

10. A 100-mL volume of the diprotic acid H_2X, 0.1000 M concentration, was titrated with 0.1000 M NaOH. At exactly 50.0 mL of added NaOH the pH was 4.80 and at the first equivalence point, 100.0 mL of added NaOH, the pH was 7.15. The value of K_{a2} was (a) 8.40×10^{-9}, (b) 5.62×10^{-10}, (c) 7.08×10^{-8}, (d) 1.32×10^{-10}, (e) 3.16×10^{-10}.

11. The weak acid HY has a K_a of 8.50×10^{-6}. The number of milliliters of 0.1000 M NaOH that must be added to 50.00 mL of this acid to give a pH 5.00 buffer is (a) 50.0, (b) 23.0, (c) 27.0, (d) 19.5, (e) 31.6.

12. A weak acid HZ, 0.0800 M in concentration, was titrated with 0.500 M NaOH. The value of K_a of HZ was 9.00×10^{-6}. The pH at the equivalence point was (a) 7.05, (b) 8.32, (c) 9.16, (d) 10.90, (e) 8.94.

6

Nonaqueous Solvents and Acid-Base Titrations in Nonaqueous Solvents

L E A R N I N G   G O A L S

1. Major ways in which nonaqueous solvents are classified.
2. Meaning of self-dissociation of solvents.
3. Distinction between aprotic and amphiprotic solvents.
4. Distinction between acidic solvents and basic solvents.
5. How dielectric constant affects solvent properties and how it it influences ionization and the existence of ion pairs in solution.
6. Differences in leveling solvents and differentiating solvents.

6.1 Advantages of Nonaqueous Solvents

Examination of titration curves for various strength acids (Figure 5.11) and bases (Figure 5.13) shows that acids or bases with dissociation constants less than about 1×10^{-8} cannot be titrated successfully in water at about $0.1\ M$ concentration. This is because at lower values of the dissociation constants the breaks in the titration curves are too small to enable observation of the end point, reflecting incomplete titration reactions. Furthermore, some acids and bases are not soluble enough in water to produce a $0.1\ M$ solution, requiring dissociation constants higher than 1×10^{-8} to enable end point measurement.

These problems can sometimes be overcome by using a solvent other than water—that is, a nonaqueous solvent. Nonaqueous solvents can be chosen that emphasize the acidity or basicity of analytes so that the analytes can be titrated. Furthermore, a number of acids and bases that are not sufficiently soluble in water to be titrated are soluble enough to titrate in properly chosen nonaqueous solvents. Nonpolar, non-ionic species, usually organic in nature, may be much more soluble in a water-immiscible organic nonaqueous solvent than in water, so that these species can be dissolved and titrated at significant concentrations in an organic solvent.

6.2   **Classification of Nonaqueous Solvents**

Self-Dissociation and the Role of Proton Exchange

Nonaqueous solvents are most conveniently classified on the basis of their relative acidity or basicity and whether or not they undergo **self-dissociation**, as shown below for the familiar example of water:

$$H_2O + H_2O \rightleftharpoons H_3O^+ + OH^- \tag{6.1}$$

The simplest type of solvents are those that are **aprotic**, meaning that they neither accept nor donate protons (H^+ ions). Sometimes these solvents are termed **inert**. Common examples are shown in Figure 6.1.

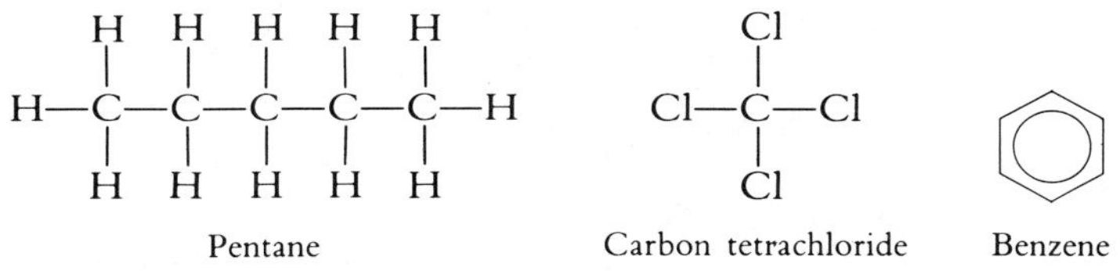

Pentane   Carbon tetrachloride   Benzene

FIGURE 6.1 Common aprotic solvents

At the opposite end of the scale are those solvents that undergo self-dissociation through exchange of protons between solvent molecules. These solvents are said to undergo **autoprotolysis** and are classified as **amphiprotic** solvents. As is the case with water, the general reaction for the autoprotolysis of a protonated solvent, HSolv, is

$$HSolv + HSolv \rightleftharpoons H_2Solv^+ + Solv^- \tag{6.2}$$

The extent of this reaction can be described by a **self-dissociation constant, K_s**

$$K_s = [H_2Solv^+][Solv^-] \tag{6.3}$$

In addition to water, these solvents include liquid ammonia

$$NH_3 + NH_3 \rightleftharpoons NH_4^+ + NH_2^- \tag{6.4}$$

methanol (methyl alcohol)

$$CH_3OH + CH_3OH \rightleftharpoons CH_3OH_2^+ + CH_3O^- \tag{6.5}$$

and acetic acid

$$HOAc + HOAc \rightleftharpoons H_2OAc^+ + OAc^- \tag{6.6}$$

Acidic and Basic Solvents

Water and the alcohols have approximately the same tendency to donate as to accept protons. However, some solvents, such as acetic acid, have a greater tendency to donate than to accept protons and are called **acidic solvents**. Other solvents, like ethylenediamine

$$
\begin{array}{c}
\overset{\displaystyle H}{|}\ \ \overset{\displaystyle H}{|} \\
\overset{\displaystyle H}{\underset{\displaystyle H}{}}\!\!>\!\!N\!-\!\overset{|}{\underset{|}{C}}\!-\!\overset{|}{\underset{|}{C}}\!-\!N\!<\!\!\overset{\displaystyle H}{\underset{\displaystyle H}{}} \\
\underset{\displaystyle H}{|}\ \ \underset{\displaystyle H}{|}
\end{array}
$$

have a large affinity for protons and little tendency to donate them; such solvents are **basic solvents**. As will be discussed, the relative acidity or basicity of a solvent determines the kinds of solutes that can be titrated in the solvent.

Dielectric Constant and Boiling Point

In addition to relative acidity or basicity, three major properties that determine the uses of a solvent are **boiling point**, **dielectric constant**, and self-dissociation constant (see Equation 6.3). The boiling point determines to a large extent the convenience of a solvent for use in the laboratory. With a boiling point of $64°C$, methanol will have some tendency to evaporate in laboratory use. Boiling at $-78°C$, liquid ammonia must obviously be handled with special equipment and precautions in the laboratory. Dielectric constant largely determines the ability of a solvent to dissolve ionic solutes; a high dielectric constant enables separation of oppositely charged ions in solution. Water's dielectric constant of 78.5 at $25°C$ is about the highest of all common liquids, explaining water's excellent solvent properties for ionic solutes. By comparison, the dielectric constant of benzene is only 2.3 at $25°C$, and it is a very poor solvent for ionic compounds. In general, a high dielectric constant is desirable in solvents used as titration media.

6.3   Autoprotolysis and Acid-Base Reactions in Nonaqueous Solvents

Titration of Weak Acid

Commonly, the titration of a weak acid in a nonaqueous solvent is performed with a base in which the anion is the same as that formed by the autoprotolysis of solvent. For example, a weak acid HA can be titrated in methanol by a solution of sodium methoxide, $Na^+CH_3O^-$. The titration reaction is

$$HA + CH_3O^- \rightarrow CH_3OH + A^- \tag{6.7}$$

$$K = \frac{[A^-]}{[HA][CH_3O^-]} \tag{6.8}$$

In the expression for K of the titration reaction, $[CH_3OH]$ is omitted because it is the solvent present at a high constant concentration. The titration reaction, 6.7, may be viewed as the acid dissociation reaction from which is subtracted the autoprotolysis reaction of the solvent:

$$HA + CH_3OH \rightleftharpoons CH_3OH_2^+ + A^- \tag{6.9}$$

$$K_a = \frac{[CH_3OH_2^+][A^-]}{[HA]} \tag{6.10}$$

$$-(CH_3OH + CH_3OH \rightleftharpoons CH_3OH_2^+ + CH_3O^-) \tag{6.11}$$

$$K_s = [CH_3OH_2^+][CH_3O^-]$$

$$\overline{HA + CH_3O^- \rightleftharpoons CH_3OH + A^-} \tag{6.12}$$

$$K = \frac{[A^-]}{[HA][CH_3O^-]} = \frac{K_a}{K_s} \tag{6.13}$$

From the above it is seen that the equilibrium constant K of the titration reaction increases with decreasing values of the autoprotolysis constant K_s. Therefore, a complete titration reaction is favored by a low value of K_s.

Titration of Weak Base

The preceding showed the titration of a weak acid by a strong base, sodium methoxide. An analogous reaction can be shown for the titration of a weak base B with a strong acid in a nonaqueous solvent. For example, a strong acid titrant may be prepared in acetic acid solvent by adding perchloric acid, $HClO_4$, to the solvent to form an acetic acid solution of $H_2OAc^+ClO_4^-$. The protonated acetic acid molecule reacts with a weak base as follows:

$$H_2OAc^+ + B \rightarrow HB^+ + HOAc \tag{6.14}$$

This reaction may be obtained by subtracting the autoprotolysis reaction for acetic acid from the base dissociation reaction of the weak base B in acetic acid.

$$B + HOAc \rightleftharpoons HB^+ + OAc^- \tag{6.15}$$

$$K_b = \frac{[HB^+][OAc^-]}{[B]} \tag{6.16}$$

$$-(HOAc + HOAc \rightleftharpoons H_2OAc^+ + OAc^-) \tag{6.17}$$

$$K_s = [H_2OAc^+][OAc^-]$$

$$\overline{H_2OAc^+ + B \rightleftharpoons HB^+ + HOAc} \tag{6.18}$$

$$K = \frac{[HB^+]}{[H_2OAc^+][B]} = \frac{K_b}{K_s}$$

Again, it is seen that the equilibrium constant K of the titration reaction *increases* with increasing K_b in the solvent employed and with *decreasing* K_s.

Therefore, both weak acids and weak bases are best titrated in solvents with low autoprotolysis constants.

6.4   Acid-Base Character of Solvents

Enhancement of Acidity and Basicity

The strength of *weak bases* is amplified in *acidic* solvents, whereas the strength of *weak acids* is increased in *basic* solvents. The base "dissociation constant" of a weak base B in an acidic solvent, such as formic acid, HCO_2H, is exemplified by the reaction

$$B + HCO_2H \rightleftharpoons HB^+ + HCO_2^- \tag{6.19}$$

This is analogous to the following reaction in water:

$$B + H_2O \rightleftharpoons HB^+ + OH^- \tag{6.20}$$

It is readily seen that formate ion, HCO_2^-, in formic acid is analogous to hydroxide ion, OH^-, in water. Furthermore, K_b of the weak base in formic acid (the equilibrium constant of Reaction 6.19) is higher than K_b of the base in water. This is because formic acid is an acidic solvent with strong proton–donor qualities.

The acidity of weak acids is enhanced in basic solvents because of the tendency of such solvents to acquire protons. Consider the dissociation of a weak acid HA in the basic solvent ethylenediamine, abbreviated En. The reaction is

$$HA + En \rightleftharpoons HEn^+ + A^- \tag{6.21}$$

and its equilibrium toward the right is favored because of the strong tendency for ethylenediamine to accept protons.

Acidic Solvents to Enhance Basicity

The feasibility for the titration of a weak acid or weak base is favored by larger values of K_a and K_b, respectively. Therefore, it is possible through the use of acidic solvents to titrate weak bases that are too weak to titrate in water. Similarly, it is possible by using basic solvents to titrate acids that are too weak to titrate in water. For example, it is virtually impossible to titrate pyridine, C_6H_5N, in water because the compound is such a weak base in that solvent with a K_b of only 1.7×10^{-9}. However, in acidic acetic acid solvent, the K_b value of pyridine is greatly increased, due to the reaction

$$C_6H_5N + HOAc \rightleftharpoons C_6H_5NH^+ + OAc^- \tag{6.22}$$

which has a relatively strong tendency to go to the right because of the strong proton-donor tendencies of HOAc. Thus, pyridine can be titrated in anhydrous acetic acid with perchloric acid titrant:

$$C_6H_5N + H_2OAc^+ + ClO_4^- \rightleftarrows C_6H_5NH^+ + HOAc + ClO_4^- \qquad (6.23)$$

Basic Solvents to Enhance Acidity

Just as bases that are very weak in water can be titrated successfully in acidic solvents, *acids* that are very weak in water can be titrated in *basic* solvents. Such an acid is phenol,

which has a K_a of only 1.00×10^{-10} in water and cannot be titrated in that solvent. However, use of ethylenediamine (En) as a basic solvent enhances the acid dissociation of phenol to such a degree that phenol is readily titrated with strong base in ethylenediamine solvent.

Leveling and Differentiating Solvents

Because of the high value of its self-dissociation constant ($K_w = 1.00 \times 10^{-14}$) and the fact that it is neither predominantly acidic nor basic, water is a leveling solvent. This quality is best seen in the strengths of acids in water. Several important acids, such as $HClO_4$, HCl, HNO_3, and the first proton from H_2SO_4, are classified as strong acids in water, meaning that the reaction

$$\underset{\substack{\text{strong} \\ \text{acid}}}{HA} + H_2O \rightarrow H_3O^+ + A^- \qquad (6.25)$$

lies completely to the right—that is, the acid is completely dissociated. In respect to these acids, however, some solvents are **differentiating solvents**, in which there are differences in the strengths of a number of acids that are all strong in water. Thus, in anhydrous acetic acid solvent, HCl is not classified as a strong acid because the acidic solvent has relatively little tendency to accept protons in the acid dissociation reaction

$$HCl + HOAc \rightleftarrows H_2OAc^+ + Cl^- \qquad (6.26)$$

An equivalent way of putting it is that in anhydrous acetic acid the conjugate base of HCl, the Cl^- ion, has a relatively strong tendency to accept protons

$$Cl^- + HOAc \rightleftarrows HCl + OAc^-$$

from the acidic proton-donor solvent.

6.5   Dielectric Constant of Solvents

Ion Pairs in Nonaqueous Solvents

In terms of a liquid's solvent properties, the dielectric constant D may be viewed as a measure of the work required to separate oppositely charged ions in the solvent. The *higher* the dielectric constant, the *less* work required to separate oppositely charged ions. Thus, in benzene, with a dielectric constant of only 2.3, much energy must be expended to separate oppositely charged ions, and such ions tend to exist as **ion pairs** held together by their opposite charges in this solvent. Water ($D = 78.5$) has a very high dielectric constant, such that oppositely charged ions in aqueous solution have a very low tendency to be paired through purely electrostatic forces of attraction. Solvents with intermediate dielectric constants, such as acetonitrile ($D = 37.4$) and ethanol ($D = 24.3$), exhibit behavior between that represented by the extremes of benzene and water insofar as ionic solutes are concerned. Liquids with high dielectric constants tend to be better solvents for ionic compounds, particularly salts. Thus, water dissolves a wide range of salts, whereas most salts are virtually insoluble in benzene.

Nature of Ionization in Low-Dielectric-Constant Solvents

Because of ion–pair formation, it is necessary to view dissociation of acids and bases in low-dielectric-constant solvents as two-step processes. For example, in acetic acid ($D = 6.2$) perchloric acid exists totally as ions of H_2OAc^+ and ClO_4^- and is, therefore, completely ionized. However, the ions are present almost entirely as the ion pair

$$H_2OAc^+ClO_4^-$$

Although this ion pair does dissociate in acetic acid

$$H_2OAc^+ClO_4^- \rightleftharpoons H_2OAc^+ + ClO_4^- \tag{6.27}$$

the dissociation constant K_d is only about 10^{-5}.

$$K_d = \frac{[H_2OAc^+][ClO_4^-]}{[H_2OAc^+ClO_4^-]} \cong 10^{-5} \tag{6.28}$$

Since the tendency to form ion pairs decreases the degree of dissociation of acids and bases, these species are relatively weaker in low-dielectric-constant solvents. This is strictly true for initially uncharged acids, HA, and initially uncharged bases, B. These react with solvents, HSolv, that can accept or donate protons by the general reactions

$$HA + HSolv \rightleftharpoons H_2Solv^+ + A^- \tag{6.29}$$

$$B + HSolv \rightleftharpoons HB^+ + Solv^- \tag{6.30}$$

which involve formation of ions. Thus, a low-dielectric-constant solvent with little tendency to separate ion pairs favors the formation of species such as $H_2Solv^+A^-$ and HB^+Solv^- with little tendency to dissociate, thus lowering the strengths of acids and bases. For example, carboxylic acids, such as benzoic acid,

typically have K_a values around 10^{-5} in water, but about five orders of magnitude lower in the lower-dielectric-constant solvent ethanol.

6.6   Solvent Choice for Nonaqueous Titrations

Solvents Commonly Used

From the preceding discussion it is obvious that the solvent is crucial in determining the suitability of a nonaqueous titration. The major solvents that have been used as media for such titrations are summarized in Table 6.1. These fall into the two broad categories of *amphiprotic*—that is, capable of both accepting and donating protons—and *aprotic*, which are solvents that do not accept or donate protons. Further subdivisions among the amphiprotic solvents are those that are acidic, basic, and neither predominantly acidic nor basic.

Methanol and Ethanol

Methanol and ethanol have been extensively investigated as solvents for acid-base titrations. They are neither predominantly acidic nor basic and have the advantage of low self-dissociation constants (see Section 6.3). Because of the relatively low dielectric constants of these solvents, undissociated ion pairs are formed, tending to weaken initially uncharged acids and bases (see Reactions 6.29 and 6.30). For weak charged acids, such as NH_4^+, however, the dissociation does not involve the formation of additional charged species—that is, charge separation. The acid dissociation reaction of NH_4^+ in ethanol is

$$NH_4^+ + C_2H_5OH \rightleftharpoons NH_3 + C_2H_5OH_2^+ \tag{6.31}$$

in which there is a +1 charge on both sides of the reaction and the low dielectric constant of the ethanol solvent has no effect on the dissociation reaction. In such a case the low self-dissociation constant of the solvent has a predominant effect, and the degree of neutralization of NH_4^+ by strong base in ethanol is much higher than the degree of reaction of NH_4^+ with strong base titrant in water. Therefore, NH_4^+ can be titrated successfully with strong base in ethanol, but not in water.

TABLE 6.1 Solvents employed as titration media*

Solvent name	Solvent formula	$K_s^\dagger$	Dielectric constant	Boiling point, °C
Amphiprotic solvents, neither acidic nor basic				
Water	H_2O	1.00×10^{-14}	78.5	100
Methanol	$H-\overset{\displaystyle H}{\underset{\displaystyle H}{C}}-OH$	2.0×10^{-17}	32.6	64
Ethanol	$H-\overset{\displaystyle H}{\underset{\displaystyle H}{C}}-\overset{\displaystyle H}{\underset{\displaystyle H}{C}}-OH$	8×10^{-20}	24.3	78
Amphiprotic solvents, basic				
Ammonia	NH_3	2×10^{-38} ($-50°\,C$)	27 ($-50°\,C$)	-78
Ethylenediamine	$\overset{\displaystyle H}{\underset{\displaystyle H}{N}}-\overset{\displaystyle H}{\underset{\displaystyle H}{C}}-\overset{\displaystyle H}{\underset{\displaystyle H}{C}}-\overset{\displaystyle H}{\underset{\displaystyle H}{N}}$	5×10^{-16}	12.9	117
Amphiprotic solvents, acidic				
Formic acid	$H-\overset{\displaystyle O}{\overset{\displaystyle \|}{C}}-OH$	6.3×10^{-7}	58.5	101
Acetic acid	$H-\overset{\displaystyle H}{\underset{\displaystyle H}{C}}-\overset{\displaystyle O}{\overset{\displaystyle \|}{C}}-OH$	3.5×10^{-15}	6.13	118
Sulfuric acid	H_2SO_4	2.5×10^{-4}	100	340
Aprotic solvents				
Acetonitrile	$H-\overset{\displaystyle H}{\underset{\displaystyle H}{C}}-C\equiv N$	—	36.7	82
Dimethylsulfoxide	$H-\overset{\displaystyle H}{\underset{\displaystyle H}{C}}-\overset{\displaystyle O}{\overset{\displaystyle \|}{S}}-\overset{\displaystyle H}{\underset{\displaystyle H}{C}}-H$	—	46	Decomposes ~ 100°C
Methyl isobutyl ketone	$H-\overset{\displaystyle H}{\underset{\displaystyle H}{C}}-\overset{\displaystyle O}{\overset{\displaystyle \|}{C}}-\overset{\displaystyle H}{\underset{\displaystyle H}{C}}-\overset{\displaystyle H-CH_3}{\underset{\displaystyle CH_3}{C}}-H$	—	13.1	119

* At 25°C unless otherwise specified
† Equilibrium constant for the reaction, $HSolv + HSolv \rightleftarrows H_2Solv^+ + HSolv^-$

Basic Solvents

Although the solvent properties of liquid ammonia have been investigated extensively, the difficulties of handling this solvent at the low temperatures required to keep it in the liquid state have prevented its use for routine acid–base titrations. The basic solvent of choice is ethylenediamine, which has been used as a solvent for the titration of phenols, carboxylic acids, imides, amine salts, sulfa drugs, and some inorganic salts. The acidities of these chemical species are enhanced in ethylenediamine. Several bases have proven useful as titrants in ethylenediamine media. Tetrabutylammonium hydroxide, $(C_4H_9)_4NOH$, dissolved in ethanol, isopropanol, or a mixed solvent of benzene and methanol has been employed as a base titrant. Sodium amino-ethoxide, $Na^{+-}OCH_2CH_2NH_2$, a product of the reaction of ethanolamine, $HOCH_2CH_2NH_2$, with Na metal is a good titrant base in ethylenediamine solution. Sodium methoxide can be prepared by the reaction of sodium metal with methanol

$$2CH_3OH + 2Na \rightarrow 2Na^+ + 2CH_3O^- + H_2(g) \tag{6.32}$$

and can be used as a standard base in a solvent consisting of a mixture of methanol and benzene.

Dimethylformamide, $H-\overset{\overset{\displaystyle O}{\|}}{C}-N(CH_3)_2$, is a very weakly basic solvent. Its dielectric constant of 38 is higher than that of ethylenediamine. In general it can be used as the medium to titrate materials for which ethylenediamine is used as a solvent, using the same basic reagents as are employed in ethylenediamine. Because of dimethylformamide's relatively high dielectric constant, it dissolves a number of salts, of which ammonium salts are among the most important. It is also a good solvent for titratable organics and polymers.

Titrations in Acetic Acid

Acetic acid is the most extensively investigated acidic solvent employed in titrations. Perchloric acid, $HClO_4$, is the acid of choice for a titrant in this medium. Sodium acetate, Na^+OAc^-, can be used as a base. The presence of water interferes in the titration of very weak bases. However, acetic anhydride may be added to react with the water and produce acetic acid:

$$
\underset{\overset{\displaystyle |}{\displaystyle H}}{\overset{\overset{\displaystyle H}{|}}{H-C}}-\underset{}{\overset{\overset{\displaystyle O}{\|}}{C}}-O-\underset{\overset{\displaystyle |}{\displaystyle H}}{\overset{\overset{\displaystyle O}{\|}}{C}}-\underset{}{\overset{\overset{\displaystyle H}{|}}{C}}-H \;+\; H_2O \;\rightarrow\; 2\underset{\overset{\displaystyle |}{\displaystyle H}}{\overset{\overset{\displaystyle H}{|}}{H-C}}-\overset{\overset{\displaystyle O}{\|}}{C}-OH \tag{6.33}
$$

One practical problem with the use of acetic acid in volumetric work is its high temperature coefficient of expansion, which is about four times that of water. The titrant volume can be corrected for temperature changes using known correction factors.

Glacial (pure) acetic acid has been used as a solvent for the titration of very weak nitrogen bases, amines, amides, amino acids, and ureas. Many salts, particularly ammonium and sodium salts, can be titrated as bases in acetic acid. This is because anions that are not at all basic in water behave as bases in acetic acids and undergo the following titration reaction with H^+ from $HClO_4$:

$$Na^+X^- + H_2OAc^+ \rightarrow Na^+ + HX + HOAc \qquad (6.34)$$

Even the chloride ion of NaCl can be titrated with H^+ provided by $HClO_4$ in glacial acetic acid, where the titration reaction is

$$Na^+Cl^- + H_2OAc^+ \rightarrow Na^+ + HCl + HOAc \qquad (6.35)$$

Figure 6.2 illustrates the advantages of glacial acetic acid as a solvent for the titration of weak base, in this case the acetate ion in sodium acetate. In aqueous media there is a barely discernible break in the titration curve, which would not be suitable for a quantitative determination. In glacial acetic acid, however, the break is very pronounced and a good end point is readily obtained.

Normally for titrations in acetic acid it is best to employ a glass electrode (see Chapter 13) to measure pH as a function of reagent added and calculate the end point volume from the titration curve obtained. In favorable cases crystal violet and methyl violet indicators can be employed. The end point with methyl violet is indicated by a change in color from violet to green; further addition of acid results in a yellow color. End points with both indicators are hard to discern, and there is little reason to use these indicators since glass electrodes function well in an acetic acid medium.

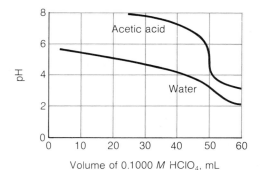

FIGURE 6.2   Curves for the titration of 100 mL of 0.0500 M sodium acetate in aqueous solution and in glacial acetic acid solution titrated with 0.1000 M $HClO_4$ in water and glacial acetic acid, respectively. The pH values were measured with a glass electrode (see Chapter 13); those values obtained in glacial acetic acid are not directly comparable to those obtained in water.

Aprotic Solvents

Aprotic or inert solvents, such as those listed at the bottom of Table 6.1, have been used as media for the titration of acids, bases, and amines. In addition to the compounds listed in Table 6.1, ethers and esters (ethyl acetate) have been employed as nonaqueous titration media.

Programmed Summary of Chapter 6

This summary contains in a programmed format the major terms and concepts introduced in this chapter. To get the most out of it, you should fill in the blanks for each question and then check the answers at the end of the book to see if you are correct.

The two major advantages offered by nonaqueous solvents for the titration of acids or bases are increased (1) _____ of the solute and increased (2) _____. Nonaqueous solvents used for acid–base titrations are most conveniently classified on the basis of their relative (3) _____ and whether or not they undergo (4) _____. Solvents that neither accept nor donate protons are called (5) _____ or (6) _____. The process of solvent self-dissociation through the exchange of protons is called (7) _____, and such solvents are said to be (8) _____. The general reaction for a solvent HSolv undergoing the preceding reaction is (9) _____ and its equilibrium is described by a (10) _____. Three common solvents that have about the same tendency to accept as to donate electrons are (11) _____. Solvents that have a greater tendency to donate than to accept protons are called (12) _____, and those that have the greater tendency to accept protons are called (13) _____. In addition to relative acidity or basicity, three major properties that determine the uses of a solvent are (14) _____. The equilibrium constant K for the titration of an acid HA by a base in a nonaqueous solvent in terms of the dissociation constant of the acid and the self-dissociation constant of the solvent is (15) _____. This shows that a complete titration is favored by a low level of (16) _____ of the solvent. The strengths of weak acids are increased in (17) _____ solvents, and the strengths of weak bases are increased in (18) _____ solvents. An example of an acid that is too weak to titrate in water but strong enough in basic ethylenediamine to be titrated is (19) _____, and an example of a base that is too weak to titrate in water but can be titrated in acidic acetic acid is (20) _____. A solvent in which several acids are strong acids is probably classified as a (21) _____, whereas a solvent that tends to accentuate differences in acidic strength is a (22) _____. Dielectric constant is a measure of the work required to (23) _____ in solution. In low-dielectric-constant solvents, ions tend to exist as (24) _____, which tend to (25) _____ the degree of dissociation of acids and bases. Low solvent dielectric constant does not greatly affect the dissociation constant of an initially charged acid such as NH_4^+ because dissociation of the acid does not involve (26) _____. The most commonly used basic solvent is (27) _____, and the most used acidic solvent is (28) _____. The strong acid of choice for uses in acidic nonaqueous solvents, such as acetic acid, is (29) _____. For titrations in acetic acid it is normally best to determine the end point from (30) _____. Two visual indicators that can be employed in acetic acid are (31) _____.

Questions

1. Liquid H_2S has been used experimentally as a nonaqueous solvent. Suggest the autoprotolysis or self-dissociation reaction of that solvent.
2. Name six examples of aprotic solvents.
3. The electron–dot structure of ethylenediamine is

    H H H H
    :N̈:C̈:C̈:N̈:
    Ḧ Ḧ Ḧ Ḧ

    How does this suggest that ethylenediamine should function as a basic solvent?
4. What base in ethanol, C_2H_5OH, is analogous to NaOH in water? (It will be helpful to know about the common bases in methanol.)
5. Perchloric acid, $HClO_4$, is the strong acid of choice for use in nonaqueous solvents. Concentrated solutions of perchloric acid contain about 30% water by mass. Suggest the nature of perchloric acid solute species in the amphiprotic solvent formic acid, HCO_2H, and the aprotic solvent acetonitrile, $CH_3C{\equiv}N$. Consider also the dielectric constants of the solvents.
6. What factor, or factors, would favor acetate ion, OAc^-, from sodium acetate as a base in the relatively low dielectric con-

stant solvent, ethanol, as compared to NH_3 base in the same solvent?
7. Show with a chemical reaction or two how the relatively low tendency of ethanol, C_2H_5OH, to acquire protons increases the acidic strength of NH_4^+ in ethanol.
8. Explain the statement that the strongest base that can exist in a solvent is the solvent's conjugate base.
9. Although acetic acid is a more effective *differentiating solvent* for *acids* than is water, acetic acid is a better *leveling solvent* for *bases* than water. Explain the latter observation.
10. What are some typical base titrants used in ethylenediamine?
11. Of the following, the *least* likely reason to choose a nonaqueous solvent for acid-base titration is (a) to emphasize the acidity of an analyte, (b) to emphasize the basicity of an analyte, (c) to dissolve nonpolar non-ionic species, (d) to increase the magnitude of the break in a titration curve for some analytes, (e) to increase the solubility of highly ionic chemical species.
12. Match the solvent formula (designated by a letter) with the description of its general properties (designated by a number).

a. H_2O

b. 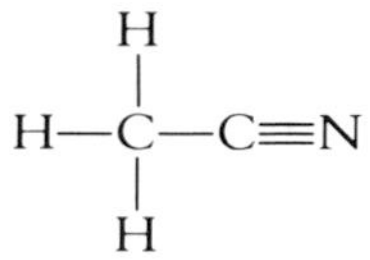
$$
\begin{array}{c}
H \\
| \\
H-C-C\equiv N \\
| \\
H
\end{array}
$$

c. NH_3

d. 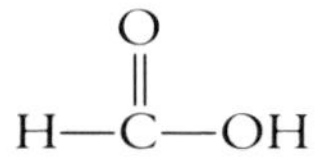
$$
\begin{array}{c}
O \\
\parallel \\
H-C-OH
\end{array}
$$

1. Amphiprotic solvent, acidic
2. Aprotic solvent

3. Amphiprotic solvent, neither acidic nor basic
4. Amphiprotic solvent, basic

13. Which of the following statements is true, pertaining to a low-dielectric-constant solvent such as benzene: (a) salts are especially soluble, (b) oppositely charged ions tend to exist as ion pairs, (c) most acids are strong, (d) most bases are strong, (e) less work is required to separate oppositely charged ions.

Problems

1. Assuming that the dissociation constant of the ion pair $H_2OAc^+ClO_4^-$ is 1.00×10^{-6} in acetic acid, calculate the percentage of perchloric acid dissociated in a 2.00×10^{-3} M solution of $HClO_4$ in acetic acid solvent.

2. The density of pure acetic acid is 1.049 g at 20° C. What is the molarity of acetic acid under those conditions?

3. From Table 6.1 it is seen that the autoprotolysis constant K_s of methanol solvent has a value of 2.0×10^{-17}. How would you suggest that acid-base neutrality is defined in methanol?

7 Solubility Equilibria

L E A R N I N G   G O A L S

1. Classes of soluble and insoluble inorganic compounds most likely to be encountered in quantitative analysis.
2. Simple solubility equilibria calculations based upon solubility product expressions.
3. Activities and activity coefficients and how they are incorporated into chemical equilibrium calculations.
4. An introduction to how ions common to a slightly soluble compound affect its solubility in a medium with which the compound is equilibrated.
5. Calculation of solubilities in media in which there are species that react with the anion of the slightly soluble compound.
6. The effects of cation reactions on solubilities.
7. Solubilities of metal hydroxides and their uses in quantitative analysis.
8. The concept of intrinsic solubility and how it contributes to solubility not accounted for by solubility product calculations.
9. Separations of metal ions with sulfide precipitation.
10. Henry's law as applied to gas solubilities.

7.1   Applications of Solubility to Chemical Analysis

Importance of Precipitation and Slightly Soluble Compounds

Although reference is often made to insoluble compounds, every compound dissolves to at least a minuscule extent. Therefore, it is more proper to refer to *slightly soluble* compounds. The difference between soluble and slightly soluble is somewhat indistinct, and an exact distinction is of little importance. For uses in chemical analysis, a slightly soluble compound is one that dissolves to an extent of $0.01\ M$ or less when saturated in a particular medium; most slightly soluble compounds of use in chemical analysis have lower solubilities than this.

The process of producing a slightly soluble compound from solution by a chemical or physical process in the solution is called **precipitation**, and the product is called a **precipitate**. Figure 7.1 shows the precipitation of silver chloride from solutions of sodium chloride and silver nitrate.

The formation of slightly soluble substances has several uses in analytical chemistry. The most obvious such use is to form a precipitate of the analyte

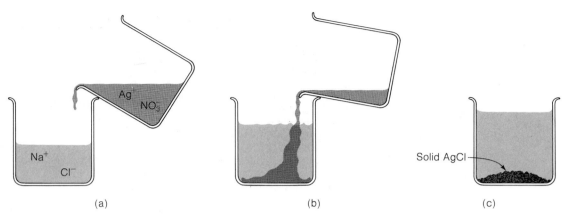

FIGURE 7.1 (a) Addition of a solution of $AgNO_3$ to one of NaCl, results in (b) the precipitation reaction $Ag^+ + NO_3^- \rightarrow AgCl(s)$, leaving (c) a precipitate of solid silver chloride, AgCl

stoichiometrically and to weigh the precipitate or a modified form of it. This is the basis of **gravimetric analysis** (see Chapter 8). In cases where a form of the analyte suitable for weighing is not obtained, it is often possible to remove the analyte from impurities in a sample solution by precipitation, followed by determination of the analyte by nongravimetric analysis. Conversely, interfering substances may be separated from the sample by precipitation, allowing the determination of the remaining analyte by a variety of methods. In most such applications, the precipitate should be reasonably pure and have a particle size that permits filtration.

7.2 Nature and Classes of Slightly Soluble Substances

The solubility of compounds depends largely upon the type of compound. Table 7.1 gives solubility rules for ionic compounds, primarily metal salts. Use is made of some of these compounds in chemical analysis. For example, some metals can be precipitated as hydroxides, as shown below for a divalent metal ion

$$M^{2+}(aq) + 2OH^-(aq) \rightarrow M(OH)_2(s) \tag{7.1}$$

then heated (ignited) to give a weighable form of the metal oxide:

$$M(OH)_2 \xrightarrow{\text{heat}} MO(s) + H_2O(g) \tag{7.2}$$

Chloride can be analyzed with very great accuracy by precipitating silver chloride with silver nitrate reagent

$$Ag^+ + Cl^- \rightarrow AgCl(s) \tag{7.3}$$

TABLE 7.1   Solubilities of inorganic ionic compounds

Anion of compound	Soluble compounds	Insoluble compounds
OH^-	Hydroxides of 1A elements—for example, NaOH–Sr(OH)$_2$, Ba(OH)$_2$; Ca(OH)$_2$ is moderately soluble.	All except those listed on the left
NO_3^-	Essentially all compounds	No common ones
ClO_4^-	All except KClO$_4$	KClO$_4$ is moderately insoluble
Cl^-	All, except those listed on the right	AgCl, Hg$_2$Cl$_2$; PbCl$_2$ is marginally insoluble
SO_4^{2-}	All, except those listed on the right	Hg$_2$SO$_4$, HgSO$_4$, Ag$_2$SO$_4$, PbSO$_4$, CaSO$_4$, SrSO$_4$, BaSO$_4$
S^{2-}	Salts of 1A elements—for example, Na$_2$S; salts of 2A elements—for example, CaS; NH$_4$HS, (NH$_4$)$_2$S	All except those listed on the left
CO_3^{2-}	Salts of 1A elements, except for Li$_2$CO$_3$; salt of NH$_4^+$	All except those listed on the left

and weighing the dried silver chloride product. Metals can be separated from solution with some selectivity by precipitation as sulfides from hydrogen sulfide at a controlled pH:

$$H_2S + M^{2+} \rightarrow MS(s) + 2H^+ \tag{7.4}$$

Controlling the pH regulates the degree of dissociation of H_2S and HS^- to S^{2-}, so that the less-soluble metal sulfides precipitate at lower pH values.

7.3   Simple Solubility Equilibria

Calculation of Precipitate Solubility

To determine the feasibility of an analysis involving a precipitate, it is necessary to calculate the solubility of the precipitate. In its simplest form, such a calculation involves the concentrations of the cation and anion, along with a solubility product. However, the anion is frequently the conjugate base of a weak acid, so pH may play a role. The cation, which is usually a metal, may form complex ion species—for example,

$$Cu^{2+} + 4NH_3 \rightleftharpoons Cu(NH_3)_4^{2+} \tag{7.5}$$

with the metal ions. There may be competing ions present that form undesired precipitates. All of these factors must be considered in the solubility calculations involved with precipitate formation. The first such calculations to be considered will be those involving only the ions forming the desired precipitate.

Solubility from Solubility Product

In the simplest case, solubility is expressed for a reaction of the type

$$BaSO_4(s) \rightleftarrows Ba^{2+}(aq) + SO_4^{2-}(aq) \tag{7.6}$$

by a solubility product K_{sp} in the form

$$K_{sp} = [Ba^{2+}][SO_4^{2-}] \tag{7.7}$$

It is necessary to know how to calculate K_{sp} from solubility. For example, given the mass solubility of $BaSO_4$ of 2.59 mg/L, 233.4 for the formula weight of $BaSO_4$, and using the symbol S for solubility in molar units, the value of S for $BaSO_4$ may be calculated.

$$S = \frac{2.59 \text{ mg}}{L} \times \frac{1 \text{ L}}{1000 \text{ mL}} \times \frac{1 \text{ mmol}}{233.4 \text{ mg}} = 1.11 \times 10^{-5} \text{ mmol/mL} \tag{7.8}$$

From Reaction 7.6,

$$[Ba^{2+}] = [SO_4^{2-}] = 1.11 \times 10^{-5} = S \tag{7.9}$$

$$K_{sp} = [Ba^{2+}][SO_4^{2-}] = 1.11 \times 10^{-5} \times 1.11 \times 10^{-5} = 1.23 \times 10^{-10} \tag{7.10}$$

For a slightly more complicated case, calculate the solubility product of $PbCl_2$ if the solubility of this salt is 4.41 mg/mL.

$$S = \frac{4.41 \text{ mg}}{mL} \times \frac{1 \text{ mmol}}{278 \text{ mg}} = 1.59 \times 10^{-2} \text{ mmol/mL} \tag{7.11}$$

From the reaction

$$PbCl_2(s) \rightleftarrows Pb^{2+} + 2Cl^- \tag{7.12}$$

$$[Pb^{2+}] = S \qquad [Cl^-] = 2S \tag{7.13}$$

The solubility product is given by

$$K_{sp} = [Pb^{2+}][Cl^-]^2 = S(2S)^2 = 1.59 \times 10^{-2}(2 \times 1.59 \times 10^{-2})^2 = 1.61 \times 10^{-5} \tag{7.14}$$

Consider next the reverse calculation, that of solubility from solubility product under conditions such that neither the cation nor the anion undergoes further reactions in solution. For example, the solubility product of silver chromate, Ag_2CrO_4, fw = 331.7, is 1.29×10^{-12}. What is the molar solubility and solubility in milligrams per liter of Ag_2CrO_4? The dissolution reaction is

$$Ag_2CrO_4(s) \rightleftarrows 2Ag^+ + CrO_4^{2-} \tag{7.15}$$

from which it is seen that the solubility product expression is

$$K_{sp} = [Ag^+]^2[CrO_4^{2-}] = 1.29 \times 10^{-12} \tag{7.16}$$

and the ion concentrations in terms of S are the following:

$$[Ag^+] = 2S \qquad [CrO_4^{2-}] = S \tag{7.17}$$

With this information, the solubility is calculated as follows:

$$[Ag^+]^2[CrO_4^{2-}] = (2S)^2 S = K_{sp} = 1.29 \times 10^{-12} \qquad (7.18)$$

$$4S^3 = 1.29 \times 10^{-12} \qquad (7.19)$$

$$S = \left(\frac{1.29 \times 10^{-12}}{4}\right)^{1/3} = 6.86 \times 10^{-5} \ M \qquad (7.20)$$

$$S(\text{mg/L}) = 6.86 \times 10^{-5} \ \text{mol/L} \times 331.7 \ \text{g/mol} \times 1000 \ \text{mg/g} = 22.8 \ \text{mg/L} \qquad (7.21)$$

To check the value of S used, calculate $[Ag^+]$ and $[CrO_4^{2-}]$ and substitute back into the expression for K_{sp}.

$$[Ag^+] = 2S = 2 \times 6.86 \times 10^{-5} \ M = 1.37 \times 10^{-4} \ M \qquad (7.22)$$

$$[CrO_4^{2-}] = S = 6.86 \times 10^{-5} \ M \qquad (7.23)$$

$$[Ag^+]^2[CrO_4^{2-}] = (1.37 \times 10^{-4})^2 \times 6.86 \times 10^{-5} = 1.29 \times 10^{-12} = K_{sp} \qquad (7.24)$$

EXAMPLE

The value of K_{sp} of CaF_2 is 4.9×10^{-11}. What is the molar solubility S of CaF_2 in water sufficiently basic to prevent appreciable binding of F^- ion by H^+ ion?[1]

The reaction for CaF_2 going into solution is

$$CaF_2(s) \rightleftharpoons Ca^{2+} + 2F^-$$

from which the following are obtained:

$$K_{sp} = [Ca^{2+}][F^-]^2 = 4.9 \times 10^{-11}$$

$$[Ca^{2+}] = S \qquad [F^-] = 2S$$

Substituting into the K_{sp} expression gives

$$[Ca^{2+}][F^-]^2 = S(2S)^2 = 4S^3 = K_{sp}$$

and solving for S gives the following:

$$S = \left(\frac{K_{sp}}{4}\right)^{1/3} = \left(\frac{4.9 \times 10^{-11}}{4}\right)^{1/3} = 2.31 \times 10^{-4} \ M$$

[1] If the pH is near pK_a of HF, or below, the reaction $F^- + H^+ \rightarrow HF$ increases the solubility of CaF_2.

7.4   Activity and Activity Coefficients

The Activity Concept

In some cases the use of ion concentrations alone in the calculation of solution concentrations involving equilibrium constants yields results that are too far from the true value. This occurs particularly when ions with charges other than $+1$ or -1 are involved (for example, Ca^{2+}, $Fe(CN)_6^{4-}$) and when ionic solutes other than those involved in the equilibrium reaction are present. In some cases such ionic solutes are deliberately added to solution to provide a constant solution concentration of ions and thus minimize effects from *variation* of ion concentration. Under those circumstances an added ionic solute is called a **supporting electrolyte**.

Account may be taken of the influence upon equilibrium reactions of ions that do not participate directly in the equilibrium reaction but still influence the position of the equilibrium (extent of ionization of weak acid, solubility of a slightly soluble salt) by substituting the **activity** for concentrations. Activity is defined as

$$a_X = [X] f_X \tag{7.25}$$

where a_X, $[X]$, and f_X are the activity, molar concentration, and activity coefficient of species X. The **activity coefficient** f_X is a number without units which yields activity when multiplied by concentration, and is in a sense a correction factor that converts concentration to an *effective concentration* (activity). The activity coefficient (as well as activity) varies with the concentration of ions present.

The use of activities in place of concentrations in equilibrium calculations enables the use of the same equilibrium constant value over a reasonably wide range of electrolyte concentration. Consider the reaction

$$NH_3 + H_2O \rightleftharpoons NH_4^+ + OH^- \tag{7.26}$$

for which the equilibrium constant in terms of activities is the following:

$$K_b = \frac{a_{NH_4^+} a_{OH^-}}{a_{NH_3}} = \frac{[NH_4^+] f_{NH_4^+} [OH^-] f_{OH^-}}{[NH_3] f_{NH_3}} = \frac{[NH_4^+][OH^-]}{[NH_3]} \times \frac{f_{NH_4^+} f_{OH^-}}{f_{NH_3}} \tag{7.27}$$

In this expression K_b is constant, whereas the activity coefficients vary with electrolyte concentration. Therefore, different values of $[NH_4^+]$ and $[OH^-]$ may be obtained at the same analytical concentration (nominal molarity) of NH_3 for different total electrolyte concentrations. When expressed in terms of activities, the equilibrium constant K_b has the same value over a wide range of electrolyte concentrations and is called a **thermodynamic equilibrium constant**. The expression for the **apparent equilibrium constant**, designated K_b' in this case to distinguish it from the thermodynamic equilibrium constant, is the following:

$$K_b' = \frac{[NH_4^+][OH^-]}{[NH_3]} = K_b \frac{f_{NH_3}}{f_{NH_4^+} f_{OH^-}} \tag{7.28}$$

The value of the apparent equilibrium constant does vary with electrolyte concentration.

Ionic Strength

As discussed above, activity coefficients vary with electrolyte concentration. The effect is due to the density of ion charge in the solution. The effect of electrolytes on activity coefficients is independent of the kinds of electrolytes (for example, $NaNO_3$ versus Na_2SO_4) and depends upon a parameter called **ionic strength**, μ, defined by the following equation:

$$\mu = \tfrac{1}{2}(m_1 Z_1^2 + m_2 Z_2^2 + m_3 Z_3^2 + m_4 Z_4^2 + \cdots) \qquad (7.29)$$

In this equation the m's are molar concentrations of ions and the Z's are ionic charges as shown by the following examples.

EXAMPLE

What are the values of μ in (1) a solution of 0.0100 M $NaNO_3$ and (2) 0.0100 M Na_2SO_4?

1. $m_{Na^+} = 0.0100 \qquad Z_{Na^+} = 1$
    $m_{NO_3^-} = 0.0100 \qquad Z_{NO_3^-} = 1$
    $\mu = \tfrac{1}{2}(0.0100 \times 1^2 + 0.0100 \times 1^2) = 0.0100$
2. $m_{Na^+} = 2 \times 0.0100 = 0.0200 \qquad Z_{Na^+} = 1$
    $m_{SO_4^{2-}} = 0.0100 \qquad Z_{SO_4^{2-}} = 2$
    $\mu = \tfrac{1}{2}(0.0200 \times 1^2 + 0.0200 \times 2^2) = 0.0500$

Note that the ionic strength of the Na_2SO_4 solution is five times that of the $NaNO_3$ solution with the same molar concentration.

EXAMPLE

What is the ionic strength of a solution 0.0100 M in $NaCl$, 0.00500 M in $MgSO_4$, and 0.00250 M in K_2SO_4?

$m_{Na^+} = 0.0100 \qquad Z_{Na^+} = 1$

$m_{Mg^{2+}} = 0.00500 \qquad Z_{Mg^{2+}} = 2$

$m_{K^+} = 2 \times 0.00250 = 0.00500 \qquad Z_{K^+} = 1$

$m_{Cl^-} = 0.0100 \qquad Z_{Cl^-} = 1$

$m_{SO_4^{2-}} = 0.00500 + 0.00250 = 0.00750 \qquad Z_{SO_4^{2-}} = 2$

(Note that there are two sources of SO_4^{2-} ion.)

$\mu = \tfrac{1}{2}(0.0100 \times 1^2 + 0.00500 \times 2^2 + 0.00500 \times 1^2 + 0.0100 \times 1^2$
$\qquad + 0.00750 \times 2^2)$

$\qquad = 0.0375$

Determination of Mean Activity Coefficients

The activity coefficients mentioned previously, such as $f_{NH_4^+}$, are *individual activity coefficients* that apply to single ions. *Individual ion activity coefficients cannot be determined experimentally.* It is possible to determine *mean activity coefficients,* which for an electrolyte $X_m Y_n$ are defined as

$$f_{\pm} = (f_A^m f_B^n)^{\frac{1}{m+n}} \tag{7.30}$$

where $f_{\pm}$ is the mean activity coefficient and f_A and f_B are individual activity coefficients. In the case of a slightly soluble salt, the mean activity coefficient may be used to relate the *thermodynamic solubility product* to the *apparent solubility product* as shown for $BaSO_4$.

$$K_{sp} = [Ba^{2+}][SO_4^{2-}] f_{\pm}^2 = K'_{sp} f_{\pm}^2 \tag{7.31}$$

In this case the apparent equilibrium constant (solubility product) is designated K'_{sp} to distinguish it from the thermodynamic solubility product. The value of K'_{sp} may be determined at various ionic strengths. At zero ionic strength the mean activity coefficient is equal to 1. Thus extrapolating K'_{sp} values to zero ionic strength gives the value of K_{sp}. With that value established it is possible to calculate $f_{\pm}$ at a particular ionic strength by substituting the value of K'_{sp} determined at that ionic strength into Equation 7.31.

Calculation of Activity Coefficients from the Debye-Hückel Equation

Activity coefficients for single ions can be calculated from the Debye-Hückel relationship

$$-\log f_X = \frac{0.5085 Z_X^2 \sqrt{\mu}}{1 + 3.281 \times 10^{-3} a_X \sqrt{\mu}} \tag{7.32}$$

in which f_X is the activity coefficient of ion X, Z_X is the ion's charge, μ is ionic strength, a_X is the effective diameter of the hydrated ion in picometers ($1 \text{ pm} = 10^{-12} \text{ m} = 10^{-2}$ Ångstroms), and the two numerical constants in the equation are valid at 25° C. For many singly charged ions (see Table 7.2), the effective diameter of the ion is estimated to be about 300 pm, in which case the denominator of the Debye-Hückel relationship may be approximated by $1 + \sqrt{\mu}$. For ionic strengths less than about 0.01, the second term of the denominator becomes negligible relative to 1, yielding the simplified equation

$$-\log f_{\pm} = 0.5085 Z_X^2 \sqrt{\mu} \tag{7.33}$$

Table 7.2 gives the calculated values of some activity coefficients as a function of ionic strength. The greatest uncertainty in calculating activity coefficients is knowledge of a_X. Nevertheless, the Debye-Hückel equation and simplifications of it give satisfactory values for ionic strengths up to about 0.10, a sufficient range for most applications in quantitative analysis.

TABLE 7.2   Activity coefficients at different ionic strengths at 25° C

Ion	Effective ion diameter, a_X, (pm)[1]	Activity coefficient at designated ionic strength				
		0.001	0.005	0.010	0.050	0.10
H^+ (strongly hydrated)	900	0.967	0.933	0.914	0.86	0.83
Li^+	600	0.965	0.929	0.907	0.84	0.80
Na^+, HSO_3^-, HCO_3^-, $H_2PO_4^-$, $C_2H_3O_2^-$, IO_3^-	400–450	0.964	0.928	0.902	0.82	0.78
OH^-, F^-, ClO_3^-, ClO_4^-, HS^-, SCN^-, MnO_4^-	350	0.964	0.926	0.900	0.81	0.76
K^+, NO_2^-, NO_3^-, Cl^-, Br^-, I^-, CN^-	300	0.964	0.925	0.899	0.80	0.76
NH_4^+, Ag^+, Rb^+, Cs^+	250	0.964	0.924	0.898	0.80	0.75
Mg^{2+}, Be^{2+}	800	0.872	0.755	0.69	0.52	0.45
Ca^{2+}, Mn^{2+}, Fe^{2+}, Co^{2+}, Ni^{2+}, Cu^{2+}, Zn^{2+}	600	0.870	0.749	0.675	0.48	0.40
Ba^{2+}, Cd^{2+}, Hg^{2+}, Sr^{2+}, S^{2-}	500	0.868	0.744	0.67	0.46	0.38
CO_3^{2-}, $C_2O_4^{2-}$, SO_3^{2-}, Pb^{2+}	450	0.868	0.742	0.665	0.46	0.37
CrO_4^{2-}, Hg_2^{2+}, SO_4^{2-}, HPO_4^{2-}	400	0.867	0.740	0.660	0.44	0.36
Al^{3+}, Cr^{3+}, La^{3+}, Ce^{3+}, Fe^{3+}	900	0.738	0.54	0.44	0.24	0.18
PO_4^{3-}, $Fe(CN)_6^{3-}$	400	0.725	0.50	0.40	0.16	0.095

[1] The values of effective diameter are still commonly found in the chemical literature in units of Ångstroms, Å, where 1 Å = 100 pm.

EXAMPLE

Calculate the activity coefficient of OH^- ion in 0.0700 M NaOH.

$$\mu = \frac{1}{2}(0.0700 \times 1^2 + 0.0700 \times 1^2) = 0.0700$$

$$a_{OH^-} = 350 \text{ pm}$$

$$-\log f_{OH^-} = \frac{0.5085 \times 1^2 \sqrt{0.0700}}{1 + 3.281 \times 10^{-3} \times 350 \sqrt{0.0700}} = 0.103$$

$$f_{OH^-} = 10^{-0.103} = 0.79$$

Activity Coefficients in Solubility Calculations

For most of the solubility calculations in this chapter, concentrations are used in place of ion activities. However, particularly at higher ionic strengths and for multiply charged ions, activity coefficients may be significantly less than 1, so that activities are often appreciably lower than concentrations. Approximating activities by concentrations usually introduces essentially no error if the slightly soluble substance is dissolved in pure water, because the ion concentrations and, therefore, the ionic strengths are so low. However, if an electrolyte that reacts with neither the cation nor the anion is present to raise the ionic strength, the effect upon solubility can be to increase the solubility significantly by decreasing the activity coefficients.

To show the effects of changing activity coefficients upon solubilities, consider the effect of an electrolyte upon the solubility of lanthanum iodate, $La(IO_3)_3$. The solubility product expression for this salt is

$$K_{sp} = a_{La^{3+}}(a_{IO_3^-})^3 = 6.2 \times 10^{-12} \qquad (7.34)$$

where a is the activity of the ion in the subscript. If activities are approximated by concentrations, as can be done without appreciable error for the calculation of the solubility of $La(IO_3)_3$ in pure water, the solubility S is calculated as follows:

$$La(IO_3)_3(s) \rightleftharpoons La^{3+} + 3IO_3^- \qquad (7.35)$$

$$[La^{3+}] = S \qquad [IO_3^-] = 3S \qquad (7.36)$$

$$S(3S)^3 = [La^{3+}][IO_3^-]^3 = K_{sp} = 6.2 \times 10^{-12} \qquad (7.37)$$

$$S = \left(\frac{6.2 \times 10^{-12}}{27}\right)^{1/4} = 6.92 \times 10^{-4} \ M \qquad (7.38)$$

Next calculate the solubility of $La(IO_3)_3$ at an ionic strength of 0.10, such as would be the case in a solution of 0.10 M KNO_3. At this ionic strength

$$f_{La^{3+}} = 0.18 \qquad f_{IO_3^-} = 0.78 \qquad (7.39)$$

where f is the activity coefficient of the species designated. Recall that activities and concentrations are related as follows:

$$a_{La^{3+}} = [La^{3+}]f_{La^{3+}} \qquad a_{IO_3^-} = [IO_3^-]f_{IO_3^-} \qquad (7.40)$$

Considering activity coefficients at an ionic strength of 0.10 gives the following for S of $La(IO_3)_3$:

$$a_{La^{3+}}(a_{IO_3^-})^3 = K_{sp} = [La^{3+}]f_{La^{3+}}([IO_3^-]f_{IO_3^-})^3 \qquad (7.41)$$

$$[La^{3+}][IO_3^-]^3 = \frac{K_{sp}}{(f_{La^{3+}})(f_{IO_3^-})^3} = \frac{6.2 \times 10^{-12}}{0.18 \times (0.78)^3} = 7.3 \times 10^{-11} \qquad (7.42)$$

$$S(3S)^3 = 7.3 \times 10^{-11} \qquad (7.43)$$

$$S = \left(\frac{7.3 \times 10^{-11}}{27}\right)^{1/4} = 1.28 \times 10^{-3} \ M \qquad (7.44)$$

It is seen that the higher ionic strength gives an appreciably higher concentration of ions from the slightly soluble salt. This is because the "effectiveness" of the La^{3+} and IO_3^- ions is lessened in the presence of other ions, so that they do not have such a great tendency to precipitate $La(IO_3)_3$.

EXAMPLE

Earlier in this chapter, S of Ag_2CrO_4 in pure water was calculated as 6.86×10^{-5} M. What is S at an ionic strength of 0.05 where $f_{Ag^+} = 0.80$ and $f_{CrO_4^{2-}} = 0.44$?

$$(a_{Ag^+})^2 a_{CrO_4^{2-}} = ([Ag^+]f_{Ag^+})^2 [CrO_4^{2-}]f_{CrO_4^{2-}} = K_s$$

$$[Ag^+]^2[CrO_4^{2-}] = \frac{K_{sp}}{(f_{Ag^+})^2 f_{CrO_4^{2-}}} = \frac{1.29 \times 10^{-12}}{(0.80)^2 0.44} = 4.58 \times 10^{-12}$$

$$[Ag^+] = 2S \qquad [CrO_4^{2-}] = S$$

$$(2S)^2 S = 4.58 \times 10^{-12}$$

$$S = \left(\frac{4.58 \times 10^{-12}}{4}\right)^{1/3} = (1.14 \times 10^{-12})^{1/3} = 1.05 \times 10^{-4} \ M$$

Again it is seen that the solubility is significantly higher in the medium with the higher ionic strength.

7.5   Solubility in the Presence of a Common Cation or Anion

Large Excess of Common Ion

Sometimes it is necessary to calculate the solubility of a slightly soluble salt in a medium that already contains the cation or anion of the salt. As an example consider the solubility of AgCl, $K_{sp} = 1.82 \times 10^{-10}$ in several media. In pure water, the solubility is given very simply by the following:

$$AgCl(s) \rightleftharpoons Ag^+ + Cl^- \tag{7.45}$$

$$S = [Ag^+] = [Cl^-] = \sqrt{K_{sp}} = \sqrt{1.82 \times 10^{-10}} = 1.35 \times 10^{-5} \ M \tag{7.46}$$

To see the effects of an excess of a common ion, we may calculate the solubility of AgCl in a solution initially 1.00×10^{-3} M in Ag^+ because of the presence of dissolved $AgNO_3$. In this case the concentration of Ag^+ already present is about 100 times what would be obtained by the dissolution of AgCl in pure water; the latter is in turn suppressed by the Ag^+ already present. Therefore, it may be assumed that at equilibrium

$$[Ag^+] = [Ag^+]_{AgNO_3} + [Ag^+]_{AgCl} = 1.00 \times 10^{-3} \ M + \cancel{S} = 1.00 \times 10^{-3} \ M \tag{7.47}$$

$$[Cl^-] = S \tag{7.48}$$

$$[Ag^+][Cl^-] = (1.00 \times 10^{-3} \ M + S) \, S = K_{sp} \tag{7.49}$$

$$S = \frac{K_{sp}}{1.00 \times 10^{-3}} = \frac{1.82 \times 10^{-10}}{1.00 \times 10^{-3}} = 1.82 \times 10^{-7} \ M \tag{7.50}$$

Common Ion Concentration Close to Solubility

Next consider the solubility of AgCl in a medium that is initially 8.00×10^{-6} M in Ag^+. This value is too large to ignore relative to the contribution of dissolved AgCl to $[Ag^+]$, but too small to consider as the sole contributor to $[Ag^+]$. In this case the calculations below give the value of S.

$$[Ag^+] = [Ag^+]_{AgNO_3} + [Ag^+]_{AgCl} = 8.00 \times 10^{-6} \ M + S \qquad (7.51)$$

$$[Cl^-] = S \qquad (7.48)$$

$$[Ag^+][Cl^-] = K_{sp} = (8.00 \times 10^{-6} + S)S = 1.82 \times 10^{-10} \qquad (7.52)$$

$$S^2 + 8.00 \times 10^{-6}S - 1.82 \times 10^{-10} = 0 \qquad (7.53)$$

$$S = 1.01 \times 10^{-5} \ M \qquad (7.54)$$

Here it is seen that both the Ag^+ originally present (8.00×10^{-6} M) and that contributed from AgCl (1.01×10^{-5} M) are of comparable magnitude, and neither may be neglected. Therefore it is necessary to solve a quadratic equation to get the correct answer to this problem.

EXAMPLE

Calculate S of AgCl in a medium that is initially 7.50×10^{-6} M in Ag^+ and 6.00×10^{-6} M in Cl^-.

Before addition of AgCl,

$$[Ag^+][Cl^-] = 7.50 \times 10^{-6} \times 6.00 \times 10^{-6} = 4.5 \times 10^{-11} < K_{sp}$$

The solubility product has not been exceeded. After equilibration with AgCl, the concentrations of Ag^+ and Cl^- are given by the following:

$$[Ag^+] = 7.50 \times 10^{-6} + S \qquad [Cl^-] = 6.00 \times 10^{-6} + S$$

$$(7.50 \times 10^{-6} + S)(6.00 \times 10^{-6} + S) = K_{sp} = 1.82 \times 10^{-10}$$

$$S^2 + 1.35 \times 10^{-5}S - 1.37 \times 10^{-10} = 0$$

$$S = 6.71 \times 10^{-6} \ M$$

A somewhat more challenging problem is the calculation of the solubility of Ag_2CrO_4 in a solution of $AgNO_3$ that is initially 5.00×10^{-5} M in Ag^+. This concentration is of comparable magnitude to the value of 1.37×10^{-4} M Ag^+ in a solution of pure water saturated with Ag_2CrO_4 discussed in Section 7.3 ($S = 6.86 \times 10^{-5}$ M). An equation solvable for S is derived as follows:

$$Ag_2CrO_4(s) \rightleftharpoons 2Ag^+ + CrO_4^{2-} \qquad (7.55)$$

$$[Ag^+] = [Ag^+]_{AgNO_3} + [Ag^+]_{Ag_2CrO_4} = 5.00 \times 10^{-5} + 2S \qquad (7.56)$$

$$[CrO_4^{2-}] = S$$

$$[Ag^+]^2[CrO_4^{2-}] = K_{sp} = (5.00 \times 10^{-5} + 2S)^2 S = 1.29 \times 10^{-12} \qquad (7.57)$$

$$S^3 + 5.00 \times 10^{-5} S^2 + 6.25 \times 10^{-10} S - 3.23 \times 10^{-13} = 0 \qquad (7.58)$$

The value of S can be obtained by Newton's approximation method (see Appendix). To do this, simply set Equation 7.58 equal to $f(S)$, giving

$$f'(S) = S^3 + 5.00 \times 10^{-5} S^2 + 6.25 \times 10^{-10} S - 3.23 \times 10^{-13} \qquad (7.59)$$

and take the first derivative of $f(S)$ to give the following:

$$f'(S) = 3S^2 + 1.00 \times 10^{-4} S + 6.25 \times 10^{-10} \qquad (7.60)$$

According to Newton's approximation, if S_0 is somewhat close to the solution of S, S_1 is even closer, where

$$S_1 = S_0 - \frac{f(S_0)}{f'(S_0)} \qquad (7.61)$$

By a method of successive approximations, when a value of S is chosen such that $S_1 = S_0$—that is, $f(S_0) = 0$—the polynomial is solved. The initial value of S should be chosen such that it is relatively close to the correct value. At this point it may be recalled that S in pure water was calculated to be $6.86 \times 10^{-5} M$; this should be considered a maximum value of S because the presence of Ag^+ from a source other than Ag_2CrO_4 decreases the solubility. If S were $6.86 \times 10^{-5} M$, in a solution initially $5.00 \times 10^{-5} M$ in Ag^+, the value of $[Ag^+]$ after equilibration with solid Ag_2CrO_4 would be the following, calculated from Equation 7.56:

$$[Ag^+] = 5.00 \times 10^{-5} + 2 \times 6.86 \times 10^{-5} = 1.87 \times 10^{-4} M \qquad (7.62)$$

Substituting this value into the K_{sp} expression gives

$$S = [CrO_4^{2-}] = \frac{K_{sp}}{[Ag^+]^2} = \frac{1.29 \times 10^{-12}}{(1.87 \times 10^{-4})^2} = 3.69 \times 10^{-5} \qquad (7.63)$$

It is readily reasoned that S would not be this low, because the value of $[Ag^+]$ employed in calculating it was too high; therefore, the minimum value of S, S_{min}, is $3.69 \times 10^{-5} M$. A logical starting value of S to use in Newton's approximation is

$$S_0 = \frac{S_{max} + S_{min}}{2} = \frac{6.86 \times 10^{-5} + 3.69 \times 10^{-5}}{2} = 5.28 \times 10^{-5} M \qquad (7.64)$$

Using this value for the initial "guess" in Newton's approximation leads in one iteration to the correct value of $S = 5.30 \times 10^{-5} M$. The value of $5.30 \times 10^{-5} M$ for S may be checked by substituting into the expression for K_{sp} as follows:

$$[Ag^+] = 5.00 \times 10^{-5} + 2S = 5.00 \times 10^{-5} + 2 \times 5.30 \times 10^{-5} = 1.56 \times 10^{-4} \qquad (7.65)$$

$$[CrO_4^{2-}] = S = 5.30 \times 10^{-5} \qquad (7.66)$$

$$[Ag^+]^2[CrO_4^{2-}] = (1.56 \times 10^{-4})^2 \times 5.30 \times 10^{-5} = 1.29 \times 10^{-12} = K_{sp} \qquad (7.67)$$

7.6   Anion Reactions Affecting Solubility

Most of the anions of slightly soluble salts are conjugate bases of weak acids (see Section 4.1). When the pH of the medium is substantially less than pK_a of the conjugate acid of the salt anion, the anion goes into solution in a protonated form. This shifts the dissolution reaction to the right and raises the solubility of the salt. This phenomenon can best be illustrated by an example. Consider the solubility of $CaCO_3$ in a solution maintained at pH 8.50. The pertinent equilibrium constants are

$$K_{sp} = [Ca^{2+}][CO_3^{2-}] = 4.47 \times 10^{-9} \tag{7.68}$$

$$K_{a1} = \frac{[H^+][HCO_3^-]}{[CO_2(aq)]} = 4.45 \times 10^{-7} \qquad pK_{a1} = 6.35 \tag{7.69}$$

$$K_{a2} = \frac{[H^+][CO_3^{2-}]}{[HCO_3^-]} = 4.69 \times 10^{-11} \qquad pK_{a2} = 10.33 \tag{7.70}$$

In this case the pH is less than pK_{a2} and greater than pK_{a1}, so that virtually all the CO_3^{2-} ion goes into solution as HCO_3^-. It can be seen from the distribution of species diagram for the $CO_2(aq)-HCO_3^--CO_3^{2-}$ system (Figure 4.1) that at pH 8.50, HCO_3^- is the predominant carbonate species in solution. Therefore, the major reaction by which $CaCO_3$ goes into solution is

$$CaCO_3(s) + H^+ \rightleftarrows Ca^{2+} + HCO_3^- \tag{7.71}$$

This reaction is the sum of the two reactions below, and its equilibrium constant is the product of the equilibrium constants of these reactions.

$$CaCO_3(s) \rightleftarrows Ca^{2+} + \cancel{CO_3^{2-}} \tag{7.72}$$

$$K_{sp} = [Ca^{2+}][CO_3^{2-}] = 4.47 \times 10^{-9} \tag{7.73}$$

$$\cancel{CO_3^{2-}} + H^+ \rightleftarrows HCO_3^- \tag{7.74}$$

$$\frac{1}{K_{a2}} = \frac{[HCO_3^-]}{[H^+][CO_3^{2-}]} = \frac{1}{4.69 \times 10^{-11}} \tag{7.75}$$

$$CaCO_3(s) + H^+ \rightleftarrows Ca^{2+} + HCO_3^- \tag{7.76}$$

$$K = \frac{[Ca^{2+}][HCO_3^-]}{[H^+]} = \frac{K_{sp}}{K_{a2}} = \frac{4.47 \times 10^{-9}}{4.69 \times 10^{-11}} = 95.3 \tag{7.77}$$

At pH 8.50, $[H^+] = 10^{-8.50} = 3.16 \times 10^{-9}$. From the stoichiometry of Reaction 7.71, $[Ca^{2+}] = [HCO_3^-] = S$. Substituting these values into the expression for K gives

$$K = \frac{[Ca^{2+}][HCO_3^-]}{[H^+]} = \frac{S \times S}{3.16 \times 10^{-9}} = 95.3 \tag{7.78}$$

$$S = \sqrt{3.16 \times 10^{-9} \times 95.3} = \sqrt{3.01 \times 10^{-7}} = 5.49 \times 10^{-4}\ M \tag{7.79}$$

EXAMPLE

What is the solubility of AgCN, $K_{sp} = 7.2 \times 10^{-11}$, in water buffered at pH 6.80? K_a of HCN is 2.1×10^{-9}; $pK_a = 8.68$.

Since $pH \ll pK_a$, CN^- goes into solution as $HCN(aq)$.

$$AgCN(s) \rightleftharpoons Ag^+ + \cancel{CN^-} \qquad K_{sp} = [Ag^+][CN^-] = 7.2 \times 10^{-11}$$

$$H^+ + \cancel{CN^-} \rightleftharpoons HCN(aq) \qquad \frac{1}{K_a} = \frac{[HCN]}{[H^+][CN^-]} = \frac{1}{2.1 \times 10^{-9}}$$

$$\overline{AgCN(s) + H^+ \rightleftharpoons Ag^+ + HCN(aq)}$$

$$K = \frac{[Ag^+][HCN(aq)]}{[H^+]} = \frac{K_{sp}}{K_a} \quad \frac{7.2 \times 10^{-11}}{2.1 \times 10^{-9}} = 3.43 \times 10^{-2}$$

$$[Ag^+] = [HCN(aq)] = S \qquad [H^+] = 10^{-6.80} = 1.58 \times 10^{-7}\ M$$

$$\frac{[Ag^+][HCN(aq)]}{[H^+]} = \frac{S \times S}{1.58 \times 10^{-7}} = K = 3.43 \times 10^{-2}$$

$$S = \sqrt{1.58 \times 10^{-7} \times 3.43 \times 10^{-2}}$$

$$= \sqrt{5.42 \times 10^{-9}} = 7.4 \times 10^{-5}\ M$$

7.7   **Cation Reactions Affecting Solubility**

Formation of Complex Ions by Cations

Just as anions of a slightly soluble salt may be bound to H^+ ions, increasing the solubility of the salt, cations may be bound to chemical species in solution to increase the solubility. As discussed in more detail in Chapter 10, the species that react with cations are **complexing agents**, or **ligands**. These are electron-pair donor Lewis bases, either neutral molecules or anions, that react with metal ions to form **complex ions**. Most ligands are conjugate bases of weak acids and therefore bond to H^+ at lower pH values. Ammonia, NH_3,

$$\text{Unshared electron pair} \rightarrow \overset{\displaystyle H}{\underset{\displaystyle H}{:N-H}}$$

is a typical ligand with a great deal of importance in analytical chemistry. In water, ammonia undergoes the following reaction with silver ion, the equilibrium constant for which is designated by β_2.

$$Ag^+ + 2NH_3 \rightleftarrows Ag(NH_3)_2^+ \tag{7.80}$$

$$\beta_2 = \frac{[Ag(NH_3)_2^+]}{[Ag^+][NH_3]^2} = 1.59 \times 10^7 \tag{7.81}$$

Solubility of AgI in NH₃ Solution

Because of reaction 7.80, the presence of ammonia increases the solubility of silver salts. To see how this is so, consider the solubility of AgI, $K_{sp} = 1.00 \times 10^{-16}$. In pure water the solubility of this salt is given simply by the following:

$$AgI(s) \rightleftarrows Ag^+ + I^- \tag{7.82}$$

$$[Ag^+] = [I^-] = S \tag{7.83}$$

$$[Ag^+][I^-] = S \times S = K_{sp} = 1.00 \times 10^{-16} \tag{7.84}$$

$$S = \sqrt{1.00 \times 10^{-16}} = 1.00 \times 10^{-8} \ M \tag{7.85}$$

Next consider the solubility of AgI in water that is 0.750 M in NH₃. One might first ask whether silver ion in such a medium would be present as Ag^+ or $Ag(NH_3)_2^+$. That is readily answered by using the β_2 expression (Equation 7.81) to calculate the ratio $[Ag(NH_3)_2^+]/[Ag^+]$ in 0.750 M NH₃.

$$\frac{[Ag(NH_3)_2^+]}{[Ag^+]} = \beta_2[NH_3]^2 = 1.59 \times 10^7 (0.750)^2 = 8.94 \times 10^6 \tag{7.86}$$

These results show that essentially all silver in solution is in the complexed form. Therefore, AgI goes into solution according to the reaction

$$AgI(s) + 2NH_3 \rightleftarrows Ag(NH_3)_2^+ + I^- \tag{7.87}$$

which can be obtained by adding the two following reactions:

$$AgI(s) \rightleftarrows \cancel{Ag^+} + I^- \tag{7.82}$$

$$K_{sp} = [Ag^+][I^-] = 1.00 \times 10^{-16} \tag{7.84}$$

$$\cancel{Ag^+} + 2NH_3 \rightleftarrows Ag(NH_3)_2^+ \tag{7.80}$$

$$\beta_2 = \frac{[Ag(NH_3)_2^+]}{[Ag^+][NH_3]^2} = 1.59 \times 10^7 \tag{7.81}$$

$$\overline{AgI(s) + 2NH_3 \rightleftarrows Ag(NH_3)_2^+ + I^-} \tag{7.88}$$

$$K = \frac{[Ag(NH_3)_2^+][I^-]}{[NH_3]^2} = K_{sp}\beta_2 = 1.59 \times 10^{-9} \tag{7.89}$$

The value of S is obtained by the following calculation:

$$[Ag(NH_3)_2^+] = [I^-] = S \tag{7.90}$$

$$\frac{[Ag(NH_3)_2^+][I^-]}{[NH_3]^2} = \frac{S \times S}{(0.750)^2} = K = 1.59 \times 10^{-9} \tag{7.91}$$

$$S = \sqrt{(0.750)^2 \times 1.59 \times 10^{-9}} = 2.99 \times 10^{-5} \ M \tag{7.92}$$

From this calculation it is seen that the presence of $0.750 \ M$ NH$_3$ increases the solubility of AgI by more than three orders of magnitude compared to its solubility in water.

Solubility of AgCl in NH$_3$ Solution

Next consider the somewhat more complicated example of the solubility of AgCl, $K_{sp} = 1.82 \times 10^{-10}$, in a solution *initially* $5.00 \ M$ in NH$_3$. Here the solubility product of AgCl is sufficiently high that an appreciable fraction of the NH$_3$ is bound as Ag(NH$_3$)$_2^+$, and [NH$_3$] is less than $5.00 \ M$ at equilibrium. The calculation is the following:

$$AgCl(s) \rightleftarrows Ag^+ + Cl^- \tag{7.93}$$

$$K_{sp} = [Ag^+][Cl^-] = 1.82 \times 10^{-10} \tag{7.94}$$

$$Ag^+ + 2NH_3 \rightleftarrows Ag(NH_3)_2^+ \tag{7.80}$$

$$\beta_2 = \frac{[Ag(NH_3)_2^+]}{[Ag^+][NH_3]^2} = 1.59 \times 10^7 \tag{7.81}$$

$$AgCl(s) + 2NH_3 \rightleftarrows Ag(NH_3)_2^+ + Cl^- \tag{7.95}$$

$$K = \frac{[Ag(NH_3)_2^+][Cl^-]}{[NH_3]^2} = K_{sp}\beta_2 \tag{7.96}$$

$$= 2.89 \times 10^{-3} \tag{7.97}$$

$$[Ag(NH_3)_2^+] = [Cl^-] = S \tag{7.98}$$

$$[NH_3] = 5.00 - 2S \tag{7.99}$$

$$\frac{[Ag(NH_3)_2^+][Cl^-]}{[NH_3]^2} = \frac{S \times S}{(5.00 - 2S)^2} = K \tag{7.100}$$

Taking the square root of both sides yields

$$\frac{S}{5.00 - 2S} = \sqrt{K} = \sqrt{2.89 \times 10^{-3}} = 5.38 \times 10^{-2} \tag{7.101}$$

$$S = 0.243 \ M \tag{7.102}$$

At equilibrium

$$[NH_3] = 5.00 - 2S = 5.00 - 2 \times 0.243 = 4.51 \ M \tag{7.103}$$

EXAMPLE

Calculate the solubility of AgBr, $K_{sp} = 5.2 \times 10^{-13}$, in a solution that, at equilibrium, contains $0.100 \ M$ NH$_3$.

$$AgBr(s) + 2NH_3 \rightleftharpoons Ag(NH_3)_2^+ + Br^-$$

$$K = \frac{[Ag(NH_3)_2^+][Br^-]}{[NH_3]^2} = K_{sp}\beta_2$$

$$= 5.2 \times 10^{-13} \times 1.59 \times 10^7 = 8.3 \times 10^{-6}$$

$$[Ag(NH_3)_2^+] = [Br^-] = S$$

$$\frac{[Ag(NH_3)_2^+][Br^-]}{[NH_3]^2} = \frac{S \times S}{(0.100)^2} = 8.3 \times 10^{-6}$$

$$S = \sqrt{(0.100)^2 \times 8.3 \times 10^{-6}} = 2.9 \times 10^{-4} \ M$$

How much higher is S than the solubility of AgBr in pure water?

7.8   Complexes of Slightly Soluble Salt with Salt Anion

Complexes of Ag(I) by Cl$^-$

In many cases metal ions in a slightly soluble compound form complex ions with the anion of the compound. This is especially true in the case of some metal hydroxides, which may be redissolved in excess base by reactions of the type

$$M(OH)_2(s) + OH^- \rightleftharpoons M(OH)_3^- \ (aq) \tag{7.104}$$

A classic example of increased solubility of a compound by addition of excess anion is provided by AgCl(s). In solution, this compound forms the soluble species Ag^+, Cl^-, AgCl(aq), and $AgCl_2^-$ (aq). It also forms $AgCl_3^{2-}$, but this species is normally not a significant fraction of the soluble species for $[Cl^-] < 0.1$ and will not be considered in these calculations. The study of this system requires consideration of the following equilibrium reactions and their corresponding equilibrium constants:

$$AgCl(s) \rightleftharpoons Ag^+ + Cl^- \qquad K_{sp} = [Ag^+][Cl^-] = 1.82 \times 10^{-10} \tag{7.105}$$

$$Ag^+ + Cl^- \rightleftharpoons AgCl(aq) \qquad K_1 = \frac{[AgCl(aq)]}{[Ag^+][Cl^-]} = 2.04 \times 10^3 \tag{7.106}$$

$$AgCl(aq) + Cl^- \rightleftharpoons AgCl_2^- \qquad K_2 = \frac{[AgCl_2^-]}{[AgCl(aq)][Cl^-]} = 93 \tag{7.107}$$

Considering the above, the total solubility S of AgCl(s) in a medium containing a particular chloride ion concentration at equilibrium is

$$S = [Ag^+] + [AgCl(aq)] + [AgCl_2^-] \tag{7.108}$$

In terms of equilibrium constants and $[Cl^-]$, this expression becomes

$$S = \frac{K_{sp}}{[Cl^-]} + K_1[Ag^+][Cl^-] + K_2[AgCl(aq)][Cl^-] \tag{7.109}$$

Substituting K_{sp} for $[Ag^+][Cl^-]$ in the second term and $K_{sp}K_1$ for $[AgCl(aq)]$ in the third term on the right side of the equation yields

$$S = \frac{K_{sp}}{[Cl^-]} + K_{sp}K_1 + K_{sp}K_1K_2[Cl^-] \tag{7.110}$$

With numerical values substituted for the equilibrium constants, the preceding equation becomes

$$S = \frac{1.82 \times 10^{-10}}{[Cl^-]} + 3.71 \times 10^{-7} + 3.45 \times 10^{-5}[Cl^-] \tag{7.111}$$

Solubility of AgCl(s) in Excess Cl$^-$

Recall that the solubility of AgCl in water calculated on the basis of the solubility product of AgCl is

$$[Ag^+] = [Cl^-] = S = \sqrt{K_{sp}} = \sqrt{1.82 \times 10^{-10}} = 1.35 \times 10^{-5} \ M \tag{7.112}$$

With this value in mind we may calculate S in media containing various concentrations of Cl^-, considering the formation of AgCl(aq) and $AgCl_2^-$. Equation 7.111 can be used to calculate S of AgCl in a medium that is $1.00 \times 10^{-3} \ M$ in Cl^- as follows:

$$S = \frac{1.82 \times 10^{-10}}{1.00 \times 10^{-3}} + 3.71 \times 10^{-7} + 3.45 \times 10^{-5} \times 1.00 \times 10^{-3} \tag{7.113}$$

$$= 5.88 \times 10^{-7} \ M \tag{7.114}$$

If $[Cl^-]$ is $5.00 \times 10^{-2} \ M$, S is

$$S = \frac{1.82 \times 10^{-10}}{5.00 \times 10^{-2}} + 3.71 \times 10^{-7} + 3.45 \times 10^{-5} \times 5.00 \times 10^{-2} \tag{7.115}$$

$$= 2.10 \times 10^{-6} \tag{7.116}$$

The two preceding calculations show that the solubility of AgCl is actually higher in $5.00 \times 10^{-2} \ M$ in Cl^- than in $1.00 \times 10^{-3} \ M \ Cl^-$.

EXAMPLE

What is S of $AgCl(s)$ in pure water considering the contribution of $AgCl(aq)$?

$$[Ag^+] = [Cl^-] = \sqrt{K_{sp}} = 1.35 \times 10^{-5} \ M$$

The second term on the right of Equation 7.111 gives the contribution to S from $[AgCl(aq)]$, a constant value of $3.71 \times 10^{-7} \ M$. Thus, total S is given by

$$S = [Ag^+] + [AgCl(aq)] = 1.35 \times 10^{-5} + 0.04 \times 10^{-5} = 1.39 \times 10^{-5}$$

The third term on the right of Equation 7.111 gives the contribution of $AgCl_2^-$ to the solubility of $AgCl$. In this case the contribution is negligible as shown by the following calculation:

$$[AgCl_2^-] = 3.45 \times 10^{-5}[Cl^-] = 3.45 \times 10^{-5} \times 1.35 \times 10^{-5} = 4.66 \times 10^{-10}$$

Minimum Solubility of AgCl

The value of $[Cl^-]$ for the minimum solubility of $AgCl$ is found by calculating $dS/d[Cl^-]$ from Equation 7.111 and setting the derivative equal to zero:

$$\frac{dS}{d[Cl^-]} = -\frac{1.82 \times 10^{-10}}{[Cl^-]^2} + 3.45 \times 10^{-5} = 0 \tag{7.117}$$

$$[Cl^-] = \sqrt{\frac{1.82 \times 10^{-10}}{3.45 \times 10^{-5}}} = 2.30 \times 10^{-3} \ M \tag{7.118}$$

At this value of $[Cl^-]$, the concentrations of each of the silver-containing soluble species and their sum S are given by substitution into Equation 7.111:

$$S = \frac{1.82 \times 10^{-10}}{2.30 \times 10^{-3}} + 3.71 \times 10^{-7} + 3.45 \times 10^{-5} \times 2.30 \times 10^{-3} \tag{7.119}$$

$$= \underbrace{0.79 \times 10^{-7}}_{[Ag^+]} + \underbrace{3.71 \times 10^{-7}}_{[AgCl(aq)]} + \underbrace{0.79 \times 10^{-7}}_{[AgCl_2^-]} = 5.29 \times 10^{-7} \ M \tag{7.120}$$

It should be noted that at the point of minimum S, $AgCl(aq)$ is the predominant silver-containing soluble species and the values of $[Ag^+]$ and $[AgCl_2^-]$ are equal.

Logarithmic Concentration Diagram of the Ag(I)–Cl⁻ System

Figure 7.2 contains plots of log $[Ag^+]$, log $[AgCl(aq)]$, and log $[AgCl_2^-]$ versus log $[Cl^-]$. Such a plot is one form of a **logarithmic concentration diagram** and is useful for showing concentrations over many orders of magnitude of several species on the same plot. Furthermore, as explained in the figure caption, a logarithmic concentration diagram shows significant points and regions pertaining to the system that it represents.

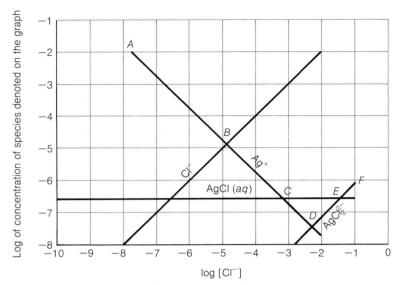

FIGURE 7.2 Logarithmic concentration diagram for the $Cl^- - Ag^+ - AgCl(aq)-$ $AgCl_2^-$ system. Line AB shows excess Ag^+; point B denotes $AgCl(s)$ equilibrated with pure water; along line BCD solubility decreases because $[Ag^+]$ decreases with added Cl^-; in the region CDE, $AgCl(aq)$ is the predominant soluble $Ag(I)$ species; at point D the solubility of $AgCl$ is at a minimum value; along line DEF the solubility increases due to the formation of $AgCl_2^-$; and along line EF, $AgCl_2^-$ is the predominant $Ag(I)$ species.

7.9   Metal Hydroxide Solubilities

The Nature of Metal Hydroxides

Metal hydroxides make up a large general class of slightly soluble compounds that are important in analytical chemistry. In some cases analytes are precipitated as hydroxides for further analysis, or interfering metals may be precipitated as hydroxides. The formation of metal hydroxides may interfere with a chemical analysis. For example, for the titration of metal ions with chelating agents (see Chapter 10), the pH must often be so high that the metal ion analyte would precipitate as a hydroxide if measures were not taken to prevent it.

It should be noted that the chemistry of metal ions in basic solution is often quite complex. The formation of pure metal hydroxides is difficult to achieve, and **basic salts** containing OH^- along with salt anions often form when base is added to a solution of a salt of a metal ion. Typical of such basic salts is basic copper(II) sulfate, $CuSO_4 \cdot 3Cu(OH)_2$. Most metals other than those in Group 1A form hydroxy complexes, such as $FeOH^{2+}$ and $Fe(OH)_2^+$ formed by iron(III). Two or more metal ions in solution may be joined together by hydroxide ion bridging groups to form species such as

In this chapter metal hydroxide solubilities will be treated in a simplified fashion, but it is well to keep the complexities mentioned above in mind for those who may have to deal with such compounds later.

Solubility of Magnesium Hydroxide

As an example of metal hydroxide solubility, consider that of $Mg(OH)_2$, $K_{sp} = 1.8 \times 10^{-11}$. Treating this like any other K_{sp} calculation, the following results are obtained:

$$Mg(OH)_2(s) \rightleftarrows Mg^{2+} + 2OH^- \tag{7.121}$$

$$[Mg^{2+}] = S \qquad [OH^-] = 2S \tag{7.122}$$

$$[Mg^{2+}][OH^-]^2 = K_{sp} = 1.8 \times 10^{-11} = S(2S)^2 \tag{7.123}$$

$$S = \left(\frac{1.8 \times 10^{-11}}{4}\right)^{1/3} = 1.6 \times 10^{-4} \ M \tag{7.124}$$

A possible complication to be considered here is the contribution of the autoionization of water

$$H_2O \rightleftarrows H^+ + OH^-$$

$$K_w = [H^+][OH^-] = 1.00 \times 10^{-14} \tag{7.125}$$

to the OH^- concentration. After the solid $Mg(OH)_2$ has equilibrated with water, the total OH^- concentration is the sum

$$[OH^-] = [OH^-]_{water} + [OH^-]_{Mg(OH)_2} \tag{7.126}$$

From the autoionization of water it is seen that the contribution to OH^- from this source is equal to $[H^+]$. If $Mg(OH)_2$ is the only significant contributor to $[OH^-]$, then $[OH^-] = 1.6 \times 10^{-4} \ M$ (Equation 7.124), and

$$[OH^-]_{water} = [H^+] = \frac{K_w}{[OH^-]} = \frac{1.00 \times 10^{-14}}{1.6 \times 10^{-4}} = 6.3 \times 10^{-11} \ M \tag{7.127}$$

Clearly, in this case

$$[OH^-]_{water} \ll [OH^-]_{Mg(OH)_2} \tag{7.128}$$

so that the contribution to the OH^- ion concentration from water can be ignored.

Contribution of Water to OH$^-$ Concentration

As another example of metal hydroxide solubility, consider the hypothetical solid $M(OH)_2$, with a K_{sp} value of 1.00×10^{-21}. First, its solubility in pure water can be calculated ignoring the contribution of water to [OH$^-$]:

$$M(OH)_2(s) \rightleftharpoons M^{2+} + 2OH^- \qquad (7.129)$$

$$[M^{2+}] = S \qquad [OH^-] = 2S \qquad (7.130)$$

$$K_{sp} = [M^{2+}][OH^-]^2 = S(2S)^2 = 1.00 \times 10^{-21} \qquad (7.131)$$

$$S = \left(\frac{1.00 \times 10^{-21}}{4}\right)^{1/3} = 6.30 \times 10^{-8} \ M = [M^{2+}] \qquad (7.132)$$

$$[OH^-] = 2S = 2 \times 6.30 \times 10^{-8} = 1.26 \times 10^{-7} \ M \qquad (7.133)$$

In this case it is seen that the value of [OH$^-$] from the dissolution of the salt is of the same order of magnitude as that from the autoionization of pure water ($1.00 \times 10^{-7} \ M$); the latter must definitely be considered as follows:

$$M(OH)_2(s) \rightleftharpoons M^{2+} + 2OH^- \qquad (7.129)$$

$$[M^{2+}] = S \qquad (7.134)$$

$$[OH^-] = [OH^-]_{M(OH)_2} + [OH^-]_{H_2O} \qquad (7.135)$$

$$H_2O \rightleftharpoons H^+ + OH^-$$

$$[OH^-]_{H_2O} = [H^+]_{H_2O} = \frac{K_w}{[OH^-]} = \frac{1.00 \times 10^{-14}}{[OH^-]} \qquad (7.136)$$

$$[OH^-] = [OH^-]_{M(OH)_2} + [OH^-]_{H_2O} = 2S + \frac{1.00 \times 10^{-14}}{[OH^-]} \qquad (7.137)$$

$$K_{sp} = [M^{2+}][OH^-]^2 = S[OH^-]^2 \qquad (7.138)$$

$$S = \frac{K_{sp}}{[OH^-]^2} \qquad (7.139)$$

Substituting Equation 7.139 into Equation 7.137 leads to

$$[OH^-] = 2\frac{K_{sp}}{[OH^-]^2} + \frac{1.00 \times 10^{-14}}{[OH^-]} \qquad (7.140)$$

which rearranges to the following:

$$[OH^-]^3 - 1.00 \times 10^{-14}[OH^-] - 2K_{sp} = 0 \qquad (7.141)$$

This equation can be solved using Newton's approximation (see Section 7.5). As an initial value of [OH$^-$], the value obtained assuming no contribution from water, $1.26 \times 10^{-7} \ M$ (see Equation 7.133), can be used. Several iterations give the correct value of $1.52 \times 10^{-7} \ M$ for [OH$^-$], and substitution into Equation 7.139 gives the following value of S:

$$S = \frac{K_{sp}}{[OH^-]^2} = \frac{1.00 \times 10^{-21}}{(1.52 \times 10^{-7})^2} = 4.34 \times 10^{-8} \ M \qquad (7.142)$$

These results may be checked by substituting for S and $[OH^-]$ on the right side of Equation 7.137.

$$2S + \frac{1.00 \times 10^{-14}}{[OH^-]} = 2 \times 4.34 \times 10^{-8} + \frac{1.00 \times 10^{-14}}{1.52 \times 10^{-7}} = 1.52 \times 10^{-7} = [OH^-] \qquad (7.143)$$

which gives the same value of $[OH^-]$.

EXAMPLE

What is the solubility of $Zn(OH)_2$, $K_{sp} = 1.2 \times 10^{-17}$, in a solution buffered at pH 8.30?

$$pOH = 14.00 - 8.30 = 5.70$$

$$[OH^-] = 10^{-5.70} = 2.00 \times 10^{-6}$$

$$Zn(OH)_2(s) \rightleftarrows Zn^{2+} + 2OH^-$$

$$[Zn^{2+}] = S$$

$$S = [Zn^{2+}] = \frac{K_{sp}}{[OH^-]^2} = \frac{1.2 \times 10^{-17}}{(2.00 \times 10^{-6})^2} = 3.0 \times 10^{-6} \ M$$

Some metal hydroxides have increased solubilities at elevated OH^- concentrations because of reactions of the type

$$M(OH)_2(s) + xOH^- \rightleftarrows M(OH)_{2+x}(aq)^{-x} \qquad (7.144)$$

in which the hydroxide ion acts both as the anion of the slightly soluble metal hydroxide and as a complexing agent in charged soluble complex ions formed from the metal hydroxide. An example is provided by $Pb(OH)_2$, $K_{sp} = 2.5 \times 10^{-16}$. The solubility of this compound calculated from its K_{sp} is $4.0 \times 10^{-6} \ M$. In the presence of excess OH^-, however, solid $Pb(OH)_2$ reacts as follows:

$$Pb(OH)_2(s) + OH^- \rightleftarrows Pb(OH)_3^- \qquad (7.145)$$

$$K_{s3} = \frac{[Pb(OH)_3^-]}{[OH^-]} = 5.0 \times 10^{-2} \qquad (7.146)$$

In the symbol for the equilibrium expression above, the subscript s denotes that the *solid* compound is a reactant and that the charged soluble complex ion has three ligands (OH^-) attached. (Recall that complexation of a metal by the anion of one of its slightly soluble compounds was discussed for AgCl in Section 7.8.)

For an example of the influence of excess OH^- on the solubility of solid $Pb(OH)_2$, consider the calculation of the solubility of $Pb(OH)_2$ in a medium in which $[OH^-]$ is 1.00×10^{-2} M at equilibrium. By substitution into the K_{s3} expression, the value of $[Pb(OH)_3^-]$ is given by the following:

$$[Pb(OH)_3^-] = K_{s3}[OH^-] = 5.0 \times 10^{-2} \times 1.00 \times 10^{-2} = 5.0 \times 10^{-4} \ M \qquad (7.147)$$

Since this value is about 100 times greater than the solubility of $Pb(OH)_2$ in pure water (calculated as 4.0×10^{-6} M from K_{sp}), $Pb(OH)_3^-$ is the predominant $Pb(II)$ species in 1.00×10^{-2} M OH^-, and the solubility of $Pb(OH)_2$ in such a medium is 5.0×10^{-4} M.

7.10   Intrinsic Solubility

Definition of Intrinsic Solubility

Often the actual solubility of a slightly soluble compound is much greater than that calculated from the solubility product. As an extreme example, the solubility product of $HgCl_2$ is 2×10^{-14}, and the solubility calculated from K_{sp} alone is 1.7×10^{-5}, whereas the molar solubility of this compound is about 0.25 M. The high solubility of the compound results from its existence in solution as undissociated $HgCl_2$. The solubility of such a neutral species is known as its **intrinsic solubility**. In Section 7.8 another example of intrinsic solubility was seen from the solubility of $AgCl(s)$, part of which was due to the presence in solution of soluble $AgCl(aq)$. It was observed in Figure 7.2 that $[AgCl(aq)]$ was independent of $[Cl^-]$, and that for the point of minimum solubility of $AgCl(s)$, $AgCl(aq)$ was the predominant silver-containing species in solution.

Solubility of CaSO$_4$ Considering Intrinsic Solubility

A good example of intrinsic solubility is provided by calcium sulfate, $CaSO_4$. The dissolution of this compound in water can be viewed as the undissociated salt going into solution

$$CaSO_4(s) \rightleftarrows CaSO_4(aq) \qquad (7.148)$$

$$[CaSO_4(aq)] = K_s = 5.0 \times 10^{-3} \ M \qquad (7.149)$$

where K_s is the intrinsic solubility of $CaSO_4$. The dissociation of $CaSO_4(aq)$ is

$$CaSO_4(aq) \rightleftarrows Ca^{2+} + SO_4^{2-} \qquad (7.150)$$

$$K_d = \frac{[Ca^{2+}][SO_4^{2-}]}{[CaSO_4(aq)]} = 5.2 \times 10^{-3} \qquad (7.151)$$

where K_d is the dissociation constant of $CaSO_4(aq)$. It should be noted that the intrinsic solubility of $CaSO_4$, $[CaSO_4(aq)]$, is constant and does not depend upon the values of $[Ca^{2+}]$ or $[SO_4^{2-}]$.

The total solubility of $CaSO_4$ in water is given by the sum

$$S = [Ca^{2+}] + [CaSO_4(aq)] = [Ca^{2+}] + 5.0 \times 10^{-3} \ M \tag{7.152}$$

For $CaSO_4$ dissolved in initially pure water, Reaction 7.150 shows that

$$[Ca^{2+}] = [SO_4^{2-}] \tag{7.153}$$

Furthermore, the value of $[Ca^{2+}]$ can be calculated from the solubility product expression of $CaSO_4$.

$$[Ca^{2+}][SO_4^{2-}] = K_{sp} = 2.6 \times 10^{-5} \tag{7.154}$$

$$[Ca^{2+}] = \sqrt{K_{sp}} = \sqrt{2.6 \times 10^{-5}} = 5.1 \times 10^{-3} \ M \tag{7.155}$$

$$S = [Ca^{2+}] + 5.0 \times 10^{-3} \ M = 5.1 \times 10^{-3} \ M + 5.0 \times 10^{-3} \ M = 1.01 \times 10^{-2} \ M \tag{7.156}$$

It is seen that in this case the undissociated solute makes up half of the total $CaSO_4$ in solution.

7.11 Separations with Sulfide Precipitates

Acidity of H_2S

Differences in solubility can serve as a means of separating metal ions in solution. Sulfide precipitates have been very extensively investigated in that regard. Sulfide is particularly useful for this purpose because it forms salts of very low solubility with a number of metal ions, and the concentration of the sulfide ion is accurately regulated over several orders of magnitude by protonation. Dissolved H_2S is a weak acid as shown by the following equilibria:

$$H_2S(aq) \rightleftharpoons H^+ + HS^- \qquad K_{a1} = \frac{[H^+][HS^-]}{[H_2S(aq)]} = 5.7 \times 10^{-8} \tag{7.157}$$

$$HS^- \rightleftharpoons H^+ + S^{2-} \qquad K_{a2} = \frac{[H^+][S^{2-}]}{[HS^-]} = 1.2 \times 10^{-15} \tag{7.158}$$

$$\overline{H_2S(aq) \rightleftharpoons 2H^+ + S^{2-} \qquad K_{a1}K_{a2} = \frac{[H^+]^2[S^{2-}]}{[H_2S(aq)]} = 6.8 \times 10^{-23}} \tag{7.159}$$

Metal Separations by Sulfide Precipitation

Sulfide separations are normally carried out in solutions saturated with H_2S gas; for such a solution $[H_2S]$ is approximately 0.1 M. If a concentration of 0.10 M is assumed for $[H_2S(aq)]$, the value of $[S^{2-}]$ as a function of $[H^+]$ is given by

$$[S^{2-}] = \frac{K_{a1}K_{a2}[H_2S(aq)]}{[H^+]^2} = \frac{5.7 \times 10^{-8} \times 1.2 \times 10^{-15} \times 0.10}{[H^+]^2} = \frac{6.8 \times 10^{-24}}{[H^+]^2} \qquad (7.160)$$

As an example of the kind of separation that might be performed with H_2S in a pH-controlled medium, consider the sulfide precipitation of CdS, $K_{sp} = 2 \times 10^{-28}$, in a medium initially 0.010 M in both Cd^{2+} and Zn^{2+}, where K_{sp} of ZnS is 4.5×10^{-24}. The maximum separation will be achieved when $[S^{2-}]$ is raised to a level at which the more soluble salt, ZnS, is just ready to precipitate. This concentration is calculated from the K_{sp} expression for ZnS as

$$[S^{2-}] = \frac{K_{sp}}{[Zn^{2+}]} = \frac{4.5 \times 10^{-24}}{0.010} = 4.5 \times 10^{-22} \qquad (7.161)$$

Assuming that $[H_2S(aq)]$ is maintained at 0.10 M and substituting into Equation 7.160 gives

$$[H^+] = \sqrt{\frac{6.8 \times 10^{-24}}{[S^{2-}]}} = \sqrt{\frac{6.8 \times 10^{-24}}{4.5 \times 10^{-22}}} = 0.12 \ M \qquad (7.162)$$

$$pH = 0.92$$

If the value of $[H^+]$ were raised to exactly 0.12 M, where ZnS is still in solution, the concentration of Cd^{2+} remaining in solution is

$$[Cd^{2+}] = \frac{K_{sp}}{[S^{2-}]} = \frac{2 \times 10^{-28}}{4.5 \times 10^{-22}} = 4 \times 10^{-7} \ M \qquad (7.163)$$

Thus in a solution 0.10 M in $H_2S(aq)$ with $[H^+]$ of 0.12 M, the concentration of Cd^{2+} is reduced about five orders of magnitude from its original value of 0.010 M. In practice, the value of $[H^+]$ would be maintained slightly higher than 0.12 M to prevent the possibility of some precipitation (coprecipitation) of contaminant ZnS, but the separation is still quite feasible.

7.12   Solubilities of Gases

Henry's Law

Gases dissolve reversibly in water according to the equilibrium reaction

$$X(g) \rightleftharpoons X(aq) \qquad (7.164)$$

The solubility of a gas is expressed in terms of **Henry's law**, which states that *at constant temperature the solubility of a gas in a liquid is proportional to the partial pressure of that gas in contact with the liquid.* Henry's law applies only to the simple equilibrium shown in Reaction 7.164 and does not account for solution

TABLE 7.3   Henry's law constants for some gases in water at 25° C

Gas	K, mol $\times$ L^{-1} $\times$ atm^{-1}
O_2	1.28×10^{-3}
CO_2	3.38×10^{-2}
H_2	7.90×10^{-4}
CH_4	1.34×10^{-3}

reactions of the gas, such as

$$NH_3 + H_2O \rightleftharpoons NH_4^+ + OH^- \tag{7.165}$$

$$CO_2 + H_2O \rightleftharpoons H^+ + HCO_3^- \tag{7.166}$$

Such reactions may result in the uptake of vastly more gas by water than is predicted by Henry's law alone.

The mathematical expression of Henry's law is

$$[X(aq)] = KP_X \tag{7.167}$$

where $[X(aq)]$ is the aqueous concentration of gas X, P_X is the partial pressure of the gas, and K is the Henry's law constant. If the concentration of the gas in solution is in units of moles per liter and P_X is in atmospheres, the units of K are moles per liter-atmosphere. The Henry's law constants for several gases at 25° C are given in Table 7.3.

Effects of Water Vapor Pressure and Temperature

In calculating gas solubilities, a correction is required for the vapor pressure of water (0.03126 atm at 25° C) by subtracting it from the total pressure of the gas. This correction is shown in the example below. Water vapor pressures at temperatures other than 25° C are readily obtained from handbooks of chemical and engineering data.

The effects of temperature upon gas solubility may be calculated from the **Clausius–Clapeyron equation**,

$$\log \frac{[X(aq)]_2}{[X(aq)]_1} = \frac{\Delta H}{2.303 R} \left(\frac{1}{T_1} - \frac{1}{T_2} \right) \tag{7.168}$$

where $[X(aq)]_1$ is the gas concentration at absolute temperature T_1, $[X(aq)]_2$ is the gas concentration at absolute temperature T_2, ΔH is the heat of solution in calories per mole, and R is the gas constant with a value of 1.987 cal/mol-deg.

EXAMPLE

Pure oxygen gas is equilibrated with water at a total pressure of 1.000 atm at 25° C. What is the value of $[O_2(aq)]$?

$$P_{O_2} = 1.000 \text{ atm} - P_{H_2O} = 1.000 \text{ atm} - 0.031 \text{ atm} = 0.969 \text{ atm}$$

$$
\begin{aligned}
[O_2(aq)] &= KP_{O_2} \\
&= 1.28 \times 10^{-3} \text{ mol/L-atm} \times 0.969 \text{ atm} \\
&= 1.24 \times 10^{-3} \text{ mol/L}
\end{aligned}
$$

Programmed Summary of Chapter 7

This summary contains in a programmed format the major terms and concepts introduced in this chapter. To get the most out of it, you should fill in the blanks for each question and then check the answers at the end of the book to see if you are correct.

Three major uses of precipitation in quantitative analysis are (1) _____. Of the common chlorides, those that are insoluble are (2) _____. Of the common sulfates, those that are insoluble are (3) _____. For a slightly soluble salt MX_2 with a solubility in pure water of S, the value of K_{sp} in terms of S is $K_{sp} = $ (4) _____. For a slightly soluble salt with the general formula M_2X_3, the expression for K_{sp} in terms of the salt's solubility in water S is $K_{sp} = $ (5) _____. To take account of the influence of electrolytes in solution upon ionic equilibrium, (6) _____ are substituted for concentrations. Activity is the product of concentration times an (7) _____. At constant temperature, equilibrium constants expressed in terms of activity coefficients are independent of the (8) _____ of the medium. Where m_X is equal to the molar concentration of an ion of charge Z_X, the formula for the ionic strength of a solution is $\mu = $ (9) _____. The kind of activity coefficient that can be determined experimentally is the (10) _____. The Debye-Hückel relationship is used for the calculation of (11) _____. The mathematical expression of the Debye-Hückel relationship is (12) _____, where the numerical constants 0.5085 and 3.281×10^{-3} are valid at (13) _____, Z_X is (14) _____, and (15) _____ is the effective (16) _____ in units of (17) _____. Using concentrations in place of activities for the calculation of solubility in pure water from K_{sp} introduces little error because (18) _____. The solubility of a slightly soluble ionic compound is (19) _____ in a high-ionic-strength medium than in water with very low ion concentrations because the (20) _____ of the ions composing the slightly soluble compound are (21) _____ in the higher-ionic-strength medium. When two chemical reactions are added to give a third reaction, their equilibrium constants are (22) _____ to give the equilibrium constant of the third reaction. The solubility of a slightly soluble ionic compound can be increased by reaction of the metal ion with (23) _____ in solution. The solubility of silver salts is higher in NH_3 solutions than in pure water because of the formation of the species (24) _____. Of the three species Ag^+, $AgCl(aq)$, and $AgCl_2^-$, the one whose concentration is independent of $[Cl^-]$ is (25) _____. According to the logarithmic concentration diagram for the $Cl^- - Ag^+ - AgCl(aq) - AgCl_2^-$ system, the predominant silver-containing species at the value of $\log[Cl^-]$ where S of $AgCl(s)$ has its minimum value is (26) _____. The solubility of some metal hydroxides is increased in a basic medium because of the formation of species such as (27) _____ in the case of $Pb(OH)_2$.

Questions

1. What general laboratory procedure is involved in gravimetric analysis?
2. Designate which of the following compounds are soluble (S) and insoluble (I): (a) $AgNO_3$, (b) Na_2S, (c) $BaSO_4$, (d) Hg_2Cl_2, (e) $FeCO_3$, (f) $Mn(OH)_2$, (g) $(NH_4)_2S$, (h) $NaOH$, (i) NH_4Cl, (j) Li_2CO_3.
3. The solubility of the slightly soluble salt M_mX_x is S mmol/mL. What is the expression for K_{sp} in terms of S?
4. Why are slightly soluble salts more soluble in a high-ionic-strength medium than in water?
5. Under what circumstances may increasing the concentration of the anion of a slightly soluble compound *increase* the compound's solubility?
6. What is the expression of Newton's approximation used to solve a polynomial with the general formula $X^3 + bX^2 + cX + d = 0$?
7. Why is the solubility of AgCN increased by lowering the pH of the solvent medium, whereas the solubility of AgCl is not affected?
8. What is meant by the intrinsic solubility of a slightly soluble compound?

Problems

1. Water saturated with silver oxalate contains 62.6 mg/L $Ag_2C_2O_4$, fw = 303.8. What is K_{sp} of $Ag_2C_2O_4$?
2. The solubility of $Ba(IO_3)_2$, fw = 487, is 356 mg/L. What is the value of K_{sp}?
3. The slightly soluble salt M_2Y_3, fw = 432, has a solubility of 78 mg/L. What is the value of K_{sp}?
4. The slightly soluble salt M_2X_3 has a solubility product of 2.68×10^{-29}. What is the salt's solubility in moles per liter?
5. The slightly soluble salt MZ_3 has a solubility product of 8.41×10^{-31}. What is its solubility in moles per liter?
6. The slightly soluble salt Me_3R has a solubility product of 5.71×10^{-28}. What is its solubility in moles per liter?
7. The solubility of Ag_2CrO_4 in pure water is 6.86×10^{-5} M. Using activities for the calculation, determine the solubility of this salt at an ionic strength of 0.005, where f_{Ag^+} is 0.924, and $f_{CrO_4^{2-}}$ is 0.740.

8. What is the solubility of Ag_2CrO_4 at an ionic strength of 0.01, where f_{Ag^+} is 0.898 and $f_{CrO_4^{2-}}$ is 0.660 (see Problem 7)?

9. What is the solubility of Ag_2CrO_4 at an ionic strength of 0.05, where f_{Ag^+} is 0.80 and $f_{CrO_4^{2-}}$ is 0.44 (see Problem 7)?

10. Calculate the solubility of AgI, $K_{sp} = 1.00 \times 10^{-16}$, in a medium initially 3.68×10^{-9} M in Ag^+ and 5.13×10^{-9} M in I^-.

11. What is S of Ag_2CrO_4 in a solution initially 6.45×10^{-5} M in CrO_4^{2-}?

12. What is S of AgCl in a solution initially 8.65×10^{-6} M in Ag^+?

13. When $CaCO_3$ dissolves in pure water, a significant fraction, but by no means all, of the CO_3^{2-} ion hydrolyzes to HCO_3^-. Calculate the exact solubility of $CaCO_3$ in pure water isolated from atmospheric CO_2, taking account of the partial hydrolysis of CO_3^{2-} ion. (An exact solution to this problem is mathematically rather complicated.)

14. Given that K_a of HCN is 2.1×10^{-9}, what is the solubility of AgCN, $K_{sp} = 7.2 \times 10^{-11}$, in water buffered at pH 6.45?

15. What is the activity coefficient of CO_3^{2-} ion at an ionic strength of 0.0200?

16. What is the solubility of AgBr, $K_{sp} = 5.2 \times 10^{-13}$, in 0.200 M NH_3? The value of β_2, the overall formation constant of $Ag(NH_3)_2^+$, is 1.59×10^7.

17. Ag_2CrO_4, $K_{sp} = 1.29 \times 10^{-12}$, was equilibrated with a solution of NH_3. At equilibrium $[NH_3] = 0.0500$ M. What was S of Ag_2CrO_4?

18. What is the solubility of AgI, $K_{sp} = 1.00 \times 10^{-16}$ in 0.465 M NH_3?

19. Considering the formation of $AgCl(aq)$ and $AgCl_2^-$, what is S of AgCl in a solution that has $[Cl^-] = 3.00 \times 10^{-3}$ M at equilibrium?

20. What is the percentage composed by $AgCl(aq)$ of all soluble silver species for a solution in equilibrium with solid AgCl when $[Cl^-]$ is such that the solubility of AgCl has its minimum value?

21. Considering the formation of $AgCl(aq)$ and $AgCl_2^-$, what is the solubility of AgCl in 1.00×10^{-2} M Cl^-?

22. The solubility product of MA is 3.75×10^{-13} and K_a of HA is 4.60×10^{-6}. The solubility S in moles per liter of MA in a solution buffered at pH 3.00 is (a) 2.89×10^{-4}, (b) 9.03×10^{-6} (c) 5.61×10^{-4}, (d) 8.13×10^{-4}, (e) 6.12×10^{-7}?

8 Analysis Based upon Solubility: Gravimetric Analysis

LEARNING GOALS

1. Characteristics of precipitates suitable for gravimetric analysis.
2. The process of precipitate formation.
3. Colloidal precipitates, the processes by which they are formed, and the terms pertaining to them.
4. Crystalline precipitates, the processes by which they are formed, and the major terms used to describe their characteristics.
5. Principles and uses of precipitation from homogeneous solution.
6. The processes, particularly ignition, used to convert precipitates to a weighable form.
7. Major inorganic and organic precipitants.
8. Gravimetric analysis by volatilization.
9. Basic steps in gravimetric calculations.
10. Gravimetric factor and how it is used in calculations.

8.1 Introduction

Gravimetric analysis consists of isolating in a pure form the analyte species or, more commonly, a compound formed from the analyte, weighing the isolated species in a pure form, and calculating the percentage of analyte in the known mass of sample from the mass of the isolated species. Although the principles of gravimetry are the simplest of all methods of analysis, obtaining a weighable product of known composition is often a complicated, sophisticated, and time-consuming process. The method does have the advantage that the element or compound weighed at the end of the analytical process can be examined for purity as a check on the adequacy of the analysis.

A number of general methods can be used to isolate an element or a compound derived from it for weighing. The most important of these is the formation of a precipitate, which will be considered in this chapter. Other methods include evolution of a volatile product by heat or a chemical reaction, extraction methods (see Chapter 16), and electrodeposition (see Chapter 15).

This chapter deals with those quantitative analysis techniques that depend upon weighing as the final measure of quantity of analyte. Most gravimetric procedures involve precipitation of a slightly soluble substance from solution as a key element of the process, although other means of isolating a substance that lead to a weighable material may be used.

8.2   Precipitation in Gravimetric Analysis

Precipitate Formation and Ignition

Precipitation (see Chapter 7) consists of reacting the analyte in solution with a reagent to form a substance, called the precipitate, of such low solubility that a negligible amount remains in solution. The medium in which the precipitate is formed is called the **mother liquor**. The precipitate should not retain significant amounts of mother liquor, must be filterable, and be either of known composition or capable of conversion to a species of known composition.

One of the simplest and inherently most accurate gravimetric analyses known is that of Cl^- by the addition of Ag^+ to form solid AgCl by the reaction

$$Ag^+ + Cl^- \rightleftharpoons AgCl(s) \tag{8.1}$$

After the precipitate is formed, it is washed to remove soluble salt impurities, dried at $140 \pm 10°\,C$, cooled, and weighed.

An example of an analysis in which the precipitate is not in a weighable form is that in which iron(III) is precipitated as iron(III) hydroxide, followed by conversion of the hydroxide to a weighable oxide. The precipitation may be accomplished by the addition of NH_3 to an acidic solution of Fe^{3+}, giving the reaction

$$Fe^{3+} + 3NH_3 + (x + 3)H_2O \rightleftharpoons Fe(OH)_3 \cdot xH_2O(s) + 3NH_4^+ \tag{8.2}$$

The gelatinous precipitate actually consists of a mixed oxide-hydroxide of iron(III) with variable amounts of water and is totally unsuited for weighing. In order to get a form suitable for weighing, it is heated to a high temperature ($1000°\,C$), a process called **ignition** in gravimetric analysis. In this case ignition yields Fe_2O_3 as follows:

$$2Fe(OH)_3 \cdot xH_2O \xrightarrow[\text{heat, } 1000°C]{} Fe_2O_3(s) + (2x + 3)H_2O \tag{8.3}$$

The iron(III) oxide is formed stoichiometrically and is suitable for weighing.

Desirable Precipitate Characteristics

The preceding discussion has mentioned several factors that are required in a precipitate suitable for gravimetric analysis. These are summarized in the accompanying box.

Characteristics of a Precipitate Suitable for Gravimetric Analysis

1. Very low solubility to avoid loss of the precipitate with the mother liquor or from washing during filtration. No more than about 0.1% of the precipitate should be lost by this route.
2. No contamination by impurities that cannot be removed by washing or by heating the precipitate after filtration.
3. A solid state that is readily separable from the mother liquor by filtration. The precipitate particles must be large enough to be retained by the filter during removal of the mother liquor and washing, and the particles should not clog the filter medium unduly.
4. It must be possible by physical or chemical means, usually heating at a specified temperature, to convert the precipitate to pure material of well-defined chemical composition.

The principles of solubility of precipitates were discussed in Chapter 7. However, regardless of their solubilities, the physical properties of precipitates vary a great deal. These properties are greatly influenced by the precipitation process and colloidal properties, which will be considered next.

8.3   **Physical Properties of Precipitates**

The isolation of a precipitate is a crucial step in most gravimetric determinations; its success depends strongly upon the physical properties of the precipitate. Precipitates are usually removed from the solutions in which they are formed by *filtration* (see Figure 8.1), employing filter crucibles having bottoms fitted with glass frits, on filter paper held in a funnel, or by other means.

Particle size is the most important factor in determining filterability. A very porous filtering medium with a rapid liquid flow rate can be used for coarse precipitates, enabling rapid, complete filtration. On the other hand, small precipitate particles require a fine filtering medium. Not only is the flow rate slower through a fine filtering medium, but such a medium tends to clog, further slowing the rate of filtration.

Particle size is a function of the chemical nature of a precipitate and the conditions under which it was formed. There are great extremes in particle size. At the lower end of the scale are **colloidal particles**, which range from 0.001 to 1 micrometer (μm) in diameter. Such particles remain in suspension and cannot be filtered out by ordinary means because they are so small. At the other end of the scale, **crystalline particles** have dimensions of the order of several tenths of a millimeter and are rapidly filtered over coarse filtration media. Between these two extremes there exists a gradation of particle sizes.

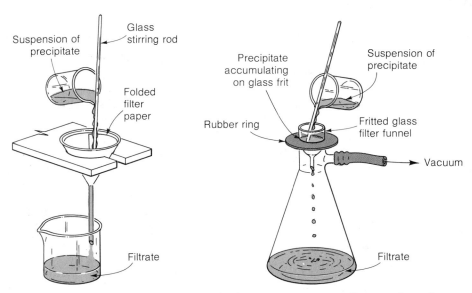

FIGURE 8.1   Two common means of isolating precipitates are filtration through filter paper (*left*) or through a fritted glass filter crucible (*right*).

8.4   The Formation of Precipitates

Relative Supersaturation in Precipitate Formation

The size and purity of precipitates are influenced by a number of factors, including the chemical nature of the precipitate and its solubility. Also important are temperature, the concentration of the analyte in solution, the concentration of the precipitating reagent, and the rate of mixing of the reagent solution with the analyte solution.

Precipitate particle size is strongly influenced by **relative supersaturation**, defined by the equation

$$\text{Relative supersaturation} = \frac{Q - S}{S} \tag{8.4}$$

where S is the equilibrium solubility of the precipitate in the precipitating medium and Q is its concentration at a particular time and place within a medium. As the precipitating reagent is added to a particular place in the medium, there are times during the process when $Q > S$ immediately upon addition of the reagent, and particularly at the point of addition. As the precipitate forms, Q approaches S. It has been found that both a longer time and greater degree of supersaturation ($Q > S$) lead to smaller, less desirable colloidal particles. On the other hand, very slow addition of precipitating reagent accompanied by vigorous mixing leads to a low degree of supersaturation favoring the formation of larger crystalline precipitates.

Nucleation and Particle Growth Processes

The effect of supersaturation upon the nature of precipitates can be explained by a mechanism of precipitate formation that involves the two phenomena of **nucleation** and **particle growth**. Nucleation consists of the formation of a new particle by the uniting of several formula units* to form a particle of solid. Since as few as four or five formula units may be involved, these particles, called **nuclei**, are extremely small. Additional precipitate can form by further nucleation. Alternatively, additional formula units of precipitate can condense on existing nuclei through the process of particle growth. Obviously the nucleation route leads to a huge number of very small particles, whereas particle growth favors fewer, larger particles.

   The relative predominance of nucleation or particle growth depends upon the degree of supersaturation. Both increase with supersaturation—nucleation exponentially and particle growth in an approximately linear fashion. A low degree of supersaturation favors particle growth and better-quality precipitates. This may be achieved by the conditions summarized in the accompanying box.

Conditions Favoring Large, Readily Filterable Particles

1. An elevated temperature increases S for most precipitates.
2. Dilute solutions favor minimum Q.
3. Slow addition of precipitating reagent accompanied by vigorous stirring tends to keep Q low.
4. When the precipitating agent is the conjugate base of a weak acid, it may be added to an acidic medium in which S is high, and S decreased slowly by the gradual addition of base. For example, oxalate precipitating reagent can be added to an acidic solution of Ca^{2+} analyte, where the equilibrium of the reaction

$$Ca^{2+} + HC_2O_4^- \rightleftharpoons CaC_2O_4(s) + H^+$$

does not lie strongly to the right. Slow addition of NH_3 takes up H^+, resulting in precipitation of most of the CaC_2O_4 as large, well-formed crystals.

---

* A formula unit is a molecule of a nonionic compound or the minimum number of ions required to make up the smallest possible quantity of an ionic compound; for example, two Ag^+ ions and one SO_4^{2-} ion form a formula unit of Ag_2SO_4.

8.5   Formation and Handling of Colloidal Particles

Formation, Dispersion, and Coagulation of Colloids

Under some conditions the formation of a colloidal precipitate, rather than crystalline solids, cannot be avoided. This occurs, for example, in the precipitation of $Fe(OH)_3 \cdot xH_2O$, which is so insoluble that a high degree of supersaturation cannot be avoided. It also occurs with relatively more soluble salts, such as AgCl, for reasons that are not completely understood. However, a settleable, filterable mass usually can be obtained from a colloidal precipitate through the process of **coagulation**, also called **agglomeration**. This phenomenon is favored by *heated solutions*, *stirring*, and the presence of *electrolytes* in solution. Why these factors favor coagulation is explained by the nature of colloids and colloidal suspensions.

Figure 8.2 represents a colloidal particle of AgCl(s) suspended in a solution of NaCl and $NaNO_3$ from which the Cl^- has been partially precipitated by the addition of $AgNO_3$ solution. Ions in solution that are common to the precipitate are preferentially adsorbed at the precipitate

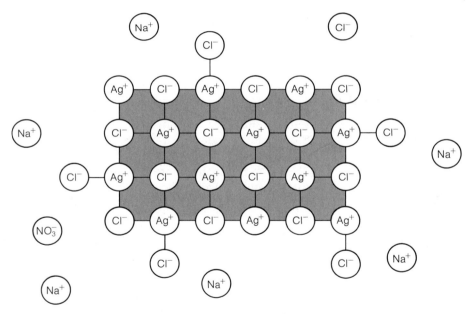

FIGURE 8.2   Colloidal AgCl particle (shaded) suspended in a solution of NaCl and $NaNO_3$. The primary adsorption layer consists of negatively charged Cl^- ions bonded to Ag^+ ions on the surface of the particle, giving the particle a net negative charge. Beyond the primary adsorption layer is a counter-ion layer consisting predominantly of positively charged ions (Na^+, H^+), neutralizing the charge of the negatively charged particle. Beyond the counter-ion layer is the bulk of the solution (not shown) in which the positive and negative charges from ions in solution are in balance.

surface. For the case under discussion, the excess Cl^- ions in solution bond to the surface Ag^+ ions on the AgCl colloidal particle, making up a *primary adsorption layer* and imparting a net negative charge to the colloidal particle. This is balanced by a *counter-ion layer* of positive ions (Na^+, H^+) which balance the negative charge of the colloidal particle. These two layers make up an *electrical double layer* that repels similar colloidal particles and overcomes the cohesive force by which particles of identical composition normally are attracted to each other. These forces must be overcome to enable the coagulation of colloidal particles.

Some colloidal particles lose their charges and coagulate as a normal consequence of mixing stoichiometric amounts of the positive and negative ions making up the particle. For example, when $AgNO_3$ precipitating reagent is first added to a solution containing Cl^- analyte, there is a large excess of Cl^- ions available to bond to the AgCl colloidal particles, giving them a double-layer charge as shown in Figure 8.2. At this point the colloidal suspension of AgCl will have a milky opaque appearance with little or no evidence of solid AgCl having precipitated from the solution. As more $AgNO_3$ is added, there are fewer excess Cl^- ions to bond to the colloidal AgCl particles; a point is reached near the stoichiometric amount of $AgNO_3$ where a curdy white precipitate of AgCl will settle from the suspension as the colloidal AgCl particles coagulate. If too much excess $AgNO_3$ solution is added, there is a tendency for the precipitate to go back into colloidal suspension, this time as colloidal AgCl particles having a positive charge because of Ag^+ ions bonded to surface Cl^- ions.

Factors Affecting Colloidal Precipitate Quality

Heating causes colloidal particles to coagulate because it decreases the surface concentration of adsorbed ions in the primary adsorption layer and imparts sufficient kinetic energy to the particles to enable them to overcome the double-layer charge repulsion and thus stick together. Stirring likewise aids this process. The addition of a soluble electrolyte is particularly effective in coagulating suspensions of charged particles. This is because additional ions in solution lessen the volume of the solution containing enough counter-ions to balance the colloidal particle charge, thus shrinking the double layer and enabling the ions to come close enough for the forces of attraction to predominate. Since the coagulation process takes time, it is often aided by allowing the precipitate to stand in contact with the mother liquor for a while, a process called **digestion**. Digestion also results in the loss of weakly bound water and gives a denser, more filterable precipitate.

The maintenance of electrical neutrality requires that when charged colloidal particles coagulate, they retain a layer of solution containing the adsorbed primary layer and counter-ion layer ions, as well as those ions added to reduce the forces of electrical repulsion between the colloidal particles. Therefore, colloidal precipitates consist of curdy or gelatinous masses in which are entrained ions of otherwise soluble salts, such as $NaNO_3$. The precipitation of such salts that would otherwise be soluble is known as **coprecipitation**, and is a source of error in gravimetric analysis.

Washing a precipitate of coagulated colloidal particles with water can remove the ions responsible for coagulation from the surface layers of solution on the particles. When this happens, some or all of the precipitate may become redispersed as a colloidal suspension. This phenomenon is called **peptization**, and it introduces error into gravimetric analysis. Peptization may be prevented by washing the coagulated precipitate with a solution of a volatile electrolyte—for example, dilute HNO_3 in the washing of a precipitate of AgCl. Since it contains ions, the wash solution prevents peptization and displaces a large fraction of the nonvolatile salt ions coprecipitated with the desired precipitate. Upon heating, the volatile electrolyte is evolved as a vapor, leaving the pure precipitate behind.

8.6   Formation and Handling of Crystalline Precipitates

Formation of Crystalline Precipitates

The formation of large, easily handled, readily filtered crystalline precipitates is favored by minimizing Q and maximizing S in Equation 8.4—that is, by avoiding excessive degrees of supersaturation. This is accomplished by precipitation from dilute, hot solutions and adding the precipitating reagent slowly with vigorous mixing. A better product of higher purity is obtained by digesting the precipitate for some time in contact with the mother liquor. In the case of crystalline products, it appears that digestion causes precipitate in smaller, less-well-formed crystals to dissolve and reprecipitate in a way that favors growth of larger, more nearly perfect crystals. During the process, pockets of impurities in the crystals are exposed, and some impurities are lost to the solution, improving precipitate quality.

Impurities in Crystalline Precipitates

The exposed surface area per unit weight of a crystalline precipitate is much less than that of a colloidal precipitate, so crystalline precipitates are not troubled with surface adsorption of impurities nearly so much as colloidal precipitates. Impurities can be incorporated in crystalline precipitates by two major processes, however. The first of these is **inclusion**, in which impurity ions and molecules are scattered throughout the interior of the crystals. This may occur through **isomorphic substitution**, in which the contaminant molecules or ions are sufficiently similar in size and properties to those of the precipitate to allow their substitution into the crystalline lattice. For example, $PbSO_4$ and $BaSO_4$ are isomorphic, and precipitation of $BaSO_4$ by the addition of SO_4^{2-} to a Ba^{2+} solution containing a trace of Pb^{2+} results in the inclusion of

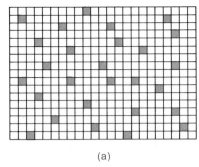
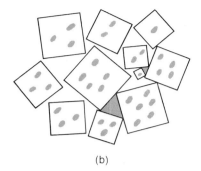

(a)   (b)

FIGURE 8.3   Impurities in crystalline precipitates (darkened areas). (a) Isomorphic substitution or solubility showing homogeneous substitution in a well-formed crystal; (b) occlusion in imperfectly formed crystals.

$PbSO_4$ in the $BaSO_4$ precipitate, even though the solubility product of $PbSO_4$ is not exceeded. Inclusions resulting from isomorphic substitution are almost impossible to eliminate by processes such as digestion or even reprecipitation; chemical removal of the interference is usually necessary. Some inclusions result from *solubility* of the impurity in the precipitate. This occurs when the two compounds in question have the same crystal form (for example, both monoclinic) but may have considerably different lattice spacings in the crystal. Like isomorphic substitution, this results in a homogeneous distribution of the impurity throughout the crystal.

The second major way in which impurities may be trapped in a crystalline precipitate is **occlusion** (Figure 8.3), which involves trapping of small quantities of the mother liquor within cavities produced during the formation of rapidly growing, imperfectly formed crystals. Slow precipitation helps prevent occlusions from forming, and digestion eliminates occluded impurities by allowing larger, more ideally formed crystals to be produced from smaller, less-well-shaped ones; in the process pockets of occluded impurities are released to the mother liquor.

8.7   **Homogeneous Precipitation**

Nature of Homogeneous Precipitation

The ideal way to achieve minimum supersaturation in the formation of a precipitate from an analyte solution would be to add precipitating reagent throughout the solution at a very slow rate, thus yielding the most pure and filterable precipitate. This can be accomplished chemically in some cases by the process known as **precipitation from homogeneous solution**. (*Homogeneous precipitation* is a term sometimes used to describe this process.) The most commonly cited example of a reagent used for this purpose is urea,

$(H_2N)_2CO$, which hydrolyzes in a solution near the boiling point by the reaction

$$(H_2N)_2CO + 3H_2O \rightarrow CO_2(g) + 2NH_4^+ + 2OH^- \qquad (8.5)$$

to produce base in the form of OH^- ion. This reaction is directly useful for the production of metal-hydroxide precipitates, such as $Al(OH)_3 \cdot xH_2O$ and $Fe(OH)_3 \cdot xH_2O$. Formation of such precipitates by direct addition of base results in the formation of a highly gelatinous mass laden with water and impurities, difficult to handle and almost impossible to filter. However, homogeneously precipitated from a hot urea solution, these hydroxides are much more pure, dense, and readily filtered. Crystalline solids formed by precipitation from homogeneous solution tend to have larger crystals with fewer impurities than those produced by direct addition of reagent.

Examples of Homogeneous Precipitation

The use of urea as a homogeneous precipitating reagent is not confined to the production of OH^-. It can be used as a source of base to convert weak acids to anions that form precipitates with various metals. The most common example of this is the use of urea to raise the pH of a solution of protonated oxalate and calcium to the point where calcium oxalate is precipitated:

$$Ca^{2+} + HC_2O_4^- + OH_{urea}^- \rightarrow CaC_2O_4(s) + H_2O \qquad (8.6)$$

There are numerous reagents available for precipitation from homogeneous solution. Oxalate for the precipitation of Ca^{2+}, Mg^{2+}, and Zn^{2+} can be produced from the hydrolysis of ethyl oxalate:

$$(C_2H_5)_2C_2O_4 + 2H_2O \rightarrow H_2C_2O_4 + 2C_2H_5OH \qquad (8.7)$$

Sulfate for the precipitation of Ba^{2+}, Ca^{2+}, Sr^{2+}, and Pb^{2+} can be formed by the hydrolysis of dimethyl sulfate:

$$(CH_3O)_2SO_2 + 2H_2O \rightarrow 2H^+ + SO_4^{2-} + 2CH_3OH \qquad (8.8)$$

A large number of metal ions can be precipitated homogeneously by sulfide generated by the hydrolysis of thioacetamide:

$$\underset{\substack{\| \\ S}}{CH_3CNH_2} + H_2O \rightarrow \underset{\substack{\| \\ O}}{CH_3CNH_2} + H_2S \qquad (8.9)$$

8.8   Treatment of Precipitates after Filtration

Heating Precipitates

A precipitate is almost invariably heated after it is formed. This serves the purposes of (1) drying, (2) removal of volatile electrolytes, (3) changing the composition of the precipitate to a weighable form, and (4) in cases where

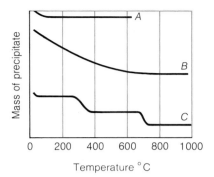

FIGURE 8.4 Mass versus temperature plots for three different types of precipitates. *A* loses water at a low temperature and does not undergo chemical change or volatilization at somewhat higher temperatures; *B* loses large quantities of water over a wide temperature range, finally reaching a stable mass; *C* loses water at low temperatures and undergoes abrupt changes in composition at two higher temperatures.

the precipitate is collected on ashless filter paper, the filter is burned, so that its mass does not add error to the mass of the precipitate.

Figure 8.4 shows three general types of curves obtained by heating various precipitates. Curve *A* is from a compound that readily loses water and volatile coprecipitate electrolytes at a low temperature, then maintains a stable mass with increased heating. Such a precipitate is AgCl containing some HNO_3, which is dried to a stable mass by heating to 110 to 120°C. Curve *B* is for a rather strongly hydrated precipitate that loses large quantities of water over a wide temperature range. A precipitate of $Al(OH)_3 \cdot xH_2O$ formed homogeneously in a medium of hydrolyzing urea would behave in this fashion, attaining a stable mass at around 650°C. Curve *C* is for a precipitate that undergoes two changes in composition with heating. The classic example of a precipitate that undergoes several changes in composition during heating is calcium oxalate monohydrate, $CaC_2O_4 \cdot H_2O$, which undergoes the following reactions:

$$CaC_2O_4 \cdot H_2O \xrightarrow[225°C]{} CaC_2O_4(s) + H_2O(g) \tag{8.10}$$

$$CaC_2O_4 + \tfrac{1}{2}O_2 \xrightarrow[450°C]{} CaCO_3(s) + CO_2(g) \tag{8.11}$$

$$CaCO_3 \xrightarrow[600-800°C]{} CaO(s) + CO_2(g) \tag{8.12}$$

The term ignition (see Section 8.2) is often applied to the heating of precipitates. Ignition is more specifically the strong heating of precipitates to drive off bound water (about 500°C is required for strongly hydrated precipitates), to bring about changes in precipitate composition, or to burn off filter paper on which precipitates have collected. The high temperatures used in ignition ruin fritted glass filter crucibles, so that porcelain, silica, or platinum crucibles are required.

8.9   **Precipitation with Inorganic Reagents**

Hydroxide and Sulfide Precipitants

As indicated by the solubility rules in Table 7.1 the most obvious precipitant to use for a number of metal cations is hydroxide ion, OH^-. Some degree of selectivity may be achieved because various metal hydroxides precipitate at different pH values, some of which are quite low. Some pH values at which major metal ions precipitate are the following: pH 3, Fe^{3+}, Sn^{2+}; pH 4, Th^{4+}; pH 5, Al^{3+}; pH 6, Cr^{3+}, Cu^{2+}, Zn^{2+}; pH 7, Fe^{2+}; pH 8, Cd^{2+}, Co^{2+}, Ni^{2+}; pH 9, Ag^+, Hg^{2+}, Mn^{2+}; pH 11, Mg^{2+}. Coprecipitation is a major problem with hydroxide precipitates but can be partially overcome by the slow, homogeneous generation of OH^- from urea. Normally the hydroxides are ignited to their corresponding oxides for weighing.

Next to hydroxide, the most widely applicable inorganic species for the precipitation of metal ions is H_2S (see Table 7.1 and Section 7.11). Among the species precipitated by H_2S for chemical analysis are As(III), Bi(III), Cu(II), Ge(IV), Sb(III), Sn(IV), and Cu(II). With the exception of bismuth, which is weighed as Bi_2S_3, these precipitates are ignited to their oxides—for example, As_2O_3, GeO_2—and weighed in that form.

Miscellaneous Precipitants

A number of other inorganic precipitants are used to precipitate metal ions. Among the most common such precipitants are the following, where the ignited product that is weighed is listed in parentheses: HCl for Ag (AgCl), HNO_3 for Sn (SnO_2), H_2SO_4 for Ba ($BaSO_4$), $(NH_4)_2HPO_4$ for Mg ($Mg_2P_2O_7$), $H_2C_2O_4$ for Th (ThO_2), and H_2PtCl_6 for K ($KPtCl_6$).

Several reagents are commonly used to precipitate anions. These include $AgNO_3$ for Cl (AgCl), $MgCl_2$ and NH_4Cl for PO_4^{3-} ($Mg_2P_2O_7$), and $BaCl_2$ for SO_4^{2-} ($BaSO_4$).

Conversion to the Elemental Form

The most fundamental way to determine an element gravimetrically is to convert it to the elemental form. The simplest reagent for doing that is the electron provided at a cathode in electrogravimetric analysis (see Chapter 15), which reduces metals according to the following type of half-reaction:

$$Cu^{2+} + 2e^- \xrightarrow[\text{cathode}]{} Cu(s) \tag{8.13}$$

In addition to copper, this technique can also be used for the electrodeposition of Ag, Bi, Cd, Co, In, Ni, Re, Sb, and Sn. In addition, a number of chemical reductants have been used to reduce elements to their elemental forms. These include SO_2 for Se and Au, $TiCl_2$ for Rh, H_2 for Re and Ir, SO_2 and H_2NOH for Te, and $SnCl_2$ for Hg. The use of $SnCl_2$ for the reduction of mercury salts to elemental mercury is especially useful for the determination of mercury by flameless atomic absorption analysis (see Chapter 22) in quantities that are far too low to weigh.

8.10   Organic Precipitants

The Nature and Advantages of Organic Precipitants

Because of the enormous variety of compounds possible through organic synthesis, it has been possible to develop a number of **organic precipitants** for inorganic species, particularly cations. These compounds are usually relatively high molecular weight chelating agents (see Chapter 10) that bond to a metal cation in at least two places. To be useful for gravimetric analysis, the product must be a slightly soluble metal chelate (see Chapter 10), such as that formed between 8-hydroxyquinoline and Zn^{2+}:

$$2 \quad + Zn^{2+} \rightarrow \quad O-Zn-O \quad + 2H^+ \tag{8.14}$$

8-Hydroxyquinoline   Precipitate of the 8-hydroxyquinolate
                            chelate of zinc(II)

As seen in Reaction 8.14, each of the 8-hydroxyquinolate anions bonds to Zn(II) in two places, through the electron-donor N and O atoms, respectively. The result is 2 five-membered rings, each containing a Zn, O, N, and two C atoms. (Most chelates have either five- or six-membered rings.) Many metals form insoluble 8-hydroxyquinolates.

Organic precipitants offer several advantages. Normally the precipitates have high formula weights, thus reducing relative errors in weighing. The compounds can be synthesized with a certain degree of selectivity for various analytes. Selectivity can be increased by varying the pH and concentration of the analyte, as well as by the use of masking agents (see Chapter 10) that complex interfering ions and prevent them from forming a precipitate with the reagent.

Specificity of Organic Precipitants

Although much of the early work on organic precipitants was directed toward the development of reagents specific for various metals, this ideal has been largely unrealized. The nearest thing to it is dimethylglyoxime used for nickel analysis. It forms a bright red precipitate with nickel(II) according to the reaction

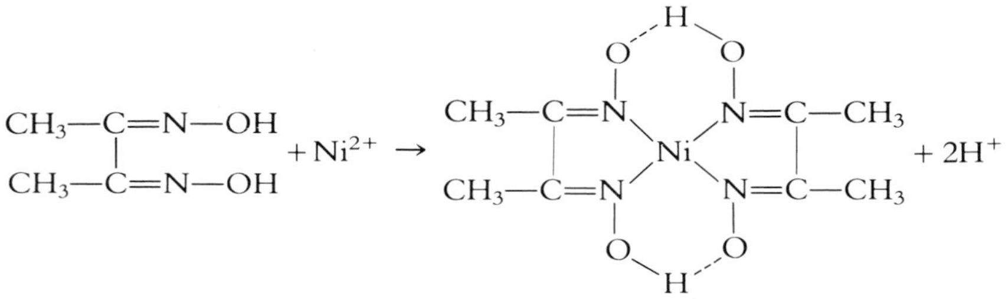

$$CH_3-C=N-OH \atop CH_3-C=N-OH \quad +Ni^{2+} \rightarrow \quad \text{[structure]} \quad +2H^+ \quad (8.15)$$

The precipitate is produced in a medium consisting of a buffer of ammonium acetate and acetic acid. Dimethylglyoxime is only sparingly soluble in water, but it can be added as a solution in ethanol (ethyl alcohol). It can also be employed as a solution of the sodium salt $Na_2C_4H_6N_2O_2 \cdot 8H_2O$, which is soluble to the extent of about 3% in aqueous solution. The nickel dimethylglyoxime precipitate can be weighed after drying at 105 to 110°C. The nickel salt of dimethylglyoxime is essentially the only precipitate formed by this compound in weakly alkaline solution. The reagent can also be used to precipitate a palladium(II) salt from acid solution.

In contrast to the specificity of dimethylglyoxime for nickel, approximately 25 cations can be precipitated by 8-hydroxyquinoline (see Reaction 8.14), commonly known as oxine. The solubilities of these precipitates vary and their formation is pH-dependent, because oxine is a weak acid. Therefore, by controlling reagent concentration and pH, it is possible to get some degree of selectivity with this reagent.

It is possible to generate 8-hydroxyquinoline homogeneously (see Section 8.7) by hydrolysis of 8-acetoxyquinoline, as shown below:

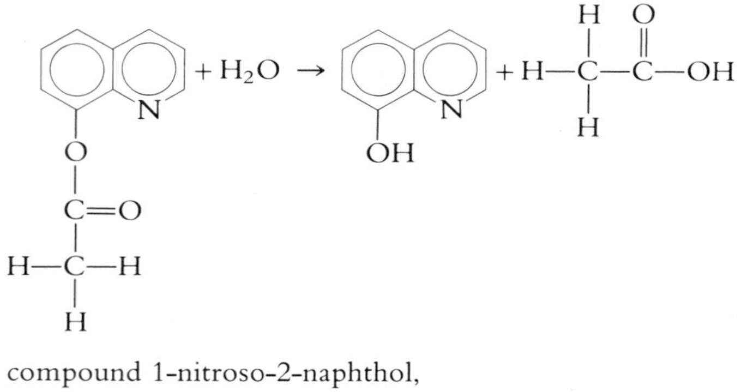

$$\text{[structure]} \quad +H_2O \rightarrow \text{[structure]} \quad +H-\overset{H}{\underset{H}{C}}-\overset{O}{C}-OH$$

(8.16)

The compound 1-nitroso-2-naphthol,

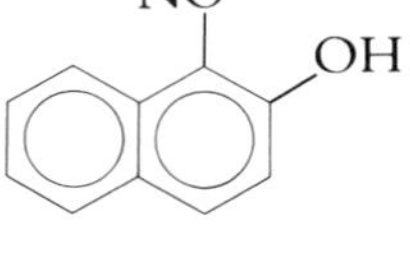

quantitatively precipitates cobalt, iron(III), palladium, and zirconium from slightly acidic media. A number of metals are partially precipitated, and many more not at all. This precipitant is used primarily as a reagent for Co(III), with which it forms a bulky, red-brown precipitate with the formula $Co(C_{10}H_6O_2N)_3$. It is hard to get this precipitate in a pure form, however, and ignition to the oxide Co_3O_4 is difficult. A major use of 1-nitroso-2-naphthol is to separate cobalt from nickel. This may be done in the analysis of alloys containing iron, nickel, and cobalt after the removal of iron.

Tetraphenylarsonium chloride, $[(C_6H_5)_4As]^+Cl^-$, forms precipitates with large, singly charged anions. This reagent can be used for the determination of thallium by the precipitation of $[(C_6H_5)_4As]^+TlCl_4^-$ salt from an approximately 1 M HCl medium. Tetraphenylarsonium chloride is also used in the reaction

$$[(C_6H_5)_4As]^+ + ClO_4^- \rightarrow [(C_6H_5)_4As]^+ClO_4^- \, (s) \qquad (8.17)$$

for the gravimetric determination of perchlorate anion.

8.11 Gravimetric Analysis by Volatilization

Determination of Water, Carbonate, Hydrogen, and Carbon

A sample constituent that forms a gas or vapor by heating or by a chemical reaction can be determined by **volatilization**. The volatile constituent can be determined *directly* by collecting it and weighing it, or *indirectly* by measuring the loss of mass of the sample as a result of the loss of the volatile constituent.

Water that is superficially adsorbed onto the surface of sample particles, water that is adsorbed within sample particles, and water of hydration can be determined by heating a sample to a suitable temperature and measuring the loss of mass of the sample. This is very commonly done with commercial products. Alternatively, the water driven off may be carried to a trap of absorbent material in a stream of predried air or nitrogen, and the mass of water collected by the trap measured by weighing. Anhydrous calcium chloride, $CaCl_2$, or magnesium perchlorate, $Mg(ClO_4)_2$, make good absorbents for water in this application.

Carbon dioxide gas can be released from carbonates (represented in general by MCO_3) in a sample by the addition of strong acids as shown in the following reaction:

$$MCO_3 + 2H^+ \rightarrow M^{2+} + H_2O + CO_2(g) \qquad (8.18)$$

The CO_2 is completely removed from the analyte solution by heating and sweeping with a stream of CO_2-free air. Water is removed in a drying tube and the CO_2 collected in a trap of solid alkali such as Na_2CO_3, or Na_2CO_3 and NaOH on a solid support material (Ascarite-II) (see Figure 8.5). The gain in weight of the carbon dioxide trap is equal to the mass of carbon dioxide released from the sample.

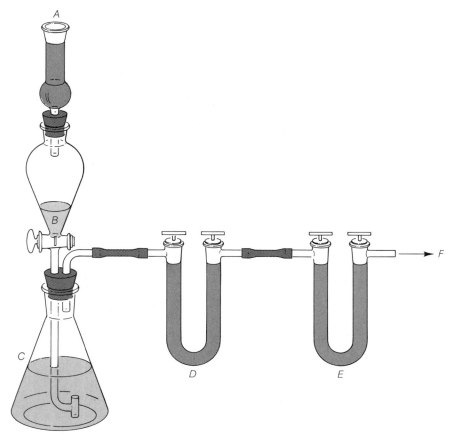

FIGURE 8.5   Simplified train for the measurement of CO_2 gas evolved from a sample. *A*, air into Ascarite tube, which removes CO_2; *B*, H_2SO_4 solution; *C*, reaction vessel; *D*, drying tube containing $MgClO_4$; *E*, Ascarite-containing tube for CO_2 removal; *F*, to vacuum.

Carbon and hydrogen can be determined in organic samples by burning the sample in pure oxygen. The water vapor produced by the hydrogen is retained in a water-absorbing trap, and the carbon dioxide produced by the carbon is retained in a trap designed to absorb CO_2. Carbon in steels and alloys can be determined by burning the metal in oxygen, a reaction promoted by a catalyst, to produce CO_2. Interfering gases such as SO_2 are removed from the gas stream before the CO_2 produced is trapped and weighed.

Miscellaneous Determinations by Volatilization

Silica, SiO_2, can be determined in some kinds of samples by treating a weighed sample in a platinum crucible with a mixture of sulfuric and hydrofluoric acids. The reaction is

$$SiO_2 + 4HF \rightarrow SiF_4(g) + 2H_2O(g) \tag{8.19}$$

in which silicon is lost as the volatile fluoride. The reagents are evaporated off at high temperature. Constituents other than silica that form volatile products with HF or acid must be absent. The silica content is measured from the loss in weight of the sample. This analysis is particularly applicable to impure ignited silica residues, such as those obtained from high-silica-content minerals.

A standard volatilization analysis recommended by the American Society for Testing and Materials (ASTM) is the hydrogen loss test for copper, tungsten, and iron powders. This determination is based upon the reaction of bound oxygen in the metal with H_2 in a gas stream:

$$H_2 + \{O\} \rightarrow H_2O(g) \tag{8.20}$$

It can be used for measuring percent mass losses of 0.05 to 3.0% in copper and iron powders and 0.01 to 0.50% in tungsten powder. These losses are used to calculate the oxygen content of the metal. The basic procedure is to heat a weighed sample of the powdered metal in a stream of H_2 in a tube contained in a tube furnace. A temperature of $875°C$ is employed for copper and tungsten powders and $1150°C$ for iron powder. Some oxides, particularly Al_2O_3, CaO, and SiO_2, are not reduced under these conditions and do not respond to the hydrogen loss test.

8.12   Gravimetric Calculations

Mass of Analyte from Mass of Precipitate

In practically all gravimetric analyses, the product that is weighed is not the same as the analyte. Therefore, it is necessary to convert the mass of product to that of analyte for substitution into the equation

$$\text{Percent analyte} = \frac{\text{mass of analyte}}{\text{mass of sample}} \times 100 \tag{8.21}$$

The calculation of the mass of analyte is accomplished by the general equation

$$\begin{array}{c}\text{Mass of}\\\text{analyte}\end{array} = \begin{pmatrix}\text{mass of}\\\text{product}\end{pmatrix}\begin{pmatrix}\text{conversion to}\\\text{moles of product}\end{pmatrix}\begin{pmatrix}\text{conversion to}\\\text{moles of analyte}\end{pmatrix}\begin{pmatrix}\text{conversion to}\\\text{mass of analyte}\end{pmatrix} \tag{8.22}$$

These kinds of calculations are best illustrated by examples. Consider the determination of aluminum in a sample containing filter alum, $Al_2(SO_4)_3 \cdot 18H_2O$ (a water-treatment chemical that removes suspended matter from water by forming gelatinous $Al(OH)_3$ that settles from suspension) and sodium sulfate filler by the precipitation and weighing of aluminum

8-hydroxyquinolate (see Section 8.10), $Al(C_9H_6ON)_3$, fw = 459.4. The formula weight of filter alum is 666.4. If a 0.9057-g sample of the filter alum contaminated with sodium sulfate yields 1.1783 g of $Al(C_9H_6ON)_3$, what was the percentage of $Al_2(SO_4)_3 \cdot 18H_2O$ in the sample? After it is dissolved in water, the aluminum sulfate undergoes the following overall reaction:

$$Al_2(SO_4)_3(aq) + 6C_9H_7ON \rightarrow 2Al(C_9H_6ON)_3(s) + 6H^+ + 3SO_4^{2-} \quad (8.23)$$

For each mole of $Al_2(SO_4)_3 \cdot 18H_2O$ originally present in the sample, 2 moles of aluminum 8-hydroxyquinolate are produced. The calculation of the mass of $Al_2(SO_4)_3 \cdot 18H_2O$ is the following:

Mass of $Al_2(SO_4)_3 \cdot 18H_2O$

$$= \underbrace{1.1783 \text{ g } Al(C_9H_6ON)_3}_{\text{mass of product}} \times \underbrace{\frac{1 \text{ mol } Al(C_9H_6ON)_3}{459.4 \text{ g } Al(C_9H_6ON)_3}}_{\substack{\text{conversion to moles} \\ \text{of product}}} \times \underbrace{\frac{1 \text{ mol } Al_2(SO_4)_3 \cdot 18H_2O}{2 \text{ mol } Al(C_9H_6ON)_3}}_{\substack{\text{conversion to moles} \\ \text{of analyte}}}$$

$$\times \underbrace{\frac{666.4 \text{ g } Al_2(SO_4)_3 \cdot 18H_2O}{\text{mol } Al_2(SO_4)_3 \cdot 18H_2O}}_{\substack{\text{conversion to mass} \\ \text{of analyte}}} = 0.8546 \text{ g } Al_2(SO_4)_3 \cdot 18H_2O \quad (8.24)$$

$$\text{Percent analyte} = \frac{\text{mass } Al_2(SO_4)_3 \cdot 18H_2O}{\text{mass sample}} \times 100 = \frac{0.8546 \text{ g}}{0.9057 \text{ g}} \times 100 = 94.36\% \quad (8.25)$$

Mass of Analyte from Mass of Ignited Precipitate

As another example, consider the determination of magnesium as the pyrophosphate $Mg_2P_2O_7$. The precipitation reaction is

$$Mg^{2+} + HPO_4^{2-} + NH_3 + 6H_2O \rightarrow MgNH_4PO_4 \cdot 6H_2O(s) \quad (8.26)$$

resulting from the addition to the magnesium sample of excess diammonium hydrogen phosphate followed by the addition of excess ammonia. The precipitate is next carefully ignited

$$2MgNH_4PO_4 \cdot 6H_2O \xrightarrow{1000°C} Mg_2P_2O_7 + 2NH_3(g) + 7H_2O(g) \quad (8.27)$$

to form the weighed product. Suppose that solid mineral magnesium carbonate is leached with acid from a 2.5378-g sample of commercial cattle feed supplement, the leachate taken through the gravimetric analysis procedure outlined above, and 0.1803 g of $Mg_2P_2O_7$, fw = 222.57, is obtained. What is the percentage of MgO in the sample? (In commercial products elemental content is often expressed as percentage of the oxide, even though that is not the form of the element in the product.) The formula weight of MgO is 40.31.

The calculation is the following:

$$\text{Mass of MgO} = \underbrace{0.1803 \text{ g Mg}_2\text{P}_2\text{O}_7}_{\text{mass of product}} \times \underbrace{\frac{1 \text{ mol Mg}_2\text{P}_2\text{O}_7}{222.57 \text{ g Mg}_2\text{P}_2\text{O}_7}}_{\substack{\text{conversion to moles} \\ \text{of product}}} \times \underbrace{\frac{2 \text{ mol MgO}}{1 \text{ mol Mg}_2\text{P}_2\text{O}_7}}_{\substack{\text{conversion to} \\ \text{moles of analyte}}}$$

$$\times \underbrace{\frac{40.31 \text{ g MgO}}{\text{mol MgO}}}_{\substack{\text{conversion to mass} \\ \text{of analyte}}} = 0.06531 \text{ g MgO} \qquad (8.28)$$

$$\text{Percent MgO} = \frac{0.06531 \text{ g}}{2.5378 \text{ g}} \times 100 = 2.573\%$$

EXAMPLE

A 0.8237-g sample containing anhydrous $\text{Al}_2(\text{SO}_4)_3$, fw = 342.16, and NaCl yielded 1.0384 g BaSO_4, fw = 233.40, when dissolved and treated with excess BaCl_2. What was the percentage of $\text{Al}_2(\text{SO}_4)_3$ in the sample?

$$\text{Mass Al}_2(\text{SO}_4)_3 = 1.0384 \text{ g BaSO}_4 \times \frac{1 \text{ mol BaSO}_4}{233.40 \text{ g BaSO}_4} \times \frac{1 \text{ mol Al}_2(\text{SO}_4)_3}{3 \text{ mol BaSO}_4}$$

$$\times \frac{342.16 \text{ g Al}_2(\text{SO}_4)_3}{\text{mol Al}_2(\text{SO}_4)_3} = 0.5074 \text{ g Al}_2(\text{SO}_4)_3$$

$$\text{Percent Al}_2(\text{SO}_4)_3 = \frac{0.5074 \text{ g}}{0.8237 \text{ g}} \times 100 = 61.60$$

8.13   Gravimetric Factor

Derivation of Gravimetric Factor

Examination of the examples shown in Section 8.12 reveals that for a specific precipitate produced from a specific analyte, three of the factors in the calculation of the mass of analyte remain the same regardless of the mass of sample, analyte, or precipitate. In Equations 8.22, 8.24, and 8.28, these are labeled *conversion to moles of product*, *conversion to moles of analyte*, and *conversion to mass of analyte*. The product of these conversion factors

$$\left(\begin{array}{c}\text{conversion to} \\ \text{moles of product}\end{array}\right)\left(\begin{array}{c}\text{conversion to} \\ \text{moles of analyte}\end{array}\right)\left(\begin{array}{c}\text{conversion to} \\ \text{mass moles of analyte}\end{array}\right) \qquad (8.29)$$

is the same for the calculation of the mass of a specific analyte corresponding to the production of a mass of a specific weighing form of the precipitate. Thus for the mass of $\text{Al}_2(\text{SO}_4)_3 \cdot 18\text{H}_2\text{O}$ analyte corresponding to a particular mass of $Al(C_9H_6ON)_3$ precipitate (Equation 8.24), the product above is

$$\underbrace{\frac{1 \text{ mol } Al(C_9H_6ON)_3}{459.4 \text{ g } Al(C_9H_6ON)_3}}_{\substack{\text{conversion to moles} \\ \text{of product}}} \times \underbrace{\frac{1 \text{ mol } Al_2(SO_4)_3 \cdot 18H_2O}{2 \text{ mol } Al(C_9H_6ON)_3}}_{\substack{\text{conversion to moles} \\ \text{of analyte}}} \times \underbrace{\frac{666.4 \text{ g } Al_2(SO_4)_3 \cdot 18H_2O}{\text{mol } Al_2(SO_4)_3 \cdot 18H_2O}}_{\substack{\text{conversion to mass} \\ \text{of analyte}}}$$

$$= 0.7253 \frac{\text{g } Al_2(SO_4)_3 \cdot 18H_2O}{\text{g } Al(C_9H_6ON)_3} \tag{8.30}$$

The result of this calculation is a **gravimetric factor**, in this case with a value of 0.7253. When a gravimetric factor is known, all that is necessary to find the mass of analyte is to multiply the mass of product (precipitate weighed) by the gravimetric factor to find the mass of analyte. Thus for the analysis of filter alum by the precipitation of aluminum 8-hydroxyquinolate cited in Section 8.12, the mass of analyte is given by the calculation

$$\begin{array}{l} \text{Mass of} \\ Al_2(SO_4)_3 \cdot 18H_2O \end{array} = 1.1783 \text{ g } Al(C_9H_6ON)_3 \times 0.7253 \frac{\text{g } Al_2(SO_4)_3 \cdot 18H_2O}{\text{g } Al(C_9H_6ON)_3}$$

$$= 0.8546 \text{ g } Al_2(SO_4)_3 \cdot 18H_2O \tag{8.31}$$

With the use of a gravimetric factor, the total calculation of the percent of analyte in a sample becomes simply

$$\text{Percent analyte} = \frac{(\text{mass of product})(\text{gravimetric factor})}{\text{mass of sample}} \times 100 \tag{8.32}$$

EXAMPLE

Calculate the gravimetric factor giving the mass of MgO that produces a particular mass of $Mg_2P_2O_7$.

$$\frac{1 \text{ mol } Mg_2P_2O_7}{222.57 \text{ g } Mg_2P_2O_7} \times \frac{2 \text{ mol MgO}}{1 \text{ mol } Mg_2P_2O_7} \times \frac{40.31 \text{ g MgO}}{1 \text{ mol MgO}}$$

$$= 0.3622 \frac{\text{g MgO}}{\text{g } Mg_2P_2O_7}$$

Calculate the gravimetric factor for converting the mass of Fe_2O_3, fw $= 159.7$, obtained from igniting a precipitate of ferric hydroxide to the mass of Fe_3O_4, fw $= 231.5$, in iron ore.

$$\frac{1 \text{ mol } Fe_2O_3}{159.7 \text{ g } Fe_2O_3} \times \frac{2 \text{ mol } Fe_3O_4}{3 \text{ mol } Fe_2O_3} \times \frac{231.5 \text{ g } Fe_3O_4}{1 \text{ mol } Fe_3O_4}$$

$$= 0.9664 \frac{\text{g } Fe_3O_4}{\text{g } Fe_2O_3}$$

Some typical gravimetric factors are given in Table 8.1.

TABLE 8.1   Gravimetric factors for some typical analytes

Analyte	Form of precipitate weighed	Gravimetric factor, g analyte/g precipitate
Cr_2O_3	$BaCrO_4$	0.3000
CoO	$Co_2P_2O_7$	0.5136
Fe_2O_3	Fe	1.4297
$(PbCO_3)_2 \cdot Pb(OH)_2$	Pb	1.2478
$MgCl_2$	$Mg_2P_2O_7$	0.8556
Hg_2Cl_2	Hg	1.1767
Ni	Ni-bisdimethylglyoximate	0.2032
P_2O_5	$Mg_2P_2O_7$	0.6377
S	$BaSO_4$	0.1374
U_3O_8	$UO_2(NO_3)_2 \cdot 6H_2O$	0.5590

Use of Gravimetric Factor

As an example of the use of the gravimetric factors in Table 8.1, suppose that extractable S (mostly pyrite, FeS_2) is leached from a 2.516-g coal sample with HCl and the S oxidized to SO_4^{2-} by hydrogen peroxide, H_2O_2. The resulting solution yields 0.1925 g of $BaSO_4$. The appropriate gravimetric factor from Table 8.1 is 0.1374 g S/g $BaSO_4$. What is the percent extractable S, % Ext S, in the coal sample?

$$\% \text{ Ext S} = \frac{0.1925 \text{ g } BaSO_4 \times 0.1374 \text{ g S/g } BaSO_4}{2.516\text{-g coal}} \times 100 = 1.051\% \text{ S} \qquad (8.33)$$

Programmed Summary of Chapter 8

This summary contains in a programmed format the major terms and concepts introduced in this chapter. To get the most out of it, you should fill in the blanks for each question and then check the answers at the end of the book to see if you are correct.

Four general ways of isolating a species to be weighed for gravimetric analysis are (1) _____. An example of a precipitate that can be dried and weighed directly is (2) _____. An example of a chemical reaction in which a precipitate is heated to 1000° C to produce a weighable form is (3) _____. A solution from which a precipitate has been formed is called the (4) _____. Four characteristics of a precipitate suitable for gravimetric analysis are (5) _____. Two extremes of precipitate particle size are (6) _____. Relative supersaturation is defined as (7) _____. The two major processes involved in the formation and growth of precipitate particles are (8) _____. Four conditions favoring large, readily filterable precipitate particles are (9) _____. A settleable, filterable mass usually can be obtained from a colloidal precipitate through the process of (10) _____, which is favored by (11) _____. An electrically charged colloidal precipitate particle, such as AgCl in suspension in NaCl solution, is chemically bonded to solution ions that compose the (12) _____ layer, beyond which is the (13) _____ layer; these two layers together make up the (14) _____. Letting a precipitate stand in contact with its mother liquor for a while to improve its quality is a process called (15) _____. The precipitation, along with a precipitate of salts whose solubility product has not been exceeded, is called (16) _____. Redispersion of a colloidal precipitate during washing is called (17) _____ and is prevented by washing with a solution of a (18) _____. A phenomenon by which impurity ions and molecules are homogeneously distributed in crystals of a crystalline precipitate is called (19) _____. Entrapment of small quantities of mother liquor and impurities in cavities formed by the rapid growth of crystalline precipitates is called (20) _____. The most common example of a reaction used for precipitation from homogeneous solution is (21) _____. The four main purposes served by heating or ignition of a precipitate after it is formed are (22) _____. A plot of mass of precipitate versus temperature in the general form

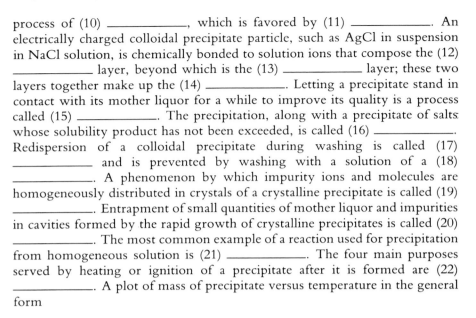

indicates that the precipitate undergoes (23) _____. The correct arrangement of the ions Ag^+, Cu^{2+}, Fe^{2+}, Fe^{3+}, and Cd^{2+} in order of increasing pH of precipitation as hydroxides is (24) _____. The two most widely used inorganic anions for the formation of precipitates from metal cations are (25) _____. Two advantages of organic precipitants for gravimetric analysis are (26) _____. Perhaps the most specific of the common organic precipitants for metals is (27) _____, which is commonly used to determine (28) _____. The chemical name for oxine is (29) _____. A cationic organic precipitant used to precipitate anions is (30) _____. Two good chemical absorbents for the gravimetric determination of volatilized water are (31) _____. Carbon dioxide volatilized from a sample may be collected on (32) _____. Carbon and hydrogen can be determined in organic samples by (33) _____. The hydrogen loss test is used for (34) _____. The mass of analyte in the gravimetric analysis of a sample is the product of the four terms (35) _____. Three of these factors may be combined to a single term called the (36) _____. The equation for calculating percent analyte from mass of sample, mass of product, and gravimetric factor is (37) _____.

Questions

1. What methods are used to isolate a weighable species for gravimetric analysis?
2. How are colloidal particles best filtered from suspension?
3. How do nucleation and particle growth vary with the degree of supersaturation in a solution from which a precipitate forms?
4. What ions predominate in (a) the primary adsorption layer and (b) the counter-ion layer of a colloidal particle suspended in a solution consisting originally of KCl from which a portion of the Cl^- has been precipitated by the addition of $AgNO_3$ solution.
5. Match the following:
   a. Isomorphic substitution
   b. Digestion
   c. Coprecipitation
   d. Peptization

   1. Allowing the precipitate to remain in contact with the mother liquor
   2. A form of inclusion
   3. Results in removal from solution of normally soluble salts
   4. Redispersion of a colloidal suspension
6. With chemical reactions illustrate the phenomenon of precipitation from homogeneous solution by (a) production of precipitant OH^-, (b) production of a precipitant anion by raising pH, and (c) direct production of a precipitant ion other than OH^-.
7. Study the mass-temperature plot of a heated precipitate. Explain the shape of the plot.

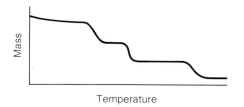

Temperature

8. A precipitate heated at $250°C$ in the atmosphere yielded a mass-time curve as shown. How might such behavior be explained?

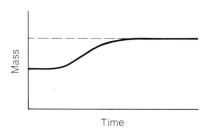

Time

9. Match the following:
   a. 8-Hydroxyquinoline
   b. 1-Nitroso-2-naphthol
   c. Tetraphenylarsonium cation
   d. Dimethylglyoxime

   1. One of the few precipitants for anions
   2. Virtually the only essentially specific precipitant
   3. Used primarily to precipitate Co(III)
   4. Precipitates a wide spectrum of metal ions
10. List the vapor species that are commonly collected in gravimetric analysis by volatilization.
11. Of what three factors or "conversions" does a gravimetric factor consist?
12. What advantage is there to using a gravimetric factor in calculations of gravimetric analysis?
13. Of the following, the factor or process most likely to cause difficulties in gravimetric analysis is (a) weighing with sufficient accuracy, (b) getting a precipitate to form, (c) obtaining a weighable product of known composition, (d) filtration, (e) calculating the solubility product.
14. A precipitate of AgCl was formed by mixing stoichiometric amounts of NaCl and $AgNO_3$ solutions. The precipitate

was isolated by filtration and washed carefully to remove all soluble ions. Then the precipitate was resuspended in the colloidal form in a solution of NaCl. The colloidal particle is represented below. Fill in each of the two open circles with the formula of the ion that each formula represents.

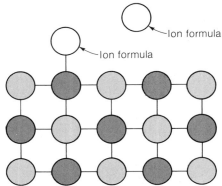

15. The reaction

$$(H_2N)_2CO(aq) + 3H_2O \xrightarrow[\text{near boiling}]{}$$

$$CO_2(g) + 2NH_4^+ + 2OH^-$$

(a) shows undesired decomposition of a precipitating agent, (b) illustrates the use of a homogeneous precipitating reagent, (c) is, itself, a precipitation reaction, (d) was not mentioned as being significant in analytical chemistry, (e) is confined to the formation of hydroxide precipitates.

16. Match the following:
   a. Hydrolyzes in hot solution
   b. About the closest thing to a metal-specific reagent
   c. A precipitant for metals that is not noted for specificity

1. $\underset{H}{\overset{H}{>}}N-\overset{\overset{\displaystyle O}{\|}}{C}-N\underset{H}{\overset{H}{<}}$

2. $CH_3-C=N-OH$
   $CH_3-\overset{|}{C}=N-OH$

3. (quinolin-8-ol structure with OH)

Problems

1. Suppose that a 2.136-g sample of $CaC_2O_4 \cdot H_2O$ was heated to a constant weight at $800°C$ and that the CO_2 and H_2O evolved were collected in suitable traps (see Figure 8.5). Calculate the weight gained by the CO_2 trap, that gained by the H_2O trap, and the equilibrium weight of the ignited residue.

2. A low-quality sample of quicklime consisting of CaO substantially contaminated with $Ca(OH)_2$ and $CaCO_3$ was analyzed for these constituents by ignition to CaO and collection and weighing of CO_2 and H_2O evolved. A 3.072-g sample of the

quicklime yielded 106.0 mg H_2O and 272.0 mg CO_2. Calculate the percentages of $Ca(OH)_2$ and $CaCO_3$. If these values are correct, what should the solid residue weigh?

3. A 3.471-g sample of a compound containing C, H, and O was ignited in a stream of O_2 and the CO_2 and H_2O produced were collected. The weight of CO_2 collected was 8.758 g and that of H_2O was 1.537 g. Calculate the percentages of C and H in the compound.

4. Dissolved Fe^{2+} in waste steel-pickling liquor (a solution of HCl and H_2SO_4 used

to dissolve coatings of rust from steel) was determined by oxidizing the Fe^{2+} to Fe^{3+} with hydrogen peroxide, homogeneously precipitating $Fe(OH)_3$ from solution by the slow hydrolysis of urea, and igniting the precipitate to Fe_2O_3, which was weighed. What is the value of the gravimetric factor for calculating the weight of Fe from the weight of Fe_2O_3.

5. Using the appropriate formula weights, atomic weights, and mole ratios, verify the gravimetric factors in Table 8.1.

6. Dust from a uranium ore grinding operation was collected in a baghouse and analyzed for U_3O_8 by dissolving the dust in appropriate solvents and precipitating and weighing $UO_2(NO_3)_2 \cdot 6H_2O$. Using the gravimetric factor given in Table 8.1, calculate the percentage of U_3O_8 in a 4.851-g sample of dust that yielded 107.3 mg $UO_2(NO_3)_2 \cdot 6H_2O$.

7. Chloride in a brine solution is determined by precipitating the chloride as $AgCl(s)$ from V mL of brine and weighing the AgCl in grams (Wt). Derive an equation that gives the molarity of chloride, M, from V, Wt, and the appropriate gravimetric factor.

8. A 12.731-g sample of coal was burned in a stream of air and the combustion products were bubbled through aqueous hydrogen peroxide oxidant. The sulfate in the peroxide solution was precipitated as $BaSO_4$. This product was found to weigh 0.1893 g. Representing the sulfur in coal as S(coal), list the reactions involved in the procedure and calculate the percent S in the coal.

9. A fertilizer sample weighing 0.6379 g was dissolved and treated with $NaClO_4$ solution, yielding 0.3816 g of $KClO_4$ precipitate. What was the percent K in the fertilizer sample?

10. A corrosion product scraped from the surface of a lightweight metal alloy was analyzed to determine the percentage of $MgCO_3$ in the product by precipitation of magnesium ammonium phosphate and ignition to $Mg_2P_2O_7$. A 0.5626-g sample of the corrosion product yielded 0.3982 g of $Mg_2P_2O_7$. What was the percent $MgCO_3$ in the sample?

11. A sample consisting of a mixture of Fe_3O_4 and Fe_2O_3 was heated in an oxygen atmosphere to convert all of the sample to Fe_2O_3. Before heating, the sample weighed 1.2389 g and after heating it weighed 1.2734 g. What was the percent Fe_2O_3 in the sample initially?

12. A sample of peat was taken from 1 m below the surface of a peat bog and sealed tightly in a bottle to prevent loss of water. Part of the sample was weighed and analyzed for calcium before drying, yielding a value of 0.907% Ca. The remainder of the sample was dried under vacuum to prevent oxidation, and again analyzed for Ca; this time a value of 8.13% Ca was obtained. What was the percent H_2O in the undried sample?

13. A 1.3072-g sample of rocket fuel yielded 1.2318 g of $(C_6H_5)_4AsClO_4$ precipitate when dissolved and treated with excess tetraphenylarsonium chloride solution. What was the percent ClO_4 oxidant in the fuel?

9 Precipitation Titrations

9.1   Definition and Examples of Precipitation Titrations

A **precipitation titration** is one in which the titrant forms a precipitate with the analyte. There is an abrupt decrease in analyte concentration and increase in titrant concentration at the equivalence point. Therefore, a plot of log [analyte] or log [titrant] versus volume of titrant added typically gives an **S**-shaped titration curve with a break at the equivalence point. The end point (experimental determination of the equivalence point) is found by graphical analysis of the break or by indicators.

Table 9.1 gives some examples of precipitation titrations. The most commonly used precipitation titrations involve Ag^+ ion and are called **argentometric titrations**. Silver ion itself can be determined by titration with Cl^- or SCN^-. Silver nitrate may be employed as a reagent for the determination of Cl^-, Br^-, I^-, or SCN^-. Tetraphenylarsonium chloride reagent (see Section 8.10) can be used to precipitate ClO_4^- ion and the titration followed potentiometrically with a perchlorate ion–selective electrode (see Chapter 13). Lead ion can be titrated with sulfate in a medium that is partially nonaqueous to decrease the solubility of $PbSO_4$.

TABLE 9.1   Examples of precipitation titrations

Analyte	Reagent	Precipitate product	Means of equivalence-point detection
Ag^+	Cl^-	$AgCl(s)$	Potentiometric with Ag or Ag–AgCl electrode producing titration curve
Cl^-	Ag^+	$AgCl(s)$	Potentiometric titration curve, adsorption indicators, formation of a colored salt, formation of a colored complex (Volhard method)
Br^-	Ag^+	$AgBr(s)$	Volhard method, potentiometric
I^-	Ag^+	$AgI(s)$	Volhard method, potentiometric
SCN^-	Ag^+	$AgI(s)$	Volhard method, potentiometric
ClO_4^-	ϕ_4As^{+*}	$\phi_4As^+ClO_4^-(s)$	Potentiometric with perchlorate-selective electrode
Pb^{2+}	SO_4^{2-}	$PbSO_4(s)$	Potentiometric with lead-selective electrode in nonaqueous media to partially reduce lead sulfate solubility

* Tetraphenylarsonium ion from tetraphenylarsonium chloride.

9.2

The Nature of a Precipitation Titration Curve

Calculation of Points for a Precipitation Titration Curve

A precipitation titration curve has the same general shape as that of a strong acid titrated by a strong base (see Section 5.2) in the case where the precipitate has a very low solubility. It is shaped like the curve for the titration of a weak acid with a strong base (see Section 5.5) when a more soluble precipitate is formed. As an example of the former, consider the titration of 100.0 mL of $0.01000\ M\ I^-$ with $0.1000\ M\ Ag^+$ titrant. The reaction is

$$Ag^+ + I^- \rightarrow AgI(s) \tag{9.1}$$

and the AgI product has a relatively low solubility product,

$$K_{sp} = [Ag^+][I^-] = 1.00 \times 10^{-16} \tag{9.2}$$

In 100.0 mL of $0.01000\ M\ I^-$ there is initially

$$100.0\ \text{mL} \times 0.01000\ \text{mmol/mL} = 1.000\ \text{mmol of}\ I^- \tag{9.3}$$

This requires the addition of 1.000 mmol of Ag^+ to precipitate all the I^-, corresponding to a total of 10.00 mL of Ag^+ to reach the equivalence point. The titration curve may be plotted in terms of $\log[Ag^+]$ versus milliliters of Ag^+ titrant added for the volumes of titrant listed in Table 9.2.

Major Regions of a Precipitation Titration Curve

There are three major regions of the titration curve for which the data shown in Table 9.2 were calculated: (1) prior to the equivalence point, (2) at the

TABLE 9.2   Data for plotting the curve for the titration of 100.0 mL of 0.0100 M I^- with 0.1000 M Ag^+ titrant

Ag^+, mL	Ag^+ excess, mmol	I^- excess, mmol	$[Ag^+]$, mmol/mL	log $[Ag^+]$	pAg*
0	—	1.00	No Ag^+ present	—	—
1.00	—	0.90	1.12×10^{-14}	-13.95	13.95
5.00	—	0.50	2.10×10^{-14}	-13.68	13.68
9.00	—	0.10	1.09×10^{-13}	-12.96	12.96
9.90	—	0.01	1.10×10^{-12}	-11.96	11.96
9.99	—	0.001	1.10×10^{-11}	-10.96	10.96
10.00	—	—	1.00×10^{-8}	-8.00	8.00
10.01	0.001	—	9.09×10^{-6}	-5.04	5.04
10.10	0.01	—	9.08×10^{-5}	-4.04	4.04
11.00	0.100	—	9.01×10^{-4}	-3.04	3.04
15.00	0.500	—	4.35×10^{-3}	-2.36	2.36

* By analogy with pH, pAg is $-\log[Ag^+]$. For any metal ion in solution, M^{z+}, $pM = -\log [M^{z+}]$. Some chemists prefer to plot pM rather than log $[M^{z+}]$.

equivalence point where exactly 10.00 mL of Ag^+ titrant have been added, and (3) beyond the equivalence point where Ag^+ is in excess. *Initially* at 0 mL added titrant, no Ag^+ has been added and it is not possible to calculate a meaningful value of log $[Ag^+]$.

The addition of 1.00 mL of Ag^+ titrant corresponds to

$$(1.00 \text{ mL } Ag^+)(0.1000 \text{ mmol/mL}) = 0.1000 \text{ mmol added } Ag^+ \tag{9.4}$$

According to Reaction 9.1, this precipitates 0.1000 mmol of I^-, leaving 0.900 mmol I^- in $100.0 + 1.0 = 101.0$ mL of solution. The concentration of I^- is

$$[I^-] = \frac{0.900 \text{ mmol}}{101.0 \text{ mL}} = 8.91 \times 10^{-3} \; M \tag{9.5}$$

The values of $[Ag^+]$ and log $[Ag^+]$ are then calculated as follows:

$$[Ag^+] = \frac{K_{sp}}{[I^-]} = \frac{1.00 \times 10^{-16}}{8.91 \times 10^{-3}} = 1.12 \times 10^{-14} \; M \tag{9.6}$$

$$\log [Ag^+] = \log 1.12 \times 10^{-14} = -13.95 \tag{9.7}$$

All the values of log $[Ag^+]$ prior to the equivalence point are calculated in an identical manner.

At the equivalence point (exactly 10 mL of added titrant) there is no stoichiometric excess of either Ag^+ or I^-. The very low concentrations of these ions in solution come from the dissolution of $AgI(s)$, and their concentrations are equal. This enables calculation of log $[Ag^+]$ as follows:

$$[Ag^+] = [I^-] = \sqrt{K_{sp}} = 1.00 \times 10^{-8} \; M \tag{9.8}$$

$$\log [Ag^+] = -8.00 \tag{9.9}$$

By referring to the table, note the tremendous jump in log [Ag$^+$] from adding just 0.01 mL of titrant from 9.99 mL to 10.00 mL.

The rest of the values of log [Ag$^+$] are determined on the basis of the presence of excess Ag$^+$. For example, after the addition of 11.00 mL of titrant, a total of 1.100 mmol of Ag$^+$ has been added, 1.000 mmol of which was precipitated by I$^-$, leaving 0.100 mmol of excess Ag$^+$ in a solution having a volume of 100.0 mL + 11.00 mL = 111.0 mL. The calculation of log [Ag$^+$] is

$$[Ag^+] = \frac{0.100 \text{ mmol}}{111.1 \text{ mL}} = 9.01 \times 10^{-4} \ M \tag{9.10}$$

$$\log [Ag^+] = \log 9.01 \times 10^{-4} = -3.04 \tag{9.11}$$

Other values of log [Ag$^+$] beyond the equivalence point are calculated in an identical manner.

Plot of a Precipitation Titration Curve

The titration curve plotted from the data in Table 9.2 is shown in Figure 9.1. The equivalence-point volume corresponds to the point half way up on the titration curve break. A vertical line drawn down from this point to the volume axis would intersect it at the equivalence volume of exactly 10.00 mL. This point could be determined graphically by the method shown for the

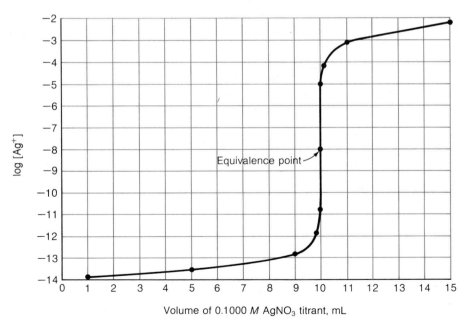

FIGURE 9.1   Curve for the titration of 100.0 mL of 0.0100 M I$^-$ with 0.1000 M Ag$^+$

TABLE 9.3   Data for plotting the curve for the titration of 100.0 mL of 0.01000 M Cl$^-$ with 0.1000 M Ag$^+$ titrant

Ag$^+$, mL	Ag$^+$ excess, mmol	Cl$^-$, mmol	[Ag$^+$], mmol/mL	log[Ag$^+$]
0	—	1.00	—	—
1.00	—	0.90	2.04×10^{-8}	-7.69
5.00	—	0.50	3.82×10^{-8}	-7.42
8.00	—	0.20	9.83×10^{-8}	-7.01
9.00	—	0.10	1.98×10^{-7}	-6.70
9.50	—	0.05	3.98×10^{-7}	-6.40
10.00	—	—	1.35×10^{-5}	-4.87
10.50	0.050	—	4.52×10^{-4}	-3.34
11.00	0.100	—	9.01×10^{-4}	-3.04
12.00	0.200	—	1.78×10^{-3}	-2.75
15.00	0.500	—	4.35×10^{-3}	-2.36

titration of an acid in Figure 5.6. Note that the break in this titration curve is extremely steep, and the equivalence point could be readily found. Furthermore, if the general shape of the titration curve is known, only three points, commonly those at 50%, 100%, and 150% titrated (5, 10, and 15 mL titrant in Figure 9.1) are enough to produce a reasonable sketch of the titration curve.

Titration of Cl$^-$ with Ag$^+$

Consider next a precipitation titration in which a much more soluble salt, AgCl, $K_{sp} = 1.82 \times 10^{-10}$, is formed. Suppose that 100.0 mL of 0.01000 M Cl$^-$ is titrated with 0.1000 M AgNO$_3$ titrant. The volumes of added Ag$^+$ titrant for which log[Ag$^+$] is calculated are given in Table 9.3. The calculations are exactly like those used to calculate the points used in constructing the curve for the titration of I$^-$ with Ag$^+$. Initially there is no Ag$^+$ present, so a value of log[Ag$^+$] cannot be calculated. Example calculations before, at, and beyond the equivalence point are the following:

Before the equivalence point, 5.00 mL added Ag$^+$

$$\text{mmol Ag}^+ \text{ added} = (5.00 \text{ mL})(0.100 \text{ mmol/mL}) = 0.500 \text{ mmol} \tag{9.12}$$

$$\text{mmol Cl}^- \text{ remaining} = 1.000 \text{ mmol} - 0.500 \text{ mmol} = 0.500 \text{ mmol} \tag{9.13}$$

$$[\text{Cl}^-] = \frac{0.500 \text{ mmol}}{105 \text{ mL}} = 4.76 \times 10^{-3} \text{ } M \tag{9.14}$$

$$[\text{Ag}^+] = \frac{K_{sp}}{[\text{Cl}^-]} = \frac{1.82 \times 10^{-10}}{4.76 \times 10^{-3}} = 3.82 \times 10^{-8} \text{ } M \tag{9.15}$$

$$\log[\text{Ag}^+] = \log 3.82 \times 10^{-8} = -7.42 \tag{9.16}$$

At the equivalence point

$$[Ag^+] = [Cl^-] = \sqrt{K_{sp}} = \sqrt{1.82 \times 10^{-10}} = 1.35 \times 10^{-5}\,M \qquad (9.17)$$

$$\log[Ag^+] = -4.87 \qquad (9.18)$$

Beyond the equivalence point, 15.00 mL added Ag^+

$$[Ag^+] \text{ in excess} = \frac{5.00\text{ mL} \times 0.1000\text{ mmol/mL}}{100.0\text{ mL} + 15.0\text{ mL}} = \frac{0.50\text{ mmol}}{115.0\text{ mL}} = 4.35 \times 10^{-3}\,M \qquad (9.19)$$

$$\log[Ag^+] = -2.36 \qquad (9.20)$$

The curve for the titration of 100.0 mL of 0.01000 M Cl^- with 0.1000 M $AgNO_3$ is shown in Figure 9.2. In comparison to the titration curve for I^-, the curve for Cl^- shows a much smaller, more gradual, sloping break. It may be inferred from the more distinct break in the I^- titration curve resulting from the lower solubility of AgI than AgCl that more dilute solutions of I^- can be titrated compared to those of AgCl.

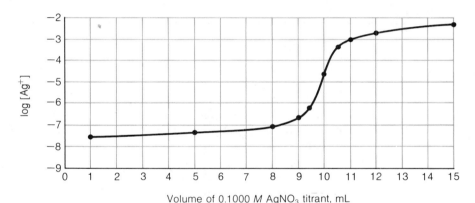

FIGURE 9.2 Curve for the titration of 100.0 mL of 0.0100 M Cl^- with 0.1000 M $AgNO_3$ titrant

9.3   Titrations with Two Different Precipitates

Titration of Iodide–Chloride Mixture with Silver Ion

When two different ions react with the same ion of opposite charge to form two different salts with greatly different K_{sp}'s, the two ions give a precipitation titration curve with two different equivalence points, much like a mixture of strong and weak monoprotic acids titrated with base (see Figure 5.13). This is

TABLE 9.4   Data for plotting the curve for the titration of 100.0 mL of a solution 0.01000 M in I⁻ and 0.01000 M in Cl⁻ with 0.1000 M Ag⁺ titrant

Ag⁺, mL	Ag⁺ excess, mmol	I⁻ excess, mmol	Cl⁻ excess, mmol	[Ag⁺], mmol/mL	log [Ag⁺]
0	—	1.00	1.00	No Ag⁺ present	
1.00	—	0.90	1.00	1.12×10^{-14}	−13.95
5.00	—	0.50	1.00	2.10×10^{-14}	−13.68
9.00	—	0.10	1.00	1.09×10^{-13}	−12.96
9.90	—	0.01	1.00	1.10×10^{-12}	−11.96
9.99	—	0.001	1.00	1.10×10^{-11}	−10.96
10.00	—	—	1.00	1.00×10^{-8}	−8.00
10.10	—	—	0.99	2.02×10^{-8}	−7.69
11.00	—	—	0.90	2.24×10^{-8}	−7.65
15.00	—	—	0.50	4.19×10^{-8}	−7.38
18.00	—	—	0.20	1.07×10^{-7}	−6.97
19.00	—	—	0.10	2.17×10^{-7}	−6.66
19.50	—	—	0.05	4.35×10^{-7}	−6.36
20.00	—	—	—	1.35×10^{-5}	−4.87
20.50	0.05	—	—	4.15×10^{-4}	−3.38
21.00	0.10	—	—	8.26×10^{-4}	−3.08

best illustrated for a mixture of I⁻ ($K_{sp}^{AgI} = 1.00 \times 10^{-16}$) and Cl⁻ ($K_{sp}^{AgCl} = 1.82 \times 10^{-16}$) titrated with AgNO₃. To show this example consider the titration with 0.1000 M AgNO₃ of a solution that is 0.01000 M in both I⁻ and Cl⁻. The volumes of added titrant to be considered are those given in Table 9.4. Since AgI has a much lower solubility than AgCl, the initial reaction that occurs as Ag⁺ is first added is

$$Ag^+ + I^- \rightarrow AgI(s) \tag{9.21}$$

Initially there are 1.000 mmol of I⁻ and 1.000 mmol of Cl⁻ present, and no Cl⁻ precipitates until 1.000 mmol of Ag⁺ has been added. Therefore, the values of log [Ag⁺] as a function of milliliters of added 0.1000 M AgNO₃ titrant are the same as those below 10.00 mL for the titration of I⁻ alone, given in Table 9.2. For this to be true the K_{sp} of Cl⁻ must not be exceeded substantially before the addition of exactly 10 mL of titrant. Consider the point at the addition of 9.99 mL of AgNO₃ titrant. According to Table 9.4, the value of [Ag⁺] at this point is 1.10×10^{-11}, as controlled by the solubility of AgI. The product

$$[Ag^+][Cl^-] = 1.10 \times 10^{-11} \times \frac{1.000}{109.9} = 1.00 \times 10^{-13} < K_{sp} \text{ of AgCl} \tag{9.22}$$

is less than K_{sp} of AgCl, so no AgCl has been precipitated up to that point. Even at the first equivalence point where 10.00 mL of 0.1000 M Ag⁺ solution have been added, the value of [Ag⁺] is given simply by the following:

$$[Ag^+] = [I^-] = \sqrt{K_{sp} \text{ of AgI}} = \sqrt{1.00 \times 10^{-16}} = 1.00 \times 10^{-8} \tag{9.23}$$

Note that the product of silver and chloride ion concentrations,

$$[Ag^+][Cl^-] = 1.00 \times 10^{-8} \times \frac{1.000}{110.0} = 9.09 \times 10^{-11} < K_{sp} \text{ of AgCl} \qquad (9.24)$$

is still slightly less than K_{sp} of AgCl. (If the initial Cl^- concentration had been somewhat higher, the solubility product of AgCl would be exceeded at the first equivalence point, and $[Cl^-]$ would limit $[Ag^+]$ at this point.)

Just beyond the first equivalence point—that is, at 10.10 mL of Ag^+ reagent, there is an excess of 0.10 mmol × 0.1000 mmol/mL = 0.010 mmol of Ag^+ beyond that required to precipitate all the I^-. This excess Ag^+ precipitates 0.010 mmol of Cl^-, leaving 0.99 mmol of Cl^- in 110.1 mL of solution. At this point $[Ag^+]$ is limited by K_{sp} of AgCl and is given by the following:

$$[Ag^+] = \frac{K_{sp}}{[Cl^-]} = \frac{1.82 \times 10^{-10}}{0.99/110.1} = 2.02 \times 10^{-8} \text{ } M1 \qquad (9.25)$$

Exactly the same approach is used to calculate all the $[Ag^+]$ values up to the second equivalence point at 20.00 mL added Ag^+ reagent.

At the second equivalence point,

$$[Ag^+] = [Cl^-] = \sqrt{K_{sp}} = \sqrt{1.82 \times 10^{-10}} = 1.35 \times 10^{-5} \text{ } M \qquad (9.26)$$

Beyond the second equivalence point the value of $[Ag^+]$ is calculated from millimoles of excess Ag^+ and the total volume of the solution.

The plot of the titration curve for 100.0 mL of 0.01000 M I^-, 0.01000 M Cl^- titrated with 0.1000 M Ag^+ is shown in Figure 9.3. A unique feature of this curve is that the first equivalence point comes at a "corner" (a *cusp* in mathematical terms) formed at the point where the solubility product of AgCl is exceeded. When that happens, precipitation of AgCl stops the steep rise in $[Ag^+]$. This occurs abruptly as I^- is stoichiometrically exhausted.

As shown in Figure 9.3, the second equivalence point occurs on a conventional S-shaped portion of the titration curve. This equivalence point, corresponding to stoichiometric precipitation of Cl^- as solid AgCl, occurs at the midpoint of the second break.

Experimental Measurement of Titration Curves with Ag^+ Titrant

The titration curve in Figure 9.3 is readily reproduced experimentally (see Chapter 25) by using a silver metal electrode, the electrical potential of which is measured against a suitable reference electrode (see Chapter 13). The electrical potential E in volts at 25° C is given by a form of the Nernst equation (see Chapters 11 and 13), as follows:

$$E = E_a + 0.0591 \log [Ag^+] \qquad (9.27)$$

where E_a is a constant for the electrode system. Plots of E versus mL $AgNO_3$ give titration curves identical in shape to those in Figures 9.1 to 9.3.

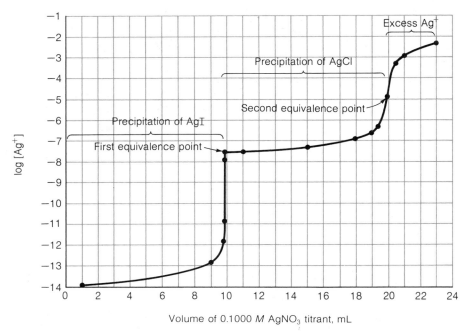

FIGURE 9.3   Curve for the titration of 100.0 mL of $0.01000\ M\ Cl^-$ and $0.01000\ M\ I^-$ with $0.1000\ M\ Ag^+$

9.4   Determination of End Point in Precipitation Titrations

The titration curves just described show an abrupt rise in titrant concentration, accompanied by an abrupt fall in analyte concentration at the equivalence points. Obviously, a reagent that responds to the abrupt rise in analyte concentration at the equivalence point can be employed as an indicator to show the end point. Several methods have been developed for showing end point in precipitation titrations. The better known of these have been developed for titrations involving Ag^+; they are based on formation of a colored precipitate, turbidity, formation of a colored complex ion, and adsorption of dyes onto charged precipitate particles. These methods are described in the following sections. In addition, precipitation titration curves may be plotted from potentiometric data (see Equation 9.27), and the end points calculated by graphical means.

9.5   Turbidimetric End Points

Turbidimetry (turbidity) refers to observation of cloudiness due to particulate matter suspended in a liquid as the result of a titration reaction. The observation of turbidity, or the lack thereof, can be taken as an indication of

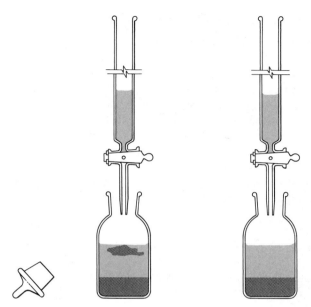

FIGURE 9.4   Addition of Cl^- titrant to a clear solution above a settled AgCl precipitate in the titration of Ag^+ with Cl^- results in formation of additional precipitate (left) before the end point and no precipitate (right) after the end point.

end point in some precipitation titrations. The Gay-Lussac–Stas analysis is one such method that, though primitive, is capable of very good accuracy. This method consists of titrating a solution containing Ag^+, HNO_3, and $Ba(NO_3)_2$ with a chloride solution and observing the point at which no more precipitate is formed. The barium nitrate is present to assist the coagulation of the precipitate (see Section 8.5). The solution to be titrated is contained in a stoppered bottle, and the Cl^- titrant is added in a volume known to be less than the equivalence-point volume (see Figure 9.4). At this point the AgCl suspension is shaken vigorously and allowed to settle from the solution until a clear (nonturbid) solution is present over the settled precipitate. The titrant is then added in 1 mL increments and an observation is made after each addition as to whether or not a precipitate forms after the addition of the titrant as indicated by the liquid becoming turbid. After each 1 mL of titrant is added, the suspension is again shaken and the precipitate allowed to settle. A point is finally reached beyond the equivalence point where no more precipitate forms and the volume of titrant is noted at that point. Next, an identical aliquot of the Ag^+ analyte is taken, and the procedure repeated, starting with the next-to-the-last volume of titrant at which fresh precipitate was observed to form in the preliminary titration. In the second titration, the reagent is added dropwise with shaking and settling until the addition of 1 drop produces no more turbidity in the clear solution above the settled precipitate. This volume is taken as the end point, and it can be determined within 1 drop (0.05 mL).

9.6   Formation of a Colored Precipitate at the End Point

The Mohr Chloride Determination

The formation of a second colored precipitate at the end point can be used for the determination of end point. The second precipitate should have a solubility slightly greater than that of the precipitate formed by the reaction of the analyte and titrant. Brick-red silver chromate, Ag_2CrO_4, meets the criterion for a second precipitate to show the end point in the titration of Cl^- with $AgNO_3$. The colored silver chromate is somewhat more soluble than the white AgCl product of this titration. The formation of Ag_2CrO_4 by reaction at the end point of Ag^+ with a low concentration of CrO_4^{2-} added to the solution titrated forms the basis of the **Mohr determination**.

Calculation of End Point Equilibria in the Mohr Titration

Prior to the end point in the Mohr titration, the Cl^- analyte is precipitated by the Ag^+ titrant according to the following reaction:

$$Ag^+ + Cl^- \rightarrow AgCl(s) \tag{9.28}$$

white precipitate

As the end point is approached, the concentration of Cl^- in the solution becomes quite low, and the value of $[Ag^+]$, limited up to the end point by the K_{sp} of AgCl, rises steeply, as shown in Figure 9.2. As this occurs, the solubility product of Ag_2CrO_4 is exceeded, and the following reaction takes place:

$$2Ag^+ + CrO_4^{2-} \rightarrow Ag_2CrO_4(s) \tag{9.29}$$

The appearance of the brick-red color Ag_2CrO_4 on the background of white AgCl precipitate particles denotes the end point.

The amount of chromate added to the solution is critical for the success of the titration. As a first estimate, consider the calculation of the initial concentration of CrO_4^{2-} that must be present so that Ag_2CrO_4 just begins to form at the equivalence point in the titration of $0.0500\ M$ Cl^- with $0.0500\ M$ Ag^+. The value of the solubility product for AgCl is 1.82×10^{-10}, so that at the equivalence point

$$[Ag^+] = [Cl^-] = \sqrt{1.82 \times 10^{-10}} = 1.35 \times 10^{-5}\ M \tag{9.30}$$

The value of $[CrO_4^{2-}]$ *at the equivalence point* required for Ag_2CrO_4 to just begin forming is given by the following:

$$K_{sp} = [Ag^+]^2[CrO_4^{2-}] = 1.29 \times 10^{-12} \tag{9.31}$$

$$[CrO_4^{2-}] = \frac{K_{sp}}{[Ag^+]^2} = \frac{1.29 \times 10^{-12}}{(1.35 \times 10^{-5})^2} = 7.08 \times 10^{-3}\ M \tag{9.32}$$

However, since a volume of titrant equal to the volume of analyte solution was added, the *original* concentration of CrO_4^{2-} was halved, so that its concentration of CrO_4^{2-} was

$$\text{Original } [CrO_4^{2-}] = 2 \times 7.08 \times 10^{-3} \, M = 0.0142 \, M \tag{9.33}$$

In practice, a concentration of the order of $7 \times 10^{-3} \, M$ CrO_4^{2-} at the equivalence point is not practical because the yellow color of the chromate ion masks the color change at the end point. A concentration of CrO_4^{2-} at the end point of about $2.5 \times 10^{-3} \, M$ is generally used for the Mohr titration. This introduces a positive titration error (see Section 5.8) because a concentration of Ag^+ higher than $1.35 \times 10^{-5} \, M$ must be reached, and a finite amount of Ag_2CrO_4 must be precipitated before the color of the silver chromate can be observed.

End Point Error in the Mohr Chloride Titration

To calculate the magnitude of the end point error obtained in a Mohr chloride titration, again consider the titration of $0.0500 \, M$ Cl^- with $0.0500 \, M$ Ag^+, starting with $5.00 \times 10^{-3} \, M$ CrO_4^{2-} in the analyte solution, so that the CrO_4^{2-} concentration is $2.50 \times 10^{-3} \, M$ at the end point. Assume that 25.00 mL of Cl^- are titrated, and that 1.0 mg of Ag_2CrO_4 must be precipitated for the end point to be observed. The concentration of Ag^+ required to initiate precipitation of Ag_2CrO_4 from $2.50 \times 10^{-3} \, M$ CrO_4^{2-} is

$$[Ag^+] = \sqrt{\frac{K_{sp}}{[CrO_4^{2-}]}} = \sqrt{\frac{1.29 \times 10^{-12}}{2.50 \times 10^{-3}}} = 2.27 \times 10^{-5} \, M \tag{9.34}$$

Of this concentration of Ag^+, part comes from the addition of Ag^+ reagent beyond the equivalence point and part comes from the dissolution of AgCl

$$AgCl(s) \rightleftarrows Ag^+ + Cl^- \tag{9.35}$$

The magnitude of the latter can be calculated from K_{sp} of AgCl as follows:

$$[Cl^-] = [Ag^+]_{\text{dissolution of AgCl}} \tag{9.36}$$

$$[Cl^-] = \frac{K_{sp}}{[Ag^+]} = \frac{1.82 \times 10^{-10}}{2.27 \times 10^{-5}} = 8.0 \times 10^{-6} \, M \tag{9.37}$$

The concentration of excess Ag^+ beyond that present from the dissolution of AgCl required to make $[Ag^+] = 2.27 \times 10^{-3} \, M$ is

$$[Ag^+]_{\text{excess}} = [Ag^+] - [Ag^+]_{\text{dissolution of AgCl}} \tag{9.38}$$

$$[Ag^+]_{\text{excess}} = 2.27 \times 10^{-5} \, M - 0.80 \times 10^{-5} = 1.47 \times 10^{-5} \, M \tag{9.39}$$

The number of mmol of excess Ag^+ in the 50.00 mL volume of solution at the equivalence point is

$$\text{mmol excess } Ag^+ = 1.47 \times 10^{-5} \text{ mmol/mL} \times 50.0 \text{ mL} = 7.35 \times 10^{-4} \text{ mmol} \tag{9.40}$$

The volume of $0.0500\ M$ Ag^+ titrant that must be added beyond the equivalence point to provide 7.35×10^{-4} mmol of excess Ag^+ is

$$\text{Volume } Ag^+ \text{ beyond} \atop \text{equivalence point} = \frac{7.35 \times 10^{-4}\ \text{mmol}}{5.00 \times 10^{-2}\ \text{mmol/mL}} = 1.47 \times 10^{-2}\ \text{mL} \tag{9.41}$$

Next, consider the excess Ag^+ that must be added to precipitate 1.0 mg of Ag_2CrO_4, fw = 332. The number of millimoles of Ag_2CrO_4 precipitated in order for the end point to be visible is

$$\text{mmol } Ag_2CrO_4 = \frac{1.0\ \text{mg } Ag_2CrO_4}{332\ \text{mg } Ag_2CrO_4/\text{mmol } Ag_2CrO_4} = 3.0 \times 10^{-3}\ \text{mmol } Ag_2CrO_4 \tag{9.42}$$

The number of millimoles of Ag^+ and volume of $0.0500\ M$ Ag^+ solution required to precipitate this quantity of Ag_2CrO_4 are

$$\text{mmol } Ag^+ = 2\ \frac{\text{mmol } Ag^+}{\text{mmol } Ag_2CrO_4} \times 3.0 \times 10^{-3}\ \text{mmol } Ag_2CrO_4 = 6.0 \times 10^{-3}\ \text{mmol } Ag^+ \tag{9.43}$$

$$\text{Volume } Ag^+ = \frac{6.0 \times 10^{-3}\ \text{mmol } Ag^+}{5.00 \times 10^{-2}\ \text{mmol } Ag^+/\text{mL}} = 0.12\ \text{mL} \tag{9.44}$$

Therefore, the total volume of Ag^+ titrant that must be added beyond the equivalence point to observe the end point is

$$\text{Total volume of } Ag^+ \text{ beyond} \atop \text{the equivalence point} = 0.0147\ \text{mL} + 0.12\ \text{mL} = 0.135\ \text{mL} \tag{9.45}$$

Recall that the 0.0147 mL of Ag^+ titrant was required to raise the value of $[Ag^+]$ sufficiently to precipitate Ag_2CrO_4, and 0.12 mL of Ag^+ titrant was consumed precipitating 1 mg of Ag_2CrO_4. The titration error is

$$\text{Titration} \atop \text{error, \%} = \frac{\text{volume } Ag^+ \text{ titrant added beyond equivalence point}}{\text{total volume titrant to reach equivalence point}} \times 100 \tag{9.46}$$

$$= \frac{0.135\ \text{mL}}{25.00\ \text{mL}} \times 100 = 0.54\% \tag{9.47}$$

Indicator Blank in the Mohr Chloride Titration

Compensation may be made for the Mohr chloride titration error by running an **indicator blank**, commonly used to correct for determinate end point errors. A suspension of solid, white, chloride-free calcium carbonate is made up in a solution having about the same volume as the titrated solution at the equivalence point and containing the equivalence-point concentration of CrO_4^{2-}. The Ag^+ reagent is added to this suspension until the end point color is observed. The volume of Ag^+ titrant required to match the end point color is subtracted from the end point volumes measured in the titrations of chloride.

9.7   **Formation of a Colored Complex at the End Point**

The Volhard Method

The formation of the colored complex ion $FeSCN^{2+}$ may be used to indicate the end point when Ag^+ is titrated with thiocyanate ion, SCN^-. The titration reaction is

$$Ag^+ + SCN^- \rightarrow AgSCN(s) \qquad (9.48)$$

When Fe^{3+} is present in the solution, the following reaction occurs at the end point:

$$Fe^{3+} + SCN^- \rightleftharpoons FeSCN^{2+} \qquad (9.49)$$
$$\text{red complex ion}$$

The end point is readily observed by the intense color of the $FeSCN^{2+}$ complex ion (see Section 1.9 and Chapter 10 for a discussion of complex ions and complexation equilibria). In order for this means of indicating end point to work, the complex ion must form at nearly the equivalence-point concentration of the titrant.

    The formation of $FeSCN^{2+}$ forms the basis of the **Volhard method**. The Volhard titration can, of course, be used for the determination of Ag^+ by titration with SCN^-. It can also be employed for the determination of Br^- and I^- by adding an excess of standard Ag^+ and back-titrating with standard SCN^-. The determination of Cl^- by this procedure is more difficult because AgCl is more soluble than AgSCN and the reaction

$$AgCl(s) + SCN^- \rightarrow AgSCN(s) + Cl^- \qquad (9.50)$$

occurs near the end point of the back-titration. This error is essentially eliminated by using a very high ($0.2\ M$) concentration of Fe^{3+}, so that $FeSCN^{2+}$ forms before any significant excess of SCN^- is added. Another approach involves removal of the solid AgCl by filtration, followed by titration of the filtrate with SCN^-. The fastest means of correcting for this error is to add nitrobenzene

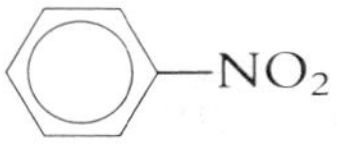

which is immiscible with water and coats the AgCl particles to prevent their redissolving. Several milliliters of nitrobenzene are added, and the titration flask is shaken vigorously before the back-titration to coat the AgCl particles.

Volhard Titration Equilibria

The Volhard titration end point is not heavily dependent upon Fe^{3+} concentration. Iron(III) concentrations between 0.002 and 1.6 M can be used with little end point error. Typically the concentration of colored $FeSCN^{2+}$ must reach

6.5×10^{-6} M before it is visually detectable by the average individual. Suppose that $[Fe^{3+}] = 0.010$ M at the end point (the concentration of Fe^{3+} commonly used); what must be the value of $[SCN^-]$ before the red $FeSCN^{2+}$ complex is observed? The formation constant for $FeSCN^{2+}$ (see Section 1.9) is given by the expression

$$K_f = \frac{[FeSCN^{2+}]}{[Fe^{3+}][SCN^-]} = 1.4 \times 10^2 \tag{9.51}$$

If $FeSCN^{2+}$ is to have a concentration of 6.5×10^{-6} M in a medium 0.010 M in Fe^{3+}, the value of $[SCN^-]$ is given by the following:

$$[SCN^-] = \frac{[FeSCN^{2+}]}{[Fe^{3+}]K_f} = \frac{6.5 \times 10^{-6}}{0.010 \times 1.4 \times 10^2} = 4.6 \times 10^{-6} \ M \tag{9.52}$$

At exactly the equivalence point the concentration of SCN^- is given by

$$AgSCN(s) \rightleftarrows Ag^+ + SCN^- \tag{9.53}$$

$$[Ag^+] = [SCN^-] = K_{sp} \text{ of AgSCN} = \sqrt{1.1 \times 10^{-12}} = 1.0 \times 10^{-6} \ M \tag{9.54}$$

Since a concentration of only 4.6×10^{-6} M SCN^- is required to form a visible concentration of the colored $FeSCN^{2+}$ product, only an insignificant excess of SCN^- need be added beyond the equivalence point for the end point to be observed.

EXAMPLE

> What should be the value of $[Fe^{3+}]$ at the equivalence point in order to ensure a visible concentration of 6.5×10^{-6} M exactly at the equivalence point? The value of K_{sp} of AgSCN is 1.1×10^{-12}.
>
> $$AgSCN(s) \rightleftarrows Ag^+ + SCN^-$$
>
> At the equivalence point
>
> $$[Ag^+] = [SCN^-] = \sqrt{K_{sp}} = \sqrt{1.1 \times 10^{-12}} = 1.0 \times 10^{-6} \ M$$
>
> From the expression for K_f of $FeSCN^{2+}$
>
> $$[Fe^{3+}] = \frac{[FeSCN^{2+}]}{K_f[SCN^-]} = \frac{6.5 \times 10^{-6}}{1.4 \times 10^2 \times 1.0 \times 10^{-6}} = 0.046 \ M$$

9.8   Adsorption Indicators for Precipitation Titrations

How Adsorption Indicators Work

Adsorption indicators are dyes that adsorb onto, or desorb from, the surface of a precipitate near the equivalence point in a precipitation titration. To be useful for end point indication, this must result in a distinct change in color, either on the surface of the precipitate or in the dye in which it is suspended. This method of end point indication is sometimes called the **Fajans method** after its originator.

The best-known adsorption indicator is *fluorescein*, which is used to indicate the equivalence point in the titration of Cl^- with Ag^+. Fluorescein is a weak acid, which partially dissociates in water to form fluoresceinate anion.

fluorescein   fluoresceinate anion (yellow-green)

The fluoresceinate anion has a yellow-green color in solution. Although fluoresceinate anion forms an intensely colored precipitate with Ag^+, it is used as an indicator at concentrations that do not exceed the solubility product of the silver fluoresceinate salt. When Cl^- is titrated with Ag^+ in the presence of fluorescein, the negatively charged fluoresceinate anions are initially repelled by the negatively charged AgCl colloidal particles, with their primary adsorption layer of Cl^- ions (see Figure 8.2). Thus the fluorescein remains in a yellow-green solution prior to the equivalence point. At the equivalence point, the colloidal AgCl particles undergo an abrupt change from a negative charge to a positive charge by virtue of Ag^+ ions adsorbed in the primary adsorption layer (Figure 9.5). The fluoresceinate ions are strongly adsorbed in the counter-ion layer of the AgCl colloids, giving these particles a red color and providing an end point color change from yellow-green to red or pink. The reaction is reversible, so that the pink color that forms at the point of addition of Ag^+ to the solution just prior to the equivalence point as a result of localized excess of silver ion disappears as the solution becomes completely mixed. Furthermore, back-titration with Cl^- solution reverses the end point color change because of desorption of the dye.

Conditions for Functioning of Adsorption Indicators

The successful application of an adsorption indicator requires a colloidal precipitate to ensure maximum adsorption of the indicator and a maximum surface area to display the color of the sorbed indicator. The colloidal particles of the precipitate must adsorb its constituent ions from solution. The adsorption indicator must be held strongly in the counter-ion layer; this almost invariably means that the indicator ion must form a precipitate with the ions

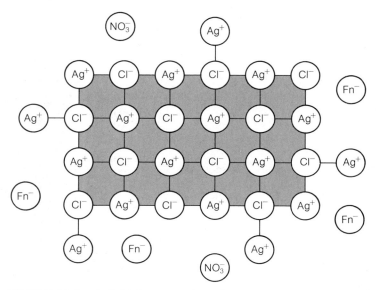

FIGURE 9.5   Colloidal particle of AgCl in a solution containing Ag^+ and fluorescein. The Ag^+ ions are shown adsorbed in the primary adsorption layer and the fluorescein-ate anions, Fn^-, are present in the counter-ion layer, imparting a red color to the particle.

adsorbed in the primary adsorption layer. However, the solubility product of this species must not be exceeded. The pH of the solution in which the titration is being carried out must be such that the indicator is largely in the ionic form—that is, the charged conjugate acid or base of the neutral indicator species. Although the active forms of some adsorption indicators are cationic, most are anionic. Furthermore, most, like fluorescein, are conjugate bases of weak acid molecules.

Programmed Summary of Chapter 9

The major terms and concepts introduced in this chapter are summarized below in a programmed format. To derive the most benefit from the summary, you should fill in the blanks for each statement and then check your answers at the end of the book to see if your choices are correct.

The steepness of a precipitation titration curve break is inversely pro-portional to the (1) ＿＿＿＿＿ of the product. The three major regions, or points, of a precipitation titration curve involving the precipitation of a single kind of analyte ion with a single kind of reagent ion are (2) ＿＿＿＿＿. For the precipitation titration of CrO_4^{2-} with Ag^+, the equivalence-point value of

$[Ag^+]$ in terms of K_{sp} is $\log[Ag^+]$ (3) _____. For a precipitation titration curve resulting from a simple 1:1 molar reaction of analyte with precipitant, $M^+ + X^- \rightarrow MX(s)$, the equivalence point on the titration curve is located (4) _____. For the precipitation titration of a mixture of I^- and Cl^- with Ag^+, the first equivalence point corresponds to complete precipitation of (5) _____ and is located at the (6) _____. The first break in the titration of a mixture of I^- and Cl^- with Ag^+ is characterized by a steep rise followed by an abrupt leveling off of the curve, forming a sharp corner. The reason that the curve becomes horizontal so abruptly is (7) _____. Five general classes of end point detection in precipitation titrations are (8) _____. Turbidimetric end points depend upon whether or not there is formation of (9) _____. The Mohr method of end point detection depends upon the formation of (10) _____ at the end point. The two factors responsible for the positive titration error inherent in the Mohr method are (11) _____. Compensation for this error is made by running (12) _____. The Volhard method depends upon formation of (13) _____ at the end point. Fluorescein is an example of an (14) _____ indicator. These operate by the mechanism of (15) _____.

Questions

1. Match the analyte in column a.–d. with its favored precipitant reagent in column 1.–4.
   a. Ag^+
   b. Ni^{2+}
   c. Pb^{2+}
   d. ClO_4^-

   1. SO_4^{2-}
   2. Tetraphenylarsonium chloride
   3. Cl^-
   4. Dimethylglyoxime

2. Two anions X^- and Y^- react with the metal ion M^+ to form MX, $K_{sp} = 1.0 \times 10^{-18}$, and MY, $K_{sp} = 1.0 \times 10^{-10}$. Roughly sketch the curves of $\log[M^+]$ versus volume of M^+ titrant for the titration of (a) a solution of X^-, (b) a solution of Y^-, and (c) a solution of a mixture of X^- and Y^-, showing approximately the relative magnitudes of the titration curve breaks.

3. Suppose that M^+ was used to titrate X^- with the formation of a precipitate of $MX(s)$, and that $\log[M^+]$ was measured from the potential of an electrode (as explained for monitoring $\log[Ag^+]$ in Section 9.3) that was immersed in solution. Draw a sketch showing how the end point would be determined from the titration curve thus generated.

4. Match the following, referring to turbidimetry for determination of end point:
   a. Gay-Lussac–Stas
   b. $Ba(NO_3)_2$
   c. Preliminary titration
   d. No turbidity formation

   1. In 1-mL increments
   2. Indicates titration should continue
   3. Coagulant
   4. Precipitant for Cl^-
   5. At end point
   6. Titration of Ag^+ with Cl^-

5. How can it be that Ag_2CrO_4 is *more soluble* than AgCl when K_{sp} of Ag_2CrO_4 (1.29×10^{-12}) is *less than* K_{sp} of AgCl (1.82×10^{-10})?

6. What is the procedure for and purpose of the indicator blank in the Mohr chloride determination?

7. Match the following:

a.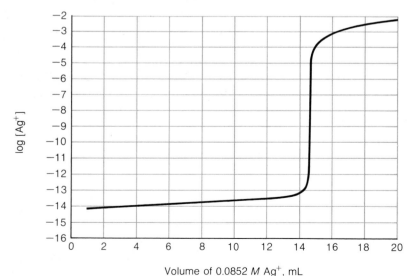

   b. Fluoresceinate anion
   c. $FeSCN^{2+}$
   d. $Ba(NO_3)_2$

   1. Required for the Volhard method
   2. Used in a modification of the Volhard method
   3. Coagulant
   4. Used in the Fajans method

8. What is the mechanism by which fluoresceinate anion indicates end point?

9. Match the following:
   a. Brick-red solid at end point
   b. Red-pink solid at end point
   c. Red solution at end point
   d. Clear solution at end point

   1. Mohr
   2. Gay-Lussac–Stas
   3. Fajans
   4. Volhard

10. Of the following, the one that is *not* used for showing end points in precipitation titrations, as exemplified by titrations involving Ag^+ ion, is (a) change of color of an organic dye indicator that remains in solution, (b) observation of turbidity, (c) formation of a colored precipitate, (d) formation of a colored complex ion, (e) adsorption of dyes onto charged precipitate particles.

Problems

1. The titration curve below is from the titration of a solution of either Cl^- or I^- with 0.0852 M standard Ag^+. Which anion was titrated and how many millimoles of it were titrated?

![Titration curve plotting log [Ag⁺] on the y-axis (from −2 to −16) against Volume of 0.0852 M Ag⁺, mL on the x-axis (from 0 to 20). The curve is nearly flat near −14 until about 14 mL, where it rises sharply to about −3.]

Volume of 0.0852 M Ag^+, mL

2. Answer the questions posed in Problem 1 for the titration curve below.

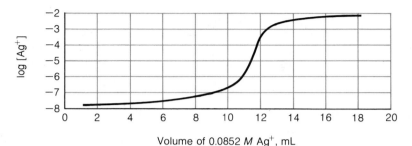

Volume of 0.0852 *M* Ag⁺, mL

3. The titration curve below is that of a mixture of Cl⁻ and I⁻ titrated with 0.0852 *M* standard Ag⁺. How many millimoles of Cl⁻ and of I⁻ were present in the mixture titrated?

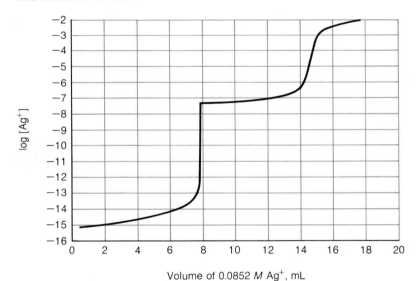

Volume of 0.0852 *M* Ag⁺, mL

4. A 2.853-g sample containing KI, NaCl, and KNO_3 was placed in a 250-mL volumetric flask, dissolved, and diluted to the mark with distilled water, and a 50-mL aliquot taken for titration with 0.0775 *M* standard $AgNO_3$. The first equivalence point came at 18.62 mL added titrant, the second at 31.36 mL total titrant. What was the percentage KI, fw = 166.0, in the sample?

5. From the information given in Problem 4, calculate the percentage of NaCl, fw = 58.4, in the total sample.

6. A 2.601-g sample containing KI, NaCl, and KNO_3 was placed in a 250-mL volumetric flask, dissolved, and diluted to the mark with distilled water. A 75.0-mL aliquot was titrated with 0.0783 *M* standard $AgNO_3$, with the first equivalence point coming at 19.37 mL and the second at 41.08 mL. What were the percentages of KI, NaCl, and KNO$_3$ in the sample?

7. The Mohr chloride titration can be used for the determination of Br$^-$ by the precipitation of AgBr, $K_{sp} = 5.2 \times 10^{-13}$, and the formation of brick-red Ag$_2$CrO$_4$, $K_{sp} = 1.29 \times 10^{-12}$, at the end point. What must be the concentration of CrO$_4^{2-}$ at the equivalence point so that precipitation of Ag$_2$CrO$_4$ is just initiated at that point?

8. Suppose for the Mohr titration described in Problem 7 that SCN$^-$ is to be determined by precipitation of AgSCN, $K_{sp} = 1.1 \times 10^{-12}$. What should be the value of [CrO$_4^{2-}$] at the equivalence point so that precipitation of Ag$_2$CrO$_4$ would just start at that point?

9. For the Mohr titration of the hypothetical anion X$^-$ to produce solid AgX, $K_{sp} = 9.8 \times 10^{-9}$, what should be the value of [CrO$_4^{2-}$] at the equivalence point for the initiation of Ag$_2$CrO$_4$ precipitation at that point?

10. A major problem in the Volhard titration is the dissolution of solid AgCl during the back-titration with SCN$^-$. Given that K_{sp} of AgSCN is 1.1×10^{-12} and K_{sp} of AgCl is 1.82×10^{-10}, what is the reaction involved and what is its equilibrium constant?

11. Suppose that 0.100 mmol of soluble NaSCN was allowed to come to equilibrium with solid AgCl in 100 mL of solution. What would be the equilibrium concentrations of Cl$^-$ and SCN$^-$?

12. Suppose that a solution of 1.00×10^{-3} M Cl$^-$ were shaken with freshly precipitated Ag$_2$CrO$_4$ until equilibrium was established with some solid silver chromate still present. What would be the reaction, its equilibrium constant, and the equilibrium values of [Cl$^-$] and [CrO$_4^{2-}$] given that K_{sp} of AgCl is 1.82×10^{-10} and K_{sp} of Ag$_2$CrO$_4$ is 1.29×10^{-12}?

13. Suppose that a solution of 7.50×10^{-4} M I$^-$ were shaken with excess freshly precipitated AgCl. Given that K_{sp} of AgCl is 1.82×10^{-10} and K_{sp} of AgI is 1.00×10^{-16}, what reaction would occur, what is its equilibrium constant, and what are the equilibrium concentrations of Cl$^-$ and I$^-$?

14. It is desired to find the percent ClO$_4^-$ in an impure perchlorate salt, the sample of which is known by the analyst to contain about 200 mg of ClO$_4^-$, fw = 99.45 (the actual perchlorate content was 223 mg). The analyst has the choice of determining the perchlorate gravimetrically by precipitation of tetraphenylarsonium perchlorate, ϕ_4AsClO$_4$, fw = 483.0, where cumulative errors in collecting and weighing the precipitate are expected to amount to 2.0 mg. The other choice is to dissolve the perchlorate salt in 50.0 mL of water and titrate it with 0.0900 M ϕ_4As$^+$ solution, following the titration with a perchlorate-selective electrode. The titration end point error for this procedure is estimated at 0.10 mL. What are the expected relative errors in parts per thousand for the two procedures?

15. A sample containing NaCl and KI was dissolved, diluted to 250 mL, and a 50.0-mL aliquot titrated with 0.1 M AgNO$_3$ solution. The first break in the titration curve came at 19.2 mL, and the second at a *total* volume of titrant of 43.6 mL. The number of millimoles of I$^-$ and Cl$^-$ in the original sample were, respectively (a) 1.60, 2.03; (b) 9.6, 12.2; (c) 12.8, 16.2; (d) 7.7, 9.8; (e) 14.7, 18.7.

10 Complexation in Quantitative Analysis

10.1   Introduction

This chapter deals with the analytical uses of complexation, in which an electron-pair-donor anion or neutral species capable of existing by itself in aqueous solution bonds to a metal ion. This phenomenon finds a number of uses in quantitative analysis. The most obvious is in the titration of metal ions, and a large number of complexation procedures have been worked out for this purpose. Complexation is useful in preventing metal ions from interfering with the determination of some analytes. Species capable of complexation, such as cyanide ion (CN^-) and NTA (see Figure 10.2), are widely used products in metal plating and other industrial applications.

This chapter deals with complexation and a subcategory of complexation, chelation, in quantitative analysis.

10.2   Complexation of Metal Ions

The Complexation Process

In the broadest sense, **complexation** is the association of two or more chemical species that are capable of independent existence. As applied to chemical analysis, this definition is generally taken to mean the bonding of a **central metal ion** capable of accepting an unshared pair of electrons with a **ligand** that can donate a pair of unshared electrons. Consider the addition of anhydrous copper(II) perchlorate to water. The salt dissolves readily according to the reaction

$$Cu(ClO_4)_2(s) + 4\:\overset{\cdot\cdot}{\underset{\overset{\cdot\cdot}{H}}{O}}:H \xrightarrow{\text{water}} Cu(H_2O)_4^{2+}(aq) + 2ClO_4^- \tag{10.1}$$

in which four water molecules, each with unshared pairs of electrons, bond to each Cu^{2+} ion. Such binding of solvent molecules to a metal ion is called **solvation** or, in the special case of solvent water, **hydration**. The $Cu(H_2O)_4^{2+}$ ion is called an **aquo complex**.

For most analytical applications, complexation occurs between a dissolved metal ion and a dissolved ligand capable of displacing water from the metal ion. This is illustrated for the reaction between a hydrated copper(II) ion and dissolved NH_3 ligand below:

$$Cu(H_2O)_4^{2+} + NH_3 \xrightarrow{\text{water}} CuNH_3(H_2O)_3^{2+} + H_2O \tag{10.2}$$

Normally for reactions that occur in water, H_2O is omitted and the complexation reaction is written simply as

$$Cu^{2+} + NH_3 \rightleftarrows CuNH_3^{2+} \tag{10.3}$$

The product, $CuNH_3^{2+}$, is called a **complex ion**, **metal complex**, or even simply a **complex**. The double arrows in the reaction indicate that the reaction is reversible.

The maximum number of electron-pair donor groups that a metal ion can accommodate in complexation is known as its **coordination number**. Typical values are 2 for Ag^+, as in $Ag(CN)_2^{2-}$; 4 for Zn^{2+}, as in $Zn(NH_3)_4^{2+}$; and 6 for Cr^{3+}, as in $Cr(NH_3)_6^{3+}$.

In addition to differences in coordination number, metal complexes differ in *geometry*, the arrangement of ligands around the central metal ion. Metal ions with a coordination number of 2, such as Cu^+, Ag^+, and Au^+, form *linear* structures. Some metals with a coordination number of 4—Cu^{2+}, Ni^{2+}, Pd^{2+}, and Pt^{2+}—have square planar structures; others with the same coordination number—Cd^{2+}, Co^{2+}, and Zn^{2+}—have tetrahedral complex ions. When N is 6 (Co^{3+}, Cr^{3+}, Fe^{3+}), octahedral structures predominate.

Names of Ligands and Metal Complexes

Ligands have special names when used in the name of a complex ion or compound. Some of the most common are listed in Table 10.1. When these ligands are involved in the formation of positively charged complex ions, a prefix is used to designate the number of complexed ligands, and Roman numerals in parentheses are used to indicate the oxidation number of the central metal ion if two or more oxidation numbers are possible. The following examples show how this terminology works: $Cu(NH_3)_4^{2+}$, tetramminecopper(II) ion; $FeOH^{2+}$, hydroxoiron(III) ion; $Zn(NH_3)_4^{2+}$, tetramminezinc ion; $BiCl_2^+$, dichlorobismuth(III) ion; $CrSO_4^+$, sulfatochromium(III) ion.

Neutral complex compounds are named very simply with the name of the cation followed by that of the anion. Therefore, the soluble complex species $AgCl(aq)$ is called simply silver chloride in solution; $Fe(NH_3)_6SO_4(aq)$ is called hexammineiron(II) sulfate, and $Pb(OH)_2(aq)$ is simply lead hydroxide.

Many complex ions, such as $Cu(CN)_4^{2-}$, are anions. Frequently the metal ion involved is given its Latin name, such as *cuprum* for copper. The ion names end in the suffix -ate, followed by the oxidation number of the metal ion in Roman numerals in parentheses. As examples, $HgCl_4^{2-}$ is called tetrachloromercurate(II) ion, $Fe(CN)_6^{4-}$ (common name ferrocyanide) is called hexacyanoferrate(II) ion; $AgCl_2^-$ is called dichloroargentate(I) ion, and $SbOHBr_5^-$ is named monohydroxopentabromoantimonate ion.

As usual for salts, those involving complex ions are designated with the name of the cation preceding that of the anion. Therefore, $K_3Fe(CN)_6$ is called potassium hexacyanoferrate(III) (common name, potassium ferricyanide). Since it is known that the $Fe(CN)_6^{3-}$ anion has an overall -3 charge, it is not necessary to specify the number of potassium ions in the compound; it can only be 3. In the compound $Cr(NH_3)_4Cl_2Cl$, two of the chlorides are complexed to Cr, and one Cl^- serves to neutralize the charge of the $Cr(NH_3)_4Cl_2^+$ ion. The name of the compound is, therefore, dichlorotetraamminechromium(III) chloride, showing that two of the chlorides are complexed to the chromium and one is not.

TABLE 10.1   Names of common ligands used in naming complex species

Ligand formula	Ligand name when uncomplexed	Ligand name when complexed
NH_3	Ammonia	Ammine-
H_2O	Water	Aquo-
Cl^-	Chloride ion	Chloro-
Br^-	Bromide ion	Bromo-
CN^-	Cyanide ion	Cyano-
F^-	Fluoride ion	Fluoro-
OH^-	Hydroxide ion	Hydroxo-
I^-	Iodide ion	Iodo-
SO_4^{2-}	Sulfate ion	Sulfato-

10.3   Chelation

Nature of Chelation

Just as metal ions may accept more than one electron-pair donor group from a ligand, a ligand may have two or more electron-pair donor groups capable of simultaneously bonding to a metal ion. This is shown in Figure 10.1 for the reaction of the amino acid glycine with Cu^{2+} ion. Here it is seen that the glycinate anion bonds simultaneously to Cu^{2+} through an N atom and an O atom, both of which are electron-pair donors. It is also noted that a five-membered ring is formed. A species such as the glycinatocopper(II) ion that is simultaneously bonded to two or more sites on a ligand is called a **metal chelate**, or **chelate**. The process of chelate formation is called **chelation**. The ligand capable of forming a chelate is called a **chelating agent** or **chelon**.

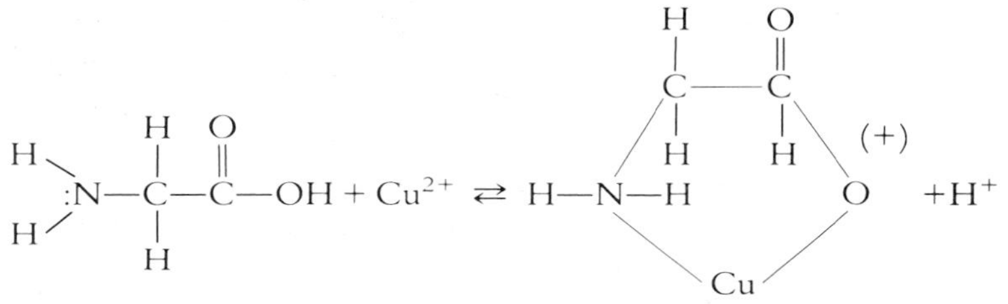

Glycine (unshared pair of
electrons shown on N)

Glycinatocopper(II) ion

FIGURE 10.1   Reaction of glycine with Cu^{2+} to form glycinatocopper(II) ion (a second glycinate anion may be chelated to the copper(II))

Names and Structures of Chelating Agents

Ligands can be classified according to the number of bonds that they form with metal ions. Ammonia, NH_3, can form only one such bond per NH_3 molecule and is called a *unidentate* ("one tooth") ligand, whereas glycinate ion, forming two bonds, is bidentate. Tridentate, tetradentate, pentadentate, and hexadentate ligands can form three through six such bonds, respectively. Table 10.2 lists some chelating agents of related structure that can form various numbers of bonds with metal ions. The structure of typical metal chelate in which nitrilotriacetate (NTA) is the ligand is shown in Figure 10.2.

10.4   Complexation Equilibria

Stepwise Formation Constant

The nature of complex species in solution, the feasibility of complexation analyses, and the shape of complexation titration curves are described by **complexation equilibria**. Consider first the simple complexation of copper-

TABLE 10.2 Chelating agents forming various numbers of bonds with metal ions

Name	Chelating agent Structure*	Maximum number of bonds
Aminoacetate (glycinate)	H, H—*N—C—C—O* with H O (bidentate structure)	2 (bidentate)
Iminodiacetate	⁻*O—C—C—*N—C—C—O*⁻ (with O, H groups)	3 (tridentate)
Nitrilotriacetate (NTA)	⁻*O—C—C—*N—C—C—O*⁻ with H—C—H, C=O, *O branch	4 (tetradentate)
Ethylenediaminetetraacetate (EDTA)	⁻*O—C—C—*N—C—C—N*—C—C—O*⁻ (with four acetate arms)	6 (hexadentate)

* Asterisks (*) designate ligand bonding sites.

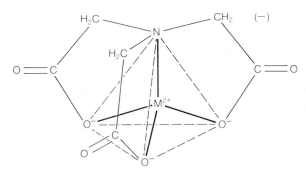

FIGURE 10.2 Nitrilotriacetate (NTA) chelate of a divalent metal ion M^{2+}, formula MT^-, where T^{3-} designates the deprotonated NTA ligand. The dashed lines outline a tetrahedron with the metal ion in the center.

(II) ion by the unidentate ligand NH_3 in water. The reaction between these two species is

$$Cu^{2+} + NH_3 \rightleftarrows CuNH_3^{2+} \tag{10.4}$$

forming monoamminecopper(II) complex ion. (Recall from Section 10.2 that the copper(II) ion is actually hydrated and NH_3 replaces H_2O. The H_2O is omitted for simplicity.) The equilibrium constant for this reaction is the **stepwise formation constant, K_1**, with the following formula:

$$K_1 = \frac{[CuNH_3^{2+}]}{[Cu^{2+}][NH_3]} = 2.0 \times 10^4 \tag{10.5}$$

The equilibrium of the addition of a second ammonia molecule,

$$CuNH_3^{2+} + NH_3 \rightleftarrows Cu(NH_3)_2^{2+} \tag{10.6}$$

is described by a second stepwise formation constant, K_2,

$$K_2 = \frac{[Cu(NH_3)_2^{2+}]}{[CuNH_3^{2+}][NH_3]} = 5.0 \times 10^3 \tag{10.7}$$

The overall process for the addition of two NH_3 molecules to a Cu^{2+} ion and the equilibrium constant for that reaction are given by the following:

$$Cu^{2+} + NH_3 \rightleftarrows CuNH_3^{2+} \qquad K_1 = \frac{[CuNH_3^{2+}]}{[Cu^{2+}][NH_3]} = 2.0 \times 10^4$$

$$\underline{CuNH_3^{2+} + NH_3 \rightleftarrows Cu(NH_3)_2^{2+} \qquad K_2 = \frac{[Cu(NH_3)_2^{2+}]}{[CuNH_3^{2+}][NH_3]} = 5.0 \times 10^3}$$

$$Cu^{2+} + 2NH_3 \rightleftarrows Cu(NH_3)_2^{2+} \tag{10.8}$$

$$\beta_2 = \frac{[Cu(NH_3)_2^{2+}]}{[Cu^{2+}][NH_3]^2} = K_1 K_2 = 1.0 \times 10^8 \tag{10.9}$$

Overall Formation Constants

The formation constant β_2 is called an **overall formation constant**. (In some calculations it may be convenient to designate K_1 as β_1; the two are identical.) Recalling that the equilibrium constant of a reaction obtained by adding two other reactions is the product of the equilibrium constants of these two reactions, β_2 is given by the following:

$$\beta_2 = K_1 K_2 \tag{10.10}$$

The stepwise and overall formation constant expressions for the complexation of a third and fourth molecule of NH_3 to copper(II) are given by the following:

$$Cu(NH_3)_2^{2+} + NH_3 \rightleftarrows Cu(NH_3)_3^{2+}$$ (10.11)

$$K_3 = \frac{[Cu(NH_3)_3^{2+}]}{[Cu(NH_3)_2^{2+}][NH_3]}$$ (10.12)

$$Cu(NH_3)_3^{2+} + NH_3 \rightleftarrows Cu(NH_3)_4^{2+}$$ (10.13)

$$K_4 = \frac{[Cu(NH_3)_4^{2+}]}{[Cu(NH_3)_3^{2+}][NH_3]}$$ (10.14)

$$Cu^{2+} + 3NH_3 \rightleftarrows Cu(NH_3)_3^{2+}$$

$$\beta_3 = \frac{[Cu(NH_3)_3^{2+}]}{[Cu^{2+}][NH_3]^3} = K_1K_2K_3$$ (10.15)

$$Cu^{2+} + 4NH_3 \rightleftarrows Cu(NH_3)_4^{2+}$$

$$\beta_4 = \frac{[Cu(NH_3)_4^{2+}]}{[Cu^{2+}][NH_3]^4} = K_1K_2K_3K_4$$ (10.16)

The values of K_3 and K_4 are, respectively, 1.0×10^3 and 2.0×10^2. Therefore, the values of β_3 and β_4 are 1.0×10^{11} and 2.0×10^{13}.

The stepwise formation constants of the ammine complexes of copper(II) are relatively close together. This means that over a wide range of NH_3 concentrations there will exist at the same time at least two, normally more, copper(II) ammine complexes in solution at significant concentrations relative to each other. This is generally true of unidentate ligands and prevents their use as titrants for the determination of metal ions in all but a few specialized cases. A major requirement for titration is a single reaction that goes essentially to completion at the equivalence point. This requirement is generally not met by unidentate ligands because of the fact that their formation constants are not very high. As explained later in the chapter, however, unidentate ligands have some important uses in analytical chemistry other than as titrants.

10.5   Influence of H$^+$ on Complex Ion Formation

Effect of H$^+$ on the Ligand

Many unidentate ligands, and practically all chelating agents, are conjugate bases of weak acids. Therefore, H$^+$ ion competes with a metal ion for the ligand and affects the equilibrium of complex formation. Consider, for example, the influence of H$^+$ on the complexation of Ag$^+$ by NH_3 as described by the following reactions and equilibrium expressions:

$$NH_4^+ \rightleftharpoons H^+ + NH_3 \tag{10.17}$$

$$K_a = \frac{[H^+][NH_3]}{[NH_4^+]} = 5.68 \times 10^{-10} \tag{10.18}$$

$$pK_a = 9.24$$

$$Ag^+ + NH_3 \rightleftharpoons AgNH_3^+ \tag{10.19}$$

$$K_1 = \frac{[AgNH_3^+]}{[Ag^+][NH_3]} = 2.04 \times 10^3 = \beta_1 \tag{10.20}$$

$$AgNH_3^+ + NH_3 \rightleftharpoons Ag(NH_3)_2^+ \tag{10.21}$$

$$K_2 = \frac{[Ag(NH_3)_2^+]}{[AgNH_3^+][NH_3]} = 7.8 \times 10^3 \tag{10.22}$$

$$Ag^+ + 2NH_3 \rightleftharpoons Ag(NH_3)_2^+ \tag{10.23}$$

$$\beta_2 = \frac{[Ag(NH_3)_2^+]}{[Ag^+][NH_3]^2} = K_1 K_2 = 1.59 \times 10^7 \tag{10.23}$$

Consider a medium that is 2.00×10^{-5} M in Ag(I) and 0.0500 M in NH_3. What is the value of $[Ag^+]$, the concentration of uncomplexed silver ion? This value can be calculated by the following treatment:

$$[Ag(NH_3)_2^+] = C_{Ag} = 2.00 \times 10^{-5} \; M \tag{10.24}$$

The above relationship is true because there is a large excess of NH_3 and essentially all of the Ag(I) is in the form of the stable $Ag(NH_3)_2^+$ complex; that is, the concentrations of Ag^+ and $AgNH_3^+$ are stoichiometrically negligible. From the expression for β_2, the following is calculated:

$$[Ag^+] = \frac{[Ag(NH_3)_2^+]}{\beta_2[NH_3]^2} = \frac{2.00 \times 10^{-5}}{1.59 \times 10^7 \times (0.0500)^2} = 5.03 \times 10^{-10} \; M \tag{10.25}$$

Consider next the same solution adjusted to pH 7.547. The pH is about 2 units below pK_a, so that most of the 0.0500 M ammonia is in the form of NH_4^+ ion. Specifically the value of $[H^+]$ is $10^{-7.547}$, which equals 2.84×10^{-8} M. At this value of $[H^+]$ the ratio

$$\frac{[NH_4^+]}{[NH_3]} = \frac{[H^+]}{K_a} = \frac{2.84 \times 10^{-8}}{5.68 \times 10^{-10}} = 50.0 \tag{10.26}$$

shows that essentially all (98%) of the ammonia is in the protonated form NH_4^+, so that the concentration of NH_4^+ can be approximated by the following:

$$[NH_4^+] = 0.0500 \; M \tag{10.27}$$

The reaction for the complexation of Ag^+ in this medium is

$$Ag^+ + 2NH_4^+ \rightleftharpoons Ag(NH_3)_2^+ + 2H^+ \tag{10.28}$$

This reaction and its equilibrium constant are given by the following:

$$Ag^+ + 2NH_3 \rightleftharpoons Ag(NH_3)_2^+ \tag{10.29}$$

$$\beta_2 = \frac{[Ag(NH_3)_2^+]}{[Ag^+][NH_3]^2} = 1.59 \times 10^7 \tag{10.30}$$

$$2NH_4^+ \rightleftharpoons 2H^+ + 2NH_3 \tag{10.31}$$

$$K_a^2 = \frac{[H^+]^2[NH_3]^2}{[NH_4^+]^2} = 3.23 \times 10^{-19} \tag{10.32}$$

$$\overline{Ag^+ + 2NH_4^+ \rightleftharpoons Ag(NH_3)_2^+ + 2H^+} \tag{10.33}$$

$$K = \frac{[Ag(NH_3)_2^+][H^+]^2}{[Ag^+][NH_4^+]^2} = \beta_2 K_a^2 = 5.13 \times 10^{-12} \tag{10.34}$$

Solving for $[Ag^+]$ gives

$$[Ag^+] = \frac{[Ag(NH_3)_2^+][H^+]^2}{K[NH_4^+]^2} = \frac{2.00 \times 10^{-5}(2.84 \times 10^{-8})^2}{5.13 \times 10^{-12}(0.0500)^2} = 1.26 \times 10^{-6}\ M \tag{10.35}$$

(Note that in solving for this value of $[Ag^+]$ it was assumed that $[Ag(NH_3)_2^+]$ was equal to the analytical concentration of Ag(I), $2.00 \times 10^{-5}\ M$. A slight correction could be made in $[Ag(NH_3)_2^+]$ for the approximately 5% of Ag(I) present as Ag^+.) Here it is seen that binding about 98% of the uncomplexed NH_3 with H^+ in the form of the NH_4^+ ion increases the value of uncomplexed Ag^+ by several orders of magnitude, indicating much less of a tendency for complexation to occur.

Effect of H^+ Concentration upon the Metal Ion

Another effect of pH upon complexation has to do with the metal ion in the complex. At higher pH values solid metal hydroxides such as $Cu(OH)_2$ may form. These tend to dissolve slowly in reacting with chelating, resulting in poor titrations. Thus methods explained later in the chapter are used to keep metal ions in solution when they are titrated with a chelating agent at relatively high pH.

10.6   Titrations Involving Unidentate Ligands

Examples of Titrations Involving Unidentate Ligands

Several titration procedures of primarily theoretical and historical interest involve unidentate ligands as either titrants or analytes. The oldest such procedure is that for the determination of I^- by titration with Hg^{2+}, dating back to 1832. The reaction is

$$Hg^{2+} + 4I^- \rightarrow HgI_4^{2-} \tag{10.36}$$

This analysis can be carried out in the presence of Cl^- or Br^- because the tetraiodomercurate(II) complex ion is much more stable than the corresponding chloride or bromide complexes. The end point can be taken as the first permanent appearance of solid red mercury(II) iodide:

$$HgI_4^{2-} + Hg^{2+} \rightleftarrows 2HgI_2(s) \tag{10.37}$$

A negative end point error is encountered. Compensation for this error may be accomplished by titrating a solution of known I^- concentration close to that of the unknown and noting the degree of the negative end point error.

Another complexometric method attributed to Liebig in 1851 is the titration of cyanide ion with Ag^+ to form the dicyanoargentate(I) complex ion:

$$Ag^+ + 2CN^- \rightarrow Ag(CN)_2^- \tag{10.38}$$

Iodide ion is used as an indicator, forming a turbid suspension of $AgI(s)$ at the end point:

$$Ag^+ + I^- \rightleftarrows AgI(s) \tag{10.39}$$

Cyanide can also be determined by titration with a solution of $NiSO_4$ to form $Ni(CN)_4^{2-}$ ion. Murexide (ammonium purpurate) can be used as an indicator.

It forms a chelate with excess Ni^{2+} having a color different from that of the unchelated dye. Cyanide itself can be used as a unidentate complexing agent for the determination of Ag^+, Cu^{2+}, Hg^{2+}, and Ni^{2+}. These form, respectively, $Ag(CN)_2^-$, $Cu(CN)_4^{2-}$, $Hg(CN)_2(aq)$, and $Ni(CN)_4^{2-}$.

Solutions of mercury(II) nitrate, $Hg(NO_3)_2$, are employed for the complexometric titration of several unidentate ligands, including Br^-, Cl^-, CN^-, and SCN^-. Where X is the ligand, these ions form $HgX_2(aq)$ when titrated with Hg^{2+}. A typical such titration consists of adding SCN^- analyte from a buret to a measured volume of standard Hg^{2+} solution containing a small quantity of iron(III). The titration reaction is

$$Hg^{2+} + 2SCN^- \rightarrow Hg(SCN)_2(aq) \tag{10.40}$$

When the Hg^{2+} is exhausted, additional SCN^- reacts with iron(III)

$$Fe^{3+} + SCN^- \rightleftharpoons FeSCN^{2+} \tag{10.41}$$

red complex ion

forming the red thiocyanatoiron(III) ion signaling the end point (note the similarity to the Volhard precipitation titration, Section 9.7).

Limitations of Titrations Involving Unidentate Ligands

Titrations like those above apply only in the rare cases in which unidentate ligands form stable complexes with a definite ligand-to-metal ratio. These requirements are met very well by several aminopolycarboxylic acid chelating agent titrants that have been adapted to the titration of a wide range of metal ions, as described in the following section. Discussed later are the uses of unidentate ligands in connection with chelation titrations with chelating agent titrants. In such an application the unidentate ligands act as auxiliary complexing agents that, for example, serve to keep metal ions in solution during titration.

10.7	**Metal Chelate Stability**

Multiple Binding Sites on Chelating Agents

A chelating agent anion with four or six electron-donor groups capable of binding to a metal ion and a structure that allows simultaneous binding to a metal ion by these groups fulfills the need for a 1:1 ligand–metal ion complex that is very stable, as required for good complexometric titrations. One such chelating agent, the NTA anion that can simultaneously bond to four sites on a metal ion, was shown in Table 10.2, and the tetrahedral structure of one of its chelates was illustrated in Figure 10.2.

For practical use as complexometric titrants, most chelating agents are aminopolycarboxylic acids containing acetic acid groups bonded to N as shown below:

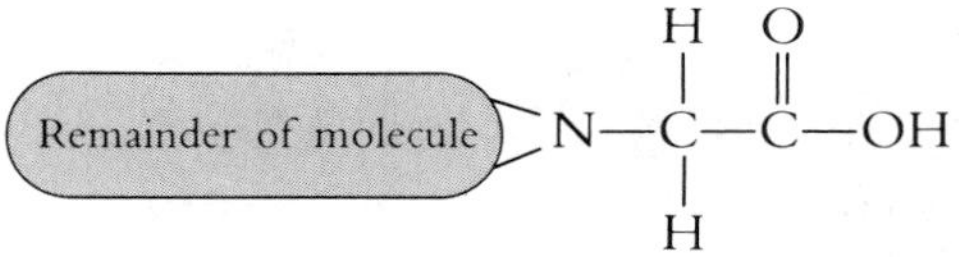

Both the carboxylate group, produced when the ionizable H is removed from the carboxylic acid group, and the N atom can bind to metal ions forming the optimum five-membered ring favored for chelate formation.

Ethylenediaminetetraacetic Acid (EDTA)

By far the most popular chelating titrant is **ethylenediaminetetraacetic acid**, also called **ethylenedinitrilotetraacetic acid**. This compound is abbreviated **EDTA**. In the fully protonated form it is commonly represented as H_4Y and with successive losses of protons as H_3Y^-, H_2Y^{2-}, HY^{3-}, and Y^{4-}. The chelating ligand is the fully deprotonated anion Y^{4-}. The structure of EDTA is shown in Figure 10.3.

The structure of an EDTA chelate of a metal ion is shown in Figure 10.4. This figure illustrates the extraordinary ability of the EDTA anion to act as a "contortionist" and bond simultaneously to six sites on the metal ion. Obviously, there is not room for another chelating agent to bond to the metal, so M^{3+} reacts with Y^{4-} to produce a 1:1 chelate, giving the desired stoichiometry for titration reactions. Furthermore, for many metals, these metal chelates are very stable, so that titration reactions go to completion at the equivalence point.

Before considering the equilibria of the reactions of EDTA with metal ions, it is necessary to take into account the fact that EDTA is protonated at all but very high pH values. The acid–base behavior of EDTA is summarized by the following chemical reactions and their corresponding equilibrium constants:

$$H_4Y \rightleftharpoons H^+ + H_3Y^- \tag{10.42}$$

$$K_{a1} = \frac{[H^+][H_3Y^-]}{[H_4Y]} = 1.0 \times 10^{-2} \tag{10.43}$$

$$pK_{a1} = 2.00$$

$$H_3Y^- \rightleftharpoons H^+ + H_2Y^{2-}$$

$$K_{a2} = \frac{[H^+][H_2Y^{2-}]}{[H_3Y^-]} = 2.1 \times 10^{-3} \tag{10.44}$$

$$pK_{a2} = 2.68$$

$$H_2Y^{2-} \rightleftharpoons H^+ + HY^{3-} \tag{10.45}$$

$$K_{a3} = \frac{[H^+][HY^{3-}]}{[H_2Y^{2-}]} = 6.9 \times 10^{-7} \tag{10.46}$$

$$pK_{a3} = 6.16$$

$$HY^{3-} \rightleftharpoons H^+ + Y^{4-} \tag{10.47}$$

$$K_{a4} = \frac{[H^+][Y^{4-}]}{[HY^{3-}]} = 5.5 \times 10^{-11} \tag{10.48}$$

$$pK_{a4} = 10.26$$

From the value of pK_{a4}, it is seen that the chelating species Y^{4-} is not the

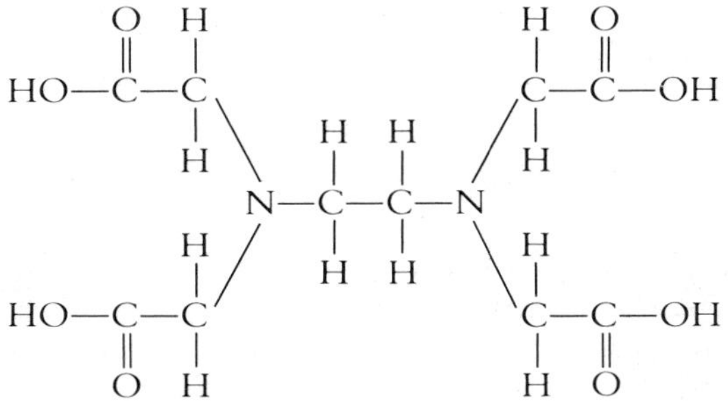

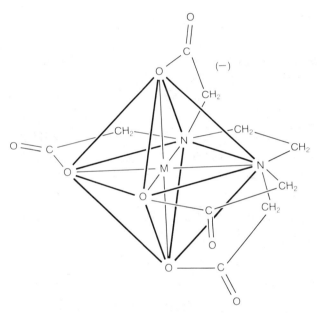

FIGURE 10.3   Structure of EDTA, H_4Y

FIGURE 10.4   EDTA anion Y^{4-} bonded to six sites on a trivalent metal ion M^{3+} to form the metal chelate MY^-. Light lines show chemical bonds (lengths not to scale) and heavy lines outline the octahedral structure of the chelate. Note that there are five rings, each composed of five atoms in the structure.

predominant form of EDTA in solution except at pH values exceeding 10.26. Since most EDTA titrations are conducted at pH's appreciably below that value, the general titration reaction is

$$H_x Y^{x-4} + M^{z+} \rightleftarrows xH^+ + MY^{z-4} \tag{10.49}$$

and the effect of pH upon the chelation equilibrium must be taken into account.

Distribution of EDTA Species as a Function of pH

The predominance of EDTA species with various degrees of protonation is best illustrated by a distribution of species diagram; the construction of these for weak polyprotic acids is discussed in Section 4.8. Such a diagram as applied to EDTA consists of a series of plots on the same graph of each of the uncomplexed EDTA species as a function of pH. These fractions are designated as α_x, where x is the unchelated EDTA species in question. For the fully protonated form H_4Y, α_{H_4Y} is given by the expression

$$\alpha_{H_4Y} = \frac{[H_4Y]}{[H_4Y] + [H_3Y^-] + [H_2Y^{2-}] + [HY^{3-}] + [Y^{4-}]} \tag{10.50}$$

Note that the sum of the five terms in the denominator is the analytical concentration of EDTA. Solving the last four of these in terms of $[H_4Y]$, $[H^+]$, and the K_a's gives the following:

$$\alpha_{H_4Y} = \frac{[H_4Y]}{[H_4Y] + \dfrac{K_{a1}[H_4Y]}{[H^+]} + \dfrac{K_{a1}K_{a2}[H_4Y]}{[H^+]^2} + \dfrac{K_{a1}K_{a2}K_{a3}[H_4Y]}{[H^+]^3} + \dfrac{K_{a1}K_{a2}K_{a3}K_{a4}[H_4Y]}{[H^+]^4}} \tag{10.51}$$

Cancelling $[H_4Y]$ in both the numerator and denominator and multiplying both numerator and denominator by $[H^+]^4$ gives the final expression

$$\alpha_{H_4Y} = \frac{[H^+]^4}{[H^+]^4 + K_{a1}[H^+]^3 + K_{a1}K_{a2}[H^+]^2 + K_{a1}K_{a2}K_{a3}[H^+] + K_{a1}K_{a2}K_{a3}K_{a4}} \tag{10.52}$$

Derived similarly, the other α values are the following:

$$\alpha_{H_3Y^-} = \frac{K_{a1}[H^+]^3}{[H^+]^4 + K_{a1}[H^+]^3 + K_{a1}K_{a2}[H^+]^2 + K_{a1}K_{a2}K_{a3}[H^+] + K_{a1}K_{a2}K_{a3}K_{a4}} \tag{10.53}$$

$$\alpha_{H_2Y^{2-}} = \frac{K_{a1}K_{a2}[H^+]^2}{[H^+]^4 + K_{a1}[H^+]^3 + K_{a1}K_{a2}[H^+]^2 + K_{a1}K_{a2}K_{a3}[H^+] + K_{a1}K_{a2}K_{a3}K_{a4}} \tag{10.54}$$

$$\alpha_{HY^{3-}} = \frac{K_{a1}K_{a2}K_{a3}[H^+]}{[H^+]^4 + K_{a1}[H^+]^3 + K_{a1}K_{a2}[H^+]^2 + K_{a1}K_{a2}K_{a3}[H^+] + K_{a1}K_{a2}K_{a3}K_{a4}} \tag{10.55}$$

$$\alpha_{Y^{4-}} = \frac{K_{a1}K_{a2}K_{a3}K_{a4}}{[H^+]^4 + K_{a1}[H^+]^3 + K_{a1}K_{a2}[H^+]^2 + K_{a1}K_{a2}K_{a3}[H^+] + K_{a1}K_{a2}K_{a3}K_{a4}} \tag{10.56}$$

Equations 10.52 through 10.56 can be used to plot a distribution of species diagram for the EDTA system. This plot is shown in Figure 10.5. It is seen that a pH of almost 12 is required before essentially all of the EDTA can be regarded as being in the complexing Y^{4-} form. Furthermore, there are no long

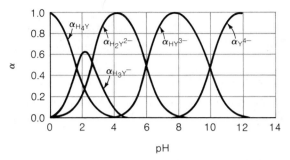

FIGURE 10.5 Distribution of uncomplexed EDTA species as a function of pH

ranges of pH in which any one form of the uncomplexed EDTA can be regarded as predominant. These considerations complicate calculations involving generalized reactions of the type shown in Reaction 10.49 (but also enable selectivity of chelation through pH control).

$$H_xY^{x-4} + M^{z+} \rightleftharpoons xH^+ + MY^{z-4} \tag{10.49}$$

Conditional Formation Constants

The problem of multiple EDTA species in stability calculations may be remedied by defining **conditional formation constants** in the form

$$K' = \frac{[MY^{z-4}]}{C_T[M^{z+}]} \tag{10.57}$$

where C_T is the total concentration of *uncomplexed* EDTA

$$C_T = [H_4Y] + [H_3Y^-] + [H_2Y^{2-}] + [HY^{3-}] + [Y^{4-}] \tag{10.58}$$

The value of C_T is related to $[Y^{4-}]$ and $\alpha_{Y^{4-}}$ by the equation

$$C_T = \frac{[Y^{4-}]}{\alpha_{Y^{4-}}} \tag{10.59}$$

Substituting this value into Equation 10.57 gives

$$K' = \underbrace{\frac{[MY^{z-4}]}{[Y^{4-}][M^{z+}]}}_{K} \alpha_{Y^{4-}} \tag{10.60}$$

The portion of the equation above corresponding to the formation constant K is so designated; the conditional formation constant is given simply by the following:

$$K' = K\alpha_{Y^{4-}} \tag{10.61}$$

EXAMPLE

The formation constant K of CuY^{2-} is 6.3×10^{18}. The value of $\alpha_{Y^{4-}}$ at pH 7.00 is 4.8×10^{-4}. What is the conditional formation constant K' of CuY^{2-} at pH 7.00?

$$K' = K\alpha_{Y^{4-}} = 6.3 \times 10^{18} \times 4.8 \times 10^{-4} = 3.0 \times 10^{15}$$

What is the value of $[Cu^{2+}]$ in a solution formed by mixing exactly 0.0100 mol of copper(II) salt and 0.0100 mol of EDTA in 1.00 L of solution adjusted to pH 7.00?

$$[CuY^{2-}] = 0.0100 \ M \text{ because } K' \text{ is very large}$$

$$[Cu^{2+}] = C_T$$

$$\frac{[CuY^{2-}]}{C_T[Cu^{2+}]} = \frac{0.0100}{[Cu^{2+}]^2} = 3.0 \times 10^{15}$$

$$[Cu^{2+}] = \sqrt{\frac{0.0100}{3.0 \times 10^{15}}} = 1.8 \times 10^{-9} \ M$$

Table 10.3 gives values of $\alpha_{Y^{4-}}$ calculated from Equation 10.56 at various pH's. Table 10.4 gives formation constants of a number of common metal-EDTA chelates. From the information contained in these two tables it is possible to calculate a number of K' values for metal ions.

Auxiliary Complexing Agents

Often the titration of a metal ion with EDTA is performed at a relatively high pH where the unchelated metal ion is not soluble. Under those conditions, an **auxiliary complexing agent** can be placed in the solution to prevent precipitation of the analyte metal ion prior to chelation by EDTA. An auxiliary complexing agent is often the conjugate base in the buffer system used to regulate the titration pH. For example, when copper(II) is titrated with EDTA using murexide indicator, the pH is buffered to 10 to enable the indicator to function properly. Furthermore, at this pH EDTA is substantially deprotonated. A buffer of NH_4Cl and NH_3 is used in which the NH_3 prevents precipitation of $Cu(OH)_2$ by binding to Cu^{2+} as an auxiliary complexing agent, forming soluble ammine complexes of copper(II). Conditional formation constants account for complexation of the metal ion with an auxiliary complexing agent. In such a case, as illustrated for the copper-EDTA chelate, the conditional formation constant is in the form

$$K'' = \frac{[CuY^{2-}]}{C_M C_T} \tag{10.62}$$

where C_M is the total concentration of copper(II) *not chelated* with EDTA:

$$C_M = [Cu^{2+}] + [CuNH_3^{2+}] + [Cu(NH_3)_2^{2+}] + [Cu(NH_3)_3^{2+}] + [Cu(NH_3)_4^{2+}] \tag{10.63}$$

TABLE 10.3   Values of $\alpha_{Y^{4-}}$ for EDTA at various pH values

pH	$\alpha_{Y^{4-}}$	pH	$\alpha_{Y^{4-}}$
2.00	3.7×10^{-14}	7.50	1.7×10^{-3}
3.00	2.5×10^{-11}	8.00	5.4×10^{-3}
4.00	3.6×10^{-9}	8.50	1.7×10^{-2}
5.00	3.5×10^{-7}	9.00	5.2×10^{-2}
6.00	2.2×10^{-5}	10.00	3.5×10^{-1}
6.50	1.7×10^{-4}	11.00	8.5×10^{-1}
7.00	4.8×10^{-4}	12.00	9.8×10^{-1}

TABLE 10.4   Formation constants of metal–EDTA chelates*

Metal cation	$K^{\dagger}$	log K	Metal cation	$K^{\dagger}$	log K
Ag^+	2.1×10^7	7.32	Cu^{2+}	6.3×10^{18}	18.80
Mg^{2+}	4.9×10^8	8.69	Zn^{2+}	3.2×10^{16}	16.50
Ca^{2+}	5.0×10^{10}	10.70	Cd^{2+}	2.9×10^{16}	16.46
Sr^{2+}	4.3×10^8	8.63	Hg^{2+}	6.3×10^{21}	21.80
Ba^{2+}	5.8×10^7	7.76	Pb^{2+}	1.1×10^{18}	18.04
Mn^{2+}	6.2×10^{13}	13.79	Al^{3+}	1.3×10^{16}	16.13
Fe^{2+}	2.1×10^{14}	14.33	Fe^{3+}	1×10^{25}	25.1
Co^{2+}	2.0×10^{16}	16.31	V^{2+}	5.0×10^{12}	12.70
Ni^{2+}	4.2×10^{18}	18.62	V^{3+}	8×10^{25}	25.9
Bi^{3+}	6.3×10^{22}	22.80	Th^{4+}	2×10^{23}	23.2
Be^{2+}	2.0×10^9	9.30	VO^{2+}	6×10^{18}	18.8
Cr^{3+}	1×10^{23}	23.0	Sn^{2+}	1.3×10^{22}	22.10

* For the reaction $M^{z+} + Y^{4-} \rightleftarrows MY^{z-4}$

$$K = \frac{[MY^{z-4}]}{[M^{z+}][Y^{4-}]}$$

† Published values of K vary with the source, ionic strength, and other factors. For research purposes, values of K that are applicable to the conditions under study should be taken from the literature.

The fraction of copper(II) not chelated by EDTA that is in the form of the Cu^{2+} ion, δ, is expressed by the following:

$$\delta = \frac{[Cu^{2+}]}{[Cu^{2+}] + [CuNH_3^{2+}] + [Cu(NH_3)_2^{2+}] + [Cu(NH_3)_3^{2+}] + [Cu(NH_3)_4^{2+}]} \quad (10.64)$$

The values of the copper ammine complex concentrations appearing in the denominator of this expression can be expressed in terms of $[Cu^{2+}]$, $[NH_3]$, and the overall formation constants of the complex ions by relationships of the general type

$$[Cu(NH_3)_i^{2+}] = \beta_i [Cu^{2+}][NH_3]^i \quad (10.65)$$

where i has values of 1 to 4. Substitution of the four equations resulting from the four values of i into the expression for δ and cancellation of $[Cu^{2+}]$ in the numerator and denominator lead to the following expression:

$$\delta = \frac{1}{1 + \beta_1[NH_3] + \beta_2[NH_3]^2 + \beta_3[NH_3]^3 + \beta_4[NH_3]^4} \tag{10.66}$$

The value of $[Cu^{2+}]$ is related to C_M and δ by the relationship

$$[Cu^{2+}] = C_M\delta \tag{10.67}$$

which rearranges to

$$C_M = \frac{[Cu^{2+}]}{\delta} \tag{10.68}$$

The formation constant expression for the EDTA chelate of copper(II) is

$$K = \frac{[CuY^{2-}]}{[Cu^{2+}][Y^{4-}]} \tag{10.69}$$

Recall from Equation 10.59 that C_T is given by

$$C_T = \frac{[Y^{4-}]}{\alpha_{Y^{4-}}} \tag{10.59}$$

Substitution into the equation for K'', Equation 10.62, yields

$$K'' = \frac{[CuY^{2-}]}{\dfrac{[Cu^{2+}]}{\delta} \dfrac{[Y^{4-}]}{\alpha_{Y4_-}}} = K\delta\alpha_{Y^{4-}} \tag{10.70}$$

10.8   Titrations with Chelating Agents

EDTA Titrant for Copper(II)

In modern practice complexometric titrations are almost invariably performed with EDTA, often in the presence of an auxiliary complexing agent such as NH_3 to keep the metal in solution prior to the equivalence point. A typical titration of this nature is that of Cu(II) in a medium containing $0.100\ M$ NH_3 adjusted to pH 10 with NH_3–NH_4^+ buffer. Under these conditions δ is given by substitution into Equation 10.66 using the β values for copper-ammine complexes as follows:

$$\delta = \frac{1}{1 + \beta_1[NH_3] + \beta_2[NH_3]^2 + \beta_3[NH_3]^3 + \beta_4[NH_3]^4}$$

$$= \frac{1}{1 + 2.0 \times 10^4(0.100) + 1.0 \times 10^8(0.100)^2 + 1.0 \times 10^{11}(0.100)^3 + 2.0 \times 10^{13}(0.100)^4}$$

$$= 4.76 \times 10^{-10} \tag{10.71}$$

Consider the titration of 25.00 mL of 0.01000 M Cu(II) with 0.01000 M EDTA in a medium maintained at pH 10 and 0.0100 M in NH_3. The curve for this titration is a plot of log [Cu^{2+}] versus milliliters of added EDTA titrant. Prior to the equivalence point, Cu^{2+} is in excess, and its concentration is calculated simply by the relationship

$$[Cu^{2+}] = \frac{\text{mmol Cu(II) not chelated by EDTA}}{\text{total volume of solution}} \delta \qquad (10.72)$$

Initially,

$$[Cu^{2+}] = 0.01000 \times 4.76 \times 10^{-10} = 4.76 \times 10^{-12} \ M \qquad (10.73)$$

$$\log [Cu^{2+}] = -11.32$$

(By analogy with pH, this value may be expressed as $pCu = -\log [Cu^{2+}]$, which has the advantage of normally giving positive values. In this case $pCu = 11.32$.)

After the addition of 5.00 mL of EDTA,

$$[Cu^{2+}] = \frac{0.2500 \ \text{mmol} - 0.0500 \ \text{mmol}}{25.00 \ \text{mL} + 5.00 \ \text{mL}} \times 4.76 \times 10^{-10} = 3.17 \times 10^{-12} \ M \qquad (10.74)$$

$$\log [Cu^{2+}] = -11.50$$

Other values of log [Cu^{2+}] prior to the equivalence point are calculated similarly and are given in Table 10.5.

At the equivalence point all but an extremely small fraction of the Cu(II) originally present and the EDTA added is in the form of CuY^{2-} so that essentially [CuY^{2-}] is given by the following:

$$[CuY^{2-}] = \frac{0.2500 \ \text{mmol}}{50.00 \ \text{mL}} = 5.00 \times 10^{-3} \ M \qquad (10.75)$$

TABLE 10.5   Log [Cu^{2+}] as a function of milliliters of added EDTA titrant for the titration of 25.00 mL 0.01000 M Cu^{2+} with 0.01000 M EDTA in a medium maintained at pH 10.00 and 0.100 M in NH_3

EDTA added, mL	Total volume, mL	Excess $Cu(NH_3)_i^{2+}$, mmol	CuY^{2-} formed, mmol	Excess EDTA, mmol	[Cu^{2+}], M	log [Cu^{2+}]
0	25.00	0.2500	0	—	4.76×10^{-12}	-11.32
5.00	30.00	0.2000	0.0500	—	3.17×10^{-12}	-11.50
12.50	37.50	0.1250	0.1250	—	1.59×10^{-12}	-11.80
20.00	45.00	0.0500	0.2000	—	5.29×10^{-13}	-12.28
24.00	49.00	0.0100	0.2400	—	9.71×10^{-14}	-13.01
25.00	50.00	—	0.2500	—	1.04×10^{-16}	-14.98
26.00	51.00	—	0.2500	0.0100	1.14×10^{-17}	-16.94
30.00	55.00	—	0.2500	0.0500	2.28×10^{-18}	-17.64
35.00	60.00	—	0.2500	0.1000	1.14×10^{-18}	-17.94

Slight dissociation of this chelate leads to the equality

$$C_M = C_T \tag{10.76}$$

The value of C_M can be obtained by substituting into the expression for K'' as follows:

$$K'' = \frac{[CuY^{2-}]}{C_M C_T} = K\delta\alpha_{Y^{4-}} = 6.3 \times 10^{18} \times 4.76 \times 10^{-10} \times 0.35 \tag{10.77}$$

$$\underset{\text{From Table 10.4}}{\uparrow} \qquad \underset{\text{From Equation 10.71}}{\uparrow} \qquad \underset{\text{From Table 10.3}}{\uparrow}$$

$$= 1.05 \times 10^9$$

$$C_M = C_T = \sqrt{\frac{[CuY^{2-}]}{K''}} = \sqrt{\frac{5.00 \times 10^{-3}}{1.05 \times 10^9}} = 2.18 \times 10^{-6} \tag{10.78}$$

The value of $[Cu^{2+}]$ is obtained from multiplying C_T by δ

$$[Cu^{2+}] = C_T\delta = 2.18 \times 10^{-6} \times 4.76 \times 10^{-10} = 1.038 \times 10^{-15} \tag{10.79}$$

$$\log[Cu^{2+}] = -14.98$$

Beyond the equivalence point there is an excess of EDTA, so that C_T as well as $[CuY^{2-}]$ can be calculated directly from stoichiometric considerations. Thus it is possible to substitute directly into the expression for K' (see Equation 10.57) to obtain the value of $[Cu^{2+}]$. For the system under consideration, K' is the following:

$$K' = \frac{[CuY^{2-}]}{C_T[Cu^{2+}]} = K\alpha_{Y^{4-}} = 6.3 \times 10^{18} \times 0.35 = 2.20 \times 10^{18} \tag{10.80}$$

At the point where a total of 30.00 mL of EDTA titrant has been added, there is a total of 0.2500 mmol CuY^{2-} and 0.0500 mmol excess EDTA in a volume of 55.00 mL of solution, so that

$$[CuY^{2-}] = \frac{0.2500 \text{ mmol}}{55.00 \text{ mL}} = 4.55 \times 10^{-3} \text{ M} \tag{10.81}$$

$$C_T = \frac{0.0500 \text{ mmol}}{55.00 \text{ mL}} = 9.09 \times 10^{-4} \text{ M} \tag{10.82}$$

The value of $[Cu^{2+}]$ is given by

$$[Cu^{2+}] = \frac{[CuY^{2-}]}{C_T K'} = \frac{4.55 \times 10^{-3}}{9.09 \times 10^{-4} \times 2.20 \times 10^{18}} = 2.28 \times 10^{-18} \text{ M} \tag{10.83}$$

$$\log[Cu^{2+}] = -17.64$$

Other values of $\log[Cu^{2+}]$ beyond the equivalence point are given in Table 10.5.

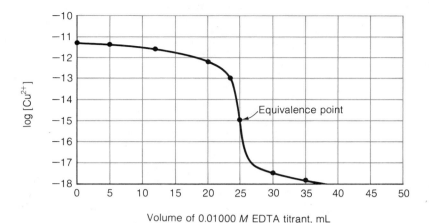

FIGURE 10.6 Curve for the titration of 25.00 mL of 0.01000 M Cu(II) with 0.01000 M EDTA in a medium maintained at pH 10.00 and 0.10 M in NH_3

EDTA Titration Curve

The titration curve constructed from the data in Table 10.5 is plotted in Figure 10.6. It is a symmetrical curve with a distinct break at the equivalence point. If log [Cu^{2+}] were monitored (by a copper ion–selective electrode, see Chapter 13) as a function of added EDTA, this titration curve could be obtained experimentally. It further suggests that the end point could be observed by the change in some solution parameter with the marked change in [Cu^{2+}] that occurs at the equivalence point. Indicators are available for this purpose, as discussed in the following section.

10.9   Metallochromic Indicators for EDTA Titrations

What Are Metallochromic Indicators?

A **metallochromic indicator** is an organic complexing agent dye that has distinctly different colors when complexed and when not complexed with a metal ion. These compounds contain light-absorbing chromophore groups such as —N=N— or >C=O, as well as electron-donor functional groups such as —CO_2H, phenolic —OH, —NH_2, or —$CH_2N(CH_2CO_2H)_2$. Normally the metallochromic indicator forms a chelate with a metal ion involving a stable five-membered ring structure in which the electron-donor functional groups bond to the metal ion. The color change resulting from the formation of an indicator-metal complex is usually greater when it results in the release of H^+ ion. The structure of one common metallochromic indicator, murexide, was shown in Section 10.6. Others are illustrated in Table 10.6.

TABLE 10.6   Some metallochromic indicators used in EDTA titrations

Indicator name	Indicator structure	Uses
Eriochrome black T		Mg, Ca, Zn, water hardness
Naphthol violet		Bi, Cd, Co(II), Cu, Mg, Mn(II), Zn
Phthalein complexone		Ba, Ca, Cd, Mg, Sr
Zincon		Zn

* These compounds contain the azo, —N=N—, chromophoric group.
† These indicators contain the sulfonic acid group, —SO$_3^-$, to increase solubility.
‡ These indicators contain the chelating —N(CH$_2$CO$_2$H)$_2$ moiety.
§ This indicator contains the phthalein chromophore (see phenolphthalein structure).

How Metallochromic Indicators Work

As shown for the titration of Cu^{2+} in Figure 10.6, there is a change in metal ion concentration of several orders of magnitude in the equivalence-point region for the titration of a metal ion with EDTA. If a small portion of the metal is complexed with a metallochromic indicator, this reduction in metal ion concentration at the equivalence point results in dissociation of the indicator-metal complex and a corresponding change in color. For example, if a divalent metal cation M^{2+} complexed with an indicator anion In^{3-} reacts with the EDTA species HY^{3-}, at the end point a reaction of the type

$$\underset{\text{color}_1}{MIn^-} + HY^{3-} \rightarrow MY^{2-} + \underset{\text{color}_2}{HIn^{2-}} \tag{10.84}$$

may occur, where MIn^- is the colored indicator-metal complex and HIn^{2-} is the free, uncomplexed indicator of a different color. The equilibrium constant of this reaction should be such that it occurs near the equivalence point.

Acid-Base Character of Metallochromic Indicators

All common metallochromic indicators are weak acids. Thus they lose hydrogen ion with increasing pH. This commonly results in different-colored species, so that metallochromic indicators can often function as acid-base indicators. In the case of Eriochrome black T, for example, the acid-base reactions are the following:

$$\underset{\text{red}}{H_3In} \rightleftarrows H^+ + \underset{\text{red}}{H_2In^-} \tag{10.85}$$

$$K_{a1} = 1.3 \times 10^{-4} \tag{10.86}$$

$$\underset{\text{red}}{H_2In^-} \rightleftarrows H^+ + \underset{\text{blue}}{HIn^{2-}} \tag{10.87}$$

$$K_{a2} = 4.0 \times 10^{-7} \tag{10.88}$$

$$\underset{\text{blue}}{HIn^{2-}} \rightleftarrows H^+ + \underset{\text{orange}}{In^{3-}} \tag{10.89}$$

$$K_{a3} = 3.2 \times 10^{-12} \tag{10.90}$$

The most readily ionizable proton (K_{a1}) is that on the sulfonic acid group, whereas the two phenolic H's (K_{a2}, K_{a3}) are much less easily ionized.

Because of the acid-base character of metallochromic indicators, proper pH adjustment is crucial for their use. For example, most Eriochrome black T complexes with metal ions are red. Therefore the pH of the titration medium in which this indicator is used should exceed 7, so that the free indicator released at the end point is a different color (blue) than the metal complex.

TABLE 10.7   Colors and pH predominance ranges of free and calcium-complexed murexide* species

Range of pH	Predominant species	Color
No Ca(II) present		
4–9.1	H_4In^-	Red-violet
9.1–10.5	H_3In^{2-}	Violet
10.5–13	H_2In^{3-}	Blue
Ca(II) present		
4–7.8	CaH_4In^+	Yellow-orange
7.8–9.6	CaH_3In	Red-orange
9.6–13	CaH_2In^-	Red

* The structure of murexide indicator is given in Section 10.6.

Consider as an example the titration of Mg^{2+} with EDTA employing Eriochrome black T as an indicator at pH 10. Examination of the distribution of species plot for EDTA, Figure 10.5, shows that at pH 10, unchelated EDTA consists of approximately equal concentrations of HY^{3-} and Y^{4-}. Thus the two end point reactions with Eriochrome black T as an indicator would be

$$MgIn^- + HY^{3-} \rightarrow MgY^{2-} + HIn^{2-} \tag{10.91}$$
$$\text{red} \qquad\qquad \text{colorless} \quad \text{blue}$$

$$MgIn^- + Y^{4-} + H_2O \rightarrow MgY^{2-} + HIn^{2-} + OH^- \tag{10.92}$$
$$\text{red} \qquad\qquad\qquad \text{colorless} \quad \text{blue}$$

The complexities of the species and the resulting colors that may be encountered at different pH levels with metallochromic indicators and their metal complexes are illustrated in Table 10.7 using, for example, various forms of uncomplexed and calcium-complexed murexide. As a function of pH and the presence or absence of Ca(II), six different species with six different colors may be present!

10.10   Color Transitions with Metallochromic Indicators

Metallochromic Indicator Color Transition Point and Range

Under specified solution conditions, including indicator concentration (commonly 1×10^{-5} M) and particularly pH, a **color transition point** can be determined or calculated for a metallochromic indicator used with a particular metal. The color transition point may be expressed in terms of the log of the hydrated metal ion concentration, above which the indicator is predominantly in the form of the metal complex, and below which the indicator is predominantly in the uncomplexed form of a different color. (The color transition point can also be expressed in terms of the negative log of metal ion concentration, pM, and is given as such in much of the literature.)

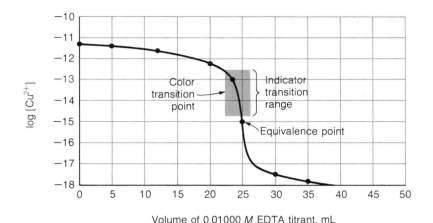

FIGURE 10.7   Curve for the titration of 25.00 mL of 0.01000 M Cu(II) with 0.01000 M EDTA in a medium maintained at pH 10.00 and 0.10 M in NH_3, showing the color transition point and the indicator transition range for murexide indicator

To illustrate the concept of the color transition point, consider the example of Cu(II) titrated with EDTA at pH 10 in 0.10 M NH_3. The titration curve is shown in Figure 10.6 and is reproduced in Figure 10.7. The published color transition point for murexide under these titration conditions is $\log [Cu^{2+}] = -13.6$. As seen in Figure 10.6 this occurs slightly before the equivalence point and would give a small, but tolerable negative end point error. By analogy with the indicator pH range for acid-base indicators (Section 5.6) there is an **indicator transition range** for metallochromic indicators. This range is normally calculated from the point at which the ratio of the indicator metal complex to uncomplexed indicator is 10:1 to the point at which the ratio is 1:10. It can be shown that this occurs over two orders of magnitude in uncomplexed metal ion concentration—that is, two units of the log of the metal ion concentration. In the case under consideration—titration of Cu(II) employing murexide as the indicator—the calculated indicator transition range would be $-12.6 < \log [Cu^{2+}] < -14.6$. This indicator transition range is shown in Figure 10.7.

Eriochrome Black T Indicator for Calcium Titration

If the formation constant of the indicator-metal complex is too low, the colored complex dissociates to metal ion and uncomplexed indicator prior to the equivalence point, giving an unacceptable negative end point error. This is the case when Eriochrome black T is used in the titration of calcium. However, an indirect method can still be used to employ this indicator in an

EDTA titration of Ca^{2+} when Mg^{2+} is present. This is because the formation constant of the calcium–Eriochrome black T complex is only approximately one-fortieth that of the magnesium complex, whereas (see Table 10.4) the formation constant of the CaY^{2-} chelate is about two orders of magnitude higher than that of the MgY^{2-} chelate. For the titration of a Ca^{2+} solution containing Eriochrome black T indicator with EDTA, a small quantity of Na_2MgY salt can be added (this does not affect the stoichiometry of the titration reaction because the salt contains equimolar quantities of Mg and EDTA). Upon the addition of Na_2MgY, the EDTA is exchanged from Mg^{2+} to Ca^{2+}

$$MgY^{2-} + Ca^{2+} \rightarrow Mg^{2+} + CaY^{2-} \tag{10.93}$$

and the Mg(II) takes up the indicator:

$$\underset{\text{red}}{CaIn^-} + Mg^{2+} \rightarrow Ca^{2+} + MgIn^- \tag{10.94}$$

As EDTA titrant is added, it first binds to all the Ca^{2+},

$$Ca^{2+} + H_xY^{x-4} \rightarrow CaY^{2-} + xH^+ \tag{10.95}$$

until at the end point EDTA displaces the less strongly bound Eriochrome black T from Mg(II)

$$\underset{\text{red}}{MgIn^-} + H_xY^{x-4} \rightarrow MgY^{2-} + \underset{\text{blue}}{HIn^{2-}} + (x-1)H^+ \tag{10.96}$$

resulting in the end point color change from red to blue.

10.11   Uses of EDTA Titrations

In order for a metal ion to be determined by direct titration with EDTA, (1) the metal-EDTA chelate must be stable, (2) the chelate must form rapidly, (3) the reaction must be stoichiometrically complete, and (4) a satisfactory indicator must be available. About 40 metal ions may be determined by direct titration.

In cases where a satisfactory indicator is not available but the metal chelate is more stable than MgY^{2-}, excess EDTA can be added to chelate the analyte metal

$$M^{2+} + H_xY^{x-4} \rightarrow MY^{2-} + xH^+ \tag{10.97}$$

and the excess chelating agent determined by back-titration with Mg^{2+} employing either Eriochrome black T or Calmagite as an indicator. This technique can be employed under conditions where the MY^{2-} chelate is formed somewhat slowly. It is also useful under solution conditions such that the analyte ion is not soluble (for example, formation of $M(OH)_2(s)$ in basic media) because the MY^{2-} chelate is a soluble species.

A variation of back-titration makes use of a standard solution of MgY^{2-} or ZnY^{2-} reagent in which the reaction with the analyte ion M^{2+} is

$$MgY^{2-} + M^{2+} \rightarrow MY^{2-} + Mg^{2+} \tag{10.98}$$

The displaced Mg^{2+} ion may then be titrated with standard EDTA using indicators mentioned previously. Analogous procedures can be used with ZnY^{2-} to provide Y^{2-} to react with the analyte ion. The excess Zn^{2+} may be titrated with EDTA employing Zincon indicator. This type of titration is called a **displacement titration**.

Analyte cations may be determined in neutral solution by adding an excess of a solution of Na_2H_2Y

$$M^{2+} + H_2Y^{2-} \rightarrow 2H^+ + MY^{2-} \tag{10.99}$$

and titrating the liberated H^+ by employing a solution of standard base. Such a determination is called an **alkalimetric titration**.

10.12   Determination of Water Hardness with EDTA

Water hardness is a measure of the total Ca^{2+} and Mg^{2+} in water; it is an important quality of water that is used for municipal or industrial purposes. A certain degree of water hardness is necessary in municipal water systems to prevent corrosion of pipes; excessive levels result in harmful deposits of solid calcium carbonate and other calcium and magnesium minerals.

Water hardness is conveniently determined by titration of total Ca^{2+} and Mg^{2+} with EDTA. The pH of the solution is adjusted to 10 to prevent excessive competition of H^+ ion for Y^{4-}; higher pH values would cause precipitation of $CaCO_3(s)$ or $Mg(OH)_2(s)$. Eriochrome black T is used as an indicator. Its use requires the presence of $Mg(II)$, which may be added as Na_2MgY (see Section 10.10).

Hardness is one of the most common water-quality parameters determined. Simple test kits are available for this determination. These usually have a container for a specific volume of water and a scoop to measure, or sealed packets containing, the correct amount of buffer. The EDTA titrant is added with a calibrated dropper. Commonly each drop of titrant corresponds to 1 grain (about 0.065 g) of $CaCO_3$ per gallon of water. Though imprecise by common laboratory standards, such determinations are rapidly performed and are quite satisfactory for monitoring water-treatment processes.

Programmed Summary of Chapter 10

The major terms and concepts introduced in this chapter are summarized below in a programmed format. To derive the most benefit from the summary, you should fill in the blanks for each question and then check the answers at the end of the book to see if your choices are correct.

In the overall process of complexation, an electron-donor species called a (1) _____ bonds reversibly to a (2) _____ to form a (3) _____. As an example of a name of a complex ion, the species $Fe(NH_3)_6^{2+}$ is called (4) _____. A chelating agent is a ligand that can (5) _____ to the same metal ion. If a chelating agent can bond simultaneously in five places to the same metal ion, it is said to be a (6) _____ chelating agent. The expression $[Cu(CN)_3^-]/[Cu(CN)_2][CN^-]$ is called a (7) _____ formation constant, whereas the expression $[Cu(CN)_3^-]/[Cu^{2+}][CN^-]^3$ is an (8) _____. The symbols, respectively, for the two preceding expressions are (9) _____. Ignoring the effect upon the metal ions, for most ligands decreasing the pH (10) _____ the stability of the complex. The unidentate ligands most commonly involved in complexometric titrations, either as titrants or analytes, are (11) _____. For successful complexometric analysis, complexing agents should react with a metal ion in a (12) _____ ratio. The most commonly used chelating titrants belong to the category of (13) _____ characterized by the group (14) _____ bonded to N atoms. Of these, the most commonly used is called (15) _____ and is abbreviated (16) _____. The compound NTA is a triprotic acid and, in the anionic form, a tetradentate chelating agent. If the completely protonated form is designated H_3T, in terms of K_{a1}, K_{a2}, K_{a3}, and $[H^+]$, $\alpha_{T^{3-}} =$ (17) _____. An equilibrium constant that compensates for protonation of a complexing agent and complexing of the metal ion with an auxiliary complexing agent is called a (18) _____. For an EDTA chelate of a metal ion in which the EDTA is partially protonated and the metal ion is complexed with an auxiliary complexing agent, using the symbols given in this chapter, the value of K'' is given by the product (19) _____. A colored complexing agent that is displaced from a metal ion by a complexing titrant near the equivalence point, resulting in a color change, is called a (20) _____. In order for Eriochrome black T to work as an indicator in the titration of Ca^{2+} with EDTA, the presence of (21) _____ is required. For a metallochromic indicator, the range of the log of the free metal ion concentration over which the indicator changes color is called the (22) _____. A titration technique in which excess EDTA is added to an analyte metal ion solution and the excess titrated with Mg^{2+} is called a (23) _____. The addition of Na_2H_2Y solution to M^{2+} analyte, followed by titration of the H^+ liberated, is called an (24) _____. The parameter in municipal or industrial water supplies most commonly determined by EDTA titration is (25) _____.

Questions

1. What is the distinction between solvation and hydration? In what sense are these phenomena complexation?

2. In the reaction

$$Cd(H_2O)_6^{2+} + Cl^- \rightleftarrows \\ Cd(H_2O)_5Cl^+ + H_2O$$

identify the aquo complex, the ligand, the central metal ion, and the complex ion.

3. What is the definition of coordination number of a metal complex ion?

4. What are the two major geometrical structures for complex ions with a coordination number of 4?

5. Name (a) $Cu(CN)_3^-$, (b) $Fe(OH)_2^+$, (c) $Mn(NH_3)_6SO_4$, (d) $Zn(OH)_2$.

6. Name (a) $Ag(CN)_2^-$, (b) $HgBr_4^{2-}$, (c) $Cr(NH_3)_5ClSO_4$.

7. What is the distinction between chelation and complexation?

8. What are the two major advantages of EDTA as a chelating titrant?

9. A metal ion M^{2+} forms the complexes ML^{2+} and ML_2^{2+} in a stepwise manner with the ligand L. Use these examples to illustrate stepwise and overall formation constants.

10. What are two effects of pH upon complexation?

11. List some unidentate ligands that have been used as complexometric titrants and the metal ions that each has been used to determine.

12. What general type of chelating agent is used for most practical complexometric titrations?

13. How many rings are formed by an EDTA anion bonded at all possible sites to a metal ion? How many atoms are in each ring?

14. From the distribution of EDTA species as a function of pH, state the predominant species at pH values of 4, 7, and 10.

15. Write the expression for the conditional formation constant K'' and define each term in the expression.

16. What is the purpose of an auxiliary complexing agent?

17. Suppose that a metal ion M^{2+} is titrated with EDTA in the presence of an auxiliary complexing agent to give a titration curve such as that shown in Figure 10.6. At the equivalence point what are the values of, or relationships between, $[MY^{2-}]$, C_M, and C_T, the parameters in the conditional formation constant K'' considering total mmol M(II) and solution volume?

18. What is the definition of a metallochromic indicator?

19. Why must Mg(II) be present in the titration of Ca(II) with EDTA when Eriochrome black T is used as an indicator.

20. An auxiliary complexing agent such as NH_3 (a) has virtually no use in chemical analysis, (b) is usually used for the direct titration of an analyte metal ion, (c) may serve to keep the metal ion in solution for titration with EDTA, (d) acts as an indicator, (e) normally is used to keep the solution acidic.

Problems

1. The table below pertains to the stepwise and overall formation constants of cadmium(II) chloro complexes. Fill in the blanks designated by letters.

Complex, $HgCl_x^{2-x}$	K_x	β_x
$HgCl^+$	5.0×10^6	5.0×10^6
$HgCl_2$	3.2×10^6	(a) _____
$HgCl_3^-$	(b) _____	1.3×10^{14}
$HgCl_4^{2-}$	1.0×10^1	(c) _____

2. For the aminopolycarboxylic acid chelating agent nitrilotriacetic acid, NTA, $K_{a1} = 2.18 \times 10^{-2}$, $K_{a2} = 1.12 \times 10^{-3}$, and $K_{a3} = 5.25 \times 10^{-11}$. Abbreviating NTA as H_3T, what is $\alpha_{T^{3-}}$ at pH 7.00?

3. For the NTA chelate of lead, PbT^-, the formation constant K is 2.45×10^{11}. What is the conditional formation constant K' at pH 7.00? (See preceding problem.)

4. What is the value of $[Pb^{2+}]$ in a pH 7.00 medium in which C_T, the concentration of unchelated NTA, is 1.00×10^{-2} M, and [Pb(II)], the total concentration of lead(II), both chelated and unchelated, is 1.00×10^{-5}?

5. Calculate $[Ag^+]$ in a medium having $[NH_4^+] + [NH_3] = 0.200$ M, pH 9.24, and $[Ag(I)] = 1.00 \times 10^{-4}$ M.

6. Calculate $\alpha_{Y^{4-}}$ for EDTA at pH 9.75.

7. What is the value of K' for the EDTA chelate of Mg(II) at pH 9.75. (See Problem 6.)

8. In a solution containing Ag(I) and NH_3, what is the value of $\delta = [Ag^+]/[Ag(I)]$ when $[NH_3] = 0.0500$ M?

9. Plot $\log [Ca^{2+}]$ versus milliliters of added EDTA for the titration of 25.00 mL of 0.01000 M Ca^{2+} with 0.01000 M EDTA in a medium buffered at pH 9.80. Calculate $\log [Ca^{2+}]$ at 0, 5.0, 12.5, 20.0, 24.0, 25.0, 26.0, 30.0, and 35.0 mL added EDTA.

10. A 2.317-g sample of soil was extracted with HCl to remove Ca^{2+} and Mg^{2+}. The extract was divided into two exactly equal portions to be titrated separately with 0.0936 M standard EDTA. One portion was adjusted to pH 10, Eriochrome black T was added as an indicator and the sample portion titrated with EDTA to an end point of 48.60 mL. Calcium was precipitated quantitatively from the remaining fraction, which required 12.37 mL of EDTA after removal of the calcium. Calculate the percentage of Ca and of Mg in the sample.

11. A 3.208-g sample of nickel ore was processed to remove interferences and 50.00 mL of 0.1200 M EDTA was added in excess to react with the Ni^{2+}. The excess EDTA was titrated with 24.17 mL of 0.0755 M standard Mg^{2+}. Calculate the percentage of Ni in the ore.

12. A 10.00-mL solution of $FeSO_4$ was added to 50.00 mL of 0.0500 M Na_2H_2Y. The H^+ released required 18.03 mL of 0.0800 M OH^- for titration. What was the molar concentration of the $FeSO_4$ solution?

13. A 50.00-mL water sample required 21.76 mL of 0.0200 M EDTA to titrate water hardness. What was the hardness in milligrams per liter of $CaCO_3$?

14. The formation constant of the $Ca^{2+} -$ EDTA chelate is

$$K = \frac{[CaY^{2-}]}{[Ca^{2+}][Y^{4-}]} = 5.0 \times 10^{10}$$

At pH 9.00, the fraction of *unchelated* EDTA that is present as the EDTA anion is $\alpha_{Y^{4-}} = 5.2 \times 10^{-2}$. Keeping in mind dilution by the volume of titrant, the value of $[Ca^{2+}]$ in moles per liter at the equivalence point in the titration of

0.0500 M Ca^{2+} with 0.0500 M EDTA in a medium buffered at pH 9.00 is (a) 7.1×10^{-7}, (b) 9.3×10^{-7}, (c) 9.7×10^{-8}, (d) 3.1×10^{-6}, (e) 5.3×10^{-6}.

15. A 0.327-g portion of a caustic mixture of solid $Ca(OH)_2$ (fw = 74.1) and Na_2CO_3 was dissolved and titrated with 0.0917 M standard EDTA, with 33.8 mL of EDTA required to reach the equivalence point (Na^+ is not chelated by EDTA). The percentage of $Ca(OH)_2$ in the sample was (a) 30.7, (b) 70.2, (c) 3.1, (d) 59.3, (e) 48.3.

11 Oxidation-Reduction Equilibria

LEARNING GOALS

1. Calculation of oxidation number or oxidation state.
2. What is involved in an oxidation-reduction reaction.
3. Understanding half-reactions.
4. Review of balancing oxidation-reduction reactions.
5. Configuration of electrochemical cells and half-cells.
6. Schematic representations of electrochemical cells.
7. Common reference electrodes, including the standard hydrogen electrode.
8. Electrode potential, standard electrode potential, and the sign convention applied to these.
9. Application of the Nernst equation to half-reactions and whole reactions.
10. Calculation of standard electrode potentials for half-reactions involving slightly soluble salts and complex ions.
11. Calculation of equilibrium constants from standard electrode potentials.
12. Consideration of the effects of liquid junction potential, internal cell resistance, and overvoltage on electrochemical cell potentials.

11.1 Oxidation-Reduction Phenomena

So far the text has discussed *acid-base reactions*, in which H^+ ion is transferred between chemical species; *precipitation reactions*, in which a slightly soluble substance is formed by the combination of a cation and an anion in aqueous solution; and *complexation reactions*, in which ligands are transferred to or from metal ions. Another very important class of reactions consisting of those in which electrons are transferred between species is the topic of this chapter. For example, when copper metal is immersed in a solution of silver nitrate, the following reaction occurs:

$$Cu + 2Ag^+ + 2NO_3^- \rightarrow 2Ag + Cu^{2+} + 2NO_3^- \tag{11.1}$$

What happens in this case is that two silver ions take two electrons (e^-) from a copper atom to yield two silver atoms and a doubly charged copper ion. A transfer of electrons has occurred; such a transfer constitutes an **oxidation-reduction reaction**.

As discussed in this and subsequent chapters, oxidation-reduction reactions are very useful in analytical chemistry. This chapter introduces oxidation-reduction phenomena and equilibria.

11.2   Oxidation-Reduction Reactions

Oxidation States or Oxidation Numbers

Various species in solution may have a variety of **oxidation states** or **oxidation numbers** (consult any general chemistry textbook for procedures used to assign oxidation state or number). The oxidation number of Fe^{3+} is clearly $+3$, and the ion is formed by the loss of three electrons from the Fe atom. The oxidation number of Cl^- ion is -1; the ion is formed by adding one negatively charged electron to a Cl atom. What is the oxidation number of C in oxalic acid, $H_2C_2O_4$? With rare exceptions, the oxidation number of each H atom is $+1$ and that of each O atom is -2 in chemical compounds. Therefore, the sum of the oxidation numbers for H and O in oxalic acid is the following:

$$\text{For H:}\quad 2\text{ H atoms} \times \frac{+1}{\text{H atom}} = +2$$

$$\text{For O:}\quad 4\text{ O atoms} \times \frac{-2}{\text{O atom}} = \underline{-8}$$

$$\text{Sum} = -6$$

Compensation must be made for this -6 by a total of $+6$ for the two C atoms because the sum of the oxidation numbers of the atoms in the $H_2C_2O_4$ molecule must be zero. Therefore, the oxidation number of each of the C atoms is

$$\frac{+6 \text{ (total for the C atoms)}}{2 \text{ C atoms}} = \frac{+3}{\text{C atom}}$$

In ions, account must be taken of the charge of the ion in assigning oxidation numbers. For example, the oxidation number or state of Mn in permanganate ion, MnO_4^-, is deduced from a knowledge that 4 O atoms, each with a -2 oxidation number, give a total of -8, which must be compensated for by $+7$ in Mn to give an ion with a net -1 charge. Therefore, Mn in MnO_4^- is in the $+7$ oxidation state.

Oxidation-Reduction Reactions

A reaction in which oxidation numbers of elements change is called an **oxidation-reduction**, or **redox**, **reaction**. The species that *gains* a higher (more positive) oxidation number is said to become **oxidized**, and the one that *loses* oxidation number is said to become **reduced**. The latter is called the **oxidizing agent**, and the former is the **reducing agent**. As a simple example of such a reaction, consider the analytically useful reaction

$$Fe^{2+} + Ce^{4+} \rightarrow Fe^{3+} + Ce^{3+} \tag{11.2}$$

Here the reducing agent, Fe^{2+}, is oxidized to Fe^{3+} by the oxidizing agent, Ce^{4+}, which simultaneously is reduced to Ce^{3+}.

Positive oxidation numbers are often designated by a Roman numeral in parentheses after the element in question. Thus, the $+7$ oxidation state of Mn in MnO_4^- can be shown as Mn(VII), the oxidation number of Ce^{3+} may be designated as Ce(III), and that of Ce^{4+} by Ce(IV). It is not yet accepted notation to use minus signs in front of Roman numerals to denote negative oxidation states. For example, Cl($-$I) is not considered proper to show the oxidation number of chloride.

EXAMPLE

For the titration of Fe(II) with dichromate in acidic solution

$$6Fe^{2+} + Cr_2O_7^{2-} + 14H^+ \rightarrow 6Fe^{3+} + 2Cr^{3+} + 7H_2O$$

indicate which are the oxidizing agents and reducing agents and show the oxidation number of each element oxidized or reduced before and after the reaction.

Oxidizing agent: $Cr_2O_7^{2-}$, Cr(VI) before reaction and Cr(III) after reaction.

Reducing agent: Fe^{2+}, Fe(II) before reaction and Fe(III) after reaction.

It is convenient to think of oxidation as a loss of electrons e^- and reduction as a gain of e^-. The *loss* of one negatively charged electron *increases* the oxidation number by 1; the *gain* of one negatively charged electron *decreases* the oxidation number by 1. Very simply, in the reaction

$$Ce^{4+} + Fe^{2+} \rightarrow Ce^{3+} + Fe^{3+} \tag{11.3}$$

1 e^- is transferred from the reducing agent, Fe^{2+}, to the oxidizing agent, Ce^{4+}. The species that undergoes *oxidation loses* electrons; the one that undergoes *reduction gains* electrons.

EXAMPLE

How many electrons are transferred in the reaction

$$2MnO_4^- + 5H_2C_2O_4 + 6H^+ \rightarrow 2Mn^{2+} + 10CO_2 + 8H_2O?$$

Each Mn(VII) in 2 MnO_4^- accepts 5 e^- for a total of 10 e^-. Each of 2 C(III) in $H_2C_2O_4$ loses 1 e^- to form CO_2. Since a total of 10 C atoms are oxidized, there is a loss of 10 e^-. Therefore, the reaction involves a net transfer of 10 electrons.

The relative tendencies of oxidizing and reducing agents to accept and lose electrons, respectively, determine the extent of an oxidation-reduction reaction—that is, its position of equilibrium. The reaction of a strong oxidizing agent with a strong reducing agent gives a very complete reaction.

11.3   Half-Reactions and Balancing Oxidation-Reduction Reactions

Oxidation-reduction reactions are conveniently broken into **half-reactions**. This is done to balance the reactions and, as is shown later in the chapter, to illustrate how an oxidation half-reaction and a reduction half-reaction can be carried out separately from each other. As an example, consider the reaction of permanganate ion, MnO_4^-, with oxalic acid, $H_2C_2O_4$, in acidic media. The half-reactions for the reduction of MnO_4^- and for the oxidation of $H_2C_2O_4$ are, respectively, the following:

$$MnO_4^- + 8H^+ + 5e^- \rightarrow Mn^{2+} + 4H_2O \tag{11.4}$$

$$H_2C_2O_4 \rightarrow 2H^+ + 2CO_2 + 2e^- \tag{11.5}$$

For a balanced oxidation-reduction reaction, *the number of electrons gained by the oxidizing agent species present must equal the number of electrons lost by the reducing agent species.* In the example under consideration, this balance can be achieved by multiplying all terms in the first half-reaction by 2

$$2(MnO_4^- + 8H^+ + 5e^- \rightarrow Mn^{2+} + 4H_2O)$$

and all the terms in the second half-reaction by 5

$$5(H_2C_2O_4 \rightarrow 2H^+ + 2CO_2 + 2e^-)$$

Addition of the two resulting half-reactions yields a balanced overall reaction:

$$2MnO_4^- + 16H^+ + 10e^- \rightarrow 2Mn^{2+} + 8H_2O \tag{11.6}$$

$$5H_2C_2O_4 \rightarrow 10H^+ + 10CO_2 + 10e^- \tag{11.7}$$

Net balanced reaction: $2MnO_4^- + 5H_2C_2O_4 + 6H^+ \rightarrow 2Mn^{2+} + 10CO_2 + 8H_2O \tag{11.8}$

EXAMPLE

The peroxodisulfate ion, $S_2O_8^{2-}$, is a powerful oxidizing agent that can even oxidize Cr^{3+} to dichromate. Balance the reaction

$$Cr^{3+} + S_2O_8^{2-} + H_2O \rightarrow Cr_2O_7^{2-} + SO_4^{2-} + H^+$$

Oxidation half-reaction:

$$2Cr^{3+} + 7H_2O \rightarrow Cr_2O_7^{2-} + 14H^+ + 6e^-$$

Reduction half-reaction:

$$S_2O_8^{2-} + 2e^- \rightarrow 2SO_4^{2-}$$

Multiplying each term in the reduction half-reaction by 3 and adding to the oxidation half-reaction yields the balanced reaction:

$$2Cr^{3+} + 3S_2O_8^{2-} + 7H_2O \rightarrow Cr_2O_7^{2-} + 6SO_4^{2-} + 14H^+$$

11.4   Physical Separation of Half-Cells: The Electrochemical Cell

Combining Half-Cells to Make an Electrochemical Cell

In many cases oxidation-reduction half-reactions can be physically separated, and the electrons exchanged through an external conductor. The device for doing this is an **electrochemical cell**, and the half-reactions occur in separate **half-cells**. Such an electrochemical cell consisting of two half-cells is shown in Figure 11.1.

The electrochemical cell in Figure 11.1 is a **galvanic cell** in which a reaction is allowed to occur spontaneously. The reaction is made up of an oxidation half-reaction that occurs at the anode and a reduction half-reaction that occurs at a cathode. (Regardless of the type of electrochemical cell, the electrode at which *oxidation* occurs is always called the *anode*, and the electrode at which *reduction* occurs is always called the *cathode*.) The two half-reactions and the overall reaction are the following:

$$\text{Cathodic half-reaction:} \quad Cu^{2+} + 2e^- \rightarrow Cu \tag{11.9}$$

$$\text{Anodic half-reaction:} \quad Pb \rightarrow Pb^{2+} + 2e^- \tag{11.10}$$

$$\overline{\text{Overall reaction:} \quad Cu^{2+} + Pb \rightarrow Cu + Pb^{2+}} \tag{11.11}$$

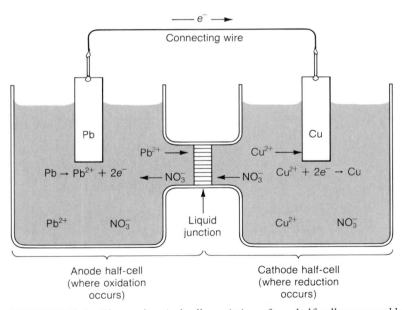

FIGURE 11.1   Electrochemical cell consisting of two half-cells separated by a porous fritted glass disk allowing transfer of ions but preventing bulk mixing of the two solutions. The overall reaction in the cell is $Pb + Cu^{2+} \rightarrow Pb^{2+} + Cu$. The electrons provided by the oxidation of Pb to Pb^{2+} are transferred via an external wire to provide e^- for the reduction of Cu^{2+} to Cu.

The electrons released to the lead anode by the oxidation of Pb are carried by an external wire to the copper cathode where they are used to reduce Cu^{2+} to copper metal. Addition of Pb^{2+} at the anodic half-cell and removal of Cu^{2+} in the cathodic half-cell results in a tendency for positive charge to accumulate in the former, and a negative charge in the latter. Since a **charge balance*** is required for any solution of an electrolyte, compensation must be made for these charge differences. That is accomplished by migration of Pb^{2+} ion from the anodic to the cathodic half-cell and migration of NO_3^- ion in the reverse direction. The junction of the two half-cells—in this case the site of a porous glass frit—is called a **liquid junction**. In some cases the two half-cells are connected by a tube (U-tube) filled with a nonreactive electrolyte such as KCl. This is called a **salt bridge**. The salt bridges illustrated in this text are shown with glass frits at both ends, which permit solution contact and transfer of ions but prevent bulk mixing of the solutions. For simplicity, the "plumbing" used to fill the tube with electrolyte solution is omitted. A salt bridge can be made of a U-tube filled with a gel of agar (resembling a harder version of gelatin) made up in the electrolyte solution. However, such salt bridges often produce high extraneous potentials called **liquid junction potentials** (Section 11.13).

Electrolytic Cells

The reactions in the galvanic cell illustrated in Figure 11.1 could be reversed (made to be nonspontaneous) by the input of electrical energy to the external circuit, forcing the electrons to flow in a reverse direction to that shown. Such an electrochemical cell in which the reactions do not occur spontaneously but are forced to occur by an external source of electricity is called an **electrolytic cell**. In this case the reactions would be

$$\text{Cathodic half-reaction:} \quad Pb^{2+} + 2e^- \rightarrow Pb \qquad (11.12)$$

$$\text{Anodic half-reaction:} \quad \underline{Cu \rightarrow Cu^{2+} + 2e^-} \qquad (11.13)$$

$$\text{Overall reaction:} \quad Pb^{2+} + Cu \rightarrow Pb + Cu^{2+} \qquad (11.14)$$

Note that, as in the galvanic cell, reduction occurs at the cathode and oxidation at the anode.

Reversibility of Cell Reactions

An electrochemical cell in which both half-reactions can be reversed by causing the electron flow (current) to go in the opposite direction with an externally applied source of electrical potential is said to be *reversible*. If a

---

* The charge balance condition for an electrolyte means that

$$\Sigma C_+ + 2\Sigma C_{2+} + 3\Sigma C_{3+} + \cdots = \Sigma C_- + 2\Sigma C_{2-} + 3\Sigma C_{3-} + \cdots$$

where ΣC_z is the sum of the analytical concentrations (nominal molarities) of all ions with charge z.

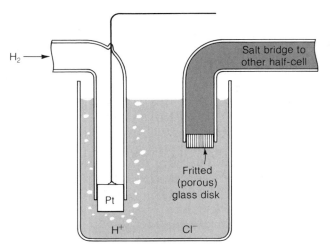

FIGURE 11.2   The hydrogen electrode

half-reaction other than the reverse of the spontaneous cathodic and/or anodic half-reaction occurs when the cell is operated as an electrolytic rather than a galvanic cell, the cell is said to be *irreversible*.

Hydrogen Electrode

Most half-reactions do not involve an elemental metal as one of the reaction constituents and thus have no metal to serve as an electrode. In such a case a noble metal, usually platinum, is employed to exchange electrons with species in solution. The most significant half-cell employing a platinum electrode is the **hydrogen electrode** shown in Figure 11.2. This half-cell has a platinum electrode in equilibrium with H_2 gas and also dipping into an acid-containing solution saturated with H_2 gas. When the electrode operates as a cathode, the half-reaction is

$$2H^+ + 2e^- \rightarrow H_2 \tag{11.15}$$

When it operates as an anode, the half-reaction is

$$H_2 \rightarrow 2H^+ + 2e^- \tag{11.16}$$

Cells without Liquid Junction

Some electrochemical cells do not have a liquid junction. Among the most studied of these is the one shown in Figure 11.3, in which one electrode is the hydrogen electrode and the other consists of Ag metal covered with solid

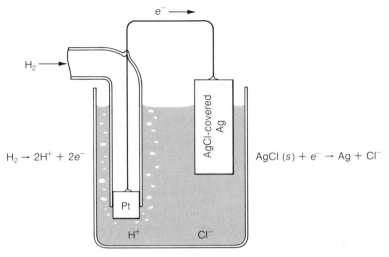

FIGURE 11.3   Electrochemical cell without liquid junction

AgCl, both electrodes dipping into a solution of HCl saturated with AgCl. The half-reactions and the overall reaction are the following:

Cathodic half-reaction: $2[AgCl(s) + e^- \rightarrow Ag + Cl^-]$ (11.17)

Anodic half-reaction: $H_2 \rightarrow 2H^+ + 2e^-$ (11.18)

Overall reaction: $2AgCl(s) + H_2 \rightarrow 2Ag + 2Cl^- + 2H^+$ (11.19)

Schematic Representation of Electrochemical Cells

A detailed description or sketch of an electrochemical cell is a rather cumbersome way to represent such a cell. A schematic representation can be employed to represent an electrochemical cell in a simplified way. *A schematic representation of an electrochemical cell lists the anode and its constituent solution on the left and the corresponding information for the cathode on the right; it employs vertical lines to show phase junctions (including liquid junctions) at which potentials may develop and denotes a salt bridge with a double vertical line.* According to this convention, the cell sketched in Figure 11.3 is represented schematically as

$Pt \mid H_2, HCl, AgCl(satd.) \mid Ag$ (11.20)

The anodic half-reaction is shown to involve the oxidation of H_2 to H^+ (see Reaction 11.18) and the cathodic half-reaction is shown as the reduction of AgCl to Ag in a medium, containing HCl and saturated with solid AgCl (see Reaction 11.17).

EXAMPLE

Give a schematic representation of the cell shown in Figure 11.1.

The anode is a lead electrode dipping into a solution of lead nitrate and is separated by a liquid junction from the cathode, which consists of a copper electrode dipping into a solution of cupric nitrate. The schematic representation of the electrochemical cell is

$$Pb \,|\, Pb(NO_3)_2(aq) \,|\, Cu(NO_3)_2(aq) \,|\, Cu$$

An electrochemical cell represented schematically as

$$Zn \,|\, ZnSO_4(aq) \,\|\, CuSO_4(aq) \,|\, Cu$$

would appear as shown in Figure 11.4. Note that the two half-cells are separated by a salt bridge, designated as $\|$ in the schematic representation. The anodic half-reaction according to the schematic representation is

$$Zn \rightarrow Zn^{2+} + 2e^- \tag{11.21}$$

and the cathodic half-reaction is

$$Cu^{2+} + 2e^- \rightarrow Cu \tag{11.22}$$

These give an overall reaction of

$$Zn + Cu^{2+} \rightarrow Zn^{2+} + Cu \tag{11.23}$$

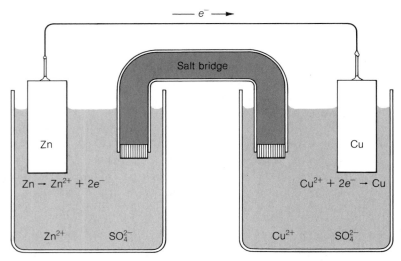

FIGURE 11.4   Electrochemical cell represented schematically as $Zn|ZnSO_4(aq)\|CuSO_4(aq)|Cu$.

Half-reactions can be represented schematically for the cell in Figure 11.4. The schematic representation of the anode is $Zn|ZnSO_4(aq)$ and that of the cathode is $CuSO_4(aq)|Cu$.

Direction of Electrochemical Cell Reactions

In all the examples cited thus far, the half-reactions do in fact go in the directions shown in either the sketches of the electrochemical cells or the schematic representations of the cells for any reasonable concentrations of reactant species in solution. Obviously, just because a cell reaction is written going in a particular direction does not necessarily mean that it does so if the cell is actually set up. The direction of a galvanic cell reaction is a function of the kinds of electrodes, the nature of the solution constituents, and the concentrations of the latter. The direction of the spontaneous electrochemical reaction in a cell is manifested by the polarity of the electrodes in the cell—that is, which electrode has a positive electrical potential relative to the other. This aspect of electrochemical cell behavior is addressed in the following section.

11.5   Electrode Potentials

Relative Half-Cell Potentials

Consider the electrochemical cell shown in Figure 11.5. It is identical to the one in Figure 11.4, except that the wire connecting the two electrodes has been replaced by a potentiometer or voltmeter that does not allow a significant current flow between electrodes. [When essentially no current flows there is no potential component due to current flowing against the resistance of the cell (IR drop) and no significant change in electroactive species concentrations at the electrode surface. Both of these possible interferences with the measurement of electrochemical cell potential are discussed in Section 11.13.] The arrangement shown in Figure 11.5 allows for the measurement of the potential of the two electrodes relative to each other, a value designated as E_{cell}. In the absence of liquid junction potentials (see Section 11.13), the value of E_{cell} is the following:

$$E_{cell} = E_{cathode} - E_{anode} \tag{11.24}$$

From the discussion in the previous section, it is known that the copper electrode is the cathode and the zinc electrode is the anode. Therefore, in this specific case

$$E_{cell} = E_{Cu} - E_{Zn} \tag{11.25}$$

where the subscripts Cu and Zn designate the metal composing the electrode.

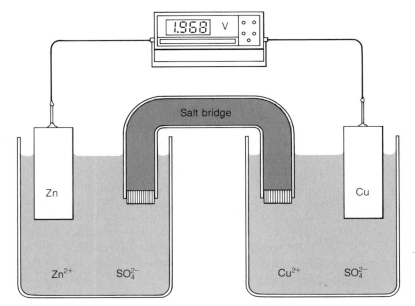

FIGURE 11.5   Electrochemical cell represented schematically as $Zn|ZnSO_4(aq)\|CuSO_4(aq)|Cu$ and set up for the measurement of relative electrode potentials

It is not possible to measure the potential of an electrode by itself. There-fore, although $E_{Cu} - E_{Zn}$ can be measured in the cell shown in Figure 11.5, it is not possible to measure either E_{Cu} or E_{Zn} by itself.

The Standard Hydrogen Electrode

Since it is impossible to measure individual half-cell potentials, relative half-cell potentials are measured instead. In order to make comparisons of relative half-cell potentials meaningful, it is necessary to arbitrarily assign a potential of exactly 0 to a specific electrode or half-cell. A potential of exactly 0 has been assigned to the **standard hydrogen electrode (SHE)**. The physical configuration of the standard hydrogen electrode is shown in Figure 11.2. The platinum foil used as a conductor in this electrode is *platinized*—that is, covered with a layer of finely divided platinum (platinum black) by the electrochemical reduction of H_2PtCl_2. This treatment greatly increases the surface area of the electrode and the amount of H_2 that it can adsorb. The solution in which the platinum electrode is immersed is kept saturated with H_2 by bubbling H_2 gas into the solution at a pressure of 1.00 atm. At this pressure of H_2 and at a concentration (more properly, an activity) of H^+ ion of exactly 1.00 mol/L, the potential of this electrode is assigned a value of 0 at all temperatures.

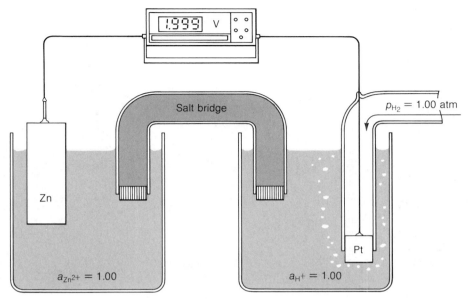

FIGURE 11.6   Electrochemical cell in which the potential of the $Zn|Zn^{2+}$ $(a = 1.00)$ half-cell can be measured versus the SHE

The half-reaction of the hydrogen electrode is

$$2H^+ + 2e^- \rightleftharpoons H_2(g) \tag{11.26}$$

This reaction is reversible and, as part of an electrochemical cell, the hydrogen electrode can function as either a cathode (half-reaction proceeding to the right) or an anode (half-reaction proceeding to the left).

Meaning and Measurement of Electrode Potential

The **electrode potential** of a particular electrode is defined as the electrical potential of an electrochemical cell in which one half-cell contains the electrode in question and the other half-cell is the standard hydrogen electrode. Used in this way, the SHE is called a **reference electrode**. Consider the electrochemical cell shown in Figure 11.6 in which the potential of a zinc metal electrode immersed in Zn^{2+} at an activity a of 1.00 can be measured. The schematic representation of this cell is the following:

$$Zn\,|\,Zn^{2+}(a = 1.00)\,\|\,H^+(a = 1.00)\,|\,H_2(1.00 \text{ atm}),\ Pt$$

The potential measured across the two electrodes of this cell would be 0.763 V, and the zinc electrode would be negative relative to the SHE. The

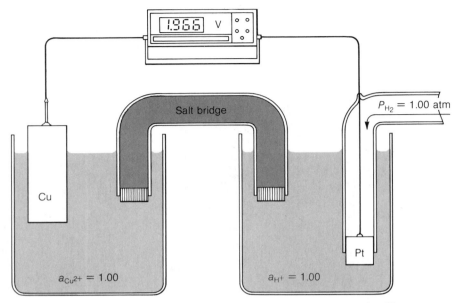

FIGURE 11.7   Electrochemical cell in which the potential of the $Cu|Cu^{2+}$ $(a = 1.00)$ half-cell can be measured versus the SHE.

negative potential of the zinc electrode relative to the SHE arises from the strong tendency of zinc metal to go into solution as zinc ion, leaving negatively charged electrons on the zinc electrode. Therefore, if the voltmeter were removed from the circuit and the two electrodes connected by a wire, the spontaneous electrochemical reaction would be

$$Zn + 2H^+ \rightarrow Zn^{2+} + H_2(g) \tag{11.27}$$

As you may recall from previous laboratory experiments and from consideration of the activity series in general chemistry, this is exactly the reaction that occurs when a piece of zinc metal is placed into a solution of acid (H^+).

Consider next the electrochemical cell in Figure 11.7. This cell is used to measure the potential of a copper electrode immersed in a solution for which $a_{Cu^{2+}} = 1.00$. In this case the potential measured would be 0.337 V, and the copper electrode would be the positive electrode. This implies a deficiency of electrons on the copper electrode relative to the platinum electrode, so that the spontaneous electrochemical reaction would be

$$Cu^{2+} + H_2(g) \rightarrow Cu + 2H^+ \tag{11.28}$$

Sign Conventions for Electrode Potentials

From the preceding discussion it is reasonable to conclude that the *sign* of the potential of an electrode versus the SHE is indicative of the direction tendency of the half-reaction at that electrode—that is, whether the electrode would act as a cathode or anode. To predict relative reaction tendencies from electrode potentials versus the SHE it is necessary to have a *sign convention for electrode potentials*. The sign convention used in analytical chemistry is that recommended by the International Union of Pure and Applied Chemistry (IUPAC), based upon a 1953 meeting of that organization held in Stockholm. According to this convention, the term **electrode potential** *applies to half-reactions written as reductions*; for example,

$$Cu^{2+} + 2e^- \rightleftarrows Cu \tag{11.29}$$

The potential for a half-reaction written in this manner is also properly called a *reduction potential*. The advantage of using the reduction potential (electrode potential) is the fact that it is measured directly by the potential versus the SHE of a half-cell having the chemical constituents of the half-reaction under consideration. The **standard electrode potentials, E^0**, are measured when all activities are exactly 1; for the reduction of Zn^{2+} and Cu^{2+} these are respectively

$$Zn^{2+} + 2e^- \rightleftarrows Zn(s) \qquad E^0 = -0.763 \text{ V} \tag{11.30}$$

$$Cu^{2+} + 2e^- \rightleftarrows Cu(s) \qquad E^0 = +0.337 \text{ V} \tag{11.31}$$

It has just been noted that the potential of a zinc electrode immersed in Zn^{2+}, $a = 1.00$ (Figure 11.6), is -0.763 V versus the SHE, and the potential of a similar copper electrode (Figure 11.7) is $+0.337$ V versus the SHE. The sign of E^0 indicates whether or not a reduction is spontaneous relative to the SHE. The negative E^0 for the overall reaction

$$Zn^{2+} + H_2(g) \rightleftarrows Zn(s) + 2H^+ \qquad E^0 = -0.763 \tag{11.32}$$

indicates that the reaction does not go spontaneously to the right and, in fact, goes in the reverse direction. However, the positive E^0 for the overall reaction

$$Cu^{2+} + H_2(g) \rightleftarrows Cu(s) + 2H^+ \qquad E^0 = +0.337 \text{ V} \tag{11.33}$$

shows that this reaction does in fact go spontaneously to the right.

11.6   The Nernst Equation and the Effect of Concentration on Electrode Potential

Effects of Changing Concentrations on Electrode Potentials

Consider the copper electrode in Figure 11.7. It is reasonable to believe that changing the concentration of Cu^{2+} in the left half-cell would affect the potential of the electrode. In fact, if the value of $[Cu^{2+}]$ is lowered to 1.00×10^{-3} M, the potential of the copper electrode is lowered from $+0.337$ to $+0.248$ V versus the SHE. The shift to more negative potentials can be justified on the basis of Le Chatelier's principle applied to the half-reaction

$$Cu^{2+} + 2e^- \rightleftarrows Cu \tag{11.34}$$

A decrease in $[Cu^{2+}]$ is accompanied by an increase in the "activity" of electrons at the electrode surface. Because electrons are negatively charged, this leads to a negative shift in potential. Another way of viewing this phenomenon is that at a lower concentration of Cu^{2+}, there is a lower demand by Cu^{2+} for electrons from the Cu electrode, leading to a more negative charge.

To cite another example, consider the zinc electrode shown in Figure 11.6. If $[Zn^{2+}]$ is lowered to 1.00×10^{-4} M, the potential of this electrode shifts from -0.763 to -0.881 V versus the SHE. This change in $[Zn^{2+}]$ to a lower value tends to shift the equilibrium of the half-reaction

$$Zn^{2+} + 2e^- \rightleftarrows Zn \tag{11.35}$$

to the left, increasing the tendency of Zn to go into solution as Zn^{2+}, leaving electrons behind, thus increasing the electron "activity" on the electrode and making its potential more negative.

The Nernst Equation

The quantitative relationship between species concentrations and electrode potentials was first explained by the nineteenth-century German chemist Nernst and is known as the **Nernst equation**. Consider a half-cell in which the generalized half-reaction is

$$aA + bB + \cdots + ne^- \rightleftarrows cC + dD + \cdots \qquad E^0 \tag{11.36}$$

As applied to the actual half-reaction

$$MnO_4^- + 8H^+ + 5e^- \rightleftarrows Mn^{2+} + 4H_2O \qquad E^0 = +1.51 \text{ V} \tag{11.37}$$

$a = 1$, $b = 8$, $n = 5$, $c = 1$, $d = 4$, and E^0, the standard electrode potential, $= +1.51$ V. The Nernst equation for the potential E of the generalized half-reaction in Reaction 11.36 is

$$E = E^0 - \frac{2.303RT}{nF} \log \frac{[C]^c [D]^d \cdots}{[A]^a [B]^b \cdots} \tag{11.38}$$

where R is the gas constant (8.316 joule/mol-deg), T is the absolute temperature (298 K at 25°C),* n is the number of electrons involved in the half-reaction (units of equivalent per mole), and F is the faraday (96,491 coulomb per equivalent). At 25° C, the term $2.303RT/F$ has a value of 0.0591,

---

* According to SI convention, absolute temperature is denoted simply by K, without the ° sign. This can cause some confusion in analytical chemistry where K is used to denote several other things, particularly equilibrium constants.

so that Equation 11.38 becomes

$$E = E^0 - \frac{0.0591}{n} \log \frac{[C]^c[D]^d \cdots}{[A]^a[B]^b \cdots} \tag{11.39}$$

The log term in the Nernst equation is sometimes expressed as the reaction quotient Q so that the Nernst equation in its general form is

$$E = E^0 - \frac{2.303RT}{nF} \log Q \tag{11.40}$$

The form of Q is the same as that of the equilibrium constant K of a reaction, except that the concentrations in the Q expression are not usually equilibrium concentrations; but when they are, $Q = K$.

For the specific case of the half-reaction given in Reaction 11.37 involving manganese species, the Nernst equation is

$$E = 1.51 \text{ V} - \frac{0.0591}{5} \log \frac{[Mn^{2+}]}{[MnO_4^-][H^+]^8} \tag{11.41}$$

Note that although H_2O appears in the half-reaction, it does not appear in the Nernst equation. This is because the activity of H_2O in aqueous solutions is constant and assigned a value of 1 by convention.

EXAMPLE

What is the Nernst equation expression for the half-reaction

$$Cr_2O_7^{2-} + 14H^+ + 6e^- \rightleftharpoons 2Cr^{3+} + 7H_2O \qquad E^0 = +1.33 \text{ V}$$

$$E = 1.33 \text{ V} - \frac{0.0591}{n} \log \frac{[Cr^{3+}]^2}{[Cr_2O_7^{2-}][H^+]^{14}}$$

Calculations with the Nernst Equation

From Equation 11.41 it is possible to calculate the potential versus the SHE generated in a solution containing specified concentrations of Mn^{2+}, MnO_4^-, and H^+. This potential must be measured at the surface of some conductor; since none participates in the half-reaction, a nonreactive conductor such as a platinum foil electrode must be used. The electrochemical cell for carrying out the measurement is shown in Figure 11.8. If, for example, $[MnO_4^-] = 1.25 \times 10^{-2} \ M$, $[Mn^{2+}] = 4.00 \times 10^{-3} \ M$, and $[H^+] = 0.100 \ M$, substitution of these values into Equation 11.41 yields

$$E = 1.51 \text{ V} - \frac{0.0591}{5} \log \frac{4.00 \times 10^{-3}}{1.25 \times 10^{-2}(0.100)^8} = 1.42 \text{ V} \tag{11.42}$$

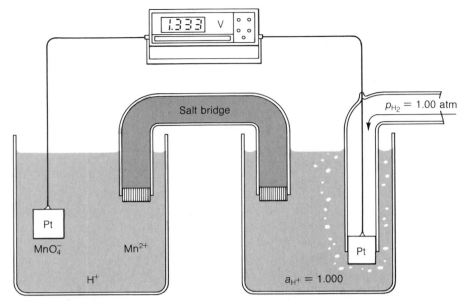

FIGURE 11.8   Cell for measuring the potential of the $Pt|Mn^{2+}$, MnO_4^-, H^+ half-cell versus the SHE

EXAMPLE

What is E calculated from Equation 11.41 when $[MnO_4^-] = 2.50 \times 10^{-2}$ M, $[Mn^{2+}] = 1.75 \times 10^{-3}$ M, and $[H^+] = 0.110$ M?

$$E = 1.51 \text{ V} - \frac{0.0591}{5} \log \frac{1.75 \times 10^{-3}}{2.50 \times 10^{-2}(0.110)^8}$$

$$= 1.51 \text{ V} - \frac{0.0591}{5} 6.51 = 1.43 \text{ V}$$

Solids and Gases in the Nernst Equation

Strictly speaking, the bracketed terms in the generalized Nernst expression, Equation 11.39, are activities; however, as in the preceding examples, these are often approximated by concentrations. When a gas is involved, the pressure of the gas in atmospheres is used in the Nernst expression. It has already been noted that the activity of H_2O in aqueous solution is taken as 1.00, so that this species does not appear in the Nernst equation. Finally, the activities of solids participating in a half-reaction, such as metal electrodes or slightly soluble salts, are constant and assigned values of exactly 1. With these stipulations in mind, it is now possible to consider some examples in which solids and gases are reaction participants.

As an example of a Nernst equation calculation involving a solid, consider the potential versus the SHE of a copper electrode immersed in an 8.75×10^{-3} M Cu^{2+} solution. The half-reaction and the corresponding Nernst equation are

$$Cu^{2+} + 2e^- \rightleftarrows Cu \qquad E^0 = +0.337 \text{ V} \tag{11.43}$$

$$E = E^0 - \frac{0.0591}{n} \log \frac{1}{[Cu^{2+}]} \tag{11.44}$$

(Note that the activity of solid Cu is taken as 1 in the Nernst expression.)

$$E = +0.337 - \frac{0.0591}{2} \log \frac{1}{8.75 \times 10^{-3}} = 0.276 \text{ V} \tag{11.45}$$

Some students may find it more convenient to handle a calculation such as that in the preceding example knowing that

$$-\log \frac{1}{X} = +\log X$$

Therefore,

$$E^0 - \frac{0.0591}{n} \log \frac{1}{[Cu^{2+}]} = E^0 + \frac{0.0591}{2} \log [Cu^{2+}]$$

In the latter form it is not necessary to take the log of a reciprocal and to multiply by -1.

As an example of a calculation with the Nernst equation involving a gas, consider the calculation of the potential versus the SHE in a half-cell in which oxygen at a pressure of 0.0900 atm is in equilibrium with a solution having $[H^+] = 1.00 \times 10^{-3}$ M. (For computational purposes it can be assumed that this potential could be measured at a noble metal electrode immersed in solution; in practice, a meaningful measurement for this system is difficult to obtain. The correct value of E is very readily calculated, though, if P_{O_2} and $[H^+]$ are known.) The reduction half-reaction, electrode potential, and corresponding Nernst equation are the following:

$$O_2(g) + 4H^+ + 4e^- \rightleftarrows 2H_2O \qquad E^0 = +1.229 \text{ V} \tag{11.46}$$

$$E = E^0 - \frac{0.0591}{4} \log \frac{1}{P_{O_2}[H^+]^4}$$

Substitution of the appropriate values leads to the solution

$$E = 1.229 - \frac{0.0591}{4} \log \frac{1}{0.0900(1.00 \times 10^{-3})^4} \tag{11.47}$$

$$= 1.036 \text{ V}$$

11.7   Practical Reference Half-Cells

Although electrode potential values are conventionally given in reference to the SHE, it should be realized that a hydrogen electrode in which a_{H^+} is exactly 1 is prohibitively difficult to make in practice. This is because it is not possible to calculate accurately the value of $[H^+]$ for which a_{H^+} is exactly 1. However, a series of values of $[H^+]$ less than 1 M can be employed, and the potential of the reference hydrogen electrode extrapolated to what it would be if a_{H^+} were exactly 1. For values of $[H^+]$ of about 0.1 and less, the Debye-Hückel relationship (see Section 7.4) is used to calculated a_{H^+}.

In practice, reference electrodes other than the SHE are used for electrode potential measurements. One of these is the **calomel electrode**, in which mercury metal is in equilibrium with a solution containing dissolved KCl and saturated with Hg_2Cl_2 (calomel); some excess solid calomel is present to ensure

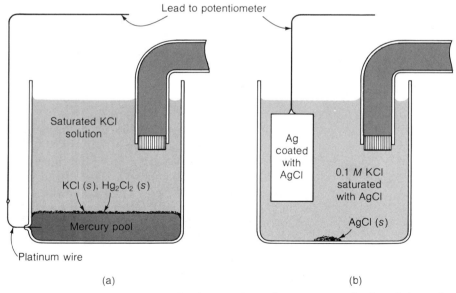

(a)   (b)

FIGURE 11.9 Commonly used reference electrodes: (a) saturated calomel electrode; (b) silver–silver chloride

saturation. The half-reaction is the following:

$$Hg_2Cl_2(s) + 2e^- \rightleftarrows 2Hg(l) + 2Cl^- \qquad E^0 = +0.268 \text{ V vs. SHE} \qquad (11.48)$$

If the solution is likewise saturated with KCl, the electrode is called a **saturated calomel electrode (SCE)**, and it has a potential of $+0.241$ V versus the SHE at $25°C$. An SCE is sketched in Figure 11.9(a). A more sophisticated commercial version is shown in Figure 11.11.

Another widely used reference electrode consists of silver metal coated with solid AgCl and immersed in a solution of KCl or NaCl saturated with AgCl as shown in Figure 11.9(b). The half-reaction and E^0 for this electrode are the following:

$$AgCl(s) + e^- \rightleftarrows Ag + Cl^- \qquad E^0 = +0.222 \text{ V vs. SHE} \qquad (11.49)$$

Both of the reference electrodes just discussed involve a reversible half-reaction which, written as a reduction, consists of the reduction of slightly soluble salt to the metal, releasing the salt anion. The potential of the half-cell depends upon the anion (chloride ion) concentration. The value of E^0 for such a half-reaction is readily derived from E^0 for the metal–metal ion half-reaction and the solubility product of the salt as shown in the next section.

11.8   Standard Electrode Potentials Involving Slightly Soluble Salts or Complexed Metals

Half-Reactions Involving Slightly Soluble Salts

For a silver electrode immersed in a solution containing Ag^+, the pertinent reduction half-reaction is

$$Ag^+ + e^- \rightleftarrows Ag \qquad E^0 = +0.799 \text{ V} \qquad (11.50)$$

and E versus the SHE is given by

$$E = E^0 - 0.0591 \log \frac{1}{[Ag^+]} \qquad (11.51)$$

The Ag^+ may come from a soluble silver salt dissolved in water or it may be present at a very low level governed by the solubility of a slightly soluble silver salt, such as AgCl. Consider a silver electrode immersed in a solution saturated with slightly soluble AgCl and containing a particular concentration of excess Cl^-. In such a case, the value of $[Ag^+]$ is determined by the solubility of AgCl in excess Cl^- as given by the solubility product relationship

$$[Ag^+] = \frac{K_{sp}}{[Cl^-]} \qquad (11.52)$$

Substituted into the Nernst equation above, this expression for $[Ag^+]$ yields

$$E = E^0 - 0.0591 \log \frac{[Cl^-]}{K_{sp}} \tag{11.53}$$

The preceding Nernst equation can be rearranged to put K_{sp} (1.82×10^{-10}) into a new standard electrode potential designated E^0_{AgCl} as follows:

$$E = \underbrace{E^0 + 0.0591 \log K_{sp}} - 0.0591 \log [Cl^-] \tag{11.54}$$

$$= \quad E^0_{AgCl} - 0.0591 \log [Cl^-] = 0.223 - 0.0591 \log [Cl^-]$$

This is the Nernst equation for the half-reaction

$$AgCl(s) + e^- \rightleftarrows Ag + Cl^- \qquad E^0_{AgCl} = +0.223 \text{ V} \tag{11.55}$$

(The value of E^0_{AgCl} just calculated varies by 0.001 V, or only 1 mv, from that given for the identical half-reaction in Equation 11.49, because of slight discrepancies in the laboratory determination of K_{sp} and E^0 values.) This calculation illustrates how E^0 for the reduction of a slightly soluble salt may be calculated from E^0 for the reduction of the metal cation of the salt and the K_{sp} of the salt. The reverse is also true—that is, if the E^0's are known for both the reduction of a slightly soluble salt and its metal cation, K_{sp} of the salt can be calculated. In general, E^0 for the n-electron reduction of a slightly soluble salt is the sum of E^0 for the n-electron reduction of the salt cation plus $0.0591/n \log K_{sp}$.

EXAMPLE

Given

$$AgI(s) + e^- \rightleftarrows Ag + I^- \qquad E^0_{AgI} = -0.151 \text{ V}$$

Calculate K_{sp} of AgI.

For a silver electrode immersed in a solution containing Ag^+ from the slight dissolution of solid AgI, the following apply:

$$Ag^+ + e^- \rightleftarrows Ag \qquad E^0 = 0.799 \text{ V}$$

$$E = E^0 - 0.0591 \log \frac{1}{[Ag^+]}$$

$$= E^0 + 0.0591 \log K_{sp} - 0.0591 \log [I^-]$$

When $[I^-] = 1.00$, $E = E^0_{AgI}$

$$E^0_{AgI} = E^0 + 0.0591 \log K_{sp} - 0.0591 \log (1)$$

$$\log K_{sp} = \frac{E^0_{AgI} - E^0}{0.0591} = \frac{-0.151 - 0.799}{0.0591}$$

$$= -16.074$$

$$K_{sp} = 8.43 \times 10^{-17}$$

Half-Reactions Involving Complex Ions

A Nernst equation expression can be written for the reduction of a complex ion. If the formation constant of a complex ion is known, it is possible to calculate E^0 for its reduction, and vice versa. For example, what is E^0 for the half-reaction

$$Ag(NH_3)_2^+ + e^- \rightleftharpoons Ag + 2NH_3 \qquad E^0_{complex} = ? \tag{11.56}$$

given

$$\beta_2 = \frac{[Ag(NH_3)_2^+]}{[Ag^+][NH_3]^2} = 1.59 \times 10^7 \tag{11.57}$$

Consider a half-cell consisting of a silver electrode immersed in a solution of $Ag(NH_3)_2^+$ and excess NH_3, for which the following apply:

$$Ag^+ + e^- \rightleftharpoons Ag \qquad E^0 = +0.799 \text{ V} \tag{11.58}$$

$$E = E^0 - 0.0591 \log \frac{1}{[Ag^+]}$$

Substituting in the expression for β_2 yields

$$E = E^0 - 0.0591 \log \frac{\beta_2[NH_3]^2}{[Ag(NH_3)_2^+]} \tag{11.59}$$

When both $[NH_3]$ and $[Ag(NH_3)_2^+]$ are exactly 1, $E = E^0_{complex}$.

$$E^0_{complex} = E = E^0 - 0.0591 \log \beta_2 = 0.799 - 0.0591 \log 1.59 \times 10^7 \tag{11.60}$$

$$= 0.373 \text{ V}$$

In general, E^0 for the n-electron reduction of a complex ion equals E^0 for the n-electron reduction of the central metal ion less $0.0591/n \log \beta_x$, where β_x is the overall formation constant of the complex with x ligands.

11.9   Compilations of Standard Electrode Potentials

Appendix Table F is a compilation of E^0 values for a number of half-reactions. Note that each of these half-reactions is written as a reduction, so that this is a table of reduction potentials. It is helpful to keep in mind that the lower the value of standard electrode potential, the less the tendency for the reduction reaction to proceed. Consider the following E^0 values, all versus SHE.

$$Ce^{4+} + e^- \rightleftharpoons Ce^{3+} \qquad E^0 = +1.70 \text{ V} \ (1 \ M \ HClO_4)$$

$$Fe^{3+} + e^- \rightleftharpoons Fe^{2+} \qquad E^0 = +0.771 \text{ V}$$

$$Cu^{2+} + 2e^- \rightleftharpoons Cu \qquad E^0 = +0.337 \text{ V}$$

$$2H^+ + 2e^- \rightleftharpoons H_2(g) \qquad E^0 = \ \ \ 0.000 \text{ V}$$

$$Pb^{2+} + 2e^- \rightleftarrows Pb \qquad E^0 = -0.126 \text{ V}$$

$$Cr^{3+} + e^- \rightleftarrows Cr^{2+} \qquad E^0 = -0.408 \text{ V}$$

$$Zn^{2+} + 2e^- \rightleftarrows Zn \qquad E^0 = -0.763 \text{ V}$$

All of the ions on the left of the above half-reactions can be viewed as oxidizing agents. Their order of strength as oxidizing agents is in increasing order of E^0—that is, $Zn^{2+} < Cr^{3+} < Pb^{2+} < H^+ < Cu^{2+} < Fe^{3+} < Ce^{4+}$. The cerium(IV) ion, Ce^{4+}, is a very strong oxidizing agent, with a very pronounced tendency to acquire electrons, whereas Zn^{2+} at the other end of the scale is an extremely weak oxidizing agent with a very weak affinity for electrons. The chemical species on the right of the above half-reactions can be viewed as reducing agents that give up electrons in oxidation-reduction reactions. The weaker the tendency of the half-reactions to go to the right, the weaker these reducing agents are, so that in increasing order of reducing strength (decreasing E^0) these species are $Ce^{3+} < Fe^{2+} < Cu < H_2 < Pb < Cr^{2+} < Zn$.

Relatively few E^0 values are determined by direct potential measurement because most of the half-reactions are not sufficiently reversible to yield accurate potential values at an electrode. Instead, E^0 values can be calculated from thermodynamic data with a knowledge that the change in free energy (ΔG) represents the maximum possible electrical work that can be done on the surroundings by a chemical reaction acting reversibly at constant temperature and pressure. This gives rise to the two following equations relating ΔG to E and ΔG^0 to E^0:

$$\Delta G = -nFE \qquad (11.61)$$

$$\Delta G^0 = -nFE^0 \qquad (11.62)$$

11.10   Electrochemical Cell Potentials

Spontaneous Electrochemical Cell Reactions

Thus far, the discussion of the calculation of potentials has centered on cases in which the SHE is the reference electrode. Consider now the electrochemical cell represented schematically as

$$Zn | Zn^{2+} (a = 1.00) \| Pb^{2+} (a = 1.00 | Pb$$

This representation implies the following reactions and potentials:

Cathodic reaction:	$Pb^{2+} + 2e^- \rightleftarrows Pb$	$E^0 = -0.126$ V	(11.63)
Anodic reaction:	$-(Zn^{2+} + 2e^- \rightleftarrows Zn)$	$E^0 = -(-0.763)$ V	(11.64)
Overall reaction:	$Zn + Pb^{2+} \rightleftarrows Zn^{2+} + Pb$	$E^0 = +0.637$ V	(11.65)

Note that the anodic half-reaction was obtained by subtracting the half-reaction for the reduction of Zn^{2+} and that its contribution to the net value of E^0 was obtained by subtraction. The overall electrochemical cell reaction,

Reaction 11.65, has a positive value of E^0, indicating that it tends to proceed to the right as written. This means that for a spontaneous reaction, Zn atoms give up electrons to the Zn electrode; making it negative relative to the Pb electrode, whereas the tendency is for Pb^{2+} ions to pick up electrons from the Pb electrode, making it relatively positive. Thus, if a voltmeter were connected across these two electrodes, it would read a potential of 0.637 V and the Zn electrode would be negative relative to the Pb electrode. This is consistent with Equation 11.24, stating

$$E_{cell} = E_{cathode} - E_{anode}$$

which, applied to this case, gives

$$E_{cell} = -0.126 \text{ V} - (-0.763 \text{ V}) = +0.637 \text{ V} \tag{11.66}$$

Reversibility of Electrochemical Cells

Consider next an electrochemical cell denoted schematically as

$$Pt|Ce^{3+}(a = 1.00), \ Ce^{4+}(a = 1.00)\|Fe^{3+}(a = 1.00), \ Fe^{2+}(a = 1.00)|Pt$$

The half-reactions and overall reaction denoted by this representation are the following:

Cathodic reaction: $\quad Fe^{3+} + e^- \rightleftarrows Fe^{2+} \qquad\qquad E^0 = \quad +0.77 \text{ V} \qquad (11.67)$

Anodic reaction: $\quad \dfrac{-(Ce^{4+} + e^- \rightleftarrows Ce^{3+})}{} \qquad \dfrac{E^0 = -(+1.70) \text{ V}}{} \qquad (11.68)$

Overall reaction: $\quad Fe^{3+} + Ce^{3+} \rightleftarrows Fe^{2+} + Ce^{4+} \qquad E^0 = \quad -0.93 \text{ V} \qquad (11.69)$

Here the negative E^0 for the overall reaction indicates that actually the spontaneous cell reaction would proceed to the left—that is, in the opposite direction from that in which it is written.

If the schematic of the electrochemical cell under discussion had been written in the opposite direction, it would appear as follows:

$$Pt|Fe^{2+}(a = 1.00), \ Fe^{3+}(a = 1.00)\|Ce^{4+}(a = 1.00), \ Ce^{3+}(a = 1.00)|Pt$$

This translates to

Cathodic reaction: $\quad Ce^{4+} + e^- \rightleftarrows Ce^{3+} \qquad\qquad E^0 = \quad +1.70 \text{ V} \qquad (11.70)$

Anodic reaction: $\quad \dfrac{-(Fe^{3+} + e^- \rightleftarrows Fe^{2+})}{} \qquad \dfrac{E^0 = -(+0.77) \text{ V}}{} \qquad (11.71)$

Overall reaction: $\quad Ce^{4+} + Fe^{2+} \rightleftarrows Ce^{3+} + Fe^{3+} \qquad E^0 = \quad +0.93 \text{ V} \qquad (11.72)$

Here the *positive value* of E^0 shows that the overall electrochemical cell reaction proceeds *to the right*. This means that Fe^{2+} is releasing electrons to the Pt electrode in contact with the Fe^{2+}–Fe^{3+} solution (anode) making it the negative electrode, whereas Ce^{4+} ions withdraw electrons from the other Pt electrode, making it the positive electrode (cathode).

The two preceding examples illustrate that an electrochemical reaction can be written in either direction. If the reaction in fact occurs in that direction, the value of E^0 is positive; if it tends to go in the reverse direction, E^0 is negative. The way in which the reaction is written, of course, has no effect upon what occurs in the cell; the same electrode is the anode and the other electrode is the cathode, regardless of how the reaction is written. The actual polarity of the electrodes is easy to figure out when it is known which way the reaction actually goes. The electrode at which a species is being oxidized and therefore donating negative electrons to the electrode is always negative relative to the other electrode where a species is being reduced and withdrawing negative electrons from the electrode.

11.11   Effect of Concentrations on Cell Potentials

Application of the Nernst Equation to Whole Reactions

In Section 11.6 the Nernst equation was applied to a half-cell, showing how the concentration of reaction participants affected the potential of that cell versus the SHE. The Nernst equation can also be applied to a total electrochemical reaction occurring in a cell, taking account of the effects of concentrations in both half-cells on the overall electrochemical cell potential. Consider the electrochemical cell discussed in the previous section, written schematically as

$$Pt|Fe^{2+}, Fe^{3+}\|Ce^{4+}, Ce^{3+}|Pt$$

for which the cell reaction is

$$Ce^{4+} + Fe^{2+} \rightleftarrows Ce^{3+} + Fe^{3+} \qquad E^0 = +0.93 \text{ V} \tag{11.73}$$

Note that in this case the ion concentrations are not specified. The Nernst equation applied to this electrochemical cell is

$$E = E^0 - 0.0591 \log \frac{[Ce^{3+}][Fe^{3+}]}{[Ce^{4+}][Fe^{2+}]} \quad \begin{array}{l}\leftarrow \text{ Product concentrations} \\ \leftarrow \text{ Reactant concentrations}\end{array} \tag{11.74}$$

$$\underset{\text{Negative sign}}{\uparrow} \quad \underset{\substack{\text{Divided} \\ \text{by 1}}}{\uparrow}$$

As an example of the use of this equation we may calculate E when $[Ce^{3+}] = 1.25 \times 10^{-3} M$, $[Ce^{4+}] = 9.00 \times 10^{-3} M$, $[Fe^{3+}] = 8.60 \times 10^{-4} M$, and $[Fe^{2+}] = 2.00 \times 10^{-2} M$. These are not equilibrium concentrations that could exist if the solutions in each half-cell were mixed in one solution, even though the concentrations in the log term of the Nernst equation are in the form of the equilibrium constant of Reaction 11.73. This point is addressed further in Section 11.12. Substituting these values into the Nernst equation yields

$$E = +0.93 \text{ V} - 0.0591 \log \frac{1.25 \times 10^{-3} \times 8.60 \times 10^{-4}}{9.00 \times 10^{-3} \times 2.00 \times 10^{-2}} \tag{11.75}$$

$$= +0.93 \text{ V} - 0.0591 \log (5.97 \times 10^{-3}) \tag{11.76}$$

$$= +0.93 \text{ V} - 0.0591 \times (-2.22) = +1.06 \text{ V} \tag{11.77}$$

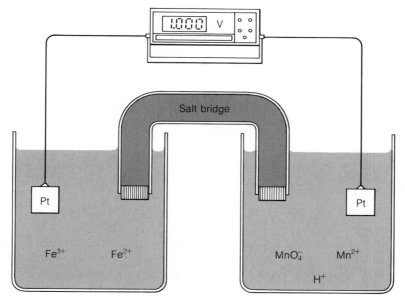

FIGURE 11.10 Electrochemical cell represented schematically as $Pt|Fe^{2+}, Fe^{3+}||MnO_4^-, Mn^{2+}, H^+|Pt$

This result shows that the potential of the electrochemical cell is $+1.06$ V. The reaction proceeds to the right as shown in Reaction 11.73. The anode containing Fe^{2+} and Fe^{3+} is, of course, negative compared to the cathode containing Ce^{4+} and Ce^{3+}. In this case the concentrations of the reaction products, Ce^{3+} and Fe^{3+}, are both lower than the concentrations of the reactants, and the potential is higher than is the case if all are at the same concentration. This is simply a reflection of Le Chatelier's principle, in that lowering the concentrations of the products tends to shift the reaction to the right.

Nernst Equation Applied to the Oxidation of Fe^{2+} by MnO_4^-

Next consider a more complicated case represented schematically as

$$Pt|Fe^{2+}, Fe^{3+}||MnO_4^-, Mn^{2+}, H^+|Pt$$

This cell is sketched in Figure 11.10. The overall cell reaction is given by the following:

Cathodic reaction:	$MnO_4^- + 8H^+ + 5e^- \rightleftharpoons Mn^{2+} + 4H_2O$	$E^0 = +1.51$ V	(11.78)
Anodic reaction:	$-5(Fe^{3+} + e^- \rightleftharpoons Fe^{2+})$	$E^0 = -(+0.77)$ V	(11.79)
Overall reaction:	$MnO_4^- + 5Fe^{2+} + 8H^+ \rightleftharpoons Mn^{2+} + 5Fe^{3+} + 4H_2O$	$E^0 = +0.74$ V	(11.80)

The Nernst equation applied to Reaction 11.80 is

$$E = E^0 - \frac{0.0591}{5} \log \frac{[Mn^{2+}][Fe^{3+}]^5}{[MnO_4^-][Fe^{2+}]^5[H^+]^8}$$

Product concentrations to the appropriate powers ←

Reactant concentrations to the appropriate powers ←

$\uparrow$
$n = 5$

(11.81)

What is the value of E if $[MnO_4^-] = 6.50 \times 10^{-3}$ M, $[Fe^{2+}] = 4.25 \times 10^{-3}$ M, $[H^+] = 0.100$, $[Mn^{2+}] = 2.00 \times 10^{-2}$ M, and $[Fe^{3+}] = 1.50 \times 10^{-3}$ M? Substituting these values into Equation 11.81 yields

$$E = +0.74 - \frac{0.0591}{5} \log \frac{2.00 \times 10^{-2} \times (1.50 \times 10^{-3})^5}{6.50 \times 10^{-3} \times (4.25 \times 10^{-3})^5 \times (0.100)^8} \quad (11.82)$$

$$= +0.74 - \frac{0.0591}{5} \log (1.69 \times 10^6) = 0.67 \text{ V} \quad (11.83)$$

The fact that E is positive indicates that the actual electrochemical cell reaction is represented by Reaction 11.80 proceeding to the right. The electrode in the half-cell containing manganese species is positive relative to the electrode immersed in the half-cell containing Fe^{3+} and Fe^{2+}.

11.12   Equilibrium Constants from E^0

Consider the last system discussed in the preceding section and the electrochemical cell shown in Figure 11.10. If the voltmeter were removed from the circuit and a wire used to connect the two electrodes, the cell reaction

$$MnO_4^- + 5Fe^{2+} + 8H^+ \rightarrow Mn^{2+} + 5Fe^{3+} + 4H_2O \quad (11.80)$$

would proceed with MnO_4^- being reduced to Mn^{2+} at the cathode and Fe^{2+} oxidized to Fe^{3+} at the anode. Eventually the reaction would stop—that is, "the battery would run down." No more electrons would flow through the wire because E of the cell would equal 0. Under these conditions the chemical species in the cell would be in a state of **chemical equilibrium**. The equilibrium constant expression for the cell reaction (Reaction 11.80) is

$$K = \frac{[Mn^{2+}][Fe^{3+}]^5}{[MnO_4^-][Fe^{2+}]^5[H^+]^8} \quad (11.84)$$

This is the log term in Equation 11.81, which is

$$E = E^0 - \frac{0.0591}{5} \log \frac{[Mn^{2+}][Fe^{3+}]^5}{[MnO_4^-][Fe^{2+}]^5[H^+]^8} \quad (11.81)$$

Substituting 0 for E at equilibrium and K for the log term yields

$$0 = E^0 - \frac{0.0591}{5} \log K \quad (11.85)$$

which rearranges to

$$\log K = \frac{5E^0}{0.0591} \tag{11.86}$$

Since in this specific case E^0 is $+0.74$ V, $\log K$ and K are

$$\log K = \frac{5 \times 0.74}{0.0591} = 62.6 \qquad K = 4 \times 10^{62} \tag{11.87}$$

The rule for calculating equilibrium constants of reversible oxidation-reduction reactions from standard electrode potentials is that *log K equals the product of the E^0 of the reaction and n divided by 0.0591 (at 25° C).* Mathematically this is simply

$$\log K = \frac{nE^0}{0.0591} \tag{11.88}$$

EXAMPLE

Calculate $\log K$ and K for the reaction

$$2Fe^{3+} + Sn^{2+} \rightleftarrows 2Fe^{2+} + Sn^{4+}$$

The value of E^0 is given by

$$\begin{array}{ll}
2(Fe^{3+} + e^- \rightleftarrows Fe^{2+}) & E^0 = \quad +0.771 \text{ V} \\
-(Sn^{4+} + 2e^- \rightleftarrows Sn^{2+}) & E^0 = -(+0.154) \text{ V} \\
\hline
2Fe^{3+} + Sn^{2+} \rightleftarrows 2Fe^{2+} + Sn^{4+} & E^0 = \quad +0.617 \text{ V}
\end{array}$$

$$\log K = \frac{2 \times 0.617}{0.0591} = 20.88$$

$$K = 10^{20.88} = 7.6 \times 10^{20}$$

EXAMPLE

Calculate $\log K$ and K for the reaction

$$MnO_4^- + 5Ce^{3+} + 8H^+ \rightleftarrows Mn^{2+} + 5Ce^{4+} + 4H_2O$$

The problem is solved as follows:

$$\begin{array}{ll}
MnO_4^- + 8H^+ + 5e^- \rightleftarrows Mn^{2+} + 4H_2O & E^0 = \quad +1.51 \text{ V} \\
-5(Ce^{4+} + e^- \rightleftarrows Ce^{3+}) & E^0 = -(+1.70) \text{ V} \\
\hline
MnO_4^- + 5Ce^{3+} + 8H^+ \rightleftarrows Mn^{2+} + 5Ce^{4+} + 4H_2O & E^0 = \quad -0.19 \text{ V}
\end{array}$$

$$\log K = \frac{5(-0.19)}{0.0591} = -16.07$$

$$K = 10^{-16.07} = 8.51 \times 10^{-17}$$

EXAMPLE

Calculate log K and K for each of the following reactions where E^0 is given.

$$Zn + Pb^{2+} \rightleftharpoons Zn^{2+} + Pb \qquad E^0 = +0.637 \text{ V}$$

$$\log K = \frac{2 \times 0.637}{0.0591} = 21.56$$

$$K = 3.6 \times 10^{21}$$

$$Cr_2O_7^{2-} + 6Fe^{2+} + 14H^+ \rightleftharpoons 2Cr^{3+} + 6Fe^{3+} + 7H_2O \qquad E^0 = +0.56 \text{ V}$$

$$\log K = \frac{6 \times 0.56}{0.0591} = 56.8$$

$$K = 6 \times 10^{56}$$

11.13   Practical Considerations with Laboratory Cells

Several factors other than electrochemical reactions at electrode surfaces affect the potential of an actual cell in the laboratory. One of these, the liquid junction potential, has been mentioned previously. It is an electrical potential that develops at the interface where two different solutions meet. For example, the electrochemical cell in Figure 11.1 shows a solution of $Pb(NO_3)_2$ and $Cu(NO_3)_2$ interfaced at a fritted glass disk that prevents bulk mixing of the solutions. However, a slight electrical potential will develop at the interface. Cells with a salt bridge (see Section 11.4) have two liquid junction potentials, one at either end of the bridge.

Junction potentials arise from different rates of movement of positive and negative ions. The junction potential effect can be lowered by employing a salt bridge that contains a high concentration of a salt for which the mobilities (migration rates) of the cations are about the same as those of the anions. This is best met in practice with a salt bridge filled with a solution of saturated KCl (concentration slightly exceeding 4 M). The mobilities of the K^+ and Cl^- ions are within 4% of each other, and this slight difference in mobility does not contribute very much to junction potentials.

Internal cell resistance, R_{cell}, can affect the measured potential of a cell through which a current I is flowing. The magnitude of the effect of internal resistance on cell potential is IR, which is subtracted from the theoretical cell potential. This IR drop lowers the measured potential of a galvanic cell and increases the measured potential of an electrolytic cell.

Another factor affecting the potential of an electrochemical cell through which current is flowing is *overvoltage*. Overvoltage arises from two phenomena that occur at either or both electrodes. The first of these is *concentration polarization* arising from the limited rates at which electroactive species diffuse to, or away from, electrodes. Consider, for example, a platinum electrode at which Fe^{3+} can be reduced to Fe^{2+}., Under conditions of negligible current, the potential of this electrode versus the SHE is

$$E = +0.771 - 0.0591 \log \frac{[Fe^{2+}]_{bulk}}{[Fe^{3+}]_{bulk}} \tag{11.89}$$

where the subscript *bulk* denotes that the concentrations of Fe^{3+} and Fe^{2+} at the electrode surface are the same as those in the bulk of the solution. If Fe^{3+} is being *reduced* to Fe^{2+} at the electrode, the concentration of Fe^{2+} is *increased* at the electrode surface and the concentration of Fe^{3+} is *decreased*; this has the net effect of making the potential of the electrode more negative. (If Fe^{2+} were being oxidized to Fe^{3+}, the potential of the electrode would shift positive.)

Another type of overvoltage, most commonly manifested when gases are involved in electrode reactions, is *kinetic polarization*. Whereas concentration polarization results from limitations in the rate of mass transfer of reaction participants to or away from the electrode surface, kinetic polarization occurs because of limitations in the rates at which electroactive species react at the electrode surface. The overvoltage associated with this phenomenon is required to overcome the kinetic barrier to the electrode reaction. In the analytically important cases of hydrogen or oxygen gas formation, kinetic polarization may introduce overvoltages of the order of 1 V.

Kinetic polarization may be lowered by lowering the current density (amperes per unit area) at the electrode surface. It is often negligible in cases involving the deposition of a metal onto an electrode surface, or the oxidation of a metal from the electrode surface. A mercury electrode has a particularly high overvoltage for the evolution of hydrogen gas, a phenomenon that makes possible the reduction of some metals without the evolution of hydrogen (see polarography, Chapter 14).

The equation for cell potential taking into account both IR drop and overvoltage is

$$E_{cell} = E_{cathode} - E_{anode} - IR - E_{overvoltage} \tag{11.90}$$

Of course, if E_{cell} is measured under conditions of little or no current flow, the last two terms are negligible.

11.14   Indicator and Reference Electrodes

Indicator Electrodes

In many applications, the potentials of electrochemical cells are used to indicate concentrations, or concentration ratios, of electroactive species. For example, a silver electrode immersed in a solution of Ag^+ reflects the concentration of this ion, shifting relatively positive with increasing $[Ag^+]$ and relatively negative with decreasing $[Ag^+]$. A platinum electrode immersed in a solution of Ce^{4+} and Ce^{3+} reflects the ratio of $[Ce^{4+}]/[Ce^{3+}]$; the potential is relatively more positive at a high ratio than it is at a low ratio. Electrodes such as the Ag and Pt electrodes just mentioned can be employed as **indicator electrodes** to measure species concentrations in accordance with the Nernst equation. Indicator electrodes form the basis of potentiometry, which is discussed in Chapter 13.

Reference Electrodes

An indicator electrode composes only one half-cell of an electrochemical cell; its potential is always measured relative to a *reference electrode* composing the other half-cell. Obviously, a reference electrode should have a very stable, reproducible potential, even when a small current is drawn from it. It should be convenient to use, readily constructed, and have a long lifetime. In Section 11.5, the SHE was designated as a reference electrode; at a defined potential of exactly 0 V, the SHE provides the basis relative to which all other half-cell potentials are expressed. Unfortunately, in routine laboratory practice, the SHE meets essentially none of the criteria needed of a reference electrode for routine use.

As discussed in Section 11.7, the calomel electrode, for which the half-reaction is

$$Hg_2Cl_2(s) + 2e^- \rightleftarrows 2Hg(l) + 2Cl^- \tag{11.91}$$

makes a very good reference electrode in routine practice. When the source of Cl^- is 1.00 M KCl, the calomel electrode has a potential of $+0.280$ V versus the SHE, and when the filling solution is 0.100 M KCl, the potential is $+0.334$ V versus the SHE. As noted in Section 11.7, a filling solution saturated with KCl gives an electrode called the *saturated calomel electrode*, SCE, with a potential of $+0.241$ V versus the SHE.

A typical commercial saturated calomel electrode is sketched in Figure 11.11. The liquid Hg and solid Hg_2Cl_2 contained in the electrode's inner tube are formed into a paste with saturated KCl solution. Electrical contact is made with this paste by means of a platinum wire connected to an electrical lead attached to one terminal of a voltmeter. The outer tube contains saturated KCl solution added through a hole in the side of the electrode, which is plugged or covered with a rubber sleeve. Contact with a solution in which an indicator electrode is immersed is made through a small asbestos fiber, fritted disk, ground glass sleeve, or other device that allows a degree of liquid contact between the reference electrode filling solution and the solution in which it is immersed. This contact constitutes a liquid junction at which a slight liquid junction potential can develop (see Section 11.13).

As discussed in Section 11.7, the half-reaction

$$AgCl(s) + e^- \rightleftarrows Ag(s) + Cl^- \tag{11.92}$$

is frequently used as the basis of a reference electrode. The silver–silver

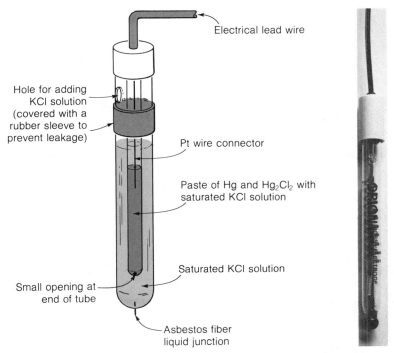

Electrical lead wire

Hole for adding
KCl solution
(covered with a
rubber sleeve to
prevent leakage)

Pt wire connector

Paste of Hg and Hg_2Cl_2 with
saturated KCl solution

Saturated KCl solution

Small opening at
end of tube

Asbestos fiber
liquid junction

FIGURE 11.11   Left: Typical configuration of a commercial saturated calomel electrode. **Right:** Ross combination glass-reference electrode for pH measurement.

chloride reference electrode has become quite popular in recent years. A commercial electrode of this type looks much like the SCE shown in Figure 11.11, with a silver chloride–coated silver wire in place of the inner tube of the calomel electrode. If the filling solution is saturated with KCl, this electrode has a potential at 25°C of +0.197 V versus the SHE.

Programmed Summary of Chapter 11

The major terms and concepts introduced in this chapter are contained in this summary in a programmed format. To derive the most benefit from the summary, you should fill in the blanks for each question and then check the answers at the end of the book to see if your choices are correct.

The species Fe^{2+}, S^{2-}, and Cr in $Cr_2O_7^{2-}$ have oxidation numbers of, respectively, (1) _____. An element that is oxidized (2) _____ in oxidation number, whereas one that is reduced (3) _____ in oxidation number. The half-reaction showing the reduction of MnO_4^- to Mn^{2+} in an acidic medium is (4) _____. A device in which an oxidation half-reaction can be allowed to occur separate from a reduction half-reaction is called an (5) _____. Two half-cells with different media (solutions in contact with the electrode) make contact at a (6) _____, which often consists of a tube filled with KCl called a (7) _____. An electrode consisting of platinum metal dipping into an acidic solution saturated with H_2 gas is called a (8) _____. The schematic representation of an electrochemical cell in which the anode is a hydrogen electrode and the cathode is platinum metal dipped into a solution of Fe^{2+} and Fe^{3+} with the anode and cathode connected via a salt bridge is (9) _____. Since it is impossible to measure individual half-cell potentials, they are all expressed relative to the (10) _____ assigned by convention a potential of (11) _____. According to the sign convention for electrode potentials, the term *electrode potential* applies to half-reactions written as (12) _____. The symbol E^0 denotes a (13) _____, which could be measured when the activities of all the species involved are (14) _____. The equation $E = 1.51 \text{ V} - (0.0591/5) \log ([Mn^{2+}]/[MnO_4^-][H^+]^8)$ is a form of the (15) _____ applied to the half-reaction (16) _____. A term for the concentration of water does not appear in the Nernst equation because the activity of H_2O in aqueous solutions is (17) _____ and assigned a value of (18) _____ by convention. The Nernst equation is used to account for differences in (19) _____ on the potential of electrochemical cells. Whereas the standard hydrogen electrode (SHE) is not practical for routine use as a reference electrode, both the (20) _____ and the (21) _____ electrode are so used. The half-reaction $M^+ + e^- \rightleftarrows M$ has a standard electrode potential of E_M^0. The slightly soluble salt MX may be reduced according to the half-reaction $MX(s) + e \rightleftarrows M(s) + X^-$, with a standard electrode potential of E_{MX}^0. The value of E_{MX}^0 can be calculated from (22) _____. A positive sign for E^0 of an overall oxidation-reduction reaction indicates that the reaction proceeds spontaneously (23) _____, whereas a negative sign indicates the opposite. For an *n*-electron oxidation-reduction reaction at 25° C, the value of the log of the equilibrium constant in terms of E^0 is $\log K = $ (24) _____. Three factors that can cause deviations from ideal potentials in electrochemical cells from which a small current is drawn are (25) _____. An electrode that is used to measure a concentration or ratio of concentrations is called an (26) _____, and its potential is always measured relative to a (27) _____.

Questions

1. Give the oxidation numbers of (a) Cr in $Cr_2O_7^{2-}$, (b) Hg in HgI_4^{2-}, (c) S in $S_2O_3^{2-}$, (d) P in H_3PO_4, (e) Fe in $Fe(NH_3)_6SO_4$, (f) Br in $KBrO_3$, (g) I in H_5IO_6, (h) C in $C_6H_8O_6$ (ascorbic acid).

2. In the following reactions, designate the element that is oxidized, the element that is reduced, the ion or compound that is

acting as the oxidizing agent, and the ion or compound that is acting as the reducing agent.

a. $2MnO_4^- + 5H_2C_2O_4 + 6H^+ \rightarrow$
$$2Mn^{2+} + 10CO_2(g) + 8H_2O$$

b. $6Fe^{2+} + Cr_2O_7^{2-} + 14H^+ \rightleftharpoons$
$$6Fe^{3+} + 2Cr^{3+} + 7H_2O$$

c. $BrO_3^- + 5Br^- + 6H^+ \rightarrow 3Br_2 + 3H_2O$

3. What is the basic requirement for a *balanced* oxidation-reduction reaction?

4. Balance the following oxidation-reduction reactions:

a. $Cu_2S + NO_3^- \rightarrow$
$$Cu^{2+} + SO_4^{2-} + NO_2$$

b. $Sn^{2+} + Cr_2O_7^{2-} + H^+ \rightarrow$
$$Sn^{4+} + Cr^{3+} + H_2O$$

c. $Cr^{3+} + MnO_4^- + H_2O \rightarrow$
$$Cr_2O_7^{2-} + Mn^{2+} + H^+$$

5. What is the device called that allows an oxidation half-reaction to occur physically separate from the reduction half-reaction, and where in this device do the half-reactions occur?

6. An electrochemical cell in which the reaction occurs spontaneously is called (a) _____, and one in which the reaction is forced to occur by application of an external electrical potential is called (b) _____.

7. What is a liquid junction, and what is its significance in an electrochemical cell?

8. In a schematic representation of an electrochemical cell, how are the following designated: (a) anode, (b) cathode, (c) phase junctions, (d) salt bridges.

9. What is an SHE? What is its physical construction? What is the half-reaction at the SHE?

10. According to the IUPAC convention, how is the electrode potential (reduction potential) related to the potential for a half-reaction actually measured versus the SHE?

11. When is an electrode potential a *standard* electrode potential?

12. What is the Nernst equation expression

applied to the half-reaction
$$BrO_3^- + 5Br^- + 6H^+ \rightleftharpoons 3Br_2(aq) + 3H_2O$$

13. What is normally used in the Nernst expression, sometimes as an approximation, for activities of (a) solutes, (b) gases, (c) solids, (d) solvent water?

14. What are the half-reactions for (a) O_2 gas being reduced to a water product, (b) the calomel reference electrode, (c) the silver–silver chloride reference electrode?

15. Where E^0 is the standard electrode potential for the reduction of Ag^+ to Ag, what is the Nernst equation for E versus the SHE of a silver electrode in a solution saturated with solid Ag_2CrO_4, solubility product K_{sp}, and a known value of excess chromate concentration, $[CrO_4^{2-}]$?

16. What should be the reduction half-reaction producing $Ag(s)$ and $C_2O_4^{2-}$ from $Ag_2C_2O_4(s)$, and what is the Nernst equation for this half-reaction is terms of its E^0 and concentrations?

17. What is the Nernst expression for the reduction of $Ag(I)$ in the complex ion $Ag(S_2O_3)_2^{3-}$ in terms of E^0 for the half-reaction $Ag^+ + e^- \rightleftharpoons Ag$, pertinent species concentrations, and β_2, the overall formation constant of the complex ion?

18. From the preceding example, what would be $E^0_{complex}$ for the half-reaction
$$Ag(S_2O_3)_2^{3-} + e^- \rightleftharpoons Ag + 2S_2O_3^{2-}$$
in terms of E^0 for the reduction of Ag^+ to Ag and β_2?

19. Given the half-reactions

$$Br_2(l) + 2e^- \rightleftharpoons 2Br^- \qquad E^0 = +1.087 \text{ V}$$
$$Cr^{3+} + e^- \rightleftharpoons Cr^{2+} \qquad E^0 = -0.408 \text{ V}$$
$$Cl_2(g) + 2e^- \rightleftharpoons 2Cl^- \qquad E^0 = +1.359 \text{ V}$$
$$Ti^{3+} + e^- \rightleftharpoons Ti^{2+} \qquad E^0 = -0.369 \text{ V}$$
$$Fe^{3+} + e^- \rightleftharpoons Fe^{2+} \qquad E^0 = +0.771 \text{ V}$$

list the oxidizing agents above in increasing order of oxidizing strength and the reducing agents in increasing order of reducing strength.

20. What is the relationship between ΔG^0 of a reaction and E^0 of the reaction?

21. What does a positive value of E^0 for an oxidation-reduction reaction indicate regarding the direction of the reaction?

22. From the information given in Question 19, what is the overall reaction and E^0 for the electrochemical cell designated schematically as

$$\text{Pt}|\text{Ti}^{2+}(a = 1.00), \quad \text{Ti}^{3+}(a = 1.00)\|\text{Br}_2(l),$$
$$\text{Br}^-(a = 1.00)|\text{Pt}$$

23. What is the Nernst equation applied to the electrochemical cell reaction in Question 22?

24. What does a negative E indicate for an oxidation-reduction reaction?

25. What is the expression (at 25° C) for log K of an oxidation-reduction reaction in terms of E^0 for the reaction?

26. What is the value of E for an oxidation-reduction reaction at equilibrium?

27. From what source do junction potentials arise?

28. What effect does internal cell resistance have on the measured potential of a cell through which a current is flowing?

29. What are the two types of overvoltage in an electrochemical cell?

30. For what purpose is an indicator electrode used?

31. What are the desirable properties of a good reference electrode?

32. What is "saturated" about the SCE?

33. What is the function of a small asbestos fiber, fritted disk, or similar device at the bottom of a commercial SCE?

Problems

1. What is the half-reaction and E versus the SHE for the half-cell designated schematically as $\text{Pt}|\text{Br}^-(a = 3.67 \times 10^{-2})$, $\text{BrO}_3^-(a = 1.05 \times 10^{-2})$, $\text{H}^+(a = 0.100)$, $E^0 = +1.44$ V?

2. What is the half-reaction and E versus the SHE for the half-cell designated as $\text{Pt}|\text{I}_2(s)$, $\text{IO}_3^-(a = 2.50 \times 10^{-2})$, $\text{H}^+(a = 0.100)$, $E^0 = 1.106$ V?

3. What is the half-reaction and E versus the SHE for the half-cell designated $\text{Pt}|\text{Cl}_2(p = 1.00$ atm$)$,   HClO $(a = 8.50 \times 10^{-3})$, H^+ $(a = 0.100)$, $E^0 = +1.63$ V?

4. Given

$$\text{Hg}_2\text{Cl}_2(s) + 2e^- \rightleftarrows 2\text{Hg}(l) + 2\text{Cl}^-$$
$$E^0 = +0.268 \text{ V}$$

and that $E = +0.241$ versus the SHE for the saturated calomel electrode, calculate

[Cl$^-$] in the SCE, approximating activities with concentrations.

5. Referring to Problem 4, what is [Cl$^-$] in a calomel electrode for which $E = +0.300$ V versus the SHE?

6. Referring to the two preceding problems, what is [Cl$^-$] in a calomel electrode having $E = +0.360$ V versus the SHE?

7. Given that $E = 0.191$ V versus the SHE for an AgBr-coated Ag electrode in a solution that is 1.00×10^{-2} M in Br$^-$, what is $K_{\text{sp}}^{\text{AgBr}}$?

8. Given that $E = 0.103$ V versus the SHE for an AgX-coated Ag electrode immersed in 7.50×10^{-3} M X$^-$, what is $K_{\text{sp}}^{\text{AgX}}$?

9. Given that $E = 0.213$ V versus the SHE for an Ag$_2$Z-coated Ag electrode in 1.45×10^{-2} M Z^{2-}, what is $K_{\text{sp}}^{\text{Ag}_2\text{Z}}$?

10. In Chapter 10, the fraction of Cu(II) present as Cu^{2+} was calculated for a medium in which $[NH_3] = 0.100\ M$. The standard electrode potential for the reduction of Cu^{2+} to Cu is $+0.337$ V. What is E versus the SHE for a Cu electrode that is $1.00 \times 10^{-3}\ M$ in Cu(II) and $0.100\ M$ in NH_3?

11. From the formation constants and α values for EDTA given in Chapter 10, what is E of a copper electrode versus the SHE in a medium in which Cu(II) concentration is $2.50 \times 10^{-3}\ M$; unchelated EDTA concentration, C_T, is $5.00 \times 10^{-2}\ M$; and pH $= 7.00$?

12. Referring to the previous two questions, what is the potential versus the SHE of a copper electrode in a solution for which $[\text{Cu(II)}] = 6.25 \times 10^{-2}\ M$, total EDTA concentration (both chelated and unchelated) $= 2.00 \times 10^{-2}\ M$, $[NH_3] = 0.100\ M$, and pH $= 10.00$?

13. Calculate E for the cell

$Pt\,|\,Fe^{2+}(a = 2.33 \times 10^{-3})$,
$Fe^{3+}(a = 9.00 \times 10^{-3})\,\|$

$\qquad\qquad Ce^{4+}(a = 4.25 \times 10^{-3})$,
$\qquad Ce^{3+}(a = 2.65 \times 10^{-3})\,|\,Pt$

Use $+1.70$ V for the standard electrode potential for the reduction of Ce^{4+}.

14. What is E for an electrochemical cell in which the left half-cell is a Pt electrode immersed in a solution in which $[Fe^{3+}] = 2.00 \times 10^{-3}\ M$ and $[Fe^{2+}] = 8.75 \times 10^{-3}\ M$, and in which the right half-cell has $[MnO_4^-] = 1.50 \times 10^{-2}\ M$, $[Mn^{2+}] = 9.00 \times 10^{-3}\ M$, pH $= 1.00$? Approximate activities with concentrations.

15. What is E for the electrochemical cell designated schematically as

$Pt\,|\,Ti^{2+}(3.75 \times 10^{-3}\ M)$,
$\qquad\qquad Ti^{3+}(1.80 \times 10^{-3}\ M)\,\|\,Br_2(l)$,
$\qquad\qquad\qquad Br^-(1.40 \times 10^{-2}\ M)\,|\,Pt$

16. What is the equilibrium constant for the reaction $Zn + Pb^{2+} \rightleftarrows Zn^{2+} + Pb$?

17. What is the equilibrium constant for the reaction $Cl_2(g) + 2Br^- \rightleftarrows Cl^- + Br_2(l)$?

18. What is the equilibrium constant for the reaction

$$Cr_2O_7^{2-} + 6Fe^{2+} + 14H^+ \rightleftarrows$$
$$2Cr^{3+} + 6Fe^{3+} + 7H_2O$$

Oxidation-Reduction Titrations

1. Calculating and plotting the potentiometric titration curve that would result from a titration given the titration's oxidation-reduction reaction.
2. End points from the titration curves of various types of potentiometric titrations.
3. Calculating and plotting titration curves involving more than one successive oxidation or reduction.
4. Types and functions of oxidation-reduction indicators.
5. Nature and uses of auxiliary oxidizing and reducing agents.
6. Common oxidizing titrants and the purposes for which they are used.
7. Common reducing titrants and the purposes for which they are used.

12.1   Introduction

Oxidation-Reduction Titrations

Oxidation-reduction reactions and equilibria were discussed in Chapter 11. This chapter addresses the use of these reactions for quantitative analysis, particularly by titration. Several important oxidation-reduction reactions meet the criteria for titration reactions by being rapid, complete, and stoichiometrically well defined. The following two general types of oxidation-reduction titration reactions are possible:

$$\text{Analyte} + \text{oxidizing agent (titrant)} \rightarrow \text{products} \tag{12.1}$$

$$\text{Analyte} + \text{reducing agent (titrant)} \rightarrow \text{products} \tag{12.2}$$

Oxidation-Reduction Titration Curves

As discussed in Chapter 5, an acid–base titration can be viewed as a plot of pH versus volume of added titrant. Similarly, an oxidation-reduction titration can be expressed graphically as a plot of electrode potential E versus volume of added titrant. Such plots for the addition of oxidant titrant and reductant titrant are shown in Figure 12.1. Figure 12.1(a) shows that E increases abruptly at the equivalence point when the titrant is a relatively strong oxidizing agent reacting with a readily reduced analyte. The increase in E reflects the fact that the medium is more oxidizing in the presence of excess oxidizing agent. In Figure 12.1(b) the value of E decreases abruptly at the equivalence point when

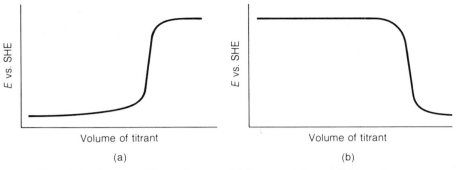

FIGURE 12.1 Curves of electrode potential E versus volume of titrant for oxidation-reduction titrations: (a) titration of an oxidizable analyte with an oxidizing agent titrant; (b) titration of a reducible analyte with a reducing agent titrant

a reducing agent is the titrant, reflecting a substantially lower electrode potential in the presence of excess reducing agent.

Figure 12.2 shows the experimental setup that could be used to collect data for titration curves involving oxidation-reduction reactions, and is the one that applies to the titration curves discussed in this chapter. The reference electrode in this figure is the SHE. In practice, an SCE or Ag–AgCl electrode (see Section 11.14) would be used for convenience. The indicator electrode is made of platinum foil and responds to changes in E. The analyte solution is placed in the titration vessel—normally a beaker—and the titrant is added through a buret. The voltmeter is used to record the potential E of the Pt indicator electrode versus the reference electrode after each addition of titrant. The whole process can be performed with autotitrators that record E as a function of volume of titrant added from a motor-driven buret. The data can be displayed on chart paper or stored for later printout by a computer.

Obviously, the end point in an oxidation-reduction titration can be determined from a plot of the titration curve. Other means are available and are discussed in this and later chapters. For routine noninstrumental titrations, the most common are oxidation-reduction indicators (see Section 12.6) that change color at the end point.

12.2   A Simple Oxidation-Reduction Titration Curve: Fe^{2+} with Ce^{4+}

Titration Conditions

As an aid to understanding oxidation-reduction titration curves, consider the simple case of Fe^{2+} analyte titrated with standard cerium(IV) oxidizing agent. The titration reaction is

$$Ce^{4+} + Fe^{2+} \rightarrow Ce^{3+} + Fe^{3+} \tag{12.3}$$

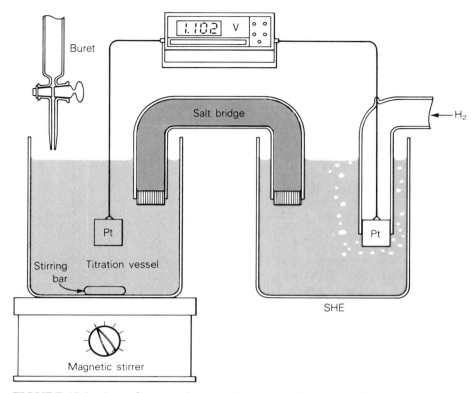

FIGURE 12.2   Setup for an oxidation–reduction titration using a platinum indicator electrode and an SHE reference electrode (in practice an SCE or Ag–AgCl reference electrode would be used for convenience)

This is a rapid 1:1 reaction with a high equilibrium constant (see Section 11.12). The reaction can be carried out by adding standard Ce^{4+} solution from a buret to the Fe^{2+} analyte solution contained in the titration vessel shown in Figure 12.2. After each addition of titrant, the potential E of the platinum electrode in the titration vessel is measured against the SHE. Because the course of the titration is followed by measuring *potential*, the titration curve is called a **potentiometric titration**. In this section a theoretical titration curve is calculated for the titration of Fe^{2+} with Ce^{4+}.

Consider the titration of 50 mL of 0.01000 M Fe^{2+} with 0.02000 M Ce^{4+}, calculating E after the addition of exactly 0, 5, 12.5, 20, 24, 25, 26, 30, and 50 mL of Ce^{4+}. During the titration, the four species involved—Ce^{4+}, Fe^{2+}, Ce^{3+}, and Fe^{3+}—can be considered at equilibrium after the addition of each portion of titrant. Therefore, the potential E versus the SHE registered by the Pt electrode in the titration vessel is given by

$$Fe^{3+} + e^- \rightleftarrows Fe^{2+} \qquad E^0 = +0.77 \text{ V} \tag{12.4}$$

$$E = 0.77 - 0.0591 \log \frac{[Fe^{2+}]}{[Fe^{3+}]} \tag{12.5}$$

or by

$$Ce^{4+} + e^- \rightleftarrows Ce^{3+} \qquad E^0 = 1.70 \text{ V} \tag{12.6}$$

$$E = 1.70 - 0.0591 \log \frac{[Ce^{3+}]}{[Ce^{4+}]} \tag{12.7}$$

In the two preceding Nernst equations, E has the same value, regardless of whether it is calculated from the ratio $[Fe^{2+}]/[Fe^{3+}]$ or from $[Ce^{3+}]/[Ce^{4+}]$. However, *before* the equivalence point, Equation 12.5 is used because both $[Fe^{2+}]$ and $[Fe^{3+}]$ can be calculated from simple stoichiometric considerations, and *after* the equivalence point, Equation 12.7 is used because both $[Ce^{3+}]$ and $[Ce^{4+}]$ can be calculated from stoichiometry. As will be seen, at the equivalence point a special relationship is used.

E before the Equivalence Point

Initially there are no cerium species present, and the solution contains $0.0100 \ M \ Fe^{2+}$. There will be a small trace of Fe^{3+} as a result of some slight oxidation of Fe^{2+} by dissolved oxygen or by water itself. However, there is no way of calculating the very low concentration of Fe^{3+}. Therefore, the initial value of E cannot be calculated.

At 5.00 mL of added $0.0200 \ M \ Ce^{4+}$,

$$\text{mmol } Fe^{3+} \text{ produced} = \text{mmol } Ce^{4+} \text{ added} = 5.00 \text{ mL} \times 0.0200 \text{ mmol/mL}$$
$$= 0.100 \text{ mmol } Fe^{3+} \tag{12.8}$$

$$\text{mmol } Fe^{2+} \text{ remaining} = \text{original mmol } Fe^{2+} - \text{mmol } Fe^{3+} \text{ produced}$$
$$= 50.0 \text{ mL} \times 0.0100 \text{ mmol/mL} - 0.100 \text{ mmol}$$
$$= 0.500 \text{ mmol} - 0.100 \text{ mmol} = 0.400 \text{ mmol} \tag{12.9}$$

The total volume of solution in the titration vessel at this point is 55.0 mL, although it cancels in the final calculation of E. The value of E can now be calculated from Equation 12.5 as follows:

$$E = 0.77 - 0.0591 \log \frac{[Fe^{2+}]}{[Fe^{3+}]} = 0.77 - 0.0591 \log \frac{0.400 \text{ mmol/55.0 mL}}{0.100 \text{ mmol/55.0 mL}}$$

$$= 0.77 \text{ V} - 0.04 \text{ V} = 0.73 \text{ V} \tag{12.10}$$

Here there may be some confusion about an aspect of the calculation of E values plotted in the preparation of oxidation–reduction titration curves. This is that the value of E is calculated from a form of the Nernst equation based upon a *reduction* half-reaction (Reaction 12.4), but the actual titration involves *oxidation* of Fe^{2+} to Fe^{3+}. In fact, a platinum electrode responds to a particular ratio of $[Fe^{2+}]$ to $[Fe^{3+}]$ regardless of whether that ratio was established by oxidation of Fe^{2+} to Fe^{3+}, reduction of Fe^{3+} to Fe^{2+}, or simply adding Fe^{2+} and Fe^{3+} salts separately to the solution. Thus the means by which Fe^{2+} and Fe^{3+} are produced in solution have no relevance to the equilibrium value of E measured.

At 12.50 mL added Ce^{4+}, E is calculated as follows:

$$\text{mmol } Fe^{3+} \text{ produced} = \text{mmol } Ce^{4+} \text{ added} = 12.50 \text{ mL} \times 0.0200 \text{ mmol/mL}$$
$$= 0.250 \text{ mmol } Fe^{3+} \tag{12.11}$$

$$\text{mmol } Fe^{2+} \text{ remaining} = \text{original mmol } Fe^{2+} - \text{mmol } Fe^{3+} \text{ produced}$$
$$= 0.500 \text{ mmol} - 0.250 \text{ mmol} = 0.250 \text{ mmol} \tag{12.12}$$

The volume of solution in the titration vessel is 62.50 mL.

$$E = 0.77 \text{ V} - 0.0591 \log \frac{0.250 \text{ mmol/62.50 mL}}{0.250 \text{ mmol/62.50 mL}} = 0.77 \tag{12.13}$$

Note that at this point, *halfway to the equivalence point*, $[Fe^{2+}]$ and $[Fe^{3+}]$ cancel in the Nernst equation, and E is exactly equal to E^0 for the half-reaction

$$Fe^{3+} + e^- \rightleftarrows Fe^{2+} \tag{12.4}$$

The remaining points prior to the equivalence point are calculated by means entirely analogous to those employed at 5.00 and 12.50 mL of Ce^{4+} titrant added. Briefly, at 20.00 mL added titrant,

$$E = 0.77 - 0.0591 \log \frac{0.100 \text{ mmol/70.0 mL}}{0.400 \text{ mmol/70.0 mL}} = 0.81 \text{ V} \tag{12.14}$$

At 24.00 mL added titrant,

$$E = 0.77 - 0.0591 \log \frac{0.020 \text{ mmol/74.00 mL}}{0.48 \text{ mmol/74.00 mL}} = 0.85 \text{ V} \tag{12.15}$$

At Equivalence Point

At the equivalence point, exactly 0.500 mmol of Fe^{3+} has been produced. Examination of the stoichiometry of the titration reaction (12.3) shows that exactly 0.500 mmol of Ce^{4+} has also been produced. Only very low concentrations of Ce^{4+} and Fe^{2+} remain unreacted. However, by the definition of equivalence point, the number of millimoles of Ce^{4+} added exactly equals the number of millimoles of Fe^{2+} originally present. Therefore, the concentration of unreacted Ce^{4+} exactly equals the concentration of unreacted Fe^{2+}, or

$$[Ce^{4+}] = [Fe^{2+}] \tag{12.16}$$

The equivalence point potential E_{eq} can be calculated by *adding* the following two reactions:

$$E_{eq} = 0.77 - 0.0591 \log \frac{[Fe^{2+}]}{[Fe^{3+}]} \tag{12.17}$$

$$E_{eq} = 1.70 - 0.0591 \log \frac{[Ce^{3+}]}{[Ce^{4+}]} \tag{12.18}$$

$$2E_{eq} = 0.77 + 1.70 - 0.0591 \log \frac{[Fe^{2+}][Ce^{3+}]}{[Fe^{3+}][Ce^{4+}]} \tag{12.19}$$

(Note that in this equation, the log term is *not* the equilibrium constant expression.)

Recall that

$$[Fe^{3+}] = [Ce^{3+}] = \frac{0.500 \text{ mmol}}{75.00 \text{ mL}} \tag{12.20}$$

and

$$[Ce^{4+}] = [Fe^{2+}] \tag{12.16}$$

at the equivalence point. Therefore,

$$2E_{eq} = 0.77 + 1.70 - 0.0591 \log \frac{[Fe^{2+}][Ce^{3+}]}{[Fe^{3+}][Ce^{4+}]} = 0.77 + 1.70 - 0.0591 \log 1$$

$$= 0.77 \text{ V} + 1.70 \text{ V} \tag{12.21}$$

$$E_{eq} = \frac{0.77 \text{ V} + 1.70 \text{ V}}{2} = 1.24 \text{ V} \tag{12.22}$$

This calculation of E_{eq} is an illustration of the general approach to calculating equivalence point potential discussed in Section 12.4.

Beyond Equivalence Point

At 26.00 mL added Ce^{4+} titrant, there is 1.00 mL of excess Ce^{4+} solution, so that

$$[Ce^{4+}] = \frac{1.00 \text{ mL} \times 0.0200 \text{ mmol/mL}}{76.00 \text{ mL}} = \frac{0.0200 \text{ mmol}}{76.00 \text{ mL}} \tag{12.23}$$

Furthermore, since 0.500 mmol of Ce^{3+} was produced up to the equivalence point,

$$[Ce^{3+}] = \frac{0.500 \text{ mmol}}{76.00 \text{ mL}} \tag{12.24}$$

At this point, and *at any volume of titrant beyond the equivalence point*, both $[Ce^{3+}]$ and $[Ce^{4+}]$ can be calculated simply from stoichiometry. Therefore, in this region E is given simply by the following Nernst expression:

$$E = 1.70 - 0.0591 \log \frac{[Ce^{3+}]}{[Ce^{4+}]} = 1.70 - 0.0591 \log \frac{0.500 \text{ mmol}/76.00 \text{ mL}}{0.0200 \text{ mmol}/76.00 \text{ mL}}$$

$$= 1.70 - 0.0591(+1.40) = 1.70 - 0.08 = 1.62 \text{ V} \tag{12.25}$$

The value of E at 30.00 mL added Ce^{4+} is calculated as in the preceding example.

$$E = 1.70 - 0.0591 \log \frac{0.500 \text{ mmol}/80.00 \text{ mL}}{0.100 \text{ mmol}/80.00 \text{ mL}} = 1.70 \text{ V} - 0.04 \text{ V} = 1.66 \text{ V} \tag{12.26}$$

The addition of 50.00 mL of standard Ce^{4+} titrant yields an excess of 0.500 mmol of Ce^{4+} beyond the equivalence point meaning that exactly twice as much Ce^{4+} has been added as is required to reach the equivalence point. This means that E equals E^0 for the half-reaction $Ce^{4+} + e^- \rightleftarrows Ce^{3+}$ as shown by the following calculation:

$$E = 1.70 - 0.0591 \log \frac{0.500 \text{ mmol}/100.0 \text{ mL}}{0.500 \text{ mmol}/100.0 \text{ mL}} = 1.70 \text{ V} \tag{12.27}$$

TABLE 12.1   Values of E versus the SHE for the titration of 50.0 mL of 0.0100 M Fe^{2+} with 0.02000 M standard Ce^{4+} solution.

Volume standard Ce^{4+} added, mL	E vs. SHE, V	Significance of point
0	—	E cannot be calculated for this point.
5.00*	0.73	
12.50*	0.77	Halfway to the equivalence point, $E = E^0$ for the half-reaction $Fe^{3+} + e^- \rightleftarrows Fe^{2+}$.
20.00*	0.81	
24.00*	0.85	
25.00	1.24	Equivalence point potential.
26.00†	1.62	
30.00†	1.66	
50.00†	1.70	Twofold amount of titrant required to reach the equivalence point, so that $[Ce^{4+}] = [Ce^{3+}]$ and $E = E^0$ for the half-reaction $Ce^{4+} + e^- \rightleftarrows Ce^{3+}$.

* At these points, prior to the equivalence point, both $[Fe^{2+}]$ and $[Fe^{3+}]$ can be calculated from simple stoichiometry, enabling calculation of E from the Nernst equation

$$E = 0.77 \text{ V} - 0.0591 \log \frac{[Fe^{2+}]}{[Fe^{3+}]}$$

† At these points, beyond the equivalence point, both $[Ce^{3+}]$ and $[Ce^{4+}]$ can be calculated from simple stoichiometry, enabling calculation of E from the Nernst equation

$$E = 1.70 \text{ V} - 0.0591 \log \frac{[Ce^{3+}]}{[Ce^{4+}]}$$

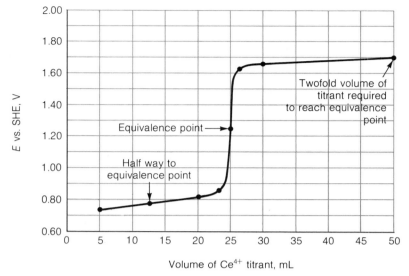

FIGURE 12.3 Curve for the potentiometric titration of 50.00 mL of 0.01000 M Fe^{2+} with 0.0200 M Ce^{4+}, E versus the SHE measured at various volumes of added titrant

Plot of the Titration Curve

The values of E corresponding to specific volumes of standard Ce^{4+} solution added are summarized in Table 12.1. The titration curve plotted from the data summarized in Table 12.1 is shown in Figure 12.3. The curve looks very much like that obtained from the titration of a strong acid with a strong base (see Section 5.2, Figure 5.2) or from the titration of a chelatable metal with EDTA, where a 1:1 metal chelate is obtained. As seen in Figure 12.3, the equivalence point occurs exactly halfway up the rising portion of the S-shaped titration curve. This holds true only in cases where the oxidizing agent species accepts the same number of electrons as are donated by each reducing agent species.

12.3   ## A More Complicated Oxidation-Reduction Titration: Fe^{2+} with MnO_4^-

Titration Conditions

Consider next the titration of 50 mL of 0.05000 M Fe^{2+} with 0.02000 M MnO_4^- in a medium maintained at $a_{H^+} = 1.00$ using the apparatus shown in Figure 12.2, and following E versus the SHE. The balanced titration reaction is

$$MnO_4^- + 5Fe^{2+} + 8H^+ \rightarrow Mn^{2+} + 5Fe^{3+} + 4H_2O \qquad (12.28)$$

The two pertinent half-reactions are the following:

$$MnO_4^- + 8H^+ + 5e^- \rightleftharpoons Mn^{2+} + 4H_2O \qquad E^0 = +1.51 \text{ V} \qquad (12.29)$$

$$Fe^{3+} + e^- \rightleftharpoons Fe^{2+} \qquad\qquad\qquad E^0 = +0.77 \text{ V} \qquad (12.30)$$

Since each MnO_4^- ion undergoes a five-electron reduction, there are five equivalents per mole of MnO_4^-. Therefore the normality, N, of the MnO_4^- solution is five times the molarity. In this case $N = 5 \times M = 5 \times 0.02000 = 0.1000$ equivalent per liter. This example is somewhat more complicated than the one discussed in the preceding section because it involves an oxidizing agent, MnO_4^-, that accepts five electrons per ion, and a reducing agent, Fe^{2+}, that loses one electron per ion. The differences will become more apparent when the equivalence point potential is calculated and the titration curve is plotted. In this case, E will be calculated after the addition of 5, 12.5, 20, 22, 24, 25, 26, 30, and 50 mL of MnO_4^- titrant.

Before the Equivalence Point

Prior to the equivalence point, both $[Fe^{2+}]$ and $[Fe^{3+}]$ can be calculated stoichiometrically, so the Nernst equation applying to these two species is used in calculating E. At 5 mL of added 0.02000 M MnO_4^-, the stoichiometry of the titration reaction shows that the number of millimoles of Fe^{3+} produced is

$$\text{mmol } Fe^{3+} = 5.00 \text{ mL } MnO_4^- \times 0.02000 \; \frac{\text{mmol } MnO_4^-}{\text{mL } MnO_4^-} \times 5 \; \frac{\text{mmol } Fe^{3+}}{\text{mmol } MnO_4^-}$$

$$= 0.500 \text{ mmol } Fe^{3+} \qquad (12.31)$$

Originally there were $50.00 \text{ mL} \times 0.05000 \text{ mmol/mL} = 2.50 \text{ mmol } Fe^{2+}$ present, but after the addition of 5.00 mL of titrant, the number of millimoles of Fe^{2+} remaining is given by the following:

$$\text{mmol } Fe^{2+} = 2.500 \text{ mmol} - 0.500 \text{ mmol} = 2.00 \text{ mmol} \qquad (12.32)$$

The value of E is readily calculated from the Nernst equation below:

$$E = 0.77 - 0.0591 \log \frac{[Fe^{2+}]}{[Fe^{3+}]} = 0.77 - 0.0591 \log \frac{2.00 \text{ mmol/55.00 mL}}{0.500 \text{ mmol/55.00 mL}}$$

$$= 0.77 - 0.0591 \times 0.602 = 0.73 \text{ V} \qquad (12.33)$$

The values of E at all volumes of added titrant less than 25 mL are calculated similarly and are summarized in Table 12.2.

At the Equivalence Point

The equivalence point is at a volume of 25.00 mL added titrant. In terms of the relevant Nernst equations, the equivalence point potential E_{eq} is

$$E_{eq} = 0.77 - 0.0591 \log \frac{[Fe^{2+}]}{[Fe^{3+}]} \tag{12.34}$$

$$E_{eq} = 1.51 - \frac{0.0591}{5} \log \frac{[Mn^{2+}]}{[MnO_4^-][H^+]^8} \tag{12.35}$$

These two equations cannot be added directly together as was done in calculating E_{eq} for the titration of Fe^{2+} with Ce^{4+} in the preceding section because the log term in the first of the equations is multiplied by 0.0591, whereas in the second it is multiplied by 0.0591/5. This problem can be remedied by multiplying the second of the preceding Nernst expressions through by 5 and performing the addition shown below:

$$5E_{eq} = 5 \times 1.51 - 5 \times \frac{0.0591}{5} \log \frac{[Mn^{2+}]}{[MnO_4^-][H^+]^8}$$

$$5E_{eq} = 5 \times 1.51 - 0.0591 \log \frac{[Mn^{2+}]}{[MnO_4^-][H^+]^8} \left. \right\} \begin{array}{l} \text{Add these two} \\ \text{equations} \end{array} \tag{12.36}$$

$$E_{eq} = 0.77 - 0.0591 \log \frac{[Fe^{2+}]}{[Fe^{3+}]} \tag{12.34}$$

$$6E_{eq} = 5 \times 1.51 + 0.77 - 0.0591 \log \frac{[Mn^{2+}][Fe^{2+}]}{[MnO_4^-][H^+]^8[Fe^{3+}]} \tag{12.37}$$

Now comes the problem of dealing with the log expression in the preceding equation. Recall that as a condition for the titration, a_{H^+} was set equal to exactly 1, so the $[H^+]^8$ term drops out. The stoichiometry of the titration reaction, Reaction 12.28, shows that *at the equivalence point*, $[Fe^{3+}] = 5[Mn^{2+}]$ and $[Fe^{2+}] = 5[MnO_4^-]$. (To some, it may appear the other way around, but realize that the stoichiometry of the titration reaction shows that $[Fe^{3+}] > [Mn^{2+}]$ and $[Fe^{2+}] > [MnO_4^-]$.) Substituting these values into Equation 12.37 yields

$$6E_{eq} = 5 \times 1.51 + 0.77 - 0.0591 \log \frac{[Mn^{2+}]5[MnO_4^-]}{[MnO_4^-](1)^8 5[Mn^{2+}]}$$

$$= 5 \times 1.51 + 0.77 - 0.0591 \log 1 = 5 \times 1.51 + 0.77$$

leading to

$$E_{eq} = \frac{5 \times 1.51 \text{ V} + 0.77 \text{ V}}{6} = +1.39 \text{ V} \tag{12.38}$$

Beyond the Equivalence Point

Beyond the equivalence point both $[MnO_4^-]$ and $[Mn^{2+}]$ can be calculated stoichiometrically. For example, after the addition of 26.00 mL of titrant, there is 1.00 mL of 0.02000 M MnO_4^- in excess. This is 1.00 mL $\times 0.02000$ mmol/mL $= 0.0200$ mmol MnO_4^-. Furthermore, the 25.00 mL $\times 0.02000$ mmol/mL $= 0.500$ mmol of MnO_4^- added to reach the equivalence point was converted to 0.500 mmol of Mn^{2+}. These values are required for the following calculation of E:

$$E = 1.51 - \frac{0.0591}{5} \log \frac{[Mn^{2+}]}{[MnO_4^-][H^+]^8}$$

$$= 1.51 - \frac{0.0591}{5} \log \frac{0.500 \text{ mmol}/76.00 \text{ mL}}{0.0200 \text{ mmol}/76.00 \text{ mL } (1)^8}$$

$$= 1.49 \text{ V} \tag{12.39}$$

All of the other potentials beyond the equivalence point are calculated in the same manner and are summarized in Table 12.2. One special point is at the addition of 50.00 mL of standard MnO_4^-. This is a total of 50.00 mL $\times 0.02000$ mmol/mL $= 1.00$ mmol of MnO_4^- added. Of this, 0.500 mmol was converted to Mn^{2+} by reduction up to the equivalence point, leaving 0.500 mmol MnO_4^-, so that

$$E = 1.51 - \frac{0.0591}{5} \log \frac{0.500 \text{ mmol}/75.00 \text{ mL}}{0.500 \text{ mmol}/75.00 \text{ mL } (1)^8} = 1.51 \text{ V} \tag{12.40}$$

At $a_{H^+} = 1$, this is where $E = E^0$ for the reduction of MnO_4^- to Mn^{2+}.

Plot and Interpretation of Titration Curve

From the potentiometric titration data summarized in Table 12.2, it is possible to construct the titration curve shown in Figure 12.4. The curve is not totally

TABLE 12.2   Data for the potentiometric titration of 50.00 mL of 0.0500 M Fe^{2+} with 0.02000 M MnO_4^- in a medium maintained at $a_{H^+} = 1.00$

MnO_4^- added, mL	Total volume, mL	Fe^{2+}, mmol	Fe^{3+}, mmol	Mn^{2+}, mmol	MnO_4^-, mmol	E vs. SHE, V
0.00	50.00	2.500	—	0.00	0.00	Undefined
5.00	55.00	2.000	0.500	0.100	—	0.73
12.50*	62.50	1.25	1.25	0.250	—	0.77
20.00	70.00	0.500	2.000	0.400	—	0.80
22.00	72.00	0.300	2.200	0.440	—	0.82
24.00	74.00	0.100	2.400	0.480	—	0.85
25.00†	75.00	$5 \times$ mmol MnO_4^-	2.500	0.500	$\frac{1}{5} \times$ mmol Fe^{2+}	1.39
26.00	76.00	—	2.500	0.500	0.0200	1.49
30.00	80.00	—	2.500	0.500	0.100	1.50
50.00‡	100.00	—	2.500	0.500	0.500	1.51

* Halfway to the equivalence point, $E = E^0$ for $Fe^{3+} + e^- \rightleftarrows Fe^{2+}$ $^{?+}$
† Equivalence point
‡ Twice as much titrant as required to reach the equivalence point, $E = E^0$ for $MnO_4^- + 8H^+ + 5e^- \rightleftarrows Mn^{2+} + 4H_2O$

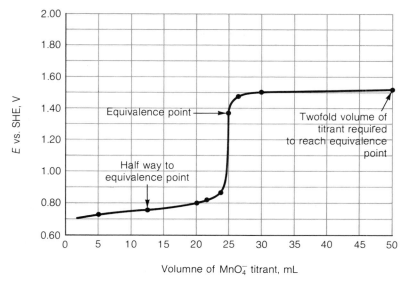

FIGURE 12.4 Curve for the potentiometric titration of 50.00 mL of 0.0500 M Fe^{2+} with 0.0200 M MnO$_4$ in a medium maintained at a_{H^+} = 1.00 M, E versus the SHE measured at various volumes of titrant

symmetrical, having a sharper bend at the top than at the bottom. The most striking feature of the titration curve is that the equivalence point is not halfway up the steeply rising portion, because E_{eq} *is closer to* E^0 *for the half-reaction involving the greater number of electrons.* In this case, E_{eq} at 1.39 V is closer to E^0 for the half-reaction involving five electrons.

$$\text{MnO}_4^- + 8\text{H}^+ + 5e^- \rightleftarrows \text{Mn}^{2+} + 4\text{H}_2\text{O} \qquad E^0 = 1.51 \text{ V} \qquad (12.29)$$

than it is for the following half-reaction involving only one electron:

$$\text{Fe}^{3+} + e^- \rightleftarrows \text{Fe}^{2+} \qquad E^0 = 0.77 \text{ V} \qquad (12.30)$$

Why this is so is obvious from the equation for calculating equivalence point potentials from E^0 values and is explained in greater detail in the following section.

12.4 Equivalence Point Potential in Oxidation-Reduction Titrations

General Case of Equivalence Point Potential

Equivalence point potentials were calculated for the titration curves constructed in the two preceding sections. The simplest type of oxidation–reduction titration is that in which a single oxidant species produces a single reduced product, a single reductant species produces a single oxidized product, and neither H^+ nor OH^- is involved in the reaction. Such a reaction can be generalized as

$$aA + bB \rightarrow a\!A + b\!B \tag{12.41}$$

where A and B are oxidized forms and A and B are reduced forms. The E^0 and the two half-reactions involved are

$$A + xe^- \rightleftarrows A \qquad E_A^0 \tag{12.42}$$

$$B + ye^- \rightleftarrows B \qquad E_B^0 \tag{12.43}$$

where x and y are the number of electrons involved in the two half-reactions, respectively. *Assuming that neither of the product species was present in the titration vessel originally*, at the equivalence point the following concentration relationships apply:

$$b[A] = a[B] \tag{12.44}$$

$$b[A] = a[B] \tag{12.45}$$

At the equivalence point $E = E_{eq}$, which is calculated after adding the two following Nernst equations:

$$xE_{eq} = xE_A^0 - 0.0591 \ \log \frac{[A]}{[A]} \tag{12.46}$$

$$yE_{eq} = yE_B^0 - 0.0591 \ \log \frac{[B]}{[B]} \tag{12.47}$$

$$(x + y)E_{eq} = xE_A^0 + yE_B^0 - 0.0591 \ \log \frac{[A][B]}{[A][B]} \tag{12.48}$$

Substituting the stoichiometric relationships expressed in Equations 12.44 and 12.45 yields

$$(x + y)E_{eq} = xE_A^0 + yE_B^0 - 0.0591 \ \log \frac{a/b[B][B]}{a/b[B][B]} \tag{12.49}$$

The terms in the log expression cancel and, since $\log 1 = 0$, the solution of E_{eq} is

$$E_{eq} = \frac{xE_A^0 + yE_B^0}{x + y} \tag{12.50}$$

Examination of the preceding equation shows that for $x = y$, E_{eq} is exactly halfway between E_A^0 and E_B^0; for $x > y$, E_{eq} is closer to E_A^0; and for $x < y$, E_{eq} is closer to E_B^0. Examples of the preceding were illustrated for the values of E_{eq} calculated at the equivalence points of the titration curves plotted in Sections 12.2 and 12.3.

EXAMPLE

What is E_{eq} for the titration of Sn^{2+} with Ce^{4+}?

The titration reaction is

$$2Ce^{4+} + Sn^{2+} \rightarrow 2Ce^{3+} + Sn^{4+}$$

$$Ce^{4+} + e^- \rightleftharpoons Ce^{3+} \qquad E^0 = +1.70 \text{ V}$$

$$Sn^{4+} + 2e^- \rightleftharpoons Sn^{2+} \qquad E^0 = +0.15 \text{ V}$$

$$E_{eq} = \frac{1.70 \text{ V} + 2(0.15)}{1 + 2} = +0.67 \text{ V}$$

Consideration of H^+ Concentration in Equivalence Point Potential

When species are involved in the titration reaction that do not undergo oxidation-reduction and are present at an activity other than 1, this species concentration must be used in calculating E_{eq}. The most commonly encountered such species is H^+. For example, calculate E_{eq} for the titration of Sn^{2+} with MnO_4^- in a medium maintained at $0.800 \ M \ H^+$. The titration reaction is

$$5Sn^{2+} + 2MnO_4^- + 16H^+ \rightarrow 5Sn^{4+} + 2Mn^{2+} + 8H_2O$$

and the concentration relationships at the equivalence point are the following:

$$2[Sn^{2+}] = 5[MnO_4^-] \tag{12.51}$$

$$2[Sn^{4+}] = 5[Mn^{2+}] \tag{12.52}$$

$$[H^+] = 0.800 \tag{12.53}$$

The two applicable half-reactions are

$$MnO_4^- + 8H^+ + 5e^- \rightleftharpoons Mn^{2+} + 4H_2O \qquad E^0 = +1.51 \text{ V} \tag{12.54}$$

$$Sn^{4+} + 2e^- \rightleftharpoons Sn^{2+} \qquad E^0 = +0.15 \text{ V} \tag{12.55}$$

The corresponding Nernst equations multiplied by the appropriate numbers and added together at E_{eq} are

$$5\left(E_{eq} = 1.51 - \frac{0.0591}{5} \log \frac{[Mn^{2+}]}{[MnO_4^-][H^+]^8}\right) \tag{12.56}$$

$$2\left(E_{eq} = 0.15 - \frac{0.0591}{2} \log \frac{[Sn^{2+}]}{[Sn^{4+}]}\right) \tag{12.57}$$

$$7E_{eq} = 5 \times 1.51 + 2 \times 0.15 - 0.0591 \log \frac{[Mn^{2+}][Sn^{2+}]}{[MnO_4^-][Sn^{4+}][H^+]^8} \tag{12.58}$$

Substituting Equations 12.51 through 12.53 into the preceding equation leads to the following:

$$7E_{eq} = 5 \times 1.51 + 2 \times 0.15 - 0.0591 \log \frac{[Mn^{2+}]^{5/2}[MnO_4^-]}{[MnO_4^-]^{5/2}[Mn^{2+}](0.800)^8} \qquad (12.59)$$

$$E_{eq} = \frac{5 \times 1.51 + 2 \times 0.15 + 0.0591 \log (0.800)^8}{7}$$

$$= \frac{7.55 + 0.30 - 0.05}{7} = 1.11 \text{ V} \qquad (12.60)$$

Formation of Two Product Species from One Reactant Species

In cases where two product species are formed from one reactant species, or vice versa, the value of E_{eq} is concentration-dependent. Consider the titration of 25.00 mL of 0.0600 M Fe^{2+} with 0.0100 M $Cr_2O_7^{2-}$ in a medium in which a_{H^+} is maintained at 1.00. The titration reaction is

$$6Fe^{2+} + Cr_2O_7^{2-} + 14H^+ \rightarrow 6Fe^{3+} + 2Cr^{3+} + 14H_2O \qquad (12.61)$$

One reactant species forms
two product species.

At the equivalence point the pertinent concentrations are

$$[Fe^{2+}] = 6[Cr_2O_7^{2-}] \qquad (12.62)$$

$$[Fe^{3+}] = \frac{\text{mmol } Fe^{2+} \text{ originally present}}{\text{total volume}} = \frac{25.00 \text{ mL} \times 0.0600 \text{ mmol/mL}}{25.00 \text{ mL} + 25.00 \text{ mL}} = 3.00 \times 10^{-2} M \qquad (12.63)$$

$$[Cr^{3+}] = \frac{2(\text{mmol } Cr_2O_7^{2-} \text{ added})}{\text{total volume}} = \frac{2 \times 25.00 \text{ mL} \times 0.0100 \text{ mmol/mL}}{25.00 \text{ mL} + 25.00 \text{ mL}} = 1.00 \times 10^{-2} M \quad (12.64)$$

The value of E_{eq} is calculated by the following treatment:

$$Fe^{3+} + e^- \rightleftarrows Fe^{2+} \qquad\qquad E^0 = +0.77 \text{ V} \qquad (12.65)$$

$$Cr_2O_7^{2-} + 14H^+ + 6e^- \rightleftarrows 2Cr^{3+} + 7H_2O \qquad E^0 = +1.33 \text{ V} \qquad (12.66)$$

$$E_{eq} = 0.77 - 0.0591 \log \frac{[Fe^{2+}]}{[Fe^{3+}]} \qquad (12.67)$$

$$6E_{eq} = 6 \times 1.33 - 0.0591 \log \frac{[Cr^{3+}]^2}{[Cr_2O_7^{2-}][H^+]^{14}} \qquad (12.68)$$

$$7E_{eq} = 0.77 + 6 \times 1.33 - 0.0591 \log \frac{[Fe^{2+}][Cr^{3+}]^2}{[Cr_2O_7^{2-}][Fe^{3+}]} \qquad (12.69)$$

(Note that the term $[H^+]^{14}$ is omitted by the titration condition that $a_{H^+} = 1.00$.)

$$7E_{eq} = 0.77 + 6 \times 1.33 - 0.0591 \log \frac{6[Cr_2O_7^{2-}](1.00 \times 10^{-2})^2}{[Cr_2O_7^{2-}] \times 3.00 \times 10^{-2}} \qquad (12.70)$$

$$E_{eq} = \frac{0.77 + 6 \times 1.33 - 0.0591 \log 2 \times 10^{-2}}{7}$$

$$= 1.26 \text{ V} \qquad (12.71)$$

12.5   Titrations Involving More Than One Oxidation or Reduction Reaction

Reduction of Two Different Oxidants by a Single Reductant

In Section 5.12 it was shown that a mixture of two acids with sufficiently different K_a values gives a titration curve with two breaks when titrated with a base. Similarly (Section 9.3), a mixture of Cl^- and I^- titrated with Ag^+ gives a curve with two different breaks. In an analogous fashion, titration of two reducing agents having E^0 values that differ by at least 0.2 V with an oxidizing titrant, or vice versa, results in a titration curve with two inflection points. For example, consider the following:

$$Ce^{4+} + e^- \rightleftarrows Ce^{3+} \qquad E^0 = +1.70 \text{ V} \qquad (12.72)$$

$$Fe^{3+} + e^- \rightleftarrows Fe^{2+} \qquad E^0 = +0.77 \text{ V} \qquad (12.73)$$

$$Cr^{3+} + e^- \rightleftarrows Cr^{2+} \qquad E^0 = -0.408 \text{ V} \qquad (12.74)$$

The relative E^0 values of these half-reactions indicate that Ce^{4+} is more readily reduced than Fe^{3+} and that Cr^{2+} is capable of reducing both Ce^{4+} and Fe^{3+}. A mixture of Ce^{4+} and Fe^{3+} titrated with Cr^{2+} would have a titration curve with two inflection points and two titration reactions, as shown in Figure 12.5. The volumes of titrant at which the two breaks occur depend upon the relative proportions of Ce^{4+} and Fe^{3+} in the solution titrated.

Two-Step Oxidation of a Reductant Species

Section 5.13 discusses the titration of a diprotic acid H_2A with a strong base. Where the K_a's of the acid are well separated, a curve with two inflection points is obtained, corresponding to the formation of HA^- and A^{2-}, respectively. An exactly analogous curve is obtained for a species that is oxidized or reduced by two one-electron steps involving well-separated E^0 values. This is best illustrated by an example—the two-step oxidation of V^{3+} to $V(OH)_4^+$. The half-reactions involved are

$$VO^{2+} + 2H^+ + e^- \rightleftarrows V^3 \qquad E^0 = +0.359 \text{ V} \qquad (12.75)$$

$$V(OH)_4^+ + 2H^+ + e^- \rightleftarrows VO^{2+} + 3H_2O \qquad E^0 = +1.00 \text{ V} \qquad (12.76)$$

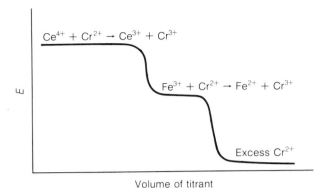

FIGURE 12.5   General shape of the curve for the titration of a mixture of Ce^{4+} and Fe^{3+} with Cr^{2+}

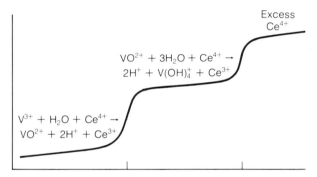

FIGURE 12.6   Titration of V^{3+} with Ce^{4+}

The oxidizing agent, Ce^{4+}, can be used to oxidize V^{3+} successively to VO^{2+} and $V(OH)_4^+$. As shown in Figure 12.6, the two inflection points are equal distances apart. The two successive titration reactions are given in the figure.

12.6   Indicators for Oxidation-Reduction Titrations

An **oxidation-reduction indicator** is a chemical species that denotes the end point in an oxidation-reduction titration through a color change involving the indicator species. Indicators are normally the handiest means of denoting end point in manual oxidation-reduction titrations, just as they are for acid-base (Section 5.6), precipitation (Sections 9.7 and 9.8), and complexation (Section 10.9) titrations.

Colored Titrant as Indicator

The simplest type of indicator is a strongly colored titrant that tints the solution when it is present in slight excess at the end point. The prime example of such a titrant is permanganate ion, MnO_4^-, which, in very dilute solution, has a light purple or pink color. Before the development of other indicators, permanganate—despite its many disadvantages as a reagent—was the oxidant titrant of choice because of its self-indicating quality. Just a slight excess of permanganate titrant imparts a visible color to the solution, thus showing the end point. Permanganate can be classified as a **specific indicator** because it reveals the presence of a specific species, in this case MnO_4^- ion in solution.

Starch Indicator for Elemental Iodine

In modern practice of chemical analysis, the most widely used specific indicator is starch, which reversibly forms an intensely colored dark blue complex compound with elemental iodine. With the use of starch indicator, titrations with titrants that consume elemental iodine (usually in the presence of I^- to increase the solubility of I_2 by forming I_3^-) have an end point manifested by the disappearance of the dark blue starch-iodine complex color formed by starch indicator added just prior to the end point. In cases where the titrant is elemental iodine in the form of I_3^- reagent, the end point in the presence of starch indicator is revealed by the appearance of the dark blue color.

Thiocyanate Complexes of Iron(III) in Titrations Involving Iron(III)

Another example of a specific indicator is provided by the use of thiocyanate ion, SCN^-, as an indicator for the titration of iron(III). The complex $FeSCN^{2+}$ is formed, which imparts a red color to the solution. Titration of the Fe(III) with Ti^{3+}, a reducing agent, yields the titration reaction

$$Fe^{3+} + Ti^{3+} + H_2O \rightarrow Fe^{2+} + TiO^{2+} + 2H^+ \qquad (12.77)$$

At the end point the small quantity of Fe(III) present in the $FeSCN^{2+}$ complex is reduced and the solution turns from red to colorless.

True Oxidation-Reduction Indicators

True oxidation–reduction indicators are those that respond to changes of E in solution. These indicators change reversibly between oxidized and reduced forms

$$In_{Ox} + e^- \rightleftharpoons In_{Red} \qquad E^0_{In} \qquad (12.78)$$

depending upon the E of the solution according to the equation

$$E = E_{In}^0 - \frac{0.0591}{n} \log \frac{[In_{Red}]}{[In_{Ox}]} \tag{12.79}$$

When the concentration ratio $[In_{Red}]/[In_{Ox}] = 10$, the color of the indicator is indistinguishable from that of a solution of the indicator in the pure reduced form; and when the ratio $[In_{Red}]/[In_{Ox}]$ is $1/10$, the color of the indicator solution is indistinguishable from that of the oxidized form of the indicator. Therefore, the indicator transition range lies between these two extremes. Examination of the Nernst equation, Equation 12.79, for indicators shows that this color change should occur over a voltage range of $\pm 0.0591/n \log 100$. For an oxidation-reduction indicator undergoing a one-electron reaction, this corresponds to an indicator color change range of 0.118 V. For many common oxidation-reduction indicators, $n = 2$, so that the color transition range is 0.0591 V.

The first widely used true oxidation-reduction indicator was diphenylamine, recommended for the titration of Fe^{2+} with $Cr_2O_7^{2-}$ in 1924. In contact with an oxidant this indicator undergoes the irreversible reaction

$$2 \, C_6H_5{-}NH{-}C_6H_5 \rightarrow C_6H_5{-}NH{-}C_6H_4{-}C_6H_4{-}NH{-}C_6H_5 + 2H^+ + 2e^- \tag{12.80}$$

followed by the reversible transition

$$C_6H_5{-}NH{-}C_6H_4{-}C_6H_4{-}NH{-}C_6H_5 \rightleftarrows$$

Diphenylbenzidine (colorless)

$$C_6H_5{-}N{=}C_6H_4{=}C_6H_4{=}N{-}C_6H_5 + 2H^+ + 2e^- \tag{12.81}$$

Diphenylbenzidine violet (violet)

so that the end point with an oxidant titrant is marked by the appearance of a violet-colored solution.

Diphenylamine suffers from a number of disadvantages, including its low solubility in water. A more satisfactory substitute is diphenylamine sulfonic acid

$$C_6H_5{-}NH{-}C_6H_4{-}SO_3H$$

used as the sodium or barium salt. The latter is only slightly soluble in water but can be put in solution in a sulfuric acid medium, where reaction of sulfate with barium produces a $BaSO_4$ precipitate, leaving the diphenylamine sulfonic acid in solution. The color change is a distinct clear to green to violet transition. This indicator finds wide application in oxidation–reduction titrations.

A number of true oxidation–reduction indicators are based upon chelates of metals, particularly iron(II) and iron(III), capable of undergoing oxidation and reduction. The best known such indicator is the 1,10-phenanthroline (o-phenanthroline) chelate of Fe(II), tris(1,10-phenanthroline) iron(II), commonly known as ferroin. The formula of 1,10-phenanthroline is $C_{12}H_8N_2$, and its structural formula is

The end point reaction for ferroin in a titration reaction with an oxidizing titrant is

$$Fe(C_{12}H_8N_2)_3^{2+} \rightleftharpoons Fe(C_{12}H_8N_2)_3^{3+} + e^- \tag{12.82}$$

resulting in a color change from red to blue. The former color is much more intense, so that in dilute solutions ferroin behaves much like a one-color indicator in that the color change appears to be from red to colorless. Ferroin changes color at about 1.11 V versus the SHE, considerably higher than the 0.80 V versus the SHE color change of diphenylaminesulfonic acid. Therefore, ferroin can be used with strongly oxidizing Ce(IV) titrant.

Chelates of iron(II) and other metals with 2,2'-bipyridine

are also employed as oxidation–reduction indicators. Oxidation of tris(2,2'-bipyridine) iron(II) to the corresponding iron(III) compound gives a color change from red to pale blue at +1.03 V.

Table 12.3 lists some oxidation–reduction indicators in decreasing order of E_{In}^0 applicable at $a_{H^+} = 1.00$, unless otherwise stated. Those indicators higher in the table are suitable for titrations in which the equivalence point comes in a relatively oxidizing medium (high E), whereas those lower in the table are appropriate for titrations having equivalence points at relatively low E values.

TABLE 12.3   Some oxidation-reduction indicators

Indicator name	Color in reduced form	Color in oxidized form	E^0 at $a_{H^+} = 1.00$, V
[Fe(II)(5-nitro-1,10-phenanthroline)$_3$](ClO$_4$)$_2$	Violet-red	Pale blue	1.25
[Fe(II)(1,10-phenanthroline)$_3$](ClO$_4$)$_2$(ferroin)	Red	Pale blue	1.11
p-Ethoxychrysoidine	Red	Pale yellow	1.0
Diphenylaminesulfonic acid	Colorless	Red-violet	0.80
Diphenylamine (diphenylbenzidine)	Colorless	Violet	0.76
2,7-Diaminophenothiazonium chloride	Colorless	Blue	0.56
1-Naphthol-2-sulfonic acid indophenol	Colorless	Red	0.54
Methylene blue	Colorless	Blue	0.36
Indigo monosulfonate	Colorless	Blue	0.26
[V(II)(1,10-phenanthroline)$_3$](ClO$_4$)$_3$	Blue-violet	Colorless	0.14

End Point from Potentiometric Titrations

The most generally applicable means of determining end points in oxidation-reduction titrations is by monitoring E with the use of a measuring (indicator) platinum electrode versus a reference electrode with the type of apparatus shown in Figure 12.2. Normally a calomel or silver–silver chloride electrode is used as a convenient reference electrode. Reference electrodes containing KCl should not be used in a medium containing perchlorate ion, ClO_4^-, because of formation of insoluble KClO$_4$ at the liquid junction of the reference electrode. Sodium chloride can be substituted in the electrode to overcome this problem. The equivalence point is taken at the appropriate point on the titration curve; as discussed in previous sections, this is not necessarily the midpoint of the titration curve.

12.7   Titrations with Standard Oxidants and Reductants

Several oxidants make excellent standard oxidant solutions. Among the most commonly used are potassium permanganate, KMnO$_4$; cerium(IV); potassium dichromate, K$_2$Cr$_2$O$_7$; and iodine, I$_2$. In general, stable solutions of these oxidants can be prepared and readily standardized, and good indicators are available for their use.

The use of standard reductant solutions is much more limited because of oxidation of the reagent with atmospheric oxygen. Frequent restandardization of these solutions is required.

Before an oxidation-reduction titration can be carried out, the analyte must all be in a single oxidation state. Various methods for doing that are discussed in the following section.

12.8   Auxiliary Oxidizing and Reducing Agents

Auxiliary Reductants

Prior to titration with either an oxidant or reductant, an analyte species must be in a single, known oxidation state. Sample processing prior to titration often does not leave an analyte in such a state. For example, dissolution of an iron alloy sample in acid results in conversion of much of the iron to iron(II):

$$Fe(s) + 2H^+ \rightarrow Fe^{2+} + H_2(g) \tag{12.83}$$

However, concurrent exposure to the atmosphere results in oxidation of a portion of the iron(II) to iron(III) by reaction with oxygen:

$$4Fe^{2+} + O_2 + 4H^+ \rightarrow 4Fe^{3+} + 2H_2O \tag{12.84}$$

Before the total dissolved iron can be determined by titration with a standard oxidant that converts iron(II) to iron(III) quantitatively, the dissolved analyte must all be in the +2 oxidation state. This is accomplished with an **auxiliary reductant**. In this case the auxiliary reductant must meet the following criteria:

1. It must be a strong enough reducing agent to reduce iron(III) to iron(II).
2. It must not be strong enough to reduce interfering species to lower oxidation states such that they would later react with the oxidizing titrant.
3. It must not remain in solution or must be capable of removal from solution.

The most obvious way to prevent excess auxiliary reductant from remaining in solution is to introduce it in the form of another phase—that is, a solid or gas. A number of solid metals, including aluminum, cadmium, copper, lead, nickel, mercury, and zinc can be used as solid reductants. In practice, metals such as these are normally contained in a **reductor**, consisting of a glass column in which the granular or powdered metal is held and through which the analyte solution flows.

Jones Reductor

The most common type of reductor is the **Jones reductor** shown in Figure 12.7. This reductor typically consists of an approximately 2-cm internal diameter, 50-cm-long glass tube packed with zinc particles coated with a layer of zinc amalgam formed by immersing the zinc in a solution of mercury(II) chloride, causing the following reaction to occur at the zinc surface:

$$Zn(s) + Hg^{2+}(aq) \rightarrow Zn(Hg) + Zn^{2+} \tag{12.85}$$

Here Zn(Hg) represents a solution of zinc metal in mercury, known as a *zinc amalgam*. The reason for amalgamating the zinc is that a mercury surface has a

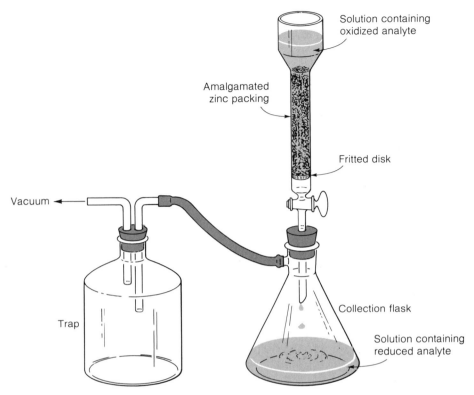

FIGURE 12.7 Jones reductor containing particles of amalgamated zinc

high overvoltage (see Section 11.13) for the half-reaction

$$2H^+ + 2e^- \rightarrow H_2(g) \tag{12.86}$$

Therefore the H^+ ion in acidic solutions poured through the column does not react with the zinc to produce H_2 gas, which would consume far more zinc than that used to reduce analyte ion, contaminate the analyte solution with too much Zn^{2+} ion, and disrupt the column packing because of bubble formation. Listed below are some of the analytically useful reactions of the Jones reductor:

$$Fe^{3+} + e^- \rightarrow Fe^{2+} \tag{12.87}$$

$$UO_2^{2+} + 4H^+ + 2e^- \rightarrow U^{4+} + 2H_2O \tag{12.88}$$

$$TiO^{2+} + 2H^+ + e^- \rightarrow Ti^{3+} + 4H_2O \tag{12.89}$$

$$Cr^{3+} + e^- \rightarrow Cr^{2+} \tag{12.90}$$

Walden Reductor

From the relatively high standard electrode potential for the reduction of Ag^+ ion

$$Ag^+ + e^- \rightleftarrows Ag \qquad E^0 = +0.799 \text{ V} \tag{12.91}$$

it may be concluded that Ag metal is a poor reducing agent. However, the reducing capability of Ag metal can be greatly enhanced by the presence of Cl^- ion, which enhances the reducing ability of silver metal by the formation of slightly soluble $AgCl(s)$. This is the basis of the **Walden reductor**, which is packed with silver coated onto copper particles and used with solutions containing chloride. Silver metal is coated onto the copper by immersing the copper particles in a solution of $AgNO_3$, resulting in the reaction

$$2Ag^+ + Cu(s) \rightarrow 2Ag(s) + Cu^{2+} \tag{12.92}$$

As a Walden reductor is used, the particles become coated with $AgCl(s)$ by the reaction

$$Ag + Cl^- \rightarrow AgCl(s) + e^- \tag{12.93}$$

which eventually reduces the efficiency of the column. The AgCl may be eliminated by immersing a zinc bar in the solution in contact with the column packing:

$$Zn + 2AgCl \rightarrow Zn^{2+} + 2Cl^- + Ag \tag{12.94}$$

When not in use the packing is kept immersed in 1 M HCl solution. Some typical reduction reactions carried out with a Walden reductor are the following:

$$Fe^{3+} + e^- \rightarrow Fe^{2+} \tag{12.95}$$

$$H_2MoO_4 + 2H^+ + e^- \rightarrow MoO_2^+ + 2H_2O \tag{12.96}$$

$$UO_2^{2+} + 4H^+ + 2e^- \rightarrow U^{4+} + 2H_2O \tag{12.97}$$

$$V(OH)_4^+ + 2H^+ + e^- \rightarrow VO^{2+} + 3H_2O \tag{12.98}$$

The Walden reductor offers some selectivity in reducing analytes. For example, unlike the Jones reductor, it does not reduce either TiO^{2+} or Cr^{3+}.

Gaseous Reductants

Gaseous reductants in the form of SO_2 and H_2S have seen limited use as auxiliary reductants. Excess reductant can be expelled from the acidified analyte solution by boiling. Slow reactions and the noxious nature of these gases have restricted their application.

Auxiliary Oxidizing Agents

The three most commonly used **auxiliary oxidizing agents** are peroxide, ammonium peroxodisulfate, and sodium bismuthate.

Peroxide can be employed as an auxiliary oxidizing agent either as sodium peroxide, Na_2O_2, or hydrogen peroxide, H_2O_2. In acidic solution, the action of peroxide as an oxidizing agent is described by the half-reaction

$$H_2O_2 + 2H^+ + 2e^- \rightleftharpoons 2H_2O \qquad E^0 = +1.78 \text{ V} \tag{12.99}$$

Peroxide undergoes autodecomposition

$$2H_2O_2 \rightarrow 2H_2O + O_2(g) \tag{12.100}$$

and this reaction can be used to eliminate excess peroxide simply by boiling the acidified solution.

The half-reaction by which ammonium peroxodisulfate, $(NH_4)_2S_2O_8$, acts as an oxidizing agent is

$$S_2O_8^{2-} + 2e^- \rightleftharpoons 2SO_4^{2-} \qquad E^0 = +2.01 \text{ V} \tag{12.101}$$

catalyzed by a very low concentration of Ag^+. This salt is a strong enough oxidizing agent to convert chromium(III) to dichromate, manganese(II) to permanganate, and even cerium(III) to cerium(IV). The oxidation of water in a boiling solution

$$2S_2O_8^{2-} + 2H_2O \rightarrow 4SO_4^{2-} + O_2(g) + 4H^+ \tag{12.102}$$

is employed to eliminate excess peroxodisulfate.

Sodium bismuthate, generally represented by the formula $NaBiO_3$, is a powerful oxidizing agent that accepts electrons in the conversion of Bi(IV) to Bi(III). It is employed by immersing the sparingly soluble solid in a boiling solution of analyte. Excess reagent can be removed by filtration.

12.9   Titration with Standard Oxidants

Oxidants are much more widely used as titrants than are reductants (solutions of reductants are very hard to store because of air oxidation). In addition to containing an oxidizing chemical species with a strong tendency to accept electrons, an oxidant solution to be used for titration should be readily standardized and relatively stable in storage. For the generalized half-reaction

$$\{Ox\} + ne^- \rightarrow \{Red\} \tag{12.103}$$

where {Ox} represents the oxidant species and {Red} its reduced form, the value of E^0 for an oxidizing titrant normally lies between 0.5 and 1.6 V. This value is a measure of the strength of the oxidizing agent and is used to determine which oxidant to use for the determination of a particular analyte. Other factors in the selection include rate of reaction, stoichiometry of reaction, availability of an indicator for the desired titration, and reagent stability.

12.10 Potassium Permanganate as a Standard Oxidant

Permanganate Reactions

Normally permanganate solutions are used as titrants in a highly acidic medium ($[H^+] > 0.1$). In such solutions the half-reaction for the reduction of permanganate is

$$MnO_4^- + 8H^+ + 5e^- \rightleftarrows Mn^{2+} + 4H_2O \qquad E^0 = +1.51 \text{ V} \qquad (12.104)$$

Although oxidations with acidic permanganate are normally rapid, heating may be required, as is done in the titration of oxalate with permanganate. Alternatively, iodine monochloride or osmium tetroxide catalysts may be needed for the titration of arsenic(III). At pH 4 and above, permanganate is normally reduced to insoluble brown $MnO_2 \cdot xH_2O$ according to the half-reaction

$$MnO_4^- + 4H^+ + 3e^- \rightleftarrows MnO_2(s) + 2H_2O \qquad E^0 = +1.695 \text{ V} \qquad (12.105)$$

The reaction of permanganate in other than highly acidic solutions is used for titration of hydrazine, N_2H_4, to N_2 gas; sulfite, thiosulfate, and sulfide to sulfate; cyanide, CN^-, to cyanate, CNO^-; and $Mn(II)$ to $Mn(IV)$.

Self-Indicating Quality of Permanganate Reagent

Even relatively dilute solutions of permanganate are intensely purple, and a slight trace of MnO_4^- ion imparts a pink color to a solution. A concentration of only about $2 \times 10^{-6} M$ MnO_4^- is required to form a perceptible color in solution. Historically, therefore, the greatest advantage of permanganate has been that it acts as its own indicator.

EXAMPLE

For the titration of 200 mL of 0.0500 M Fe^{2+} with 0.100 N (0.0200 M) MnO_4^-, what volume of titrant must be added beyond the equivalence point for the end point color to appear? How much of a percentage titration error does this cause?

$$\text{mmol excess } MnO_4^- = 2 \times 10^{-6} \text{ mmol/mL} \times 300 \text{ mL} = 6 \times 10^{-4} \text{ mmol}$$

(The total volume of solution is 300 mL at the equivalence point.)

$$\frac{\text{mL excess}}{\text{titrant}} = \frac{6 \times 10^{-4} \text{ mmol}}{0.020 \text{ mmol/mL}} = 0.03 \text{ mL}$$

$$\frac{\text{Titration}}{\text{error}} = \frac{\text{mL excess}}{\text{mL to reach equivalence point}} \times 100$$

$$= \frac{0.03 \text{ mL}}{100 \text{ mL}} \times 100 = +0.03\% \quad \text{(negligible)}$$

The end point does fade gradually because of the slow reaction

$$2MnO_4^- + 3Mn^{2+} + 2H_2O \rightarrow 5MnO_2(s) + 4H^+ \tag{12.106}$$

Permanganate in standard solutions of this reagent is consumed slowly by reactions, such as the following, in which O_2 is evolved:

$$4MnO_4^- + 4H^+ \rightarrow 4MnO_2(s) + 2H_2O + 3O_2(g) \tag{12.107}$$

The rate of decomposition is increased by excess acid or base, light, and the presence of Mn^{2+} ion or solid manganese dioxide. Substantial occurrence of MnO_4^- decomposition is evidenced by deposits of brown MnO_2 on container or buret walls. Since manganese dioxide is a product of the reaction, the reaction is said to be **autocatalytic**.

Preparation and Storage of Permanganate Solutions

Steps must be taken during the preparation and storage of standard permanganate solutions to avoid the problems just mentioned. Normally the standard solution is heated to bring about the oxidation of any residual organic matter, then filtered through a nonorganic filter to remove residual solid manganese dioxide. The solution is stored in the dark and restandardized each week it is used. If manganese dioxide residue is observed, the standard solution should be filtered and restandardized.

Standardization of Permanganate Solutions

Commonly used primary standards for permanganate solutions are sodium oxalate, $Na_2C_2O_4$; iron metal dissolved and converted to Fe^{2+} ion; or As_2O_3. When sodium oxalate is titrated with permanganate in a hot, acidic medium, the overall reaction is

$$2MnO_4^- + 5H_2C_2O_4 + 6H^+ \rightarrow 2Mn^{2+} + 10CO_2(g) + 8H_2O \tag{12.108}$$

The reaction is slow initially, and the first portions of MnO_4^- added persist for some time until an appreciable concentration of Mn^{2+} ion has built up. This product has an autocatalytic effect upon the reaction, so that in the latter parts of the titration the solution rapidly loses color with each addition of permanganate until the equivalence point is reached. The **McBride procedure** for the titration of oxalate with permanganate specifies titration at 60 to 90° C until the pink color of excess permanganate persists. This procedure gives a negative titration error of up to -0.4%, which is attributed to air oxidation of oxalate in the heated medium. A more refined method, the **Fowler–Bright procedure**, requires addition of 90 to 95% of the stoichiometric amount of permanganate to a cool oxalate solution, heating to 55 to 60° C, and finishing the titration at the elevated temperature. The Fowler–Bright procedure gives apparently exact stoichiometry but requires a rather accurate knowledge of the quantity of oxalate being titrated and the concentration of the permanganate solution. For most purposes the McBride procedure is adequate.

Determination of Iron

Historically, one of the better-known applications of permanganate as an oxidizing titrant is the determination of iron through the reaction

$$MnO_4^- + 5Fe^{2+} + 8H^+ \rightarrow Mn^{2+} + 5Fe^{3+} + 4H_2O \qquad (12.109)$$

Prior to titration the iron must be prereduced to iron(II) by a Jones or Walden reductor (see Section 12.8), or with $SnCl_2$. The oxidation of chloride ion

$$2MnO_4^- + 10Cl^- + 16H^+ \rightarrow 2Mn^{2+} + 5Cl_2(g) + 8H_2O \qquad (12.110)$$

which gives a positive titration error, is prevented by adding *Zimmerman–Reinhardt solution*, also known as preventive solution. This solution contains manganese(II) sulfate, sulfuric acid, and phosphoric acid. The Mn^{2+} ion shifts the equilibrium of the half-reaction

$$MnO_4^- + 8H^+ + 5e^- \rightleftarrows Mn^{2+} + 4H_2O \qquad (12.111)$$

to the left so that permanganate is a less powerful oxidizing agent with a lessened tendency to oxidize Cl^- ion. The phosphoric acid complexes iron(III), thus increasing the tendency toward oxidation of iron(II) to iron(III), in preference to the oxidation of Cl^- ion. In addition, complexation of the yellow Fe^{3+} ion to form less highly colored $FeHPO_4^+$ makes the end point color more readily visible.

Hydrogen peroxide can be titrated with permanganate in an acidic medium. The titration reaction is the following:

$$2MnO_4^- + 5H_2O_2 + 6H^+ \rightarrow 2Mn^{2+} + 5O_2(g) + 8H_2O \qquad (12.112)$$

Nitrite can be determined quantitatively by the following reaction with permanganate in an acidic medium:

$$2MnO_4^- + 5NO_2^- + 6H^+ \rightarrow 2Mn^{2+} + 5NO_3^- + 3H_2O \qquad (12.113)$$

In practice the nitrite is added from a buret to a measured volume of standard permanganate to prevent loss of volatile nitrous acid standing in an acidic solution. The end point is taken as the point where the permanganate solution is decolorized.

Metals that form sparingly soluble oxalates can be determined by isolating the precipitated oxalate salt by filtration, dissolving the metal oxalate in acid, and titrating the oxalic acid according to the reaction

$$2MnO_4^- + 5H_2C_2O_4 + 6H^+ \rightarrow 2Mn^{2+} + 10CO_2(g) + 8H_2O \qquad (12.114)$$

This method has been used especially for the determination of Ca^{2+}, and is also applicable to the determination of Mg^{2+}, Zn^{2+}, Pb^{2+}, Co^{2+}, and Ag^+.

12.11 Titrations with Dichromate

Properties of Dichromate

From E^0 of the half-reaction

$$Cr_2O_7^{2-} + 14H^+ + 6e^- \rightleftharpoons 2Cr^{3+} + 7H_2O \qquad E^0 = +1.33 \text{ V} \qquad (12.115)$$

it is seen that dichromate ion is not so strong an oxidizing agent as permanganate ion. However, dichromate offers some important advantages as a titrant. Potassium dichromate, $K_2Cr_2O_7$, can be obtained as a very pure, stable salt and can be weighed out as a primary standard. At room temperature dichromate does not oxidize chloride and has little tendency to oxidize organic matter, so these potential interferences are avoided. Furthermore, dichromate solutions are not affected by light and are stable in storage.

The formula weight of $K_2Cr_2O_7$ is 294.18, and the six-electron reduction shown in half-reaction 12.115 leads to

$$\frac{\text{Equivalent}}{\text{weight}} = \frac{294.18 \text{ g/mol}}{6 \text{ eq/mol}} = 49.03 \text{ g/eq} \qquad (12.116)$$

The normality of a dichromate solution in terms of its molarity is

$$N = 6M \qquad (12.117)$$

Dichromate solutions are orange-colored and solutions of the product Cr^{3+} ion are green. The latter color obscures the disappearance of the dichromate color at the equivalence point in a titration, so an oxidation–reduction indicator is required. Among the most useful indicators for dichromate titrations are N-phenylanthranilic acid and diphenylamine sulfonic acid. The latter is obtained from the commercially available sodium and barium salts and must be used in the presence of sulfuric acid.

N-phenylanthranilic acid

Diphenylamine
sulfonic acid

Applications of Dichromate Titrant

The most successful application of potassium dichromate titrant is for the determination of iron in a medium containing sulfuric and phosphoric acids. The titration reaction is

$$Cr_2O_7^{2-} + 6Fe^{2+} + 14H^+ \rightarrow 2Cr^{3+} + 6Fe^{3+} + 7H_2O \qquad (12.118)$$

If sodium diphenylamine sulfonate indicator is used, the titration is continued until the end point is nearly reached as manifested by a bluish–green to grayish–blue tint in the solution. The titration is then continued dropwise until the addition of one drop changes the solution color to blue-violet or intense purple that persists with mixing.

Standard dichromate solutions may be used for the determination of oxidizable substances (reducing agents). This is the basis of the **chemical oxygen demand (COD) test** used in the analysis of natural waters and wastewaters. The water sample is mixed with a measured volume of standard dichromate, along with sulfuric acid, Ag_2SO_4, and $HgSO_4$, and the resulting mixture is refluxed at a temperature that is elevated to enable dichromate to oxidize organic matter. The silver sulfate acts as a catalyst for the oxidation of reducing substances in the water and the mercuric sulfate complexes Cl^- ion, which otherwise would be oxidized, constituting an interference. Chemically bound organic carbon, {C}, is oxidized according to the general reaction

$$3\{C\} + 2Cr_2O_7^{2-} + 16H^+ \rightarrow 3CO_2(g) + 4Cr^{3+} + 8H_2O \qquad (12.119)$$

The excess dichromate is then titrated with a solution of standard iron(II) prepared from primary standard ferrous ammonium sulfate, $Fe(NH_4)_2(SO_4)_2 \cdot 6H_2O$. The quantity of oxidizable material in solution is expressed in terms of the equivalent amount of molecular O_2 that would be consumed in oxidizing the analytes oxidized by the dichromate. For example, if O_2 were the oxidizing agent, the reaction equivalent to Reaction 12.119 would be

$$\{C\} + O_2 \rightarrow CO_2(g) \qquad (12.120)$$

EXAMPLE

A 20.0 mL sample of water was refluxed with 10.0 mL of 0.250 N $K_2Cr_2O_7$, after which 13.60 mL of 0.100 N Fe(II) solution was required to titrate the excess dichromate. What was the COD of the water in milligrams of O_2 per liter?

meq $Cr_2O_7^{2-}$ consumed
$$= 10.0 \text{ mL} \times 0.250 \text{ meq/mL} - 13.6 \text{ mL} \times 0.100 \text{ meq/mL}$$
$$= 2.50 - 1.36 = 1.14 \text{ meq (meq is the abbreviation for}$$
$$\text{milliequivalent} = \tfrac{1}{1000} \text{ equivalent)}$$

The half-reaction for oxygen acting as an oxidizing agent is $O_2 + 4H^+ + 4e^- \rightleftarrows 2H_2O$.

Therefore, the equivalent weight of O_2 is one-fourth the molecular weight, or 8 mg/meq.

$$COD = \frac{1.14 \text{ meq} \times 8 \text{ mg/meq}}{0.020 \text{ L}} = 456 \text{ mg/L}$$

Standard dichromate can be employed indirectly for the determination of oxidizing agents by reacting the analyte with a known quantity of iron(II) in solution and titrating the excess iron(II) with standard dichromate. Among the oxidizing agents thus determined are dichromate itself, inorganic and organic peroxides, permanganate, and chlorate.

12.12   Titrations with Cerium(IV)

Properties of Cerium(IV) Oxidant

A standard solution of acidic cerium(IV) is an excellent oxidizing titrant. Cerium(IV) is a potent oxidizing agent, forming a single product, cerium(III). Solutions of the reagent are very stable and can be stored without special precautions. Cerium(IV) does not oxidize chloride ion at an appreciable rate. The major disadvantages of cerium(IV) are that it is expensive and cannot be used in neutral or basic media. The cerium(IV) ion is not intensely colored enough to serve as its own indicator.

The value of E^0 for the half-reaction

$$Ce^{4+} + e^- \rightleftarrows Ce^{3+}$$

depends upon the acidic medium in which the cerium species are contained. For 1 M solutions of acid the E^0 values are $+1.44$ V in H_2SO_4, $+1.28$ V in HCl, $+1.61$ V in HNO_3, and $+1.70$ V in $HClO_4$. Thus, cerium(IV) is a very powerful oxidizing agent in $HClO_4$ medium. The differences in these values reflect interaction of the cerium species with anions of the acids in the solution medium. In fact, there is little of the simple hydrated ion, $Ce(H_2O)_6^{4+}$, in solutions of cerium(IV). Sulfate and nitrate complexes exist in sulfuric and nitric acid solutions, respectively; $Ce(OH)^{3+}$ and $Ce(OH)_2^{2+}$ complexes exist in perchloric acid solution; and cerium(IV) dimer is present in more concentrated solution. Sparingly soluble basic salts of cerium(IV) precipitate from solutions for which $[H^+] < 0.1$ M.

Standard Solutions of Cerium(IV)

Standard solutions of cerium(IV) can be prepared from several compounds, including cerium(IV) sulfate, $Ce(SO_4)_2$; cerium(IV) hydroxide; ammonium cerium(IV) sulfate dihydrate, $(NH_4)_4[Ce(SO_4)_4] \cdot 2H_2O$; and anhydrous ammonium cerium(IV) sulfate, $(NH_4)_4[Ce(SO_4)_4]$. This last salt, whose formula weight is 548.23, can be obtained in a pure enough form to use as a primary standard. Generally, cerium(IV) standard solutions are prepared in a sulfuric acid medium.

Indicators for Cerium(IV) Titrations

The oxidation-reduction indicator most commonly used with cerium(IV) titrant is the orthophenanthroline complex of iron(II) (see Section 12.6), commonly known as ferroin, which changes color from orange-red to pale blue at the end point. Phenanthroline complexes of other metals (see Table 12.3) can also be used. Another suitable indicator is N-phenylanthranylic acid, which changes color from yellowish-green to purple at the equivalence point.

Standardization of Cerium(IV) Solutions

Cerium(IV) titrant solutions may be standardized with several primary standards, including arsenic trioxide, As_2O_3; iron; and sodium oxalate, $Na_2C_2O_4$. The reaction for the titration of arsenic trioxide in an acidic medium is

$$2Ce^{4+} + H_3AsO_3 + H_2O \rightarrow 2Ce^{3+} + H_3AsO_4 + 2H^+ \tag{12.121}$$

The reaction is catalyzed by a very small trace of osmium tetroxide. Standardization with pure iron is carried out by simply dissolving an accurately weighed quantity of A.R. (analytical reagent) iron wire in acid, reducing the dissolved iron to iron(II) with an auxiliary reducing agent (see Section 12.8) and titrating with the cerium(IV) solution using ferroin indicator

$$Ce^{4+} + Fe^{2+} \rightarrow Ce^{3+} + Fe^{3+} \tag{12.122}$$

Although cerium(IV) solutions can be standardized with sodium oxalate, the procedure is complicated by the fact that the elevated temperature required oxidizes indicators that otherwise might be employed. This problem is overcome by adding excess cerium(IV) solution to a weighed quantity of sodium oxalate in a sulfuric acid medium, heating to 60°C to oxidize the oxalate, and titrating the remaining excess Ce^{4+} with a solution of Fe^{2+}. The exact concentration of the latter solution need not be known, but it is determined relative to that of the cerium(IV) standard solution in a separate titration, giving the value of milliliters of cerium(IV) per milliliter of iron(II). Using this ratio the volume of excess cerium(IV) remaining after the oxidation of oxalate can be calculated from the volume of iron(II) solution required to titrate the excess cerium(IV).

Determinations with Cerium(IV) Titrant

Cerium(IV) solutions are useful for the titration of iron(II) from iron ore and other iron-containing samples. Other typical applications include titration of nitrites,

$$2Ce^{4+} + NO_2^- + H_2O \rightarrow 2Ce^{3+} + NO_3^- + 2H^+ \tag{12.123}$$

hydrogen peroxide,

$$2Ce^{4+} + H_2O_2 \rightarrow 2Ce^{3+} + O_2(g) + 2H^+ \qquad (12.124)$$

and uranium(IV),

$$2Ce^{4+} + U^{4+} + 2H_2O \rightarrow UO_2^{2+} + 4H^+ + 2Ce^{3+} \qquad (12.125)$$

Methods have been developed using cerium(IV) as a replacement for permanganate in most of the determinations developed earlier for use with permanganate.

12.13   Oxidation-Reduction Titrations Involving Iodine

Elemental iodine is either consumed or produced in a number of significant chemical analysis procedures. Because of the low aqueous solubility of I_2, this species is kept in solution by the triiodide ion complex formed between I_2 and I^-:

$$I_2 + I^- \rightarrow I_3^- \qquad K = 710$$

The pertinent half-reaction for analyses involving elemental iodine and iodide is

$$I_3^- + 2e^- \rightleftarrows 3I^- \qquad E^0 = +0.545 \text{ V} \qquad (12.126)$$

The two major categories of iodine oxidation–reduction reactions are **iodimetric methods** that make use of I_3^- as a weakly oxidizing titrant and **iodometric methods** in which I_3^- is liberated from a solution of KI by an oxidant, and the amount liberated is titrated with sodium thiosulfate or As(III) reagent. Both of these approaches will be discussed separately.

12.14   Iodimetric Methods

Advantages of Elemental Iodine Titrant

Solutions of I_3^- contain elemental iodine complexed to I^- ion and are referred to as **iodine solutions**. One of the greater attractions for the use of iodine titrant is the availability of **starch indicator**, which has a very high sensitivity for elemental iodine as manifested by an intense purple color; furthermore, the indicator is reversible. Because it is only a weak oxidant, iodine is a selective reagent for the determination of reducing agents.

Standard Solutions of Iodine Titrant

Standard solutions of iodine are made up in a large excess of iodide. The triodide complex forms slowly in dilute solution, so iodine solutions are made up in a concentrated form and diluted. Standard iodine solutions are somewhat unstable and require restandardization at intervals of several days. Part of the instability is due to loss of volatile I_2 to the atmosphere and air oxidation of I^- ion by the reaction

$$6I^- + O_2 + 4H^+ \rightleftharpoons 2I_3^- + 2H_2O \qquad \qquad (12.127)$$

which is catalyzed by acids, light, and heat. (This reaction causes an increase in normality, whereas other interferences cause a decrease.) Iodine reacts with organic dust and vapors, cork stoppers, rubber stoppers, and rubber tubing, so exposure to these should be avoided.

An interesting means of preparing a standard solution that functions as an iodimetric titrant is to dissolve an accurately weighed quantity of pure potassium iodate, KIO_3, in a slight excess of KI solution and dilute to a desired volume. This solution can be used as an iodimetric titrant in acidic media because it generates I_3^- by the following reaction:

$$IO_3^- + 8I^- + 6H^+ \rightarrow 3I_3^- + 3H_2O \qquad \qquad (12.128)$$

The I_3^- generated in this reaction will subsequently react with any analytes present that normally are titrated with I_3^-.

EXAMPLE

> Although the IO_3^- ion undergoes a five-electron reduction in Reaction 12.128, its equivalent weight is one-sixth its formula weight when used as a titrant in place of I_3^-. Explain.
>
> The reduction of one IO_3^- ion in the presence of I^- leads to the formation of three I_3^- ions, each of which undergoes a two-electron reduction in subsequent reactions. Another way to view it is that the one I atom in IO_3^- undergoes a five-electron reduction in Reaction 12.128, and another one-electron reduction in a subsequent reaction of I_3^- for a total six-electron reduction.

The color of I_3^- ion, visible at concentrations as low as $5 \times 10^{-6}\ M$, can be used as an approximate indicator of end point in iodimetric titrations. A more sophisticated approach involves addition of a small quantity of chloroform or carbon tetrachloride to the titration mixture; elemental iodine present in excess at the end point extracts into these solvents, producing an intense violet color in the organic phase. By far the best indicator for iodine, however, is β-amylose, a form of starch that can be isolated from arrowroot, potato, and rice starches in soluble form as **soluble starch**. The macromolecules comprising this form of starch have helical chain structures, and it is believed that elemental iodine is adsorbed to this helical structure, giving an intense blue color. The reaction is reversible, which is an important characteristic for an indicator to have. However, in the presence of moderately high concentrations of iodine, starch indicator forms an iodine complex that does not readily dissociate reversibly in a relatively large excess of iodine, so the indicator is usually added near the end point, when iodine solutions are being titrated. The approach to the end point is readily apparent from the fading of the I_3^- complex color to a light yellow tint.

Starch suspensions are subject to decomposition, particularly by bacterial action, to products that do not function properly as iodine indicators. Decomposition is slowed by heating the suspension after it is formed and storing it under sterile conditions. Bacteriostatic chloroform or mercury(II) iodide can be added to prevent decomposition. Usually it is simplest to make up a fresh starch suspension every day or two.

Sodium starch glycollate can be used as a substitute for starch as an indicator for titrations involving iodine. A white, nonhygroscopic powder, this material readily dissolves in hot water to form a solution that is stable for several months. This indicator does not form an insoluble complex with iodine at relatively high iodine concentrations. Added at the beginning of the titration of an iodine solution, this indicator is green; it turns deep blue immediately before the end point, losing color sharply at the end point.

Historically, arsenious oxide, As_2O_3, has been the primary standard of choice for the standardization of iodine solutions. Allegations that this compound is carcinogenic may restrict its use. Arsenious oxide dissolves readily in 1 M NaOH by the reaction

$$As_2O_3(s) + 4OH^- \rightarrow 2HAsO_3^{2-} + H_2O \tag{12.129}$$

Arsenic(III) in base readily oxidizes to arsenic(V), so the solution is made slightly acidic by the addition of HCl as soon as the As_2O_3 is dissolved. The reaction for the titration of arsenic(III) with iodine is

$$H_3AsO_3 + I_3^- + H_2O \rightleftharpoons H_3AsO_4 + 2H^+ + 3I^- \tag{12.130}$$

in the pH range of 4 to 9. The reaction is reversed at low pH, such that arsenic(V) oxidizes I^-. Therefore, Reaction 12.130, which lowers the pH by production of H^+ ion, must be carried out in a buffered medium. This is accomplished by adjusting the pH of the analyte solution to slightly less than 7, then saturating it with $NaHCO_3$; the resulting $CO_2(aq)$–HCO_3^- buffer consumes H^+ produced by Reaction 12.130 and maintains the pH at 7 to 8.

An alternate primary standard for iodine solutions is barium thiosulfate monohydrate, a sparingly soluble solid with the formula $BaS_2O_3 \cdot H_2O$. This is a heterogeneous reaction in which the iodine solution reacts with the solid primary standard

$$I_3^- + 2BaS_2O_3 \cdot H_2O(s) \rightarrow S_4O_6^{2-} + 2Ba^{2+} + 3I^- + 2H_2O \tag{12.131}$$

Analytes Determined Iodimetrically

A number of analytes can be determined by titration with iodine solution. Trivalent arsenic may, of course, be determined by I_3^-, as shown in Reaction

12.130; trivalent antimony as H_3SbO_3 is determined by an analogous reaction. As will be discussed in the next section, one of the most important reactions of iodine is with thiosulfate

$$2S_2O_3^{2-} + I_3^- \rightarrow S_4O_6^{2-} + 3I^- \tag{12.132}$$

Sulfite and hydrazine, both important oxygen scavengers that reduce the corrosiveness of industrial water, are determined iodimetrically by the following two reactions:

$$SO_3^{2-} + I_3^- + H_2O \rightarrow SO_4^{2-} + 2H^+ + 3I^- \tag{12.133}$$

$$N_2H_4 + 2I_3^- \rightarrow N_2(g) + 6I^- + 4H^+ \tag{12.134}$$

Tin is determined iodimetrically by oxidation of Sn(II) to Sn(IV)

$$Sn^{2+} + I_3^- \rightarrow Sn^{4+} + 3I^- \tag{12.135}$$

and H_2S by oxidation to elemental sulfur

$$H_2S + I_3^- \rightarrow S(s) + 3I^- + 2H^+ \tag{12.136}$$

12.15   Iodometric Titrations

The Nature of Iodometric Analyses

Iodometric analyses are used to determine oxidants from the amount of iodine liberated by reactions of the general type

$$3I^- + \{Ox\} \rightarrow I_3^- + \{Red\} \tag{12.137}$$

where $\{Ox\}$ is the general formula of the oxidant and $\{Red\}$ is its reduced form. The elemental iodine thus liberated is readily titrated with thiosulfate

$$2S_2O_3^{2-} + I_3^- \rightarrow S_4O_6^{2-} + 3I^- \tag{12.138}$$

to yield iodide and tetrathionate ion. This titration is normally carried out at a pH of 5 or less, with care taken to prevent air oxidation of I^-. The oxidation of thiosulfate quantitatively to tetrathionate is unique to iodine; other oxidizing agents produce at least some sulfate, SO_4^{2-}.

As discussed in Section 12.14, soluble starch is used as the indicator for titrations of iodine. It is added when the iodine is about exhausted in the solution being titrated, as evidenced by fading of the red-brown color of relatively concentrated I_3^- to the light yellow of a very dilute solution of this complex.

Preparation and Preservation of Standard Thiosulfate Solutions

Standard thiosulfate solutions are subject to several instability problems, including microbial degradation, light-catalyzed decomposition, oxidation by atmospheric oxygen, and autodecomposition by the following reaction at pH 5 or less

$$S_2O_3^{2-} + H^+ \rightarrow HSO_3^- + S(s) \tag{12.139}$$

The presence of turbidity from elemental sulfur is a clear indication of decomposition, and a solution exhibiting turbidity should be discarded. Microbial degradation can be avoided by preparing and storing thiosulfate under sterile conditions and by adding bacteriostatic agents, such as chloroform, HgI_2, and sodium benzoate. Decomposition catalyzed by light can be avoided by preventing exposure of the standard solution to strong light, particularly direct sunlight. Autodecomposition of thiosulfate by Reaction 12.139 can be avoided by raising the solution pH to 9 to 10 with a small quantity of borax, Na_2CO_3, or Na_2HPO_4. A further advantage is that this is the pH range in which bacterial action is at a minimum.

The analyte solutions of iodine that are titrated should contain enough acid to maintain a low pH throughout the titration. Higher pH values favor the formation of hypoiodite, HOI,

$$I_3^- + H_2O \rightleftarrows HOI + 2I^- + H^+ \tag{12.140}$$

which in turn causes the following reaction to occur:

$$4HOI + S_2O_3^{2-} + H_2O \rightleftarrows 2SO_4^{2-} + 4I^- + 6H^+ \tag{12.141}$$

The overall effect of these reactions is to cause a negative end point error in the titration of I_3^- with thiosulfate.

Standardization of Thiosulfate Solutions

Thiosulfate solutions can be standardized by titrating I_3^- solutions formed by adding a known quantity of primary standard oxidant to a solution of I^-. The best primary standard oxidant for this purpose is potassium iodate, KIO_3, which reacts as follows with I^-:

$$IO_3^- + 8I^- + 6H^+ \rightarrow 3I_3^- + 3H_2O \tag{12.142}$$

The reaction of each IO_3^- ion produces three I_3^- ions, each of which subsequently undergoes a two-electron reduction to I^- by reaction with thiosulfate (Reaction 12.138). Therefore, the equivalent weight of KIO_3 is one-sixth its formula weight, as shown below:

$$\frac{\text{Equivalent weight}}{\text{of } KIO_3} = \frac{214.00 \text{ g/mol}}{6 \text{ eq/mol}} = 35.67 \text{ g/eq} \tag{12.143}$$

Other oxidants used to standardize thiosulfate solutions via titration of liberated I_3^- include potassium hydrogen iodate, $KH(IO_3)_2$; potassium bromate, $KBrO_3$; and potassium dichromate, $K_2Cr_2O_7$.

Copper metal can be weighed out as a primary standard for the standardization of thiosulfate by an interesting series of reactions. The copper metal weighed on an analytical balance is dissolved in nitric acid by the following reaction:

$$Cu + 4HNO_3 \rightarrow Cu^{2+} + 2NO_3^- + 2NO_2(g) + 2H_2O \qquad (12.144)$$

Excess HNO_3, which interferes by oxidizing I^- to iodine, is eliminated by adding concentrated H_2SO_4 and heating to drive off volatile nitric acid. *Both the dissolution of copper and the heating to drive off HNO₃ must be carried out in a hood because of the danger from highly toxic NO₂ gas.* The cupric ion produced by dissolving copper metal oxidizes I^- in solution, a reaction driven by the very low solubility of copper(I) iodide:

$$2Cu^{2+} + 5I^- \rightarrow 2CuI(s) + I_3^- \qquad (12.145)$$

The I_3^- produced by the preceding reaction is titrated with thiosulfate ion using starch as an indicator.

Iodometric Determinations

A number of analyses have been developed based upon the liberation of I_3^- from I^- solution, followed by the titration of iodine with thiosulfate. This approach has been widely used in the past for the determination of copper in ores and other copper-containing samples. Iron(III) is the major interference in this determination, because it is a strong enough oxidant to oxidize I^- to the elemental state:

$$2Fe^{3+} + 3I^- \rightarrow 2Fe^{2+} + I_3^- \qquad (12.146)$$

This interference is overcome by adding ammonium hydrogen fluoride, NH_4HF_2, complexing the iron(III) as FeF_6^{3-} ion, which does not oxidize iodide. In addition, NH_4HF_2 buffers the pH of the analyte solution to the optimum value of slightly greater than 3.2.

Available chlorine can be determined in bleaching powder by an iodometric titration. The active ingredient in bleaching powder is an oxidant, calcium hypochlorite, $Ca(OCl)_2$. Other constituents of commercial bleaching powder are basic calcium chloride, $CaCl_2 \cdot Ca(OH)_2 \cdot H_2O$, and calcium hydroxide, $Ca(OH)_2$. The available chlorine in bleaching powder is the Cl_2 that would be liberated by the reaction

$$OCl^- + Cl^- + 2H^+ \rightarrow Cl_2 + H_2O \qquad (12.147)$$

going to completion; it normally has a value of about 37% by weight of the bleaching powder. Available chlorine can be determined by adding a weighed quantity of the product to an iodide solution and acidifying with acetic acid to bring about the reaction

$$OCl^- + 3I^- + 2H^+ \rightleftarrows Cl^- + I_3^- + H_2O \qquad (12.148)$$

and titrating the elemental iodine with thiosulfate.

The **Winkler determination** of dissolved oxygen in water is based upon an iodometric procedure. The dissolved oxygen is "fixed" by reaction with basic manganese(II)

$$O_2 + 4Mn(OH)_2(s) + 2H_2O \rightarrow 4Mn(OH)_3(s) \tag{12.149}$$

to produce manganese(III) hydroxide. This compound then reacts with I^- in an acidified medium

$$2Mn(OH)_3(s) + 6H^+ + 3I^- \rightarrow 2Mn^{2+} + I_3^- + 6H_2O \tag{12.150}$$

and the I_3^- titrated with thiosulfate.

Among the other species that commonly have been determined by the oxidation of iodide to iodine and subsequent titration of the latter are hydrogen peroxide,

$$H_2O_2 + 2H^+ + 3I^- \rightarrow I_3^- + 2H_2O \tag{12.151}$$

chlorates,

$$ClO_3^- + 9I^- + 6H^+ \rightarrow Cl^- + 3I_3^- + 3H_2O \tag{12.152}$$

arsenic(V) in 4 M HCl,

$$H_3AsO_4 + 2H^+ + 3I^- \rightarrow H_3AsO_3 + I_3^- + H_2O \tag{12.153}$$

and elemental Cl_2 and Br_2, both represented below as X_2

$$X_2 + 3I^- \rightarrow I_3^- + 2X^- \tag{12.154}$$

12.16   Titrations with Reducing Agents

For use in chemical analysis, by far the most useful reducing titrant is thiosulfate employed in iodometric determinations. As mentioned previously, other reductants are troublesome to use and store because of their reactions with atmospheric oxygen. This normally requires that the titrant be stored in an oxygen-free atmosphere (under N_2 or CO_2) and that titrations be carried out in a similar atmosphere. Such procedures are mandatory for the very strong reductants chromium(II) and titanium(III). Weaker reductants can sometimes be used by adding excess reductant to the analyte and back-titrating with a standard oxidant solution. The reductant is restandardized frequently with the oxidant solution to correct for air oxidation.

Iron(II) is readily prepared as a reductant solution. The half-reaction for iron(II) acting as a reducing agent is simply

$$Fe^{2+} \rightarrow Fe^{3+} + e^- \tag{12.155}$$

Solutions of iron(II) can be prepared from **Mohr's salt**, $Fe(NH_4)_2(SO_4)_2 \cdot 6H_2O$. Another convenient source of iron(II) is iron(II) ethylenediammonium sulfate, $FeC_2H_4(NH_4)_2(SO_4)_2 \cdot 4H_2O$, known as **Oesper's salt**. Prepared in approximately 0.5 M H_2SO_4 medium and restandardized daily during use, iron(II) solution is reasonably convenient to use.

An acidic solution of arsenic(III) prepared by weighing As_2O_3 as a primary standard and dissolving it in acid is a good reducing titrant that is stable indefinitely in acidic media. The half-reaction for arsenic(III) acting as a reductant is

$$H_3AsO_3 + H_2O \rightarrow H_3AsO_4 + 2H^+ + 2e^- \tag{12.156}$$

Titanium(III) is a powerful reducing agent that undergoes the following half-reaction:

$$Ti^{3+} + H_2O \rightarrow TiO^{2+} + 2H^+ + e^- \tag{12.157}$$

Solutions of titanium(III) are very unstable in the presence of oxygen and must be very carefully protected from exposure to the atmosphere.

Programmed Summary of Chapter 12

The major terms and concepts introduced in this chapter are contained in this summary in a programmed format. To derive the most benefit from the summary, you should fill in the blanks for each question and then check the answers at the end of the book to see if your choices are correct.

A titration that is measured by following E as a function of volume of added titrant is called a (1) _____. In such a titration, a Nernst equation can be written for the half-reactions of both analyte and titrant. Prior to the equivalence point, it is best to use the Nernst equation applicable to the (2) _____, and at the equivalence point it is necessary to consider (3) _____. The reaction for the titration of iron(II) with permanganate in an acidic medium is (4) _____. An oxidizing titrant undergoes a three-electron reduction as it reacts with an analyte that undergoes a one-electron oxidation. The H^+ ion is involved in the titration reaction but is maintained at an activity of exactly 1 throughout. Using E^0_{Ox} and E^0_{Red} to designate the standard electrode potentials of the titrant and analyte half-reactions, respectively, the value of the equivalence point potential is $E_{eq} = $ (5) _____. When H^+ is a titration reaction participant, in addition to knowing the E^0's of the two half-reactions involved, it is necessary to know the value of (6) _____ at the equivalence point. In an oxidation-reduction titration in which a single reactant species forms two product species, to calculate E_{eq} it is necessary to know the E^0's of the half-reactions involved, the value of $[H^+]$, and (7) _____. The titration with a strong oxidant of two reducible species with widely separated E^0 values would give a titration curve shaped such that (8) _____, and during the course of the titration E would trend (9) _____. Three specific indicators used in oxidation–reduction titrations are (10) _____. A true oxidation–reduction indicator may be described as (11) _____. For an indicator undergoing a one-electron oxidation–reduction reaction, the color transition range is approximately (12) _____. Four commonly used oxidizing titrants are (13) _____. In order to place an analyte all in the reduced form prior to titration, the analyst uses (14) _____. A Jones reductor consists of (15) _____ packed in a column; a Walden reductor is packed with (16) _____ and is used with solutions that contain (17) _____ in addition to the analyte that is reduced. Three common auxiliary oxidizing agents are (18) _____. Standard reductants are hard to store because they (19) _____. The autocatalytic decomposition product of MnO_4^- is (20) _____. The organic reagent most commonly determined by titration with permanganate is (21) _____ and its reaction with MnO_4^- in acid is (22) _____. When reduced in an acidic medium, dichromate undergoes a (23) _____ electron reduction and each dichromate ion forms (24) _____. The chemical oxidation demand determination consists of oxidation of organic matter in water by (25) _____ followed by titration of excess oxidizing agent with (26) _____. Iodimetric titrations involve titrations with elemental (27) _____ in the form of (28) _____ ion using (29) _____ as an indicator. Iodometric titrations depend upon the oxidation of I^- to (30) _____, which in turn is titrated with (31) _____. The Winkler method is used to determine (32) _____. Mohr's salt and Oesper's salt are used to prepare standard solutions of (33) _____.

Questions

1. Discuss the physical setup and data output obtained in an oxidation–reduction titration curve.

2. Consider a simple potentiometric titration in which the titrant is an oxidant, M^{3+},

$$M^{3+} + e^- \rightleftarrows M^{2+} \qquad E_M^0$$

and the analyte is a reductant, Z^+,

$$Z^{2+} + e^- \rightleftarrows Z^+ \qquad E_Z^0 \qquad E_M^0 \gg E_Z^0$$

(a) What is the titration reaction? What is the significance of the value of E (b) initially, (c) halfway to the equivalence point, (d) at the equivalence point, and (e) at a total of twice the volume of titrant required to reach the equivalence point?

3. Suppose that Z^+ in Question 2 were titrated with the oxidant YO_2^+ in a medium maintained at $1.00\ M$ in H^+, where YO_2^+ undergoes the half-reaction

$$YO_2^+ + 4H^+ + 4e^- \rightleftarrows Y^+ + 2H_2O, \quad E_Y^0.$$

Sketch the titration curve and designate the approximate location of the equivalence point on the steeply rising portion of the curve.

4. For a potentiometric titration involving the half-reactions

$$A + xe^- \rightleftarrows A \qquad E_A^0$$
$$B + ye^- \rightleftarrows B \qquad E_B^0$$

where x and y represent numbers of electrons, under what circumstances is the equivalence point potential given by the simple formula

$$E_{eq} = \frac{xE_A^0 + yE_B^0}{x + y}$$

5. In the pH region in which the half-reaction for dichromate acting as an oxidizing agent is

$$Cr_2O_7^{2-} + 14H^+ + 6e^- \rightleftarrows 2Cr^{3+} + 7H_2O$$

what should be the effect of increasing $[H^+]$ on the oxidizing strength of $Cr_2O_7^{2-}$?

6. The titration of a sample of two reducing agents A and B, both at the same molar concentration, gave the titration curve shown when Ce^{4+} was used as a titrant. What does the curve reveal about the reductant species?

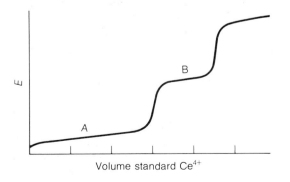

Volume standard Ce^{4+}

7. Match:

   a. Starch
   b. MnO_4^-
   c. Diphenylamine sulfonic acid
   d. $FeSCN^{2+}$

   1. True oxidation-reduction indicator
   2. Turns dark blue in the presence of $I_2(I_3^-)$
   3. Added as a specific indicator and turns colorless in the presence of reducing titrant
   4. A titrant that serves as its own indicator

8. Give the names and structures of two organic compounds which, chelated with iron(II), produce oxidation-reduction indicators.

9. What is the purpose of an auxiliary reducing agent?

10. What is the principle of a reductor, such as the Jones reductor?

11. Why is the reaction of MnO_4^- with oxalic acid slow during the initial stages of the titration of oxalic acid with permanganate?

12. How does the use of preventive solution overcome a positive titration error in the titration of Fe^{2+} with MnO_4^- oxidant?

13. What are some important advantages of dichromate over MnO_4^- as an oxidizing titrant?

14. Why is I_2 normally kept in solution along with an excess of I^-?

15. What effect does exposure to the atmosphere have upon (a) I_2 and (b) I^-?

16. What is the basic principle of iodometric determinations?

17. Soluble starch is normally added near the end of a titration of I_3^-. What physical appearance of the solution reveals that it is time to add starch indicator?

18. What special precautions are necessary with the use of strong reductants, such as $Cr(II)$, as titrants?

Problems

1. A $0.100\ M$ solution of Fe^{2+} was titrated with a $0.100\ M$ solution of an oxidizing agent M^{3+}, which is reduced to M^{2+}. When 110% of the amount of the volume of M^{3+} required to reach the equivalence point had been added, the value of E versus the SHE was $+1.52$ V. Calculate E^0 for the half-reaction $M^{3+} + e^- \rightleftharpoons M^{2+}$.

2. For the titration of Fe^{2+} with Ce^{4+} calculate E versus the saturated calomel electrode when the amount of Ce^{4+} required to reach the equivalence point has been added.

3. This titration curve was obtained from the titration of M^+ with X^{x+1} to give the products, $M^{m+1} + X^+$. The two pertinent half-reactions are

$$M^{m+1} + me^- \rightleftharpoons M^+ \qquad E^0_M$$
$$X^{x+1} + xe^- \rightleftharpoons X^+ \qquad E^0_X$$

Either m or x equals 1. From the titration curve estimate E^0_M, E^0_X, m, and x.

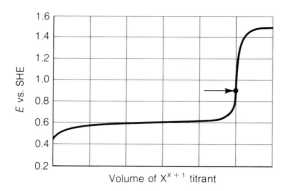

Volume of X^{x+1} titrant

4. Calculate the equivalence point potential E_{eq} for the titration of a solution $0.100\ M$ in Fe^{2+} and $1.00\ M$ in H^+ titrated with a solution $0.0200\ M$ in MnO_4^- and $0.0800\ M$ in H^+.

5. Suppose that a solution initially $0.100\ M$ in Fe^{2+} and $0.500\ M$ in Fe^{3+} is titrated with $0.0200\ M$ MnO_4^- in a medium maintained at $1.00\ M$ in H^+. What is the equivalence point potential?

6. For a particular oxidation-reduction indicator, the color is indistinguishable from that of the reduced form when $[In_{Red}]/[In_{Ox}] = 3/1$ and indistinguishable from the oxidized form when the same ratio is $1/5$. What is the indicator transition range in volts if the indicator undergoes a one-electron oxidation?

7. For the titration of Fe^{2+} with Ce^{4+} the equivalence point potential was calculated to be 1.24 V versus the SHE. Ferroin indicator changes color at 1.11 V. What is the percentage of Fe^{2+} unoxidized at the point where ferroin indicator changes color?

8. In an application of the Winkler method for the determination of oxygen in water, the I_2 liberated by the dissolved oxygen in 300 mL of water required 15.0 mL of $0.0250\ N$ thiosulfate solution for titration. What was the level of dissolved oxygen in the water in milligrams per liter?

9. A solution of dichromate titrant was prepared by dissolving 11.47 g of $K_2Cr_2O_7$ and diluting to 1 L in a medium in which $[H^+] = 1.00\ M$. How many millimoles of Fe^{2+} could be oxidized to Fe^{3+} by each milliliter of this solution?

10. A solution of MnO_4^- was standardized using iron wire as a primary standard. A total of 1.120 g of Fe wire was dissolved in acid, diluted to 250 mL, and a 50-mL aliquot taken for titration. This aliquot required 36.81 mL of permanganate solution for oxidation of the prereduced Fe(II) to Fe(III). The permanganate solution was then used to titrate oxalate obtained by

dissolving CaC_2O_4 in acid; a total of 43.17 mL of the MnO_4^- solution was required to oxidize the oxalate from the CaC_2O_4 sample. What was the weight of CaC_2O_4 in the calcium oxalate sample?

11. A 0.125 N (0.0250 M) solution of MnO_4^- was used to titrate 207 mg of $Na_2C_2O_4$ dissolved in 50.0 mL of acid. Assuming that 2×10^{-6} M MnO_4^- is required to form a perceptible color in solution, what volume of the standard permanganate should have been added beyond the equivalence point to make the end point perceptible?

12. In a chemical oxygen demand (COD) determination a 25.0-mL sample of water was refluxed with 10.0 mL of 0.250 N $K_2Cr_2O_7$, after which 16.32 mL of 0.0916 N Fe(II) solution was required to titrate the excess dichromate. What was the COD of the water in milligrams of oxygen per liter?

13 Potentiometry

LEARNING GOALS

1. Major types of indicator electrodes.
2. Principles of membrane electrodes, as compared to electrodes in which an oxidation-reduction half-reaction is involved.
3. Construction, principles, and uses of glass electrodes.
4. What is meant by ion-selective electrode.
5. Principles and major types of liquid membrane ion-selective electrodes.
6. Principles and major types of solid-state ion-selective electrodes.
7. Ion-selective electrode response and calibration.
8. Calculating ion-selective electrode error in concentration from error in potential and understanding why a small error in potential gives a large error in relative concentration.
9. Technique of standard addition potentiometry.
10. Types and construction of practical reference electrodes.
11. Gran plots.
12. Use of ion-selective electrodes for potentiometric titrations.

13.1 What Is Potentiometry?

Potentiometry is the measurement of ion activity from the electrical potential produced by the ion acting at an electrode surface. Potentiometry can even be extended to neutral species whose concentrations determine ion activities in special electrodes; common among these are acidic or basic gases whose concentrations influence H^+ ion activity (see Gas-Sensing Electrodes, Section 13.10).

In Chapter 11 it was shown that a metal ion can exchange electrons reversibly with an electrode composed of the metal, as shown below for Hg^{2+} ion

$$Hg^{2+} + 2e^- \rightleftarrows Hg \qquad E^0 \tag{13.1}$$

Furthermore, the potential of the metal electrode versus a stable reference electrode is given by the Nernst equation

$$E = E^0 - \frac{0.0591}{2} \log \frac{1}{[Hg^{2+}]} = E^0 + \frac{0.0591}{2} \log [Hg^{2+}] \tag{13.2}$$

(Recall that this equation applies to a two–electron half-reaction at 25°C.) When the presence of the metal ion in solution is due to equilibrium with a slightly soluble salt of the metal ion, the electrode may respond to the concentration of the salt anion. This is shown below for a silver electrode in a chloride solution in contact with slightly soluble solid silver chloride.

$$AgCl(s) + e^- \rightleftharpoons Ag + Cl^- \qquad E^0 \qquad\qquad (13.3)$$

$$E = E^0 - 0.0591 \log [Cl^-]$$

In a number of important cases the potential-sensing element is not a conducting metal, but a membrane on the surface of which an ion exchanges reversibly. The most important example of this kind of phenomenon is the exchange of H^+ ions on the special glass membrane of a *glass electrode* (see Sections 13.3 and 13.4). Although a reversible oxidation-reduction reaction does not occur, the electrode potential as a function of ion activity behaves in a Nernstian manner.

As a generalization for the discussion of potentiometry, it is possible to represent the potential response of an electrode by the Nernstian-type equation

$$E = E_a + \frac{2.303RT}{zF} \log a_X z \qquad\qquad (13.4)$$

where E is the potential of the *indicator electrode* responding to the ion X^z versus the *reference electrode* used in the system; E_a (analogous to E^0) is a constant for the electrode system employed; z is the signed charge of ion X, which has activity $a_X z$; and the other terms are constants for the Nernst equation as discussed in Section 11.6. Note that z has a sign as well as a charge; it is, for example, +1 for H^+ ion, +2 for Ca^{2+} ion, and −2 for S^{2-} ion. The value of E_a is a constant just for the individual electrode system employed, and in practical applications is never taken versus the difficult-to-use standard hydrogen electrode. At constant ionic strength, such as that approximated for the low concentrations of ions commonly measured potentiometrically, ion concentration can be substituted for activity (in this case, substantial differences in ionic strength will show up as differences in E_a). Taking the above considerations into account leads to simple equations in a Nernstian form that relate measured potential to ion concentration regardless of whether the measurement is taken at a metal electrode participating in a reversible oxidation-reduction half-reaction or at a membrane (there are, of course, concentration limits outside of which the Nernstian-type equation does not apply). Below are equations in a Nernstian form for electrodes that respond to Ca^{2+}, Ag^+, F^-, and S^{2-}, respectively (E is in volts at 25° C).

$$E = E_a + \frac{0.0591}{2} \log [Ca^{2+}] \qquad \text{Potential in volts} \qquad (13.5)$$

$$E = E_a + 0.0591 \log [Ag^+] \qquad\qquad (13.6)$$

$$E = E_a - 0.0591 \log [\text{F}^-] \tag{13.7}$$

$$E = E_a - \frac{0.0591}{2} \log [\text{S}^{2-}] \tag{13.8}$$

13.2   Indicator Electrodes

Electrochemical Cells for Potentiometric Measurements

Figure 13.1 is a schematic representation of a cell that could be used for the potentiometric measurement of metal ion concentration. It is seen to consist of a reference electrode, in this case a saturated calomel electrode (SCE), and an indicator or measuring electrode composed of metal M. An indicator electrode is an electrode that responds to an ion in solution by a change in potential as measured against a reference electrode. Thus an indicator electrode is a key component of a cell used for potentiometric measurements.

First-Order Electrodes

Table 13.1 summarizes the common types of indicator electrodes. **First-order electrodes** are made of a metal and produce a potential response to the ion of the metal in solution. The metals commonly used for such electrodes are cadmium, copper, lead, mercury, and silver, which respond to Cd^{2+}, Cu^{2+}, Pb^{2+}, Hg^{2+} (Hg_2^{2+}), and Ag^+ ions, respectively. The divalent metal cation (in the case of silver the monovalent metal cation) of such an electrode metal is in equilibrium with the electrode according to the half-reaction

$$\text{M}^{2+} + 2e^- \rightleftarrows \text{M} \tag{13.9}$$

The Nernstian potential of the electrode at 25°C is, therefore,

$$E = E_a + \frac{0.0591}{2} \log [\text{M}^{2+}] \tag{13.10}$$

Unfortunately, metals other than those mentioned above are not suitable for the construction of first-order electrodes. Many common metals, such as cobalt, chromium, iron, nickel, and tungsten develop oxide coatings on their surfaces that prevent their functioning as an electrode. Additional problems are produced by structural deformations and strains in the metal itself.

Second-Order Electrodes

Second-order electrodes are those in which the concentration of the metal ion to which the electrode responds is governed by a species (usually an anion) bound to the metal in a slightly soluble compound or complex. Thus a silver electrode (sometimes covered with a thin layer of AgCl) responds to [Cl⁻] in a

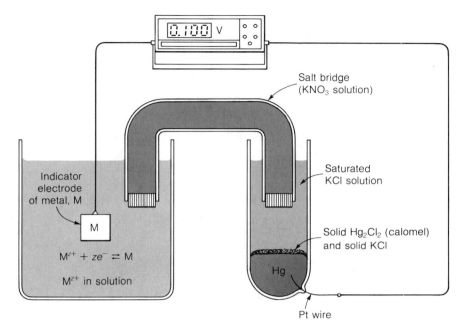

FIGURE 13.1 Cell used for the potentiometric measurement of metal ion M^{z+}

TABLE 13.1 Major types of indicator electrodes

Electrode type	Example of electrode	Example half-reaction
First-order, metal–metal cation	Cu metal	$Cu^{2+} + 2e^- \rightleftarrows Cu$
Second-order, metal–metal salt, salt anion	Ag coated with $AgCl(s)$	$AgCl + e^- \rightleftarrows Ag + Cl^-$
Second-order, metal–metal complex in solution	Mercury metal, $Hg(l)$	$HgY^{2-} + 2e^- \rightleftarrows Hg(l) + Y^{4-}$ (Y^{4-} is EDTA anion)
Metal amalgam	Solution of Cd in Hg metal, $Cd(Hg)$	$Cd^{2+} + 2e^- \overset{Hg}{\rightleftarrows} Cd(Hg)$
Inert electrode for oxidation-reduction measurement (E)	Platinum metal	$Ce^{4+} + e^- \overset{Pt}{\rightleftarrows} Ce^{3+}$
Solid membrane ion-selective electrode	Glass membrane	Most commonly responds to H^+, also designed for other monovalent cations
Liquid membrane ion-selective electrode	Organophilic membrane soaked with liquid ion exchanger	Response depends upon type of liquid ion exchanger. Designed for a variety of cations and anions, including Ca^{2+}, K^+, NO_3^-, ClO_4^-

solution saturated with $AgCl(s)$. For response to silver ion the equation for E is

$$E = E_a + 0.0591 \log [Ag^+] \tag{13.11}$$

In a solution containing Cl^- that is saturated with $AgCl(s)$

$$[Ag^+] = \frac{K_{sp}}{[Cl^-]} \tag{13.12}$$

Combination of the two preceding equations gives

$$E = E_a + 0.0591 \log K_{sp} - 0.0591 \log [Cl^-] \tag{13.13}$$

Since the first two terms on the right of the preceding equation are constants, they can be combined to give a new value of E_a as follows:

$$E = E_a - 0.0591 \log [Cl^-] \tag{a6(13.14)}$$

This form of the Nernst equation applies to the half-reaction

$$AgCl(s) + e^- \rightleftarrows Ag + Cl^- \tag{13.15}$$

Another useful second-order electrode consists of mercury metal in solution with a low concentration of HgY^{2-} chelate, a very stable species with a formation constant of 6.3×10^{21}. So long as excess Y^{4-} is present or cations forming EDTA chelates more stable than HgY^{4-} are absent, $[HgY^{2-}]$ remains constant. Under those circumstances the pertinent half-reaction is

$$HgY^{2-} + 2e^- \rightleftarrows Hg(l) + Y^{4-} \tag{13.16}$$

and the following equation relates E and $[Y^{4-}]$:

$$E = E_a - \frac{0.0591}{2} \log [Y^{4-}] \tag{13.17}$$

When a metal ion is titrated with EDTA (see Chapter 10), the value of $\log [Y^{4-}]$ undergoes an abrupt change at the equivalence point; therefore, the electrode system described above can be used to follow the titration with EDTA of a metal ion forming EDTA chelates weaker than HgY^{2-}. Figure 13.2 shows the shape of a titration curve that can be obtained with the type of electrode system described above. Note that a large break in E occurs at the equivalence point.

Metal Amalgam Electrodes

A **metal amalgam electrode** consists of a metal dissolved in mercury. The amalgam can be prepared by reducing a metal ion, such as copper, at a mercury metal cathode:

$$Cu^{2+}(aq) + 2e^- \xrightarrow{\text{Hg}} Cu(Hg) \tag{13.18}$$

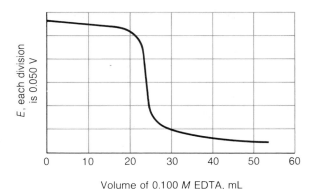

FIGURE 13.2 Titration of 50.0 mL of 0.0500 M Ca^{2+} with 0.100 M EDTA at pH 10.0

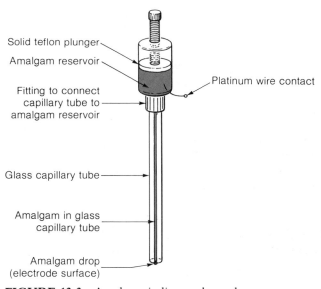

FIGURE 13.3 Amalgam indicator electrode

Often an amalgam electrode gives better potentiometric behavior than an electrode composed of the pure metal. One major advantage of an amalgam electrode is that it can be employed in the form of drops of amalgam extruded from a glass capillary as shown in Figure 13.3; a new drop and a fresh electrode surface can thus be employed for each potentiometric measurement.

For a copper amalgam electrode, the pertinent half-reaction and resulting Nernstian relationship are the following:

$$Cu^{2+}(aq) + 2e^- \overset{Hg}{\rightleftharpoons} Cu(Hg) \tag{13.19}$$

$$E = E_a + \frac{0.0591}{2} \log [Cu^{2+}] \tag{13.20}$$

Among other factors, E_a is a function of the concentration of copper in the amalgam. Analogous equations may be written for other metals that form amalgams suitable for potentiometric measurement.

Electrodes for Oxidation-Reduction Measurement

Electrodes for oxidation-reduction measurement are designed to give the electrode potential E resulting from the solutes in solution versus a reference electrode. A meaningful value of E requires the presence of the reduced and oxidized components of a reversible half-reaction, such as a solution of Fe^{3+} and Fe^{2+} in an acidic medium. Thus E versus the SHE for a mixture of Fe^{3+} and Fe^{2+} is given by

$$E = 0.77 - 0.0591 \log \frac{[Fe^{2+}]}{[Fe^{3+}]} \tag{13.21}$$

The value of E is quite low in a solution of a strong reducing agent, such as Cr^{2+}, and quite high in a solution of a strong oxidizing agent (electron acceptor), such as Ce^{4+}.

An electrode for oxidation-reduction measurement should be inert in the sense of serving only as a site for the exchange of electrons. The noble metals—palladium, platinum, and gold—can serve this purpose, as can some forms of carbon. Actually no material is completely inert in its application as an oxidation-reduction electrode. For example, a platinum electrode covered with finely divided platinum particles (a platinized electrode covered with platinum black) serves to catalyze the half-reaction

$$2H^+ + 2e^- \rightleftharpoons H_2(g) \tag{13.22}$$

in the SHE and acts as much more than a site for the exchange of electrons.

Membrane Electrodes

Membrane electrodes are indicator electrodes having a surface at which specific interaction takes place with ions resulting in a Nernstian potential response; the surface consists of a material other than the reduced form of the ion measured. Since membrane electrodes respond selectively to particular ions, they are often called **ion-selective electrodes** (or *specific-ion electrodes*, an incorrect name because no electrode is absolutely specific for a particular ion).

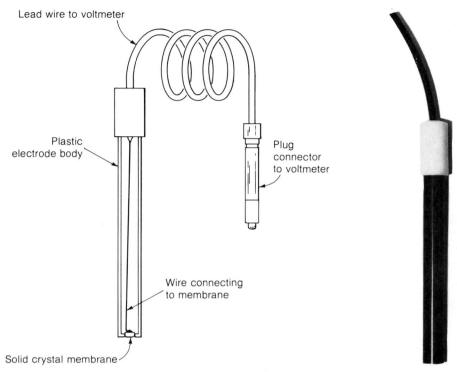

Lead wire to voltmeter

Plastic electrode body

Plug connector to voltmeter

Wire connecting to membrane

Solid crystal membrane

FIGURE 13.4 **Left:** Construction of a solid membrane ion-selective electrode with a slightly soluble crystal as the solid membrane. **Right:** Fluoride ion-selective electrode, an example of the solid-membrane type.

The membrane in a membrane electrode can be a solid, as shown in Figure 13.4. Special types of glass are used as membranes in electrodes selective for H^+ and some other monovalent cations (Sections 13.3 through 13.6). Slightly soluble crystals are used for other types of solid membrane ion-selective electrodes (Section 13.7). The other major type of membrane used in ion-selective electrodes is the "liquid membrane," in which an ion exchanger bound to the analyte ion and dissolved in an organic solvent is held by a thin layer of synthetic filter material (see Section 13.9).

13.3   **Electrodes with Glass Membrane for pH Measurement**

Construction of the Glass Electrode

Figure 13.5 illustrates an electrode system in which the indicator electrode has a glass membrane and is called a *glass electrode*. Its potential is measured against a commercial reference electrode such as the SCE (see Sections 11.7, 11.14, and 13.14).

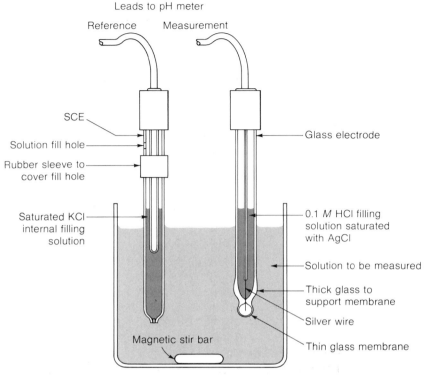

FIGURE 13.5   Construction of commercial SCE

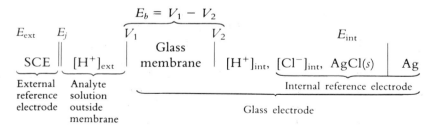

FIGURE 13.6   Schematic diagram of a glass electrode–reference electrode system for the measurement of pH. For concentrations, $[H^+]_{ext}$ is the concentration of H^+ ion in the external analyte solution, $[H^+]_{int}$ is the concentration of H^+ ion in the internal filling solution, $[Cl^-]$ is the concentration of chloride ion in the internal filling solution, and $AgCl(s)$ denotes that the internal filling solution is saturated with sparingly soluble $AgCl$. The potentials E_{ext} and E_{int} arise from differences in H^+ ion activities in contact with the external and internal surfaces of the glass membrane, respectively. Other terms are explained in the text.

The glass electrode consists of a tube on the end of which is a thin glass membrane with a strong affinity for water molecules—that is, it is **hydrated**—and sensitive to H^+ ion. An insulated wire lead through the center of the tube terminates in a short silver wire immersed in a solution of HCl saturated with $AgCl(s)$. The silver wire and the HCl solution constitute an **internal reference electrode**. The unvarying potential of the internal reference electrode and the constant H^+ ion concentration of the internal filling solution in contact with the inside surface of the glass membrane ensure that any potential variations of the glass electrode are due to interaction of H^+ ion with the *external* surface of the glass membrane. It is the concentration-dependent tendency of the H^+ ion to attach to the external surface of the glass membrane that causes the glass electrode to give a potential response with differing H^+ ion concentration and, therefore, pH.

Response of the Glass Electrode

Figure 13.6 is a schematic diagram of the electrode system shown in Figure 13.5 for the measurement of pH. The potential of the electrode system E is given by the equation

$$E = E_{int} + E_j + E_{asy} + E_b - E_{ext} \tag{13.23}$$

where E_{int} is the potential of the internal reference electrode, E_j is the junction potential where the external reference electrode contacts the analyte solution, E_{asy} is the asymmetry potential characteristic of the particular membrane, E_b is the **boundary potential** difference between the glass membrane surfaces due to differences in internal and external hydrogen ion concentrations ($[H^+]_{ext}$ and $[H^+]_{int}$), and E_{ext} is the potential of the external reference electrode (SCE). Except for V_1, a component of E_b, all of the potentials constituting E in Equation 13.23 are constant for a particular electrode system for at least limited periods of time and can be combined in a constant E_a so that

$$E = E_a + V_1 \tag{13.24}$$

The value of V_1 varies in a Nernstian manner with the activity of H^+ in the external (analyte) solution. Approximating analyte hydrogen ion activity by concentration, $[H^+]$, leads to the following equation for glass electrode potential versus a reference electrode at $25°C$:

$$E = E_a + 0.0591 \log [H^+] \tag{13.25}$$

In terms of pH the relationship is simply

$$E = E_a - 0.0591 \, pH \tag{13.26}$$

In practice, the magnitude of the slope of the two preceding equations may differ slightly from 0.0591; compensation for this can be made by calibration with two standard solutions of rather widely separated concentration.

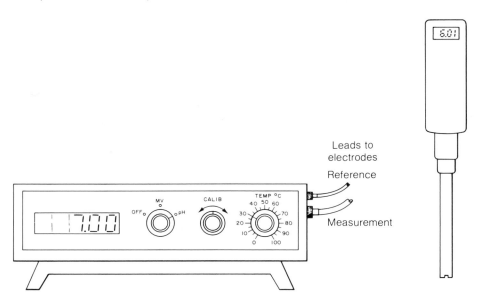

FIGURE 13.7   pH meters for measuring the potential of a glass electrode against a reference electrode, calibrated directly in pH units. **Top left:** Bench model meter. **Top right and bottom left:** Hand-held "pH wand," the handle of which is a miniaturized pH meter connected to a combination glass and reference electrode.

The pH Meter

Equation 13.26 shows that, in the absence of interferences, E is a linear function of pH. Thus the voltmeter used to measure pH can be calibrated directly in pH units as shown in Figure 13.7. In the pH measurement mode, the meter in Figure 13.7(a) is set to the known pH of a standard buffer solution in which the electrodes are immersed; this adjustment uses the "Calib." knob to compensate for variations in E_a. The temperature-adjust knob sets the slope of Equation 13.26; a setting of $25°C$ corresponds to a slope of 0.0591 V per pH unit.

13.4   Function of Glass Electrode Membrane

Nature of the Glass Electrode

An appreciation of the uses and behavior of a glass electrode requires some understanding of the principles of the glass membrane used in these electrodes and shown in Figure 13.6 as a component of a cell used for pH measurement. Figure 13.8 represents a cross section of such a membrane. Many compositions have been tried for the fabrication of glass membranes, and several are in commercial production. One of the more widely used is Corning 015 glass, with an analytical composition of approximately 22% Na_2O, 6% CaO, 72% SiO_2, responding specifically to H^+ ion up to about pH 9, with interferences from Na^+ and other singly charged cations possible above that pH.

Examination of Figure 13.8 shows that both surfaces of a glass membrane that responds to H^+ consist of a thin layer of hydrated glass gel. Thus glass membranes for pH measurement must be **hygroscopic**, containing chemically bound water molecules. The hydrated glass surface layers of the membrane may bond with H^+ from the solution in contact with the glass according to the equilibrium

$$Na^+\{glass\} + H^+ \rightleftarrows H^+\{glass\} + Na^+ \tag{13.27}$$

For a membrane that responds selectively to hydrogen ion, the equilibrium of the above reaction lies substantially to the right. Across the hydrated glass gel layer there exists a gradation of monovalent cation exchange sites occupied predominantly by H^+ ions at the surface layer in contact with water to sites occupied exclusively by Na^+ ion at the inner boundary of the hydrated layer.

Response of the Glass Membrane to H^+

As noted in the preceding section, the boundary potential E_b varies with the relative concentrations of hydrogen ion outside and inside the glass membrane, $[H^+]_{ext}$ and $[H^+]_{int}$, respectively. *For identical membrane surfaces, the*

Inside surface of glass
membrane, $[H^+]_{int}$

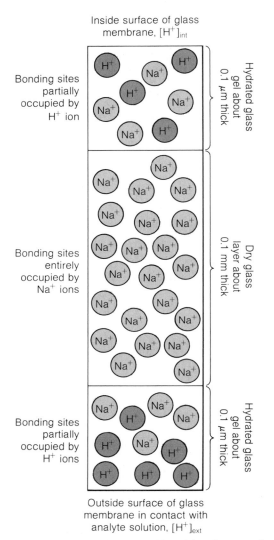

Bonding sites
partially
occupied by
H^+ ion

Hydrated glass
gel about
0.1 μm thick

Bonding sites
entirely
occupied by
Na^+ ions

Dry glass
layer about
0.1 mm thick

Bonding sites
partially
occupied by
H^+ ions

Hydrated glass
gel about
0.1 μm thick

Outside surface of glass
membrane in contact with
analyte solution, $[H^+]_{ext}$

FIGURE 13.8 Cross section of the glass membrane of a glass electrode showing hydrated and dry layers and bonding sites for monovalent cations. Note that thicknesses of layers are not drawn to scale.

value of E_b is given by

$$E_b = V_1 - V_2 = \frac{RT}{F} \ln \frac{a_{H^+_{ext}}}{a_{H^+_{int}}}$$

where $a_{H^+_{ext}}$ and $a_{H^+_{int}}$ are the hydrogen ion activities in the external analyte solution and in the internal filling solution, respectively. Although E_b would

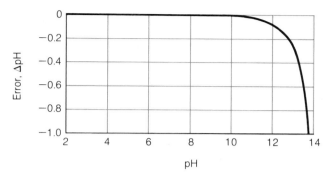

FIGURE 13.9 Alkaline error caused by Na^+ ion for a typical glass pH electrode in a medium 1.0 M in Na^+

be 0 for identical ideal membrane surfaces when $a_{H^+_{ext}} = a_{H^+_{int}}$, in fact a small **asymmetry potential** exists across the membrane; this potential is thought to arise from factors such as strains introduced during solidification of the membrane during fabrication, and mechanical and chemical changes on the surfaces after manufacture of the electrode. The asymmetry potential varies slowly with time, introducing error that is easily overcome by restandardization each time the electrode is used or every hour or two during continuous use.

Errors with Glass Electrode

At higher pH values a glass electrode designed for pH measurement may respond to monovalent cations, most commonly Na^+, introducing an **alkaline error**. Such an error is shown graphically in Figure 13.9 and mathematically by the equation

$$E = E_a + 0.0591 \log (a_{H^+} + K_{H,Na} a_{Na^+}) \tag{13.28}$$

applicable at 25° C, where a_{H^+} is the activity of the hydrogen ion, a_{Na^+} is the activity of interfering sodium ion, and $K_{H,Na}$ is the **selectivity constant**, for sodium ion interference. Selectivity constant values for interfering ions are published by manufacturers of electrodes and used to correct for alkaline error.

13.5   **Practical Measurement of pH**

A Common, Simple Laboratory Measurement

Although pH measurement is the most common, and in many respects the most simple, instrumental determination performed in the analytical laboratory, care must always be exercised to follow procedures applicable to the specific electrode, pH meter, and sample used. The operation of a properly working pH meter is just a matter of following the manufacturer's instructions. These instruments can range from small, simple, handheld devices to sophisticated computerized systems that even display instructions for the measurement of pH.

Care of Glass Electrodes

Glass electrodes for the measurement of pH have improved greatly in durability, reliability, and other respects during the last two decades. It should be kept in mind, though, that the glass membrane is a thin, fragile component that is subject to mechanical and chemical damage. Even in the absence of gross insult, glass membranes eventually wear out as a result of changes in their chemical composition caused by the media to which they are exposed. Sometimes a sluggish electrode that resists calibration can be rejuvenated by washing in 6 M HCl followed by a water rinse. If this fails a 1-min soaking in a 20% (wt/wt) aqueous solution of ammonium hydrogen fluoride, NH_4HF_2, contained in a plastic beaker followed with a water rinse will remove a surface layer from the electrode membrane, thus rejuvenating the electrode or completing its demise. (*Caution: Use care in handling this toxic solution and avoid breathing vapors from it.*) Glass electrodes should be stored with the membrane immersed in water; otherwise the electrode should be soaked in water for at least two hours before use. Of course, a glass electrode must be employed with a knowledge of its useful pH range and correction made for alkaline error (see Section 13.4) at higher pH values, if necessary.

Buffers and Standardization of Glass Electrodes

One of the most crucial operations in pH measurement is the accurate calibration of a glass electrode with a standard buffer. A **standard buffer** used for electrode calibration is a solution of accurately known pH prepared from carefully weighed acid, base, and salt components. Standard buffers are usually purchased commercially as solutions or as packets of buffer components that are dissolved and diluted to a specified volume. Directions for preparing buffers to known pH are given in standard chemical handbooks. Table 13.2 lists some examples of buffers covering a wide range of pH.

Accurate pH measurement requires that the meter be standardized with a buffer having a pH value as close as possible to that of the analyte solution. For accurate pH measurement over a wide range of values, the system should be calibrated with two buffers having pH's near either end of the range to be measured. To see how this is done, refer to the pH meter in Figure 13.7. First, with the electrodes immersed in the buffer that is the closer to pH 7, the meter is calibrated to the buffer pH with the calibration control. Next (after the electrodes are rinsed in distilled water and excess water removed by blotting

TABLE 13.2 Examples of buffers over a wide pH range

Buffer composition	pH at 25° C
0.0496 M potassium tetroxalate; prepared by dissolving 12.61 g $KHC_2O_4 \cdot H_2C_2O_4 \cdot H_2O$ in water and diluting to 1 L	1.679
Saturated potassium hydrogen tartrate (0.034 M); prepared by equilibrating excess $KHC_4H_4O_6$ with water at 25° C and filtering to remove excess solid	3.557
0.0496 M potassium hydrogen phthalate (KHP); prepared by dissolving 10.12 g $KHC_8H_4O_4$ (dried at 110° C for 2 h) in water and diluting to 1 L	4.004
pH 6.9 phosphate buffer; prepared by dissolving 3.387 g KH_2PO_4 and 3.533 g Na_2HPO_4 (salts dried at 120° C for 2 h) in water and diluting to 1 L	6.863
pH 7.4 phosphate buffer; prepared by dissolving 1.179 g KH_2PO_4 and 4.303 g Na_2HPO_4 (salts dried at 120° C for 2 h) in water and diluting to 1 L	7.415
0.01 M borax; prepared by dissolving 3.80 g $Na_2B_4O_7$ (undried)* in water and diluting to 1 L	9.183
0.025 M $NaHCO_3$ and 0.025 M Na_2CO_3; prepared by dissolving 2.092 g $NaHCO_3$ (unheated) and 2.640 g Na_2CO_3 (dried at 275° C for 2 h) in CO_2-free water and diluting to 1 L	10.014
Saturated (0.02025 M) calcium hydroxide; add freshly calcined CaO to CO_2-free water, boil, and allow to stand with excess $Ca(OH)_2$ present at 25° C, filtering buffer immediately before use	12.454

* These and higher-pH buffers must be protected from contamination by atmospheric CO_2.

with a tissue, a step that must always be performed when changing the medium in which the electrodes are immersed), the electrodes are placed in the other buffer and the meter calibrated to its pH with the temperature calibration. This step corrects the electrode response to non-Nernstian response because an actual electrode system may not give the exact Nernstian slope of −0.0591 V per pH unit in Equation 13.26. The two preceding steps are repeated, if necessary, until the meter reads the correct pH for both buffers without further adjustment.

Measurement of pH

After calibration of the pH measurement system, the electrodes are rinsed and placed in the analyte solution for pH measurement. Prior to taking the measurement, the solution should be stirred, preferably with a magnetic stirrer. In some applications it is necessary to measure the "pH" of solids, actually the pH of water equilibrated with a solid. For example, the pH of flour is measured by first equilibrating 10 g of sample with 100 mL of recently boiled water for 30 min with frequent shaking, allowing to stand 10 min, pouring off supernate into a beaker, and recording the supernate pH with a pH measuring system calibrated by pH 4.01 and pH 9.18 buffers.

Effect of Ionic Strength

Ionic strength affects pH values. The glass electrode responds to H^+ ion *activity*, which varies significantly with ionic strength at values of the latter above about 0.01. Buffer systems in which one of the constituents is doubly charged give significantly different values of a_{H^+} at different ionic strengths. For example, consider the buffer system composed of $H_2PO_4^-$ and HPO_4^{2-} ions. The hydrogen activity is governed by the equilibrium

$$H_2PO_4^- \rightleftarrows H^+ + HPO_4^{2-} \tag{13.29}$$

$$K_a = \frac{a_{H^+} a_{HPO_4^{2-}}}{a_{H_2PO_4^-}} = 6.34 \times 10^{-8} \tag{13.30}$$

At an ionic strength of 0.001, the activity coefficients of H^+, $H_2PO_4^-$, and HPO_4^{2-} are 0.967, 0.964, and 0.867, respectively. At an ionic strength of 0.1, the corresponding values are 0.83, 0.78, and 0.36, respectively. As a result, an $H_2PO_4^- - HPO_4^{2-}$ buffer exhibits different pH's at ionic strengths of 0.001 and 0.1 (see Problem 3 at the end of this chapter).

13.6   Glass Electrodes for Monovalent Cations Other Than H^+

Much of the research on glass electrode membrane formulation has involved research on minimizing *alkaline error* resulting from the presence of alkali metal ions, particularly Na^+. As a result, with commonly used pH electrodes, this is not a problem for most analytes at pH values below about 12. The existence of alkaline error, however, has suggested that membranes might be formulated that would respond selectively to monovalent cations other than H^+. For a sodium ion electrode, as an example, the *selectivity constant*, $K_{H,Na}$, is *maximized* in the following equation presented previously in Section 13.4:

$$E = E_a + 0.0591 \log (a_{H^+} + K_{H,Na} a_{Na^+}) \tag{13.28}$$

It has been shown that this can be brought about by the addition of Al_2O_3 and B_2O_3 to glass used for membranes. As a result, electrodes have been developed that respond selectively to Na^+, K^+, and Li^+ at a pH as low as 5. These electrodes give a linear response of E to pNa, pK, and pLi, which are the negative logs of the activities of the respective ions denoted. Typical responses of such electrodes are shown in Figure 13.10.

Glass electrodes selective to potassium and sodium ions are commercially available. Various types of glass membranes also respond to Rb^+, Cs^+, NH_4^+, and Ag^+.

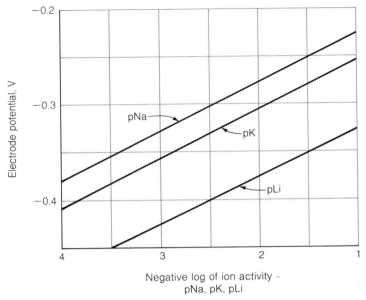

FIGURE 13.10 Response of alkali metal–sensitive glass electrode to Na^+, K^+, and Li^+ at pH 8.0

Ion-Selective Electrodes

Modern Ion-Selective Electrodes

The success of the glass electrode for pH measurement led to many attempts to develop electrodes for various kinds of ions. Except for glass electrodes useful for several monovalent ions mentioned in the previous section, these attempts were largely unsuccessful, and by the mid-1960s potentiometry had become a useful but not very exciting branch of analytical chemistry. However, the introduction of new types of electrodes optimistically called *"specific-ion"* *electrodes* caused a resurgence of interest in potentiometric research and applications in the late 1960s. The first of these new developments in potentiometry was the introduction of the calcium ion–selective electrode in 1966. This was an exciting development in two respects—it provided a potentiometric sensor for an ion that previously could not be measured potentiometrically, and it was the first example of the liquid membrane type of electrode (see Section 13.9), which was an entirely new approach to indicator electrode technology. In 1967 the fluoride ion–selective electrode was introduced (see Section 13.8). The first nonglass membrane solid-state electrode, it filled a void in analytical methodology with importance to aquatic chemistry, industrial chemistry, biological studies, and other areas. The electrode responds to fluoride ion concentrations down to about $1 \times 10^{-6}\ M$. This concentration is about as low as that reached by other methods, and the fluoride ion–selective electrode is much simpler than other accepted means of fluoride analysis. Potentiometric fluoride analysis is capable of high accuracy. The only major common interferences are hydroxide ion and metal ions complexed by fluoride; these can be eliminated by buffering and the addition of metal chelating agents.

Solid and Liquid Membrane Electrodes

Solid membrane electrodes have been marketed for bromide, cadmium, chloride, copper(II), cyanide, fluoride, iodide, lead, silver, sulfide, and thiocyanate ions. Liquid membrane electrodes include those for calcium, fluoroborate, nitrate, perchlorate, potassium, and divalent cations (a measure of water hardness). In addition, some specialty electrodes, such as those for measuring gases, have been produced; these are discussed in Section 13.10.

13.8	**Solid-State Electrodes**

Construction of Solid-State Electrodes

Figure 13.4 shows the construction of an electrode with a solid membrane. In some cases these electrodes employ an internal reference electrode, like that in a glass electrode. Figure 13.11 shows a fluoride solid-state electrode with an internal reference electrode rather than a direct connection to the inside surface of the sensing membrane.

Commercially Available Solid-State Electrodes

Table 13.3 lists commercially available solid-state electrodes. Missing from this list are electrodes for several ions, such as SO_4^{2-} and CO_3^{2-}, which commonly need to be measured in samples of commercial products, water, and soil. The lack of such electrodes is due to the fact that slightly soluble salts of these ions are not available having the qualities required of a membrane for potentiometric measurement. Generally, otherwise suitable salts do not have the requisite electrical conductivity. In addition, membrane materials must be reasonably strong and resist corrosion and abrasion.

Solid Membrane Materials

One of the few materials that meets all the criteria for solid-state membrane electrodes is lanthanum fluoride, LaF_3, used as a membrane in fluoride-

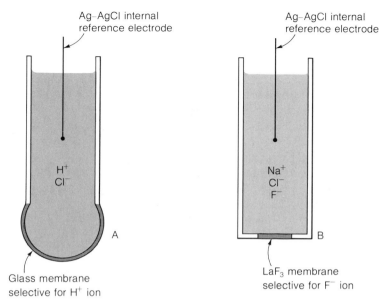

FIGURE 13.11   Solid-state fluoride electrode with an internal reference electrode (*right*) compared with a glass electrode (*left*).

TABLE 13.3   Commercially available solid-state electrodes

Analyte ion	Concentration range, M	Preferred pH	Most common interferences
Br^-	$10^0 - 5 \times 10^{-6}$	2–12	S^{2-}, I^-, Br^-
Cd^{2+}	$10^{-1} - 10^{-7}$	3–7	Ag^+, Hg^{2+}, Cu^{2+}
Cl^-	$10^0 - 5 \times 10^{-5}$	2–11	S^{2-}
Cu^{2+}	$10^0 - 10^{-8}$	3–7	S^{2-}, Ag^+, Hg^{2+}
CN^-	$10^{-2} - 10^{-6}$	11–13	S^{2-}, I^-
F^-	$10^0 - 10^{-6}$	5–8	OH^-, F^-
I^-	$10^0 - 5 \times 10^{-8}$	3–12	S^{2-}
Pb^{2+}	$10^0 - 10^{-6}$	4–7	Ag^+, Hg^{2+}, Cu^{2+}
Ag^+	$10^0 - 10^{-7}$	2–9	S^{2-}, I^-, Br^-, Cl^-
S^{2-}	$10^0 - 10^{-7}$	13–14	Hg^{2+}
Na^+	$10^0 - 10^{-6}$	9–10	Ag^+, H^+
SCN^-	$10^0 - 5 \times 10^{-6}$	2–12	OH^-, S^{2-}, I^-, Br^-

selective electrodes. Other solid membranes used in electrodes contain silver ion, Ag^+, because of the adequate electrical conductivity of silver salts. Thus pellets of AgCl, AgBr, and AgI are used in electrodes to measure Cl^-, Br^-, and I^-, respectively. Solid Ag_2S is used as a membrane material for sulfide electrodes. Mixtures of this salt with CdS, CuS, and PbS provide membranes for cadmium, cupric [copper(II)], and lead ion–selective electrodes, respectively.

13.9   Liquid Membrane Electrodes

The Nature of Liquid Membrane Electrodes

Liquid membrane electrodes are those that produce an ion-selective potential response across a membrane soaked with an organic solution of an ion exchanger capable of binding and exchanging the analyte ion. Figure 13.12 represents a liquid membrane electrode selective for nitrate ion. The disk material constituting the membrane is organophilic—that is, it attracts the organic solvent containing the ion exchanger and repels water. The liquid ion exchanger is chosen for its affinity for the ion being determined. For example, the ion exchanger used in the nitrate ion–selective electrode is a solution of a substituted phenanthroline complex of nickel(II). The calcium electrode employs a phosphate ester

$$Ca^{2+} \left\{ {}^{-}O-\overset{\overset{\textstyle O}{\uparrow}}{\underset{\underset{\textstyle OC_{10}H_{21}}{|}}{P}}-OC_{10}H_{21} \right\}_2$$

exhibiting an affinity of 10,000:1 for Ca^{2+} over Na^+. As shown in Figure 13.11, the organic solution of ion exchanger is held in a reservoir in contact with the membrane and soaks into the membrane around its edges. Contacting the center of the membrane is a reservoir of an aqueous solution containing Cl^- ion and the ion for which the electrode is selective (a solution of $NaNO_3$ and NaCl for a nitrate-selective electrode, one of $CaCl_2$ for a calcium-selective electrode). The chloride ion provides a stable potential at the Ag–AgCl internal reference electrode. The analyte ion in the aqueous internal filling solution provides a stable potential at the internal membrane surface.

Liquid membrane electrodes selective for BF_4^-, Ca^{2+}, K^+, and NO_3^- have been developed in which the ion exchanger solution is held by a semisolid polyvinylchloride gel. These electrodes look much like solid-state electrodes.

Response of Liquid Membrane Electrodes

The potential response of a liquid membrane electrode is based upon the exchange of the analyte ion in the solution external to the electrode with the ion exchanger in the organic solution held by the membrane. For the case of calcium ion, this exchange can be represented by the equilibrium

$$Ca^{2+} + 2R^- \rightleftarrows Ca^{2+}\{^-R\}_2 \tag{13.31}$$

$$\text{Aqueous Organic} \qquad \text{Organic}$$

where R represents the organic phosphate ester anion exchanger.

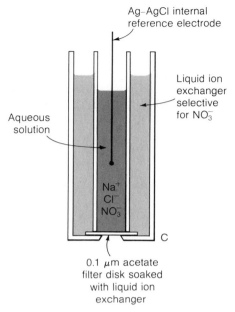

FIGURE 13.12 Liquid membrane electrode selective for nitrate ion

At 25°C, the general equation for the potential response of a liquid membrane ion-selective electrode is

$$E = E_a + \frac{0.0591}{z} \log a_X z \tag{13.32}$$

where E is the indicator electrode potential versus an *external* reference electrode, z is the signed charge of the ion, $a_X z$ is the activity of the analyte ion, and E_a is a constant for the electrode system employed. For actual chemical analyses, it is often more convenient to express potentials (E, E_a, $2.303RT/F$) in millivolts and to substitute concentrations in place of activities. With these substitutions, the potential response of a calcium-selective liquid membrane electrode is

$$E = E_a + \frac{59.1}{2} \log [Ca^{2+}] \qquad \text{Potential in millivolts} \tag{13.33}$$

that of a nitrate-selective electrode is

$$E = E_a - 59.1 \log [NO_3^-] \qquad \text{Note} - \text{sign for anion} \tag{13.34}$$

and that of a potassium-selective electrode is

$$E = E_a + 59.1 \log [K^+] \tag{13.35}$$

Interferences with Liquid Membrane Electrodes

Liquid membrane electrodes are subject to interferences from nonanalyte ions. The magnitude of this kind of interference is expressed by a *selectivity constant* such as that discussed for glass pH electrodes in Section 13.4. This constant may be denoted as $K_{X,Y}$ for the interference of ion Y with the measurement of ion Y at an X-selective electrode. The formula for the electrode potential in the presence of interfering ion is

$$E = E_a + \frac{59.1}{z} \log\left([X^z] + K_{X,Y}[Y^y]^{z/y}\right) \tag{13.36}$$

where z and y are the signed charges of ions X and Y, respectively. Note that the degree of interference is a function of the ratio z/y.

As an example of a calculation involving a selectivity constant with an ion-selective electrode, consider the interference of Na^+ with calcium electrode response. The value of $K_{Ca,Na}$ is 0.003. Suppose that E of the calcium electrode as measured against a reference electrode is -200.0 mV in 1.00×10^{-5} M standard Ca^{2+} and 218.0 mV in a solution of unknown Ca^{2+} concentration that is also 5×10^{-2} M in Na^+. What is the percentage error introduced by the presence of Na^+? The two equations for E are

$$E = -200.0 = E_a + \frac{59.1}{2} \log 1.00 \times 10^{-5} \qquad \text{In standard } Ca^{2+} \tag{13.37}$$

$$E = -218.0 = E_a + \frac{59.1}{2} \log\left[[Ca^{2+}] + 0.003(5 \times 10^{-2})^{2/1}\right] \tag{13.38}$$

In unknown Ca^{2+} with Na^+ interference

Solution of the two preceding equations for the unknown Ca^{2+} concentration yields

$$\log\left\{\frac{[Ca^{2+}] + 0.003(5 \times 10^{-2})^2}{1.00 \times 10^{-5}}\right\} = \frac{2 \times 18.0}{59.1} = 0.609 \tag{13.39}$$

$$[Ca^{2+}] = 3.32 \times 10^{-5} \, M \tag{13.40}$$

$$\% \text{ error due to } Na^+ = \frac{0.003(5 \times 10^{-2})^2}{3.32 \times 10^{-5}} \times 100 = 22.6\% \tag{13.41}$$

TABLE 13.4   Commercially available liquid membrane electrodes

Analyte ion	Concentration range, M	Preferred pH	Most common interferences
Ca^{2+}	10^0–5×10^{-7}	6–8	Zn^{2+}, Fe^{2+}, Pb^{2+}
BF_4^-	10^{-1}–3×10^{-6}	3–10	NO_3^-
NO_3^-	10^0–7×10^{-6}	3–10	ClO_4^-, I^-
ClO_4^-	10^0–2×10^{-6}	3–10	I^-
K^+	10^0–10^{-6}	3–10	Cs^+
$Ca^{2+} + Mg^{2+}$	10^{-2}–6×10^{-6}	5–8	Zn^{2+}, Fe^{2+}, Cu^{2+}, Ni^{2+}, Ba^{2+}

Commercially Available Liquid Membrane Electrodes

Commercially available liquid membrane electrodes are listed in Table 13.4. Selectivity constants are also given for common interferences.

13.10   Gas-Sensing Electrodes

Figure 13.13 shows a gas-sensing electrode. These work by virtue of a hydrophobic membrane that is not permeated by solution or ionic solutes in solution, but allows for the passage of gas molecules of various classes. For example, a CO_2-selective electrode allows CO_2 gas dissolved in the external solution to diffuse through the membrane. The internal filling solution contains HCO_3^-, and at the interface between the glass bulb of the glass

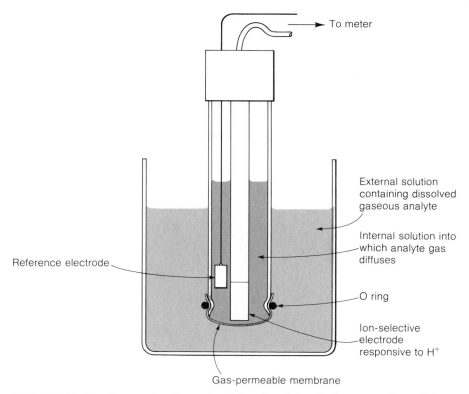

FIGURE 13.13 Gas-sensing electrode with a glass electrode that responds to pH changes caused by gases diffusing through the selective gas-permeable membrane

electrode and the gas-permeable membrane, the equilibrium of the reaction

$$CO_2(aq) + H_2O \rightleftharpoons H^+ + HCO_3^- \qquad\qquad (13.42)$$

is shifted to the right, causing a decrease in pH. This decrease is related by the appropriate conversion factors to the concentration of CO_2 in the external solution.

Gas-sensing electrodes are manufactured for the measurement of CO_2, HCN, HF, H_2S, NH_3, and SO_2. These electrodes can measure gases in the atmosphere, as well as in solution.

13.11   Ion-Selective Electrode Response and Calibration

Figure 13.14 shows a calibration plot for a fluoride electrode throughout the range over which it is useful. It is seen that the electrode gives a linear Nernstian response with a slope of -59.1 mV per unit log $[F^-]$ from about -2

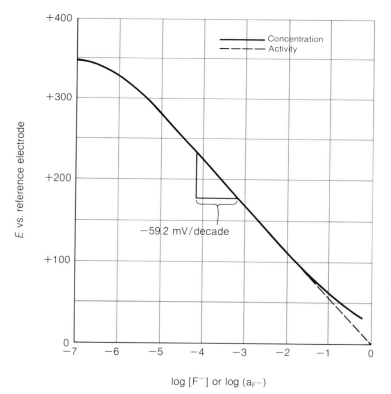

FIGURE 13.14   Potential response of fluoride electrode to log of fluoride ion concentration and activity

to almost -6 for log [F$^-$]. At concentrations below about $1 \times 10^{-6} M$ F$^-$, the electrode reaches a point where there is no longer any potential response with changes in [F$^-$]. This lower limit of electrode response is due to the slight solubility of the LaF$_3$ membrane.

In the upper fluoride concentration region, it is seen that plots for log fluoride ion activity and concentration deviate because of changes in fluoride ion activity coefficient. This does not present a problem in fluoride analysis because the concentration of analyte fluoride ion is seldom so high and the solutions can always be diluted.

Figure 13.14 illustrates an ion-selective electrode calibration curve that can be used in determining unknowns. Such a curve can be prepared over the concentration range that includes the unknown concentration values. A potential measurement in an unknown solution can be used on the plot to give the value of log [F$^-$] pertaining to that potential value. In modern practice the calibration curve is stored on a computer and the computer automatically calculates [F$^-$] for each value of E from an unknown input to it. Because of gradual shifts in the calibration curve, it is advisable to run standards frequently interspersed among a series of unknowns as a recalibration measure.

13.12 Error with Ion-Selective Electrode Measurements

An inherent error with the direct potentiometric measurement of ion concentrations arises from the fact that the signal measured, E, is a function of *log* ion concentration and is not directly proportional to ion concentration itself. Mathematically, a slight error in the log of a number (here measured by E) yields a relatively larger error in the number (in this case the concentration). This is best seen by example. Suppose that an error of 1 mV is made in the direct potentiometric measurement of nitrate ion with a nitrate-selective electrode. The percent relative error in [NO$_3^-$] is obtained by solving the two following equations, where Δ is the error in nitrate ion concentration:

$$E = E_a - 59.1 \log [\text{NO}_3^-] \qquad \text{No error in } E \tag{13.43}$$

$$E - 1 = E_a - 59.1 \log ([\text{NO}_3^-] + \Delta) \qquad \text{1-mV error in } E \tag{13.44}$$

Subtracting the second equation above from the first yields

$$1 = 59.1 \log \frac{[\text{NO}_3^-] + \Delta}{[\text{NO}_3^-]} \tag{13.45}$$

Solving for the percent relative error yields

$$\frac{[\text{NO}_3^-] + \Delta}{[\text{NO}_3^-]} = \text{antilog} \frac{1}{59.1} = 1.040 \tag{13.46}$$

$$\text{Percent relative error} = \frac{\Delta}{[\text{NO}_3^-]} \times 100 = 0.040 \times 100 = 4.0\% \tag{13.47}$$

In this case, and for any electrode measuring the concentration of a singly charged ion, an error of only 1 mV in potential results in a relative concentration error of 4.0%; for a divalent ion such as Ca^{2+}, the error is twice as great, 8.0%.

A 1-mV error occurs readily in potentiometric measurements. It can arise from changes in measuring and reference electrodes, effects of the medium upon the measuring electrode membrane, changes in liquid junction potential where the reference electrode contacts the analyte solution, and variations in the potential measuring device. Added to these causes are potential fluctuations that seem to defy logical explanation. Therefore, the direct potentiometric measurement of ion concentration should be undertaken with special care employing frequent checks with standard solutions.

13.13   Standard Addition in Potentiometric Measurement

For the most accurate possible potentiometric measurements, both the standard and unknown solutions should be in the same medium at the same temperature and measurements taken in them at the same time. Such conditions are approached by the method of **standard addition**, which consists of measuring E in a known volume of the analyte solution; adding a carefully measured, relatively small volume of standard solution with a concentration of analyte considerably greater than that of the unknown; measuring E after mixing; and calculating the original analyte ion concentration from the change in E.

As an example of a standard addition measurement, consider a 100-mL volume of unknown Ca^{2+} concentration that gives a reading of $+127.3$ mV at a calcium-selective electrode. Addition of 5.00 mL of 1.00×10^{-2} M Ca^{2+} standard shifts the potential to $+136.8$ mV. The two pertinent equations are

$$136.8 = E_a + \frac{59.1}{2} \log \left([Ca^{2+}] \frac{100}{105} + \frac{5.00 \times 1.00 \times 10^{-2}}{105} \right) \tag{13.48}$$

$$127.3 = E_a + \frac{59.1}{2} \log [Ca^{2+}] \tag{13.49}$$

Note that the addition of 5.00 mL of Ca^{2+} standard dilutes the original Ca^{2+} concentration $[Ca^{2+}]$ by $100/105$ and that the increment in the calcium ion concentration from the addition of the standard is $(5.00 \text{ mL} \times 1.00 \times 10^{-2} \text{ mmol/mL})/105$ mL. Subtraction of the second equation from the first above and solving for $[Ca^{2+}]$ yields

$$9.5 = \frac{59.1}{2} \log \left(\frac{0.952[Ca^{2+}] + 4.76 \times 10^{-4}}{[Ca^{2+}]} \right) \tag{13.50}$$

$$\frac{0.952[Ca^{2+}] + 4.76 \times 10^{-4}}{[Ca^{2+}]} = 2.096 \tag{13.51}$$

$$[Ca^{2+}] = 4.16 \times 10^{-4} \text{ M} \tag{13.52}$$

There is little likelihood of spurious shifts in E during the time required to make the two potential measurements involved in standard addition. Therefore, standard addition is the preferred technique when relatively high accuracy is required.

13.14 Reference Electrodes

Types and Construction of Reference Electrodes

The most popular types of commercial reference electrodes are the saturated calomel electrode, SCE, and the silver–silver chloride electrode. The SCE, described in Sections 11.7 and 11.14, is based upon the half-reaction

$$Hg_2Cl_2(s) + 2e^- \rightleftarrows 2Hg(l) + 2Cl^-(\text{satd. KCl}) \tag{13.53}$$

and has a potential of $+0.242$ V versus the standard hydrogen electrode. The silver–silver chloride electrode is based upon the half-reaction

$$AgCl(s) + e^- \rightleftarrows Ag + Cl^- \qquad E^0 = +0.222 \text{ V} \tag{13.54}$$

Figure 13.15 shows details of the construction of a commercial SCE. It is in the shape of a rod to be conveniently held in an electrode holder and dipped into a solution. An electrical connection is made to the reference side of the pH meter (voltmeter) through an external lead. This wire goes through an inner glass tube and connects to a short length of platinum wire dipping into a short column of mercury. The bottom of the mercury column contacts a paste of solid Hg_2Cl_2 (calomel), solid KCl, and liquid mercury moistened with an aqueous solution of saturated KCl. Near the bottom of the inner glass tube is a small hole that allows contact with the filling solution in the outer glass tube. The latter is filled with saturated KCl solution and contains crystals of solid KCl to maintain saturation. Provision is made for adding saturated KCl solution through a hole near the top of the outer glass tube; the hole can be closed with a small stopper or by a rubber sleeve around the outer glass tube that slips up over the hole. The commercial SCE contacts the analyte solution by means of a salt bridge at the bottom of the electrode that allows contact of solution between the inside and outside of the electrode without excessive leakage. This can be accomplished with an asbestos fiber through the tip of the electrode, a cracked glass bead, a ground glass sleeve covering a small hole drilled in the side of the tube near the bottom, or other means. This contact constitutes a *liquid junction*, associated with which is a *liquid junction potential* (Sections 11.4 and 11.13) that is a source of error in potentiometric measurements.

The physical appearance of a commercial silver–silver chloride electrode is much like that of the SCE. The filling solution is a solution of a chloride salt and may consist of saturated KCl. A silver wire dipping into this solution may hold solid AgCl in a loop of wire. Contact between the internal solution and analyte solution is made through a porous plug. If saturated KCl is used as the

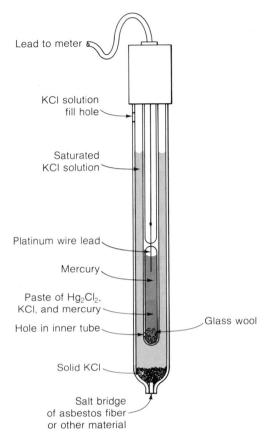

Lead to meter

KCl solution
fill hole

Saturated
KCl solution

Platinum wire lead

Mercury

Paste of Hg_2Cl_2,
KCl, and mercury

Hole in inner tube

Glass wool

Solid KCl

Salt bridge
of asbestos fiber
or other material

FIGURE 13.15   Construction of commercial SCE

filling solution, this reference electrode has a potential of $+0.197$ V versus the SHE.

Combination Electrodes

The silver–silver chloride electrode is used as an external reference in a **combination electrode** in which both the indicator electrode (usually a glass electrode for pH measurement) and the external reference electrode are mounted on the same electrode body (Figure 13.16). A combination electrode is much more convenient to use than are two separate electrodes. It is important to realize, though, that it is in fact two electrodes, an indicator and a reference electrode, mounted on the same body.

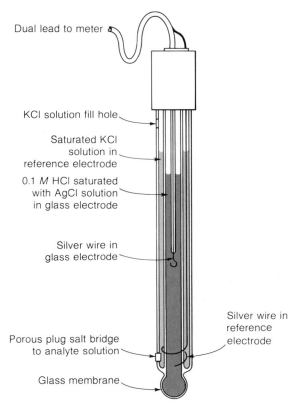

Dual lead to meter

KCl solution fill hole

Saturated KCl
solution in
reference electrode

0.1 *M* HCl saturated
with AgCl solution
in glass electrode

Silver wire in
glass electrode

Silver wire in
reference
electrode

Porous plug salt bridge
to analyte solution

Glass membrane

FIGURE 13.16 Combination electrode consisting of a glass indicator electrode and silver–silver chloride reference electrode

13.15   Potentiometric Titrations with Ion-Selective Electrodes

Nature of Potentiometric Titrations

Potentiometry works very well for following the course of titrations whenever either the analyte or titrant responds to an indicator electrode. The most common such titration is an acid-base titration in which a glass electrode is used to monitor pH as a function of either added acid or base. The resulting plot of pH versus volume of titrant gives acid-base titration curves such as those shown in Chapter 5. The course of precipitation titrations can be followed potentiometrically. This is most commonly done when Ag^+ is a reaction participant and the silver indicator electrode is employed. Titration curves of this type are shown in Chapter 9.

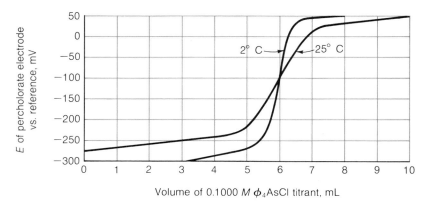

FIGURE 13.17   Precipitation titration of 50.0 mL 0.01200 M ClO_4^- with 0.1000 M ϕ_4 AsCl to form ϕ_4 $AsClO_4(s)$ precipitate at 25° and 2° C

Precipitation Titrations

The perchlorate electrode can be used to follow the titration of ClO_4^- ion with tetraphenylarsonium chloride titrant, ϕ_4 As^+, as shown in Figure 13.17. In the example shown, sensitivity was increased by lowering the titration temperature to 2° C. This decreases the solubility of the salt product; a less soluble precipitate gives a sharper titration curve break.

Complexation and Oxidation-Reduction Titrations

Complexation titrations can be followed potentiometrically by measuring the log of the concentration of metal ion titrated (for example, Cu^{2+} with a copper-selective electrode) and by monitoring the concentration of EDTA in the presence of HgY^{2-} with a mercury electrode (see Section 13.2). Oxidation-reduction titrations can be followed by monitoring E with a platinum electrode.

13.16   Gran Plots

Gran plots are a method of obtaining accurate end points from potentiometric titrations in cases where the titration curve break is not very sharp and where it is desired to take only a few points. Gran plots accomplish these goals and have the advantage of being linear. Their use is illustrated for the titration of perchlorate with tetraphenylarsonium ion described in the preceding section.

The equation for perchlorate ion response is

$$E = E_a - 59.1 \log [\text{ClO}_4^-] \qquad \text{Potentials in mV} \tag{13.55}$$

This equation can be rearranged to

$$\log [\text{ClO}_4^-] = -1.69 \times 10^{-2}E + K' \tag{13.56}$$

where 1.69×10^{-2} is the reciprocal of 59.1 and K' is $E_a/59.1$. The value of $[\text{ClO}_4^-]$ is given by antilog $(-1.69 \times 10^{-2}E + K')$.

$$[\text{ClO}_4^-] = \text{antilog} \, (-1.69 \times 10^{-2}E + K') \tag{13.57}$$

Except very close to the equivalence point where dissolution of $\phi_4\text{AsClO}_4$ may contribute significantly to $[\text{ClO}_4^-]$, the concentration of this ion is given simply by stoichiometry as

$$[\text{ClO}_4^-] = \frac{V_C M_C - V_\phi M_\phi}{V_C + V_\phi} \tag{13.58}$$

where V_C and V_ϕ are the volumes of the perchlorate and tetraphenylarsonium chloride solutions, respectively, and M_C and M_ϕ are their respective molarities. The two preceding equations can be combined and rearranged to give the following:

$$(V_C + V_\phi)\text{antilog} \, (-1.69 \times 10^{-2}E + K') = V_C M_C - V_\phi M_\phi \tag{13.59}$$

The left side of this equation can be plotted (y axis) versus the only variable on the right, V_ϕ; such a plot is shown in Figure 13.18. The equivalence point occurs where $V_C M_C = V_\phi M_\phi$—that is, where the quantity plotted on the y axis equals 0. Extrapolating the linear portion of the plot to that point gives the equivalence point value of V_ϕ.

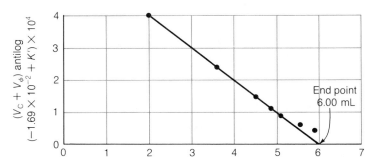

FIGURE 13.18 Gran plot for the titration of ClO_4^- with $\phi_4\text{As}^+$ (see Figure 13.17)

Programmed Summary of Chapter 13

The major terms and concepts introduced in this chapter are contained in this summary in a programmed format. To derive the most benefit from the summary, you should fill in the blanks for each question and then check the answers at the end of the book to see if your choices are correct.

The measurement of ion activity from the electrical potential produced by the ion acting at an electrode surface is known as (1) _____. For potentiometric measurement of the ion X with charge z and activity a_{Xz}, the applicable equation in the Nernstian form is (2) _____. The general type of electrode that responds to ion concentration in potentiometry is called (3) _____. Six or seven specific types of these electrodes are (4) _____. A mercury metal electrode used in a solution with a constant concentration of HgY^{2-} can be employed for the purpose of following (5) _____. The glass membrane in a glass electrode has a special affinity for water and is therefore said to be (6) _____; it serves primarily as an exchange site for (7) _____ ions. A silver wire and filling solution inside a glass electrode constitute the (8) _____. The names of the five potentials that make up the total potential of a glass and reference electrode system are (9) _____. The voltmeter used with a glass electrode to measure H^+ concentration is commonly calibrated directly in units of (10) _____. The error caused by sodium ion in the measurement of $[H^+]$ at a glass electrode is called (11) _____, and compensation is made for it by use of a constant called a (12) _____. A standard buffer is used with a glass electrode for the purpose of (13) _____. In addition to H^+ ion, glass electrodes have been developed for use with the ions (14) _____. The term *specific-ion electrode* is inaccurate because (15) _____. In place of an internal reference electrode, some solid-state electrodes have a (16) _____. Aside from the important examples of glass and LaF_3 membranes, most solid membrane materials for ion-selective electrodes are composed of salts or mixed salts of (17) _____. The component and material in a liquid membrane electrode that distinguishes it from other kinds of electrodes is (18) _____. The degree of interference by an interfering ion at an electrode designed to measure the concentration of an analyte ion is a function of the concentrations of analyte and interfering ions as well as the (19) _____ and (20) _____. The component of a gas-sensing electrode that distinguishes it from other potentiometric electrodes is a (21) _____. A calibration plot for the fluoride ion-selective electrode deviates from linearity at lower values of log $[F^-]$ because of (22) _____. A potential error of 1 mV with an ion-selective electrode for monovalent ions translates to a relative concentration error of about (23) _____. For greatest accuracy, rather than using a calibration curve with an ion-selective electrode, it is preferable to use the technique of (24) _____. The two most common reference electrodes used with ion-selective electrodes are the (25) _____ electrodes. A combination electrode normally is a combination of (26) _____. Four types of titrations that can be followed potentiometrically are (27) _____. An accurate end point can be obtained in cases where the titration curve break is not very sharp and where it is desired to take only a few points potentiometrically by the use of (28) _____.

Questions

1. The manual *Methods of Analysis of the Association of Official Analytical Chemists* lists a procedure for the measurement of the pH of peat (moss, humus, and reed-sedge types) that involves measuring the pH of water equilibrated with the solid peat. The pH thus obtained varies with the salt content of the peat, such as that from chemical fertilizers. A pH value independent of the initial salt content can be obtained by first soaking the peat in a $0.01\ M\ CaCl_2$ solution, a procedure that gives a pH 0.5 to 0.8 units lower than equilibration with water. Peat is a very complicated organic material with a high content of oxygen in carboxylic acid and phenolic—OH groups as shown.

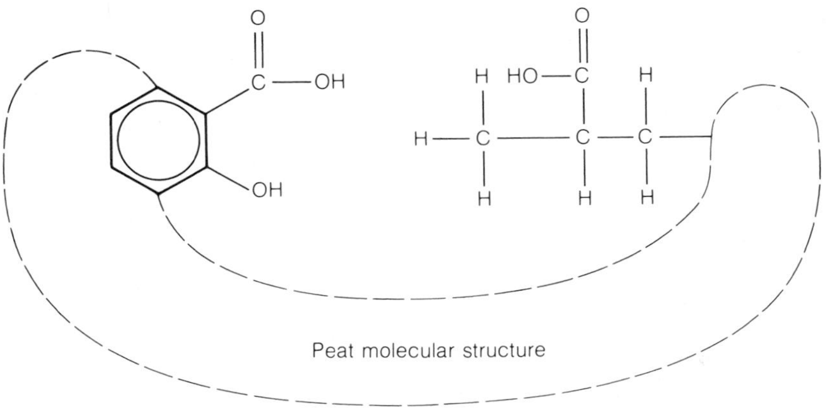

Peat molecular structure

Explain the lower pH obtained with the $CaCl_2$ solution.

2. Match:

    a. Inert electrode material
    b. First-order electrode material
    c. Membrane material for a commonly measured cation

    d. Membrane material for a commonly measured anion
    1. Glass
    2. Pt
    3. LaF_3
    4. Cd

3. An electrode that responds to an ion in

solution by a Nernstian change in potential is called (a) _____ and the electrode that must be used with it is called a (b) _____.

4. For what purpose can a mercury metal electrode be used in a solution containing a low concentration of HgY^{2-}?

5. What is a major advantage of a metal amalgam electrode in terms of electrode surface quality?

6. Why is it not totally correct to regard a platinum electrode in the SHE as an inert electrode?

7. In the simplest sense, what is the function of the membrane in a membrane electrode?

8. Why is HCl used as a filling solution in a glass electrode?

9. Where is the junction potential E_j generated in an electrode system for the measurement of pH?

10. In the equation for the potential response of a glass electrode at $25°C$, $E = E_a - 0.0591$ pH, E_a is a constant made up of a number of different potentials. (a) How are these potentials measured, or how is compensation made for them; (b) what is done to correct for deviations from the Nernstian slope of -0.0591 V per pH unit?

11. Describe the surface of the glass membrane that responds to H^+ in a glass electrode.

12. Match the following pertaining to a glass

a. Junction potential
b. Asymmetry potential
c. Negative error in pH often encountered at high pH values
d. Boundary potential
1. Can result from the presence of relatively high concentrations of Na^+
2. Occurs at external reference-analyte interface
3. Due to differences in $a_{H^+_{int}}$ and $a_{H^+_{ext}}$

4. Due to strains in the membrane and differences in membrane surfaces

13. How should a glass electrode be stored?

14. Some examples of buffers used for standardizing pH meters are listed in Table 13.2. Of these, which would you expect to have the most stable pH (the best buffer capacity), the pH 1.679 buffer composed of potassium tetroxalate or the pH 3.557 saturated potassium hydrogen tartrate buffer? Explain.

15. What is involved in measuring the pH of a solid material?

16. What constituents are added to glass membranes to increase their selectivity for monovalent ions other than H^+?

17. Which ion-selective electrodes were responsible for the renaissance of potentiometry in the mid-1960s?

18. What is the nature of the ion exchanger in liquid membrane electrodes?

19. Suggest an internal filling solution for a perchlorate-selective electrode.

20. What explains the lack of solid-state electrodes for some important ions such as SO_4^{2-} and CO_3^{2-}?

21. Why does a gas electrode respond to gases, but not other, particularly ionic, solutes in solution?

22. What determines the lower limit of electrode response for a solid-state electrode, such as the fluoride electrode?

23. Why is direct potentiometry inherently less accurate than some other common analytical techniques, such as titimetric or gravimetric analysis?

24. What method for increasing potentiometric accuracy enables measuring E for both the unknown and standard in essentially the same medium at almost the same time?

25. What is the main condition to be fulfilled if a titration is to be followed potentiometrically?

Problems

1. The formation constant of CaY^{2-} is 5.0×10^{10} and at pH 10.0 $\alpha_{Y^{4-}}$ for EDTA is 0.35. For the titration of 50.0 mL of $0.0500\ M\ Ca^{2+}$ with $0.100\ M$ EDTA in a medium maintained at pH 10 containing a very low concentration of HgY^{2-} and followed with a mercury electrode system for which $E_a = +0.200$ V, calculate E at (a) 12.5 mL, (b) 25.0 mL, and (c) 37.50 mL of added titrant.

2. Given

$$Ce^{4+} + e^- \rightleftharpoons Ce^{3+} \qquad E^0 = +1.70\ V$$

what is E versus the SHE measured at a platinum electrode in a $0.0100\ M$ solution of Ce^{4+} in which 0.025% of the Ce is present as a Ce^{3+} impurity?

3. Given the activity coefficients for H^+, $H_2PO_4^-$, and HPO_4^{2-} on page 376 and the value of K_{a2} as

$$K_{a2} = \frac{a_{H^+} a_{HPO_4^{2-}}}{a_{H_2PO_4^-}} = 6.34 \times 10^{-8}$$

calculate a_{H^+} in a buffer formed by dissolving 2.50×10^{-4} mol of NaH_2PO_4 and 2.50×10^{-4} mol of Na_2HPO_4 and diluting to 1 L (ionic strength $\mu = 0.001$) and the same solution adjusted to $\mu = 0.1$ with $NaNO_3$ electrolyte.

4. A silver electrode in a $1.00 \times 10^{-3}\ M$ solution of $AgNO_3$ had a potential E of $+0.380$ V as measured against the SCE (the SCE was isolated from the Ag^+ solution with a salt bridge filled with KNO_3 solution to prevent precipitation of AgCl). The Ag electrode was then covered with solid AgCl so that it could function as a chloride electrode of the second kind, with the following half-reaction:

$$AgCl\ (s) + e^- \rightleftharpoons Ag + Cl^-$$

The value of K_{sp} for AgCl is 1.82×10^{-10}. Calculate E_a for the Nernstian equation

$$E = E_a - 0.0591\ \log[Cl^-]$$

5. The value of E of the SCE is $+0.242$ V measured against the SHE. The following are also given:

$$Hg^{2+} + 2e^- \rightleftharpoons Hg(l)$$

$$E^0 = +0.854\ V\ vs.\ SHE$$

$$K = \frac{[HgY^{2-}]}{[Hg^{2+}][Y^{4-}]} = 6.3 \times 10^{21} \cdot$$

In Section 13.2 a second-order electrode was described consisting of mercury metal in contact with a solution containing a low concentration of HgY^{2-}. Consider such an electrode system in which $[HgY^{2-}] = 1.00 \times 10^{-5}\ M$ and in which the SCE is the reference electrode. Calculate E_a for the equation

$$E = E_a - \frac{0.0591}{2} \log\ [Y^{4-}]$$

6. The potential of a glass electrode versus a reference electrode was -0.0137 V in a pH 4.004 potassium hydrogen phthalate (KHP) buffer and -0.3038 V in a pH 9.183 borax buffer. What is the slope, $\Delta E / \Delta pH$, of the Nernstian expression for this electrode?

7. During an "acid fog" incident in a polluted urban area, fog droplets were collected amounting to an estimated 639 μL of liquid. This liquid was diluted quantitatively to 2.00 mL with CO_2-free water. The pH of this water measured 3.271. Estimate the concentration of strong acid in the acid fog in milliequivalents per milliliter.

8. This is a segment of a calibration curve for a calcium ion–selective electrode at 25° C. From the information given, estimate the value of $[Ca^{2+}]$ in a medium in which $E = -327.1$ mV.

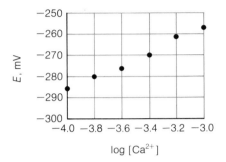

log $[Ca^{2+}]$

9. The selectivity constant $K_{Cl^-, SO_4^{2-}}$ for the interference of sulfate at a chloride-selective liquid membrane electrode is 0.14. The chloride electrode read −321.6 mV in a standard solution of 8.25×10^{-4} M Cl^- and −337.2 mV in a solution containing an unknown concentration of chloride $[Cl^-]$ and 9.00×10^{-4} M SO_4^{2-}. Calculate $[Cl^-]$ corrected for SO_4^{2-} interference.

10. What is the percent relative concentration error in calcium ion concentration resulting from a 0.4-mV error in E at a calcium ion–selective electrode?

11. What error in E corresponds to a 10.0% relative concentration error in the direct potentiometric measurement of perchlorate ion?

12. A nitrate ion-selective electrode reads −336.7 mV in 150 mL of unknown solution. Addition of 5.00 mL of 2.25×10^{-2} M standard nitrate shifts the potential to −345.1 mV. What was the potential of the unknown NO_3^- solution?

14 Voltammetry and Amperometric Titrations

LEARNING GOALS

1. The meaning of voltammetry and what a voltammogram signifies.
2. The three major types of electrodes that may be present in an electrolysis cell—that is, working electrode, reference electrode, and auxiliary electrode.
3. General characteristics of current-voltage curves and what is meant by voltammogram.
4. Polarography with a dropping-mercury electrode and the characteristics of a polarogram.
5. The mathematical expression for a polarogram in terms of current, i, and diffusion current, i_d; the significance of half-wave potential.
6. Analytical applications of polarography.
7. Principles and advantages of differential pulsed polarography.
8. The nature, advantages, and applications of stripping voltammetry at the hanging-mercury-drop electrode.
9. Principles and applications of amperometric titrations with a single polarized electrode.
10. The uses of dual polarized microelectrodes for amperometric titrations.
11. Constant-current potentiometric or bipotentiometric titration and its application to the Karl Fischer determination of water.

14.1 Current-Voltage Measurements at Microelectrodes

Microelectrodes in the context of this chapter are small electrodes with surface areas typically in the range of 1 to 10 mm^2 at which electrochemical reductions or oxidations occur. For analytical applications, the **electroactive species** that reacts at a microelectrode normally is at a low concentration (less than about 1×10^{-4} M). This low concentration is in part responsible for the fact that the current from the reaction at the microelectrode is quite small, on the order of microamperes (μA). Of course, the small size of the electrode likewise causes the current to be low. Furthermore, since a microelectrode is small relative to the volume of the solution in which it is immersed, the quantity of electroactive species oxidized or reduced at the microelectrode is negligible compared to the total quantity of the species in solution; therefore, during the course of an analysis, the analyte concentration does not change appreciably.

The reaction of an electroactive species that occurs at a microelectrode depends upon the potential applied to the electrode as measured against a reference electrode. In some cases a specific potential is applied between two microelectrodes. For other types of analysis, a constant current is caused to flow between two microelectrodes in an analyte solution and the potential difference required to maintain current flow is measured. As explained in this chapter, voltammetry and amperometric titrations are based upon **current-voltage relationships** at microelectrodes.

14.2   Voltammetry

Definition of Voltammetry

Voltammetry consists of the measurement of a current at a microelectrode (one with a small surface) as a function of the potential (versus a reference electrode) of the microelectrode. The basic components of the apparatus used for voltammetry are shown in Figure 14.1. The measurement is normally carried out in an *electrolysis cell* (Section 11.4) containing the analyte solution. A potential E is applied to a microelectrode called the **working electrode**. This potential is regulated very accurately and precisely versus a **reference electrode** by the electronic apparatus of the system. A current i flows between the working electrode and an **auxiliary electrode**; this current is normally of the order of microamperes and is measured by the appropriate electronic circuitry in the apparatus.

Current-Voltage Curve at a Microelectrode

The output from a voltammetric instrument is some form of **current-voltage curve**, a simple version of which is shown in Figure 14.2. Conventionally, E is plotted with more negative values to the right, and a cathodic current corresponding to reduction at the working electrode is plotted with increasing values upward. The general name for such a current-voltage curve is a **voltammogram**. The voltammogram illustrated in Figure 14.2 shows the current-voltage curve for the reduction of a metal ion in solution. The metal ion concentration is typically of the order of $1 \times 10^{-4}\ M$ and the solution contains $0.1\ M\ KNO_3$ or some other salt that does not react with the analyte to serve as a *supporting electrolyte* that makes the solution conducting (essentially due to the supporting electrolyte rather than the analyte). The microelectrode typically consists of a 1-mm-long piece of platinum wire extending from the end of a rotating piece of glass tubing.

The voltammogram shown in Figure 14.2 results from the reduction of the metal ion as a result of a potential applied to the working electrode.

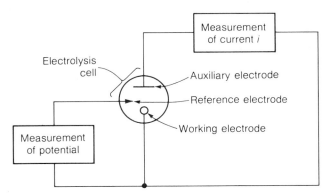

FIGURE 14.1 Generalized schematic diagram of an electrolysis cell for voltammetry

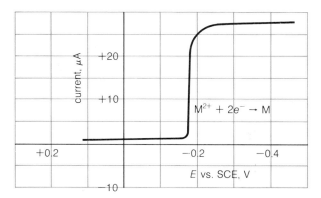

FIGURE 14.2 Voltammogram at a rotating platinum electrode for the reduction of a metal M^{2+} onto the platinum surface

Toward the left of the plot, only a small *residual current* flows because of ionic conductance and reactions of impurities in the solution. As an increasingly negative potential is applied (scanning to the right), a potential is reached at which reduction of the metal ion begins to occur, as shown for the following reaction of a divalent metal ion:

$$M^{2+} + 2e^- \rightarrow M(s) \tag{14.1}$$

The current i resulting from this reaction flows between the working and auxiliary electrodes and is measured in the circuitry connecting these electrodes (Figure 14.1). The current approaches a limiting, or plateau, value because of the limited rate at which M^{2+} ions can migrate to the surface of the working electrode to replace ions that have been reduced there. The limiting current increases with increasing concentration of reducible species, because more ions are available to migrate to the working electrode surface at higher concentrations. The potential at which the reduction occurs is a function of the identity of the reducible species; one that is readily reduced is reduced at a more positive potential than one that is more difficult to reduce. The reduction of a metal ion to the solid metal occurs at more positive potentials at higher concentrations.

Forms of Voltammetry

Voltammetry has many forms and variations. The most well known of these is polarography, which is discussed in the following section. Oxidation as well as reduction can be observed at a working electrode. Analyte can be reduced

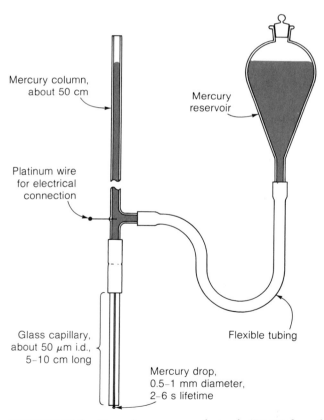

FIGURE 14.3   Dropping-mercury electrode (DME) for polarography

onto the working electrode over a period of time and then oxidized by a reversal of potential to give a relatively large anodic current for a brief period of time; this is the basis of anodic stripping voltammetry discussed in Section 14.6. Variations in the manner in which the potential is applied to the working electrode are employed for various purposes.

The development of relatively sophisticated and inexpensive electronic apparatus after World War II, and particularly since about 1960, has resulted in the development of a number of new and interesting voltammetric techniques. These include AC polarography, differential pulsed polarography, fast linear sweep voltammetry, direct anodic stripping voltammetry, and differential pulsed anodic stripping voltammetry.

14.3   Polarography

What Is Polarography?

Polarography is that branch of voltammetry in which the working electrode consists of drops of mercury issuing from the bottom of a piece of capillary glass tubing as shown in Figure 14.3. At first seeming like an unlikely sounding approach, polarography with the **dropping-mercury electrode (DME)** was the first of the practical voltammetric techniques. Polarography was developed in 1922 by the Czechoslovakian chemist Jaroslav Heyrovsky; he received the Nobel prize in chemistry in 1959 for its development.

Figure 14.4 shows a polarographic cell and associated hardware and instrumentation. A column, or "head," of mercury applies pressure to the mercury in the capillary, causing the liquid metal to flow through the capillary tube and fall in drops from its end. Provision is made for deaerating the analyte solution and the space above it with N_2 to prevent interference from the reduction of dissolved O_2. Spent mercury is collected at the bottom of the electrolysis cell.

The Polarogram

The mercury drop constituting the active surface in the DME falls off at several-second intervals, either naturally or by the action of a "drop-knocker" that taps the capillary tube at a set interval to detach the drop. Because of the regular formation and detachment of mercury drops, the current-voltage curve is discontinuous and has a sawtooth shape as shown in Figure 14.5. This characteristic voltammogram taken at the DME is called a **polarogram**.

To understand a polarogram, consider the solutions from which the polarograms shown in Figure 14.5 were obtained. The first of these, yielding polarogram A, consists of 0.1 M KCl in oxygen-free water. The KCl is the supporting electrolyte added to make the solution conducting and to prevent

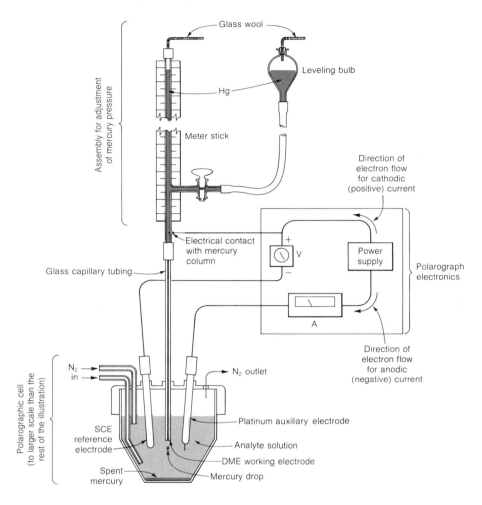

FIGURE 14.4   Above and left: Apparatus for polarography

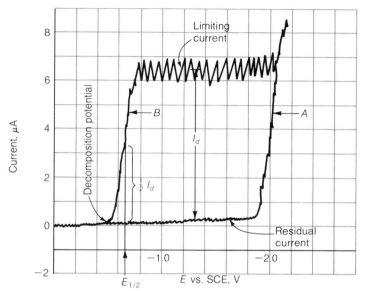

FIGURE 14.5 Polarogram consisting of a plot of current i (μA) versus potential E (V) of a dropping-mercury electrode versus an SCE. Curve A is background current for $0.1\ M$ KCl supporting electrolyte; curve B is for $5 \times 10^{-4}\ M$ CdCl$_2$ in $0.1\ M$ KCl.

electrostatic forces from acting on the electroactive species in solution, thus ensuring that only diffusion processes are involved in migration of the electroactive species to the working electrode. In a solution of supporting electrolyte only, a very small **residual current** is observed arising from traces of impurities in the solution and from a small current flow that occurs as each mercury drop acquires an electrical charge. The latter is known as a **charging current** or **condenser current**. The sharp increase in residual current at about -2.0 V results from the reduction of K$^+$ ion in the supporting electrolyte.

Polarogram B in Figure 14.5 is a classical **S**-shaped current-voltage curve obtained with the more analytically useful polarographic reactions. In this case the electrode reaction is the two-electron reduction of Cd^{2+} ion to cadmium metal dissolved in the mercury drop (cadmium amalgam) as follows:

$$\text{Cd}^{2+} + 2e^- \xrightarrow{\text{Hg}} \text{Cd(Hg)}(l) \tag{14.2}$$

This reaction begins to occur as the potential is made sufficiently negative to reach a **decomposition potential** at which i becomes perceptibly higher than the residual current. This difference is called the **limiting current**. The current is limited by the rate at which electroactive Cd^{2+} ions are transported to the electrode surface. Transport of electroactive ions to the electrode surface can occur by several mechanisms, of which it is desired to eliminate all but diffusion resulting from the random motion of electroactive species in solution. Transport of the Cd^{2+} ion to the electrode surface could occur by electrostatic attraction of the positive Cd^{2+} ion to the DME with its relatively negative charge. This is essentially eliminated by use of a large excess of supporting electrolyte, KCl in this case, the ions of which essentially mask the electrostatic attraction forces. The solution is maintained in an unstirred state and at a uniform temperature to prevent mechanical or thermal convection transport of the electroactive species to the DME. When diffusion is the only major means by which electroactive ions move to the electrode surface, the limiting current is called the **diffusion current**, i_d.

The voltage on the polarogram where i is equal to $\frac{1}{2}i_d$ is especially significant, being characteristic of the electroactive species and the medium in which it is dissolved. This voltage is called the **half-wave potential** and is given the symbol $E_{1/2}$. In cases where a metal ion is reduced to a mercury-soluble metal to form an amalgam, $E_{1/2}$ is essentially independent of the metal ion concentration. As noted in the discussion of voltammetry in Section 14.2, this is not the case when a metal ion is reduced to solid metal.

Theoretical Treatment of Polarographic Waves

The classic **S**-shaped polarographic wave can be explained by potentiometric principles and diffusion phenomena. Such a wave can result from reduction or oxidation of an electroactive species, or even from both phenomena when both the oxidized and reduced forms of the electroactive species are present. Cadmium ion, Cd^{2+}, represents a classic well-behaved polarographic species. Its reduction will be used here to illustrate the interpretation of polarographic waves.

Cadmium undergoes **reversible** oxidation-reduction processes at a DME. As shown by the double arrow in the half-reaction

$$Cd^{2+} + 2e^- \overset{Hg}{\rightleftharpoons} Cd(Hg) \qquad E_A^0 \qquad (14.3)$$

the cadmium ion can be reduced to a solution of cadmium metal in liquid mercury at the DME (a cadmium amalgam) and the cadmium in the amalgam can be oxidized to Cd^{2+} ion in solution. In order for the oxidation-reduction processes to be reversible, electron transfer must occur very rapidly at the electrode surface. Under reversible reaction conditions, the following Nernstian relationship applies at 25°C

$$E = E_A^0 - \frac{0.0591}{2} \log \frac{[Cd(Hg)]_0}{[Cd^{2+}]_0} \qquad (14.4)$$

where E is the potential (versus the reference electrode employed) of the DME, E_A^0 is the standard electrode potential (versus the aforementioned reference electrode) of the half-reaction between the metal ion in solution and *the metal in the amalgam*, $[Cd^{2+}]_0$ approximates the activity of Cd^{2+} ion *at the*

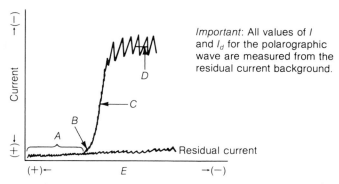

FIGURE 14.6 Polarographic wave for the reduction of Cd^{2+} at the DME

---

electrode surface, and $[Cd(Hg)]_0$ approximates the activity of cadmium metal *dissolved in mercury at the electrode surface*. Since it applies to the case where the cadmium(0) in the half-reaction is dissolved in mercury, E_A^0 *is not the same as* E^0 for the following reaction where cadmium(0) is solid cadmium metal:

$$Cd^{2+} + 2e^- \rightleftarrows Cd(s) \qquad E^0 \tag{14.5}$$

In the case of cadmium, E_A^0 is about 0.05 V more positive than E^0, reflecting greater ease of reduction of cadmium ion to a solution of Cd(0) in an amalgam than to reduction to solid cadmium metal.

The shape of the polarographic wave for the reduction of Cd^{2+} at a DME can be understood by expressing Equation 14.4 in terms of i and i_d. To understand how this is done, consider again the polarographic wave for the reduction of Cd^{2+} as shown in Figure 14.6. In region A, the potential of the DME is too positive for any significant reduction of Cd^{2+} to occur so that

$$[Cd(Hg)]_0 \cong 0 \tag{14.6}$$

and

$$[Cd^{2+}]_0 = [Cd^{2+}] \tag{14.7}$$

where $[Cd^{2+}]$ is simply the concentration of cadmium ion in the bulk of the solution. Since at this point, $[Cd^{2+}]_0$ is much higher than $[Cd(Hg)]_0$, the Nernstian relationship expressed in Equation 14.4 reflects the fact that E is too positive for significant reduction of Cd^{2+} to occur. At point B the value of E has been shifted sufficiently negative for a perceptible current to flow because of the reaction

$$Cd^{2+} + 2e^- \xrightarrow{Hg} Cd(Hg) \tag{14.8}$$

This reaction slightly dissipates the concentration of Cd^{2+} in aqueous solution immediately adjacent to the electrode surface so that

$$[Cd^{2+}]_0 < [Cd^{2+}] \tag{14.9}$$

and a relatively low concentration of cadmium metal is present as an amalgam in the mercury *at the electrode surface*.

$$[Cd(Hg)]_0 < [Cd^{2+}]_0 \qquad (14.10)$$

Diffusion Rates and Diffusion-Limited Current

If the potential were held at the value represented by point B, the current would remain constant *because i is a function of the rate of diffusion of Cd^{2+} ions from the bulk of the solution to the electrode surface*. Mathematically this relationship is expressed as

$$i = k_{Cd^{2+}}([Cd^{2+}] - [Cd^{2+}]_0) \qquad (14.11)$$

where $k_{Cd^{2+}}$ is a proportionality constant related to the rate of Cd^{2+} ion diffusion in water. It is seen that when $[Cd^{2+}]_0 = [Cd^{2+}]$, the current i is equal to 0, as in region A of Figure 14.6. This is because only residual current flows. Furthermore, i reaches a maximum value proportional to $[Cd^{2+}]$ when $[Cd^{2+}]_0 = 0$. This is the case at point D in the polarogram. Under these circumstances, Cd^{2+} ion is diffusing to the electrode at its maximum rate, all of these ions reaching the electrode are being reduced, and the magnitude of the current is as high as it can get based upon the reduction of Cd^{2+} alone. Therefore, on the upper plateau portion of the polarogram

$$i = i_d \qquad (14.12)$$

and it follows from Equation 14.11 that

$$i_d = k_{Cd^{2+}}[Cd^{2+}] \qquad (14.13)$$

which, substituted back into Equation 14.11, yields

$$[Cd^{2+}]_0 = \frac{i_d - i}{k_{Cd^{2+}}} \qquad (14.14)$$

The concentration of $Cd(Hg)$ at the electrode surface is proportional to i such that

$$[Cd(Hg)]_0 = \frac{i}{k_{Cd(Hg)}} \qquad (14.15)$$

where $k_{Cd(Hg)}$ is a proportionality constant relating to the rate of Cd atom diffusion in mercury. Substituting the two preceding equations into the Nernstian relationship

$$E = E_A^0 - \frac{0.0591}{2} \log \frac{[Cd(Hg)]_0}{[Cd^{2+}]_0} \qquad (14.4)$$

yields the following:

$$E = \underbrace{E_A^0 - \frac{0.0591}{2} \log \frac{k_{Cd^{2+}}}{k_{Cd(Hg)}}}_{E_{1/2}} - \frac{0.0591}{2} \log \frac{i}{i_d - i} \qquad (14.16)$$

Current as a Measure of Half-Wave Potential and Reversibility

The half-wave potential in terms of current is the potential value on the polarogram where i, measured from the residual current background, is equal to $\frac{1}{2}i_d$ (point C in Figure 14.6). At that point the third term in the preceding equation equals 0 and $E_{1/2}$ consists of the first two terms of the equation. This leads to the following identical versions of the Nernst equation for polarography:

$$E = E_{1/2} - \frac{0.0591}{2} \log \frac{i}{i_d - i} = E_{1/2} + \frac{0.0591}{2} \log \frac{i_d - i}{i} \qquad (14.17)$$

$$\underset{\text{Change of sign}}{\uparrow}$$

This equation applies to ideal polarographic behavior for a *reversible* process (see Equation 14.3). For a reversible *n*-electron reduction, a plot of $\log\left[(i_d - i)/i\right]$ versus E over the polarogram is a straight line with a slope of $+0.0591/n$ V at 25° C. Such a plot is used to demonstrate reversibility.

14.4   Chemical Analysis with Polarography

Qualitative and Quantitative Analysis

Polarographic analysis can be both qualitative and quantitative. The value of $E_{1/2}$ in a particular medium is indicative of the identity of the electroactive species (qualitative analysis) and i_d under controlled conditions provides a quantitative analysis. Figure 14.7 illustrates both qualitative and quantitative analysis of analytes by polarography. In the deaerated buffered complexing medium employed, the half-wave potentials for the reduction of Tl(I), Cd(II), and Zn(II) are well separated, so that these three ions could be readily distinguished in this solution. The values of i_d for the ions can be used with suitable calibration curves for the ions in the same medium to give values of the concentrations of ions present.

As with many instrumental analysis techniques, the various forms of voltammetry can be used for quantitative analysis by *standard addition*, a technique discussed for potentiometric measurement in Section 13.13. Standard addition is applicable when analyte concentration is a linear function of a parameter such as polarographic diffusion current, i_d. In this case the current is measured in a known volume of analyte, V, and again after the addition of a small volume, v, of standard solution containing a known concentration, C, of the analyte. After addition of the standard, the current due to the original analyte present is reduced slightly by the dilution factor, $V/V + v$, but increased even more by the added standard analyte, increased concentration $Cv/(V + v)$. These factors and values of the current before and after standard addition enable the calculation of the original analyte concentration.

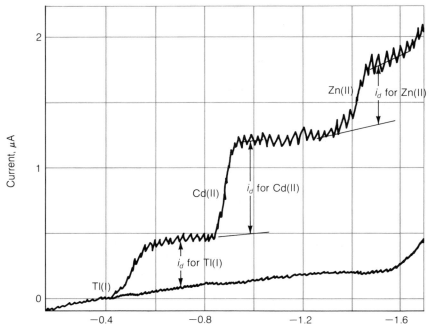

FIGURE 14.7   Polarographic reduction of a mixture of 1×10^{-4} M Tl(I), 2×10^{-4} M Cd(II), and 1×10^{-4} M Zn(II) in a supporting electrolyte 1 M in NH_3 and 1 M in NH_4Cl with 0.002% Triton X-100 maxima suppressant. The lower trace is the residual current in the supporting electrolyte and Triton X-100 alone.

Practical Considerations in Polarography

There are several important practical considerations in the technique of polarographic chemical analysis. As with many other analytical techniques, polarography is subject to interferences, which must be known and dealt with. Oxygen from the atmosphere interferes by virtue of a two-step reduction, the first to H_2O_2 at an $E_{1/2}$ of approximately -0.14 V and the second to H_2O at about -0.9 V as measured against the SCE. Both steps are irreversible, yielding drawn-out polarographic waves. Oxygen is readily eliminated by purging the system with nitrogen gas prior to analysis, and passing N_2 over the analyte solution during the scan (see Figure 14.4).

Polarographic maxima consisting of "humps" atop polarographic waves (see Figure 14.8) constitute another major interference with polarographic analysis. The origins of this phenomenon are not completely understood; fortunately, maxima can be eliminated by adding traces of some surface-active high-molecular-weight adsorptive substances. Gelatin and the

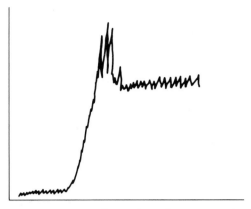

FIGURE 14.8   Illustration of current maxima on a polarographic wave

commercial surface-active agent Triton X-100 are effective maxima suppressants. Dyes such as methyl red and even carpenter's glue have been used to eliminate maxima!

Polarographic methods have been developed for the determination of a wide range of metal ions and inorganic species. These include metal ions that can be reduced to an amalgam. Also included are metal ions and inorganic species that can be reduced or oxidized at the DME to other species in solution.

Methods have been developed for the determination of organic compounds polarographically. These include peroxides; epoxides; carbonyl compounds; carboxylic acids; nitrogen compounds with nitro, nitroso, amine oxide, and azo groups; hydroquinones; mercaptans; alkenes; and organohalides.

Conventional polarography has been supplanted for the most part by other techniques of analysis, particularly spectrophotometric methods (see Chapters 20 through 22). However, more advanced voltammetric techniques are quite useful in some applications and are best understood with a knowledge of conventional polarography. These techniques are discussed in the following sections.

14.5   Differential Pulsed Polarography

Limitations of Conventional Polarography

So far the discussion has centered on classic DC (direct current) polarography with a linear voltage scan. That is, the voltage applied to the working electrode increases linearly with time as shown in Figure 14.9(a). For reversible

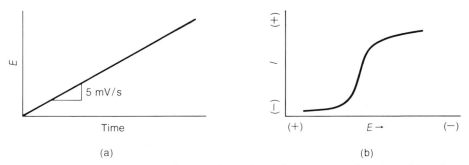

FIGURE 14.9   (a) Potential of the working electrode, DME, as a function of time for a classic DC polarographic scan; (b) S-shaped current-voltage curve from DC polarography

reduction or oxidation the result is an S-shaped current-voltage curve as shown in Figure 14.9(b). Because of the magnitude of residual current, DC polarography has a detection limit of about 1×10^{-5} M for most analytes. Such a value is not competitive with the lower detection limits of several other important analytical techniques. Furthermore, DC polarography requires that $E_{1/2}$ values differ by at least 0.2 V to resolve two different polarographic waves. Since the working range of the DME is limited to a little over 2 V, resolution of polarographic waves is a severe problem with DC polarography.

Principles of Differential Pulsed Polarography

Both of the problems just discussed are alleviated considerably by the application of **differential pulsed polarography (DPP)**, in which the voltage of the working electrode is increased in a linear fashion at about 5 mV/s with a DC pulse of an additional 20 to 100 mV applied for the last approximately 60 milliseconds (ms) of drop life as shown in Figure 14.10(a). This careful synchronization of voltage and drop lifetime is accomplished by causing the drop to detach at the end of the 60-ms voltage pulse with a mechanical drop-knocker.

The differential pulsed polarograph measures the difference in current Δi recorded just before the voltage pulse is applied and at the end of the pulse. The time interval between these two measurements is sufficient to allow for the decay of most of the capacitive current resulting from electrostatic charging of the drop by the pulsed increase in voltage. However, if there is a difference in faradaic current as a result of the oxidation or reduction of an electroactive solute between the two voltages constituting the voltage pulse (see Figure 14.11), it is recorded as a change in current Δi. Such a change in current occurs on the steeply rising portion of the conventional S-shaped

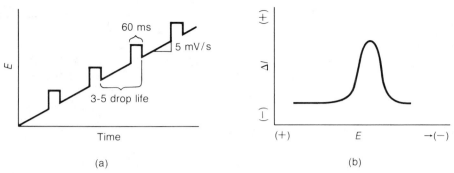

FIGURE 14.10 (a) Potential of the working electrode, DME, as a function of time for differential pulsed polarography; (b) peak-shaped plot of Δi versus E for differential pulsed polarography

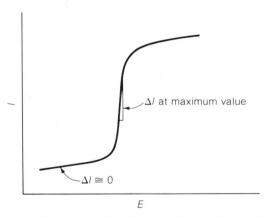

FIGURE 14.11 Differential pulsed polarography gives a signal at E values corresponding to the rising portions of a conventional DC polarogram and essentially no signal elsewhere (see Figure 14.10 b).

polarographic wave, but it is essentially zero on the level portions of the wave. The result is a peak-shaped signal as shown in Figure 14.10(b).

Advantages of Differential Pulsed Polarography

The most important advantage of DPP is that analytes can be measured at levels about three orders of magnitude lower than is the case with DC polarography. Therefore, the detection limit of DPP is of the order of $1 \times 10^{-8} \, M$, which is highly competitive with the better techniques of trace substances analysis in solution. In addition to the much more favorable detection limit, DPP is capable of resolving analytes with $E_{1/2}$ differences as small as 0.04 V. Furthermore, DPP partially alleviates problems from steeply sloping backgrounds, poorly defined polarographic waves, and peak maxima.

Differential pulse polarography has been made possible by advanced electronics developed during the last two decades. Good analytical instrumentation is now available for DPP, and relatively straightforward analyses have been developed for DPP during recent years.

14.6   Stripping Voltammetry

Principles and Advantages of Stripping Voltammetry

Stripping voltammetry is a technique in which the analyte is deposited on the surface of a mercury drop or solid electrode by means of an electrochemical reaction occurring at a selected applied potential over a period of several minutes. The analyte concentrated on the working electrode surface is then stripped off in a few seconds by the reverse of the deposition reaction caused by a rapid shift in potential. The basic principle is shown in Figure 14.12 for the determination of a metal ion M^{2+}. The conventional DC polarogram in the figure shows that the analyte metal ion is present at such a low concentration that the S-shaped polarographic curve is barely perceptible above the background. However, by using a **hanging-mercury-drop electrode** consisting of a single mercury drop employed throughout a single determination and maintained at a potential E more negative than $E_{1/2}$ for a relatively long period (5 minutes, typically), the metal can be deposited in a concentrated form on the surface layers of the drop. The electrodeposition reaction is

$$M^{2+} + 2e^- \xrightarrow{\text{Hg}} M(Hg) \tag{14.18}$$

During the deposition step the solution is stirred to increase the amount of analyte deposited; generally only a small fraction of the analyte species present is deposited on the working electrode. Control and reproducibility during this step are crucial for accurate results; all the samples and standards must be deposited on electrodes of the same size with the same stirring rate and deposition time. After the electrodeposition (preconcentration) step is completed, the potential of the working electrode is scanned to a more positive value at a relatively rapid rate of about 25 mV/s. At a potential near $E_{1/2}$, the reduced analyte begins to oxidize according to the half-reaction

$$M(Hg) \xrightarrow{\text{Hg}} M^{2+} + 2e^- \tag{14.19}$$

Since a relatively large amount of M(Hg) has been preconcentrated in the surface layers of the mercury drop, the oxidation of the metal in the amalgam results in a large anodic (negative i) current that reaches a peak value and decays back to a low background level as the M(Hg) in the mercury is exhausted. This is shown as the large inverted peak in Figure 14.12. The much higher sensitivity of anodic stripping voltammetry compared to classic DC

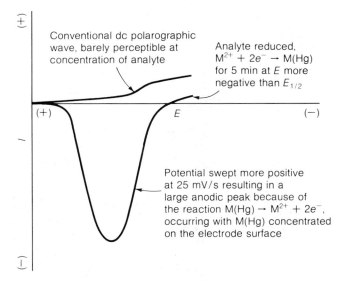

FIGURE 14.12   Representation of stripping voltammetry applied to the determination of a low concentration of metal ion M^{2+}

polarography is obvious. With the use of very long deposition times of up to 1 hour, analytes at levels as low as 1×10^{-9} M can be determined.

Apparatus for Stripping Voltammetry

An apparatus for stripping analysis is shown in Figure 14.13. One of the more crucial steps in using such an apparatus is the generation and suspension of mercury drops of a uniform size. Note that the drop is transferred to the end of a platinum wire to serve as a working electrode.

Solid electrodes have been used for anodic stripping voltammetry, with platinum the most frequent choice. Solid electrodes are easy to handle but suffer the disadvantage of not having a continually renewed surface, as is the case with each new mercury drop employed with a mercury electrode.

14.7   Amperometric Titration

Principles and Uses of Amperometric Titrations

Amperometric titration is an analytical procedure in which the titrant and/or analyte are electroactive species that react at a microelectrode to produce a current that can be plotted against volume of titrant for end point determination. The potential of the working electrode is maintained at a value versus a reference electrode such that the electroactive species is either reduced

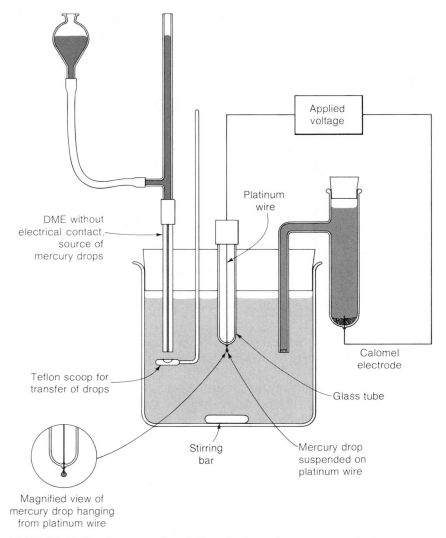

FIGURE 14.13   Apparatus for anodic stripping voltammetry employing a hanging-mercury-drop electrode

or oxidized. Under these conditions a concentration gradient exists for the electroactive species between the bulk of the solution and the electrode surface, as is the case, for example, for the polarographic reduction of Cd^{2+} to $Cd(Hg)$ at a dropping-mercury electrode, as discussed in Section 14.3. Such an electrode is said to be **polarized**. Normally the auxiliary electrode (Figure

14.1) is not polarized. However, as discussed in **Section 14.8, amperometric** titrations may be performed with two polarized electrodes.

Normally an amperometric titration curve consists of a plot of current i versus volume of titrant added. In procedures where the titrant is generated coulometrically by an electrochemical reaction (see Chapter 15), the current can be plotted versus coulombs of electricity. If the reagent is generated electrolytically at a constant rate, time can be plotted versus current at a microelectrode.

In amperometric titrations, current is a linear (not logarithmic) function of concentration. Furthermore, the end point is established by linear extrapolation from points taken well away from it so that reactions that are incomplete at the end point (those with low equilibrium constants) can be used.

Amperometric Titration Plots

The classic shapes of amperometric titration plots are illustrated in Figure 14.14. Figure 14.14(a) shows a titration plot in which the titrant is electroactive and the analyte is not. If the working electrode is held at a potential at which the titrant is either reduced or oxidized, appreciable current flows after the equivalence point as the titrant reaches excess levels. For a titration in which only the analyte is electroactive, there is a decrease in the current with added titrant, and i remains at a background level because of the flow of residual current beyond the end point. The resulting titration plot is shown in Figure 14.14(b). Figure 14.14(c) shows an amperometric plot for a titration in which both the titrant and analyte are electroactive species. Note that the plots are rounded at the end points; the rounding becomes more pronounced with lower equilibrium constants for the titration reactions. The end points are

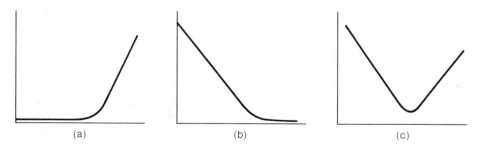

(a)	(b)	(c)

Volume of reagent or coulombs for coulometrically generated reagent

FIGURE 14.14   The most common configurations of amperometric titration curves: (a) Titrant is electroactive, analyte is not; (b) analyte is electroactive, titrant is not; (c) both analyte and titrant are electroactive.

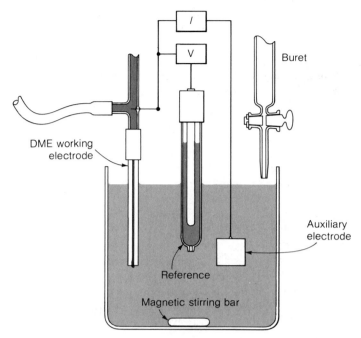

FIGURE 14.15   Amperometric titration apparatus with DME working electrode

determined by extrapolating the linear portions of the plots to their intersection points.

Advantages of Amperometric Titrations

Just as potentiometric titration is inherently more accurate and precise than direct potentiometric analysis, amperometric titration is more accurate and precise than voltammetric analysis. As compared to voltammetry, amperometric titration is less dependent upon the nature and concentration of the supporting electrolyte and upon the exact temperature employed. The major constraint on the temperature of an amperometric titration is that it remain essentially constant during the course of the titration. This can be accomplished without a temperature-regulating water bath if the titration is completed within a reasonable length of time. Unlike voltammetry, the analyte need not be electroactive in an amperometric titration as long as the titrant is electroactive.

Apparatus for Amperometric Titrations

Figure 14.15 shows the apparatus used for an amperometric titration. Precipitation titrations are most amenable to being followed amperometrically.

An example is the titration of lead ion with sulfate,

$$Pb^{2+} + SO_4^{2-} \rightarrow PbSO_4\,(s) \tag{14.20}$$

in which Pb^{2+} is reduced at a DME

$$Pb^{2+} + 2e^- \xrightarrow{\text{Hg}} Pb\,(Hg) \tag{14.21}$$

giving a titration curve similar to that in Figure 14.14(b). The solubility of $PbSO_4$ is not very low, so that an appreciable concentration of Pb^{2+} remains in solution when a stoichiometric amount of SO_4^{2-} titrant has been added. The result is that the titration curve is rounded at the end point. However, the end point is readily determined by extrapolating the linear portions of the plot to their intersection point.

If the concentration of titrant is much higher than that of the analyte, no correction need be made for dilution, because the added volume of titrant is negligible. If that is not the case, each current value can be corrected for dilution through multiplying by the factor $(V + v)/V$, where V is the original volume of the solution being titrated and v is the total volume of titrant added at each point.

14.8   Dual Polarized Microelectrodes for Amperometric Titrations

Amperometric Titrations with Two Microelectrodes

Amperometric titrations can be carried out in a system in which two microelectrodes immersed in a well-stirred solution are used to denote the end point. Usually the dual polarized microelectrodes in such a system consist of short platinum wires with a potential difference between them of 0.1 to 0.2 V. The way in which such a system provides an amperometric titration plot from which an end point can be determined is illustrated by the following example.

As an example of an amperometric titration, consider the oxidation of Fe^{2+} analyte with Ce^{4+} titrant for which the titration reaction is

$$Fe^{2+} + Ce^{4+} \rightarrow Fe^{3+} + Ce^{3+} \tag{14.22}$$

Initially, Fe^{2+} is the only electroactive species present at an appreciable concentration. It is present at a sufficient level to be oxidized at the anode, but there is no Fe^{3+} to be reduced at the cathode so no appreciable current flow is observed. Addition of the first increment of Ce^{4+} produces some Fe^{3+}, as shown by the titration reaction. This enables the two following reactions to occur:

$$\text{Anode:} \quad Fe^{2+} \rightarrow Fe^{3+} + e^- \tag{14.23}$$

$$\text{Cathode:} \quad Fe^{3+} + e^- \rightarrow Fe^{2+} \tag{14.24}$$

The cathodic reaction in which Fe^{3+} is reduced limits the current in the initial phases of the titration because of the very low concentration of Fe^{3+} available

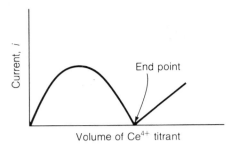

FIGURE 14.16   Amperometric titration of Fe^{2+} with Ce^{4+} employing twin polarized microelectrodes. This is the shape of the titration curve obtained when both the analyte and titrant undergo reversible voltammetric reactions

to diffuse to the cathode surface. This limits the total current flowing to the electrodes. A maximum current value occurs essentially halfway to the end point, where $[Fe^{3+}] \cong [Fe^{2+}]$. Here both the cathodic and anodic reactions limit the current equally, which results in a maximum current flow. Beyond the halfway point the current decreases with added titrant because of the decreasing concentration of Fe^{2+}, which limits the anodic reaction and thus the overall current. At the end point there is no Fe^{2+} to sustain the anodic reaction (14.23), and the voltage difference between the two microelectrodes is insufficient to bring about the oxidation of the Ce^{3+} product of the titration reaction (14.22). Beyond the end point both Ce^{3+} and excess Ce^{4+} are present. Both are reversibly electroactive species, so that at the dual polarized electrodes the following reactions occur:

$$\text{Anode:}\quad Ce^{3+} \rightarrow Ce^{4+} + e^- \tag{14.25}$$

$$\text{Cathode:}\quad Ce^{4+} + e^- \rightarrow Ce^{3+} \tag{14.26}$$

Just beyond the end point the low concentration of Ce^{4+} limits the current, which continues to increase with added titrant up to the point at which $[Ce^{4+}] \cong [Ce^{3+}]$. The resulting amperometric titration curve is shown in Figure 14.16.

Voltammetric Basis of Dual-Polarized-Electrode Amperometric Titrations

In the preceding example appreciable current flows when both the reduced and oxidized forms of a voltammetrically reversible **couple** (species that can be converted to each other by the reversible transfer of electrons) are present in appreciable quantities. Why this is so can be seen by examination of the voltammograms in Figure 14.17. When only the reduced or oxidized form is present (for example, the oxidized form, Fe^{3+}, in the preceding example), the

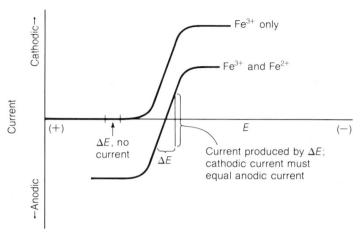

FIGURE 14.17 Voltammograms of Fe^{3+} and an $Fe^{3+}-Fe^{2+}$ mixture showing that a current is produced at twin polarized electrodes separated by a small potential difference ΔE in the latter case

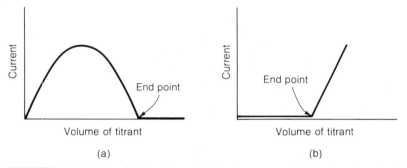

FIGURE 14.18 Dual microelectrode titration curves: (a) Analyte is reversibly electroactive, titrant is not; (b) titrant is reversibly electroactive, analyte is not.

voltage difference between the microelectrodes produces no current. However, when both the oxidized and reduced forms are present, the voltammogram has both anodic and cathodic components, so that a small potential difference between the microelectrodes results in a relatively large current.

Two other general shapes of titration curve are possible with twin polarized electrodes. These are when the analyte but not the titrant is reversibly electroactive [Figure 14.18(a)] and the opposite case [Figure 14.18(b)].

14.9   Bipotentiometric Titrations

Potential between Two Microelectrodes at Constant Current

The detection of a titration end point by two microelectrodes that have a constant potential maintained between them has just been discussed. It is possible to follow the course of some titrations with two microelectrodes between which a small current is forced; the potential difference required to maintain that current is measured. A titration followed by this means is a **constant-current potentiometric** or **bipotentiometric titration**.

The Karl Fischer Determination

The best known bipotentiometric titration is the **Karl Fischer titration** for the determination of water in nonaqueous liquids and some solids, such as crystals containing water of hydration. Given the importance of low levels of water in solvents, gasoline, and other industrial products, the Karl Fischer method is a widely employed determination. Most pH meters are equipped with two jacks in the back labeled "K-F" or "Karl Fischer," to which the leads of the bipotentiometric microelectrodes are connected. The pH meter has a special circuit for generating a constant current, normally of about 5 or 10 μA, and the potential necessary to support such a current is read on the millivolt scale of the pH meter.

The Karl Fischer determination is carried out in methanol or ethylene glycol monomethyl ether solvent.

Methanol   Ethylene glycol monomethyl ether

The classic Karl Fischer reagent* consists of I_2, pyridine, and SO_2 dissolved in one of the above solvents in a mole ratio of $1:10:3$. Water reacts with the reagent according to the following overall reaction:

Pyridine-   Sulfur   Pyridine
iodine   dioxide–pyridine
adduct   adduct

(14.27)

$$2 \, \text{(⬡)} NH^{+}I^{-} + \text{(⬡)} NH^{+ \, -}O_3SOCH_3$$

---

* Pyridine has an outrageous odor, so the classic Karl Fischer reagent is very unpleasant to work with. A propietary pyridine-free Karl Fischer reagent developed by the German scientist Eugen Scholz is now commercially available.

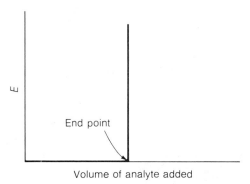

FIGURE 14.19 Plot of potential versus volume of analyte solution added for a Karl Fischer titration

Commonly analyte solution containing an unknown content of water is added from a buret into a measured volume of Karl Fischer reagent held in a container designed to exclude atmospheric water vapor and fitted with the two microelectrodes. Up to the end point the solution contains both I_2 and I^- as the pyridine adduct and pyridinium ion–iodide ion pair, respectively. Some of the latter is even present initially due to reaction of water impurity with the reagent. As long as both I_2 and I^- are present, only a very small potential difference is required between the two electrodes to maintain a current of just 5 or 10 μA because the following two half-reactions occur at the two electrodes:

$$\text{Cathode:}\quad I_2 + 2e^- \rightarrow 2I^- \tag{14.28}$$

$$\text{Anode:}\quad 2I^- \rightarrow I_2 + 2e^- \tag{14.29}$$

At the end point, however, the I_2 in the reagent is exhausted, there is no electroactive solute to readily react at the cathode, and the potential difference between the two microelectrodes must increase to a very high level to enable the reduction of solvent at the cathode by a reaction such as the following:

$$2CH_3OH + 2e^- \rightarrow 2CH_3O^- + H_2(g) \tag{14.30}$$

As a result, a Karl Fischer titration plot has the general appearance shown in Figure 14.19.

Other Forms of Bipotentiometric End Point

In some cases a bipotentiometric end point is manifested by an abrupt *decrease* in potential difference. This occurs, for example, in some titrations with I_2 (I_3^-) titrant, where all the I_2 is converted to I^- prior to the end point. At the end point excess I_2 is present and the current between the two microelectrodes can be sustained by a vastly reduced potential in the half-reactions described in Equations 14.28 and 14.29.

Programmed Summary of Chapter 14

The major terms and concepts introduced in this chapter are contained in this summary in a programmed format. To derive the most benefit from the summary, you should fill in the blanks for each question and then check the answers at the end of the book to see if your choices are correct.

The small electrodes used for voltammetry and amperometric titrations are given the general name of (1) _____. Chemical species that undergo oxidation or reduction reactions at a microelectrode are called (2) _____. The physical properties that can be measured at a microelectrode from which the analyte identity and/or quantity can be deduced are the (3) _____ relationships. Voltammetry consists of the measurement of a (4) _____ at a microelectrode as a function of a (5) _____ applied to the microelectrode. In an electrolysis cell the current flows between the (6) _____ electrode and the (7) _____, whereas a voltage is maintained or measured between the (8) _____ electrode and the (9) _____ electrode. A plot of current at a microelectrode as a function of applied potential is a current-voltage curve, more specifically a (10) _____, and, if a DME is used, the plot is called a (11) _____. A supporting electrolyte enables the solution to (12) _____ and eliminates transport of analyte ion to the working electrode because of (13) _____. In a polarogram the current that flows in the absence of added analyte species is called a (14) _____. The upper "plateau value" of current in a polarogram is called the (15) _____ and its magnitude is constrained by (16) _____. In terms of current, the half-wave potential in a polarogram or voltammogram occurs where (17) _____. In terms of reaction rate and direction, the Nernst equation applies in polarography when the reaction is (18) _____. For qualitative and quantitative analysis with polarography, (19) _____ is an indication of the identity of the electroactive species and (20) _____ is a measure of the (21) _____. A polarographic interference manifested by an irregular "hump" on the polarogram is an example of (22) _____. Near the end of the life of each mercury drop in differential pulsed polarography a (23) _____ is applied, yielding a plot of Δi versus E that is in the shape of a (24) _____ and extending the detection limit to lower concentrations by about (25) _____ orders of magnitude in comparison to conventional DC polarography. The basic principle of stripping voltammetry is the electrodeposition of analyte over a time period of at least (26) _____, followed by (27) _____ from the electrode by a reverse scan of potential over a few seconds' time. With use of stripping voltammetry using deposition times of up to 1 hour, analytes at concentrations as low as (28) _____ can be determined. A titration procedure in which the titrant and/or analyte are electroactive species that react at a microelectrode to produce a current is called (29) _____. Either a constant (30) _____ or a constant (31) _____ can be applied between dual polarized microelectrodes used for amperometric titration. The most common constant-current potentiometric titration with dual polarized electrodes is the (32) _____ determination of (33) _____.

Questions

1. Match the following terms and descriptions applied to voltammetry:

   a. Working electrode
   b. Reference electrode
   c. Auxiliary electrode
   d. Polarogram
   1. Potential stays constant
   2. Relates current and potential
   3. Normally has a small surface area
   4. In circuit for current measurement

2. Which form of voltammetry was first widely developed?

3. Why does a polarogram at the DME have a "sawtooth" appearance?

4. Is the diffusion current i_d measured from zero current; if not, from which lower value is it measured?

5. This is a measurement of i_d at a DME for a solution containing both the oxidized and reduced form of an analyte. Explain why the residual current can be ignored, or why it *should not* be ignored.

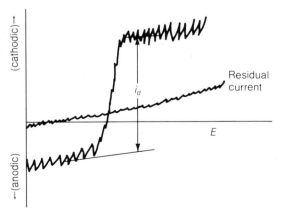

6. What is the distinction between E^0 and E_A^0 for the voltammetric reduction of the same metal ion?

7. At a potential sufficiently positive such that no faradaic current is flowing in scanning a polarogram for the reduction of Cd^{2+}, what is the relationship between $[Cd^{2+}]$ and $[Cd^{2+}]_0$?

8. In reference to the preceding question, what is the value of $[Cd^{2+}]_0$ when $i = i_d$?

9. In terms of currents, what is the definition of half-wave potential for a polarographic wave?

10. What measurable electrical parameters are useful for qualitative and quantitative analysis, respectively, in conventional DC polarography?

11. For what purpose may Triton X-100 or gelatin be employed in DC polarography?

12. What is the nature of the way in which potential is applied with respect to time in differential pulsed polarography?

13. What are the two major advantages of differential pulsed polarography over classic DC polarography?

14. Which voltammetric technique inherently involves a preconcentration step, and how is this accomplished?

15. What is the shape of the voltammetric signal measured in anodic stripping voltammetry?

16. What role might be played by voltammetry at a DME or platinum working electrode in determining the feasibility of an amperometric titration with a single working electrode?

17. What analogies can be drawn between the relationship between potentiometry and

potentiometric titration and that between voltammetry and amperometric titration.

18. How might a voltammogram be used to determine the feasibility of a titration using two polarized working electrodes?

19. For an amperometric titration with dual polarized electrodes in which only the titrant species is voltammetrically reversible, should the current continue to rise indefinitely beyond the equivalence point? If not, why not?

20. What is the principle behind the constant-current or bipotentiometric titration with two microelectrodes?

21. After working Problem 1 of this chapter, describe the differences you would expect from the polarographic reduction of two different metal ions of equal concentration and essentially identical diffusion qualities, one of which has a +1 charge, the other +2.

Problems

1. Consider the reversible polarographic reduction of two metal ions, Me^+ and Mt^{2+}, to their respective amalgams, Me(Hg) and Mt(Hg). Suppose that both have the same $E_{1/2}$ values and are at concentrations such that they have identical values of i_d. From calculations of E as a function of i, plot the polarographic reduction wave for each and explain any differences observed.

2. The metal ion M^{2+} yields a reversible polarographic wave for reduction to M(Hg). On the polarogram, at $E = -0.399$ V, $i = 3$ μA, and at $E = -0.434$ V, $i = 10$ μA. Calculate $E_{1/2}$ and i_d.

3. Suppose that two divalent ions (+2 charge) are both reduced reversibly at the DME, one at $E_{1/2} = -0.300$ V versus the SCE, the other at -0.400 V. Note that these two values are not separated sufficiently to give two distinct waves. If either ion were present separately, i_d for each would be 10 μA; i is additive at potentials where both ions are being reduced. Assuming a residual current of 0, construct a composite polarogram for the reduction of the two ions.

4. This polarogram was run on the same instrument and under exactly the same conditions as described in Figure 14.7. It

is for the reduction of Cd^{2+} to Cd(Hg). Estimate $[Cd^{2+}]$.

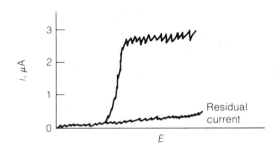

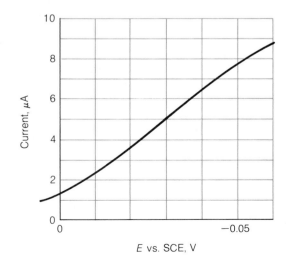

5. The polarographic wave on p. 424 is for the two-electron reduction of a metal ion M^{2+} at the DME at 25° C. Demonstrate whether or not the reduction is reversible and explain. Consider the residual current to be at $i = 0$ and that $i_d = 10.0\ \mu A$.

6. This peak is the anodic stripping curve for a prereduced metal ion analyte, M^{2+}, at a hanging-mercury-drop electrode; the upper curve is a blank plot without any prereduction. Estimate the number of moles of M(Hg) deposited on the mercury during the predeposition step. Recall that the faraday, F, is 96,500 coulombs per equivalent.

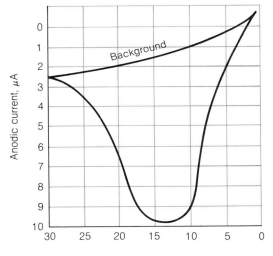

Time of scan to more positive potential, s

7. An amperometric titration using a single electrode gave the following data, where volume of titrant is given in milliliters and the corresponding current in microamperes in parentheses: 0 (0.02), 0.5 (0.02), 1.0 (0.02), 2 (0.02), 2.5 (0.02), 2.5 (0.04), 4 (0.11), 4.5 (0.2), 5.0 (0.37), 5.5 (0.60), 6.0 (0.86), 6.5 (1.18), 7.0 (1.50), 7.5 (1.88), 8.0 (2.31), 8.5 (2.79), 9.0 (3.26), 9.5 (3.72), 10.0 (4.2). Which is the elec-

troactive species? What is the end point volume?

8. A pharmaceutical compound was determined by differential pulsed polarography under specified analytical conditions. A plot of peak current in nanoamperes (nA) vs. concentration in $\mu g/mL$ yielded a straight line with a slope of 9.3 nA $\times mL \times \mu g^{-1}$. A sample solution of the compound gave a peak current of 107 nA. What was the concentration of the compound in the sample?

9. Food grade phosphoric acid was prepared by precipitating arsenic(III) sulfide with H_2S and removing the solid product by filtration with excess filter aid (diatomaceous earth). A 1.00-g sample of the product was extracted with NaOH solution to extract the arsenic(III) from the filter aid. This solution was diluted 1-to-100 and a 2.00 mL aliquot of the diluted solution was analyzed for arsenic(III) by differential pulsed polarography, giving an arsenic(III) peak current of 167 nA. Addition of 1.27 μg of As(III) with a micropipet from a relatively concentrated As(III) standard solution (negligible change in sample volume) yielded a peak current of 283 nA. Assuming linearity of current response with As(III) concentration, what was the percentage of arsenic in the original sample?

10. A 25.00 mL sulfate solution was titrated amperometrically with a 0.005000 M standard solution of M^{2+} ion, forming the slightly soluble salt $MSO_4(s)$. The amperometric reaction used to follow the course of the titration was the reduction of M^{2+} ion to M(Hg) at a mercury microelectrode. The resulting titration curve looked like the one in Figure 14.14(a), with significant rounding at the equivalence point due to appreciable solubility of $MSO_4(s)$. At the equivalence point volume of 29.51 mL added titrant, the current was 2.1 μA, and at 50.00 mL added titrant the current was $71.6\,\mu A$. From this information, estimate K_{sp} of MSO_4; list any assumptions made in the calculation.

11. For the polarographic reduction reaction,

$$M^+ + e^- \xrightarrow{\text{Hg}} M(\text{Hg})$$

the value of E_A^0 is 0.0253 V more negative than the value of $E_{1/2}$ (see Section 14.3). What is the ratio of k_{M^+} to $k_{M(\text{Hg})}$ (see Equation 14.16).

12. An oxidation–reduction titration was followed amperometrically with a single polarized electrode. The following are data in which the volume of titrant (mL) for each point is given, after which is the corresponding amperometric current (μA) in parentheses: 0 (0), 2 (6.3), 4 (10.9), 6 (14.0), 8 (15.7), 10 (16.6), 12 (16.5), 14 (15.7), 16 (14.1), 18 (10.7), 19 (7.5), 20 (4.0), 21 (1.2), 22 (1.1), 23 (2.0), 24 (4.7), 26 (8.4), 28 (12.0), 30 (15.5). From the data give the shape of the titration curve, explain the voltammetric behavior of the titration reactants and products, and give the end point volume.

15 Electrogravimetric and Coulometric Determinations

LEARNING GOALS

1. Concept of the electron or total quantity of electricity as an analytical reagent.
2. Functions of the components of an electrolysis cell used for electrolytic analysis.
3. How electrode potentials are determined for optimum electrolytic analysis.
4. Fundamentals of the analytical technique of electrogravimetry.
5. What is meant by a good adherent electrogravimetric deposit and the roles of current density and cathode depolarizers in forming such a deposit.
6. Typical electrogravimetric methods for the determination of metals.
7. Controlled electrode potential for electrogravimetric analysis.
8. Principles of constant-potential coulometry.
9. Constant-current coulometry and what is meant by a generator electrode and mediator species with this technique.
10. How coulometric techniques are used for neutralization, precipitation, and oxidation-reduction titrations.
11. Comparison of coulometric and volumetric titrations.

15.1 The Electron as a Chemical Reagent

Electrogravimetry

The electron can be used as a chemical reagent either for reduction of an analyte by addition of electrons at a cathode or oxidation of an analyte by removal of electrons at an anode. If the analyte or a well-defined product of it is deposited in a pure form on the electrode as a result of the **electrolysis reaction**, the weight of the substance deposited on the previously weighed electrode can be measured; this constitutes an **electrogravimetric** determination. In electrogravimetry the quantity of electricity required to form the deposit is not measured, and a knowledge of this quantity is not needed. An electrogravimetric determination of a metal ion, such as copper(II) ion, deposited on a platinum electrode by the half-reaction

$$\mathrm{Cu(II)} + 2e^- \xrightarrow{\mathrm{Pt}} \mathrm{Cu}(s) \qquad (15.1)$$

is conceptually the most fundamental and simple method of analysis known because it involves simply isolation and weighing of the analyte. Like its industrial cousin, electroplating, the practice of electrogravimetry is often much more complicated than might be assumed. For example, note that in the above half-reaction, the copper ion in solution was designated as Cu(II), rather than Cu^{2+}. This is because formation of a weighable metal deposit by reduction may require electroposition from a medium in which the metal ion is in some form otner than the hydrated metal ion. Often media containing several reagents made up to exacting concentrations are required for electrodeposition.

Coulometry

Coulometry is an electrolytic method of chemical analysis in which the quantity of electricity required to carry out an electrolysis reaction is measured. It is based upon the fact that 1 mol of electrons corresponds to 96,485 coulombs. A **coulomb**, abbreviated C, is the fundamental unit of electrical charge and is the quantity of electricity transferred by the flow of 1 ampere A of electrical current i for 1 second. A total of 96,485 C, usually rounded off to 96,500 C, is called the *Faraday*. There are two forms of coulometry. These are **constant-potential coulometry**, in which an electrolysis reaction occurs at an electrode set at a constant potential versus a reference electrode, and **constant–current coulometry**, or **coulometric titration**, in which a reagent is generated electrolytically at a constant rate. As with a volumetric titration, a coulometric titration can be interrupted (by stopping the current flow) near the end point to allow the system to come to equilibrium.

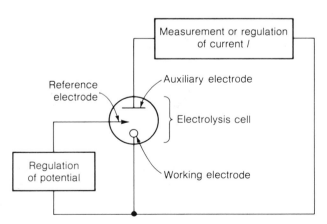

FIGURE 15.1   Generalized electrolysis apparatus for electrogravimetric or coulometric analysis

Figure 15.1 is a generalized schematic of the apparatus employed for the electrolytic methods just described. In all cases the analytical reaction, such as electrolytic deposition of copper metal, occurs at the *working electrode*. A *reference electrode* and associated circuitry for regulation of the potential of the working electrode is required for constant-potential coulometry and some forms of electrogravimetric analysis. For constant-current coulometry the current i is regulated at a constant value predetermined by the sizes of the auxiliary and working electrodes and the nature of the electrolysis reaction. For constant-potential coulometry, i is measured electronically as a function of time to give the total quantity of electricity.

Electrogravimetric and coulometric methods depend upon the measurement of the fundamental quantities, mass and electrical charge, respectively; they thus require no calibration against standards. In addition, where applicable, these methods tend to be among the more precise and accurate analytical determinations. However, they are not suitable for the determination of as low levels of analytes as are possible with voltammetric methods.

15.2   **Establishment of Potential for Electrolytic Analysis**

Feasibility of Electrolytic Analysis

A crucial factor in determining the feasibility of a constant-potential coulometric analysis or an electrogravimetric procedure with regulated potential is the establishment of the optimum potential of the working electrode versus a reference electrode, and the maintenance of that potential during the determination. In a constant-potential coulometric procedure the potential must be such that only the analyte is reduced or oxidized (100% efficient current use) and reduction or oxidation reactions of possible interfering species do not occur to a significant degree. For electrogravimetry, the potential constraint is not so severe; reactions of species other than the analyte may occur, but they must not contaminate or otherwise disturb the desired deposit on the working electrode.

Application of the Nernst Equation

Conceptually, the easiest way to visualize the regulation of potential in an electrolytic determination is in the separation of two metals in solution by the electrolytic reduction of the more readily reduced metal ion. Consider the separation of Pb from Cd in a solution that is 5.00×10^{-3} M in each of the metal ions. The pertinent half-reactions and E^0 values are given below:

$$Pb^{2+} + 2e^- \rightleftharpoons Pb(s) \qquad E^0 = -0.368 \text{ V vs. SCE} \qquad (15.2)$$

$$Cd^{2+} + 2e^- \rightleftharpoons Cd(s) \qquad E^0 = -0.645 \text{ V vs. SCE} \qquad (15.3)$$

Lead ion, Pb^{2+}, is the more readily reduced of the two metal ions, and is reduced at a more positive potential than the Cd^{2+} ion. What percentage of Pb^{2+} would remain in solution at a potential such that Cd just begins to deposit as the metal? The value of the potential of the working electrode versus an SCE reference electrode at the point where Cd would begin to deposit is given by

$$E = -0.645 + \frac{0.0591}{2} \log [Cd^{2+}]$$

$$= -0.645 + \frac{0.0591}{2} \log 5.00 \times 10^{-3} = -0.713 \text{ V} \tag{15.4}$$

The value of $[Pb^{2+}]$ at this potential is

$$E = -0.368 \text{ V} + \frac{0.0591}{2} \log [Pb^{2+}] = -0.713 \text{ V} \tag{15.5}$$

$$[Pb^{2+}] = \text{antilog}\left(\frac{-0.713 + 0.368}{\frac{0.0591}{2}}\right) = \text{antilog} -11.676 = 2.11 \times 10^{-12} \text{ } M$$

$$\tag{15.6}$$

$$\text{Percent } Pb^{2+} \text{ remaining in solution} = \frac{2.11 \times 10^{-12}}{5.00 \times 10^{-3}} \times 100 = 4.22 \times 10^{-8}\% \tag{15.7}$$

Obviously, on the basis of this calculation, a complete separation of lead from cadmium is feasible by controlled-potential electrodeposition.

Calculations such as the preceding can be used to establish the differences in E^0 needed to separate metal ions by electrodeposition. These values range from a difference of about 0.24 V for the separation of two metal ions with a $+1$ charge to only about 0.04 V for metal ions with a $+3$ charge. It should be emphasized that several potentials develop in an electrolysis cell as a current flows between the working and auxiliary electrodes. In a simplified sense, these consist of a potential as a result of the reaction at the working electrode, another potential from the reaction at the auxiliary electrode, and one generated in forcing the current i across the resistance R of the solution in the cell; this potential is the so-called iR drop of the cell. The significant potential that must be regulated to control the electrodeposition reaction is that between the working electrode and the reference electrode.

Application of Voltammetry

The similarity of the apparatus for electrolysis shown in Figure 15.1 to that for voltammetry in Figure 14.1 suggests that voltammetry might be useful for establishing the feasibility of electrogravimetric or constant-potential

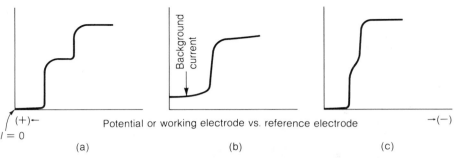

(+)←

$I = 0$

Potential or working electrode vs. reference electrode →(−)

Background current

(a)　　　　　　　　(b)　　　　　　　　(c)

FIGURE 15.2 Hypothetical voltammograms at the rotating platinum electrode that might be useful for establishing the feasibility of electrolytic determination of metal ions. (a) Two well-separated reduction waves indicate that the first metal ion reduced could be separated from the second; (b) a large, variable background current would interfere in the constant-potential coulometric determination of this metal ion; (c) two poorly separated voltammograms indicate that the two metal ions involved could not be separated by an electrolytic method.

coulometric analyses. Figure 15.2, illustrating voltammetry at a rotating platinum electrode, shows how this technique might indicate the feasibility of separating several hypothetical metal ions as explained in the figure legend.

15.3 Electrogravimetry

Quality of Electrogravimetric Deposits

Electrogravimetric procedures have been developed for the determination of several metals, including copper, cadmium, nickel, cobalt, silver, bismuth, and zinc. These are deposited from various media at the cathode as the metal. In addition, lead can be deposited as PbO_2 at the anode. Normally platinum gauze electrodes in a cylindrical configuration are employed as working and auxiliary electrodes (see Figure 15.3). The nature of the metal deposit formed is crucial to the success of electrogravimetric analysis of metals. The deposit should be strongly *adherent*—that is, it should stick firmly to the electrode. A fine-grained, smooth deposit with a metallic luster is indicative of a high-quality metal deposit that will withstand subsequent washing, drying, and weighing operations without loss or decomposition. A porous deposit or one that is powdery or flaky tends to be lost or oxidized during postdeposition handling.

Several factors influence the quality of electrogravimetric metal deposits. The stirring rate is one such factor as is the form of the metal ion in solution; normally the best-quality deposits are obtained from metals that are complexed with ligands such as NH_3 or CN^-. A low **current density**, defined as

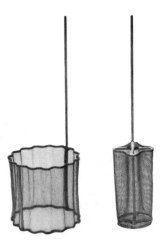

FIGURE 15.3   Platinum gauze electrodes used in electrogravimetry for the electrodeposition of analytes

current per unit electrode area (A/cm^2), is beneficial; values less than about 0.1 A/cm^2 usually give best results. Temperature influences the quality of a metal deposit, and optimum temperatures must be determined by experimentation.

The cathodic evolution of hydrogen

$$2H_2O + 2e^- \rightarrow H_2(g) + 2OH^- \tag{15.8}$$

as the metal is being deposited tends to give poor-quality deposits that do not adhere to the electrode. This can be prevented by use of a **cathode depolarizer** that is reduced at the cathode to yield a nongaseous product. One of the more useful of these is nitrate ion, which undergoes the following half-reaction at the cathode:

$$NO_3^- + 10H^+ + 8e^- \rightarrow NH_4^+ + 3H_2O \tag{15.9}$$

Nitrite ion results in the formation of a poor-quality deposit, so nitrous acid must be removed from the HNO_3 added to the electrolytic solution. This is conveniently accomplished by the addition of sulfamic acid, which reacts with nitrite ion impurity

$$H^+ + {}^-OSO_2NH_2 + NO_2^- \rightarrow N_2(g) + HSO_4^- + H_2O \tag{15.10}$$

Electrogravimetric Determination of Metal Ions

The simplest approach to the electrogravimetric determination of metal ions is the passage of a current between two electrodes in a medium containing the analyte and other chemicals. Typically the electrodes are made of platinum

TABLE 15.1 Electrogravimetric analyses of metals in a system in which the cathode potential is not regulated

Metal	Medium and conditions (product is elemental metal unless noted)
Co	Electrolyzed from a solution containing ammonium sulfate and a high concentration of NH_3 to form $Co(NH_3)_6^{2+}$ in solution. Hydrazine sulfate is added to prevent oxide formation on the deposit.
Cu	Electrolyzed from a mixture of nitric and sulfuric acids; nitrate prevents cathodic evolution of H_2 gas.
Cd	Electrolyzed from a slightly basic solution containing CN^- to put Cd(II) in the form of $Cd(CN)_4^{2-}$.
Ni	Electrolyzed from a solution containing ammonium sulfate and a relatively high concentration of NH_3 to put the nickel in the form of $Ni(NH_3)_4^{2+}$ in solution.
Ag	Excellent deposits are obtained from $Ag(CN)_2^-$ in a cyanide solution.
Zn	Electrolyzed from $Zn(OH)_3^-$ solutions with just enough excess NaOH added to prevent precipitation of $Zn(OH)_2(s)$.
Pb	Deposited at the anode as lead dioxide, $PbO_2(s)$, in a nitric acid medium by the half-reaction $Pb^{2+} + 2H_2O \rightarrow PbO_2(s) + 4H^+ + 2e^-$. Nitrate is reduced at the cathode, preventing cathodic deposition of lead metal.

gauze cylinders, with the anode rotated mechanically inside a larger-diameter cathode. A current density of less than $0.1 \ A/cm^2$ is applied across the electrodes. The reaction at the cathode is deposition of the metal and the anodic reaction is ordinarily oxidation of water.

$$2H_2O \rightarrow O_2(g) + 4H^+ + 4e^- \tag{15.11}$$

Table 15.1 summarizes the major electrogravimetric techniques for metals.

The electrogravimetric procedures just discussed are performed without cathode potential control. Most were developed and refined several decades ago when such control was possible only by tedious regulation of the cathode potential by "tuning" a rheostat by hand or with cumbersome electro-mechanical servo devices. Modern electronics used to construct **potentiostats** have made electrode potential control very easy and accurate. If these devices had been available during the times when electrogravimetry was undergoing peak development, it is likely that most of the electrogravimetric procedures would have employed electronic potentiostats to maintain constant working electrode potentials.

One element that is very difficult to determine electrogravimetrically without cathode potential control is bismuth. Bismuth can be deposited from a solution containing HCl, a small quantity of HNO_3, and oxalic acid. The purpose of the oxalic acid is to keep the bismuth in solution as an oxalate complex; otherwise it would precipitate as bismuth oxychloride (BiOCl) or other basic salts. The electrolysis is carried out at an elevated temperature of 80 to 85° C. For most of the course of the procedure, the cathode is held at -0.15 to -0.17 V versus the SCE; when the current has fallen to 10% of its initial value, the cathode potential is increased at a rate of 0.02 V/min to a value of −0.30 V. The electrolysis of 0.2 to 0.25 g of Bi requires about 20 min.

Removal of Interferences

Electrolysis can be employed at a mercury cathode to remove interfering metals from analyte solution by half-reactions such as the following for copper:

$$Cu(II) + 2e^- \xrightarrow{Hg} Cu(Hg) \tag{15.12}$$

A simple apparatus for doing this is shown in Figure 15.4. In addition to copper, the metals cadmium, cobalt, nickel, and silver are readily removed from solutions of alkali metals, alkaline earths, aluminum, and titanium.

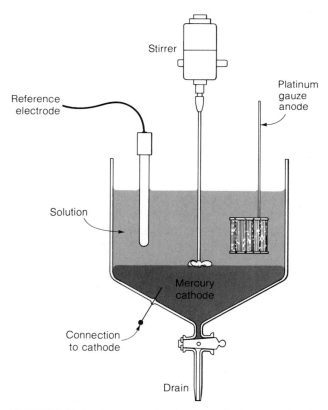

FIGURE 15.4   Apparatus for the controlled-potential reduction of metal impurities at a mercury cathode. The weights of the metals reduced are not measured, since the objective is to remove impurities. Note that the glass stirring propeller contacts the mercury surface and stirs it.

Hydrogen evolution is minimized by the high overpotential for the reduction of H_2O (or H^+) to H_2 at a mercury surface.

15.4   Constant-Potential Coulometry

Principles of Constant-Potential Coulometry

As defined in Section 15.1, constant-potential coulometry measures the species determined by the quantity of electricity that is involved in reducing or oxidizing the analyte at a working electrode set at a constant potential versus a reference electrode. The current is not constant and decreases during electrolysis in the general manner shown in Figure 15.5. The total charge C is obtained by integrating the current–time curve over the time t of the reaction.

$$C = \int_0^t i \, dt \qquad (15.13)$$

This can be done very accurately with modern electronics. At a less sophisticated level C may be calculated by visually adding the area (counting squares) beneath the plot of current versus time on recorder chart paper.

Applications of Constant-Potential Coulometry

The most straightforward application of constant-potential coulometry is the reduction of metal ions to metal amalgam or solid metal at a mercury cathode. The basic apparatus for doing so is shown in Figure 15.4. Some examples of metals that can be determined by this means are copper in a 0.5 M tartrate

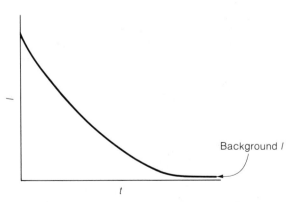

FIGURE 15.5   Plot of current versus time for a constant-potential coulometric determination

medium at pH 4.5 electrolyzed at -0.16 V versus the SCE, zinc in $1\,M$ NH_4Cl–$1\,M\,NH_3$ electrolyzed at -1.45 V versus the SCE, and Co in $1\,M$ pyridine neutralized to pH 7.0 with HCl at -1.20 V versus the SCE.

The procedure for determining a metal by constant-potential coulometry is relatively simple. The supporting electrolyte solution is first placed in the electrolysis cell and deaerated (O_2 removed) by bubbling N_2 gas through for about 5 min. The solution is next electrolyzed at the value of E to be used for the determination until the current has fallen to a very low constant background current of about 1 mA. A solution of analyte is then added and electrolyzed until the background current is again reached, usually within 1 hour. The quantity of analyte is then obtained from the value of C (see Equation 15.13) compiled electronically. If the background current is relatively large, the portion of C due to it can be subtracted in calculating the quantity of analyte.

15.5   Constant-Current Coulometry

Principles of Constant-Current Coulometry

As defined in Section 15.1, constant-current coulometry, or coulometric titration, involves an electrolytic reaction that occurs at a constant rate. It is impossible for the analyte to react with 100% current efficiency at a relatively high constant current. Therefore, it is necessary to add a **mediator** species at a relatively high concentration that can be oxidized or reduced at the working electrode to produce in turn a species that can oxidize or reduce the analyte in solution away from the working electrode. This concept is best understood by example. Suppose that it is desired to titrate coulometrically Fe^{2+} by oxidation at a platinum anode according to the half-reaction

$$Fe^{2+} \rightarrow Fe^{3+} + e^- \tag{15.14}$$

At a constant current, initially most of the electrons taken up by the anode will be taken from rather readily oxidized Fe^{2+} ions. However, as the concentration of analyte decreases, not enough Fe^{2+} ions reach the anode per unit time to depolarize it, and some other species must be oxidized to maintain the current. If a relatively high concentration of Ce^{3+} ion is present, it can be oxidized to Ce^{4+} at the anode.

$$Ce^{3+} \rightarrow Ce^{4+} + e^- \tag{15.15}$$

Recall from Section 12.2 that Ce^{4+} is an excellent oxidant for the titration of Fe^{2+}, so that the Ce^{4+} reacts with Fe^{2+} *away from the electrode*.

$$Fe^{2+} + Ce^{4+} \rightarrow Fe^{3+} + Ce^{3+} \tag{15.16}$$

The end point of the coulometric titration is reached when the Fe^{2+} is all consumed and the first trace of excess Ce^{4+} produced. Near the end point it is necessary to interrupt the flow of current several times to allow electrolytically produced Ce^{4+} to mix with the solution and react. This is completely analogous to the small increments of Ce^{4+} titrant added near the end point in a conventional volumetric titration of Fe^{2+} with Ce^{4+} titrant. As in the case of volumetric titration, at the end point there is an abrupt shift of E in the solution to more positive values. This can be observed visually in a coulometric titration with an oxidation–reduction indicator. It can be followed potentiometrically with a platinum indicator electrode, or it can be observed amperometrically.

Apparatus for Constant-Current Coulometry

Figure 15.6 shows the basic apparatus used for coulometric titration. Analyte and mediator are reduced or oxidized at the working electrode, also called a **generator electrode**. The counter electrode is physically isolated from the

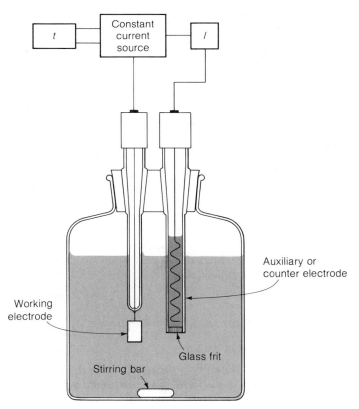

FIGURE 15.6   Apparatus for coulometric titration

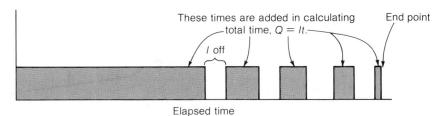

FIGURE 15.7   Plot of current versus time for a coulometric titration during the titration and near the end point. The shaded areas represent the total charge $Q = it$.

analyte solution with a glass frit, so that it does not reoxidize or re-reduce analyte that has been oxidized or reduced coulometrically. A power supply regulated to produce a constant current is connected across the two electrodes. A very accurate timing device measures the total length of time that the current passes through the analyte solution. Before the development of electronic timers, inertia in the mechanical timing device would cause error as the current was turned on and off in the vicinity of the end point; this is no longer a problem. The circuit also contains an ammeter for measuring current.

Figure 15.7 shows a typical plot of current versus time for a coulometric titration. Only the time that the current is applied, not the total elapsed time, is used for the calculation of total charge involved in the reaction.

15.6   Coulometric Acid-Base Titrations

Since it involves an electrolytic reaction, the most obvious choice for coulometric titration is in oxidation-reduction titrations. The example of the constant-current coulometric determination of iron by oxidation of Fe^{2+} was just discussed. Coulometric titrations have also been developed for acid-base, precipitation, and complexometric titrations.

Coulometric *neutralization titrations* make use of electrolytically generated H^+ or OH^- ions. The H^+ ion is generated with the working electrode in the anodic mode according to the half-reaction

$$2H_2O \rightarrow O_2(g) + 4H^+ + 4e^- \tag{15.17}$$

The H^+ generated circulates through solution neutralizing base.

$$H^+ + OH^- \rightarrow H_2O \tag{15.18}$$

A reaction such as this one occurring away from the working electrode is called a **secondary analytical reaction**. In this case, the following cathodic half-reaction occurs at the auxiliary electrode:

$$2H_2O + 2e^- \rightarrow 2OH^- + H_2(g) \tag{15.19}$$

The OH$^-$ ion generated could neutralize the H$^+$ titrant produced. This is prevented by isolating the auxiliary electrode in a glass tube with a glass frit at the bottom to prevent the OH$^-$ ion generated from mixing with the bulk of the solution (see Figure 15.6).

The coulometric titration of acids can be carried out by making the working electrode a cathode so that it generates OH$^-$ ion (Reaction 15.19). In this case H$^+$ is generated at the isolated auxiliary electrode.

End points are determined in coulometric neutralization titrations with a suitable indicator (see Section 5.6) or potentiometrically with a glass electrode (see Section 13.3). Employing the latter technique, automated coulometric titrators titrate to a predetermined pH, stop the titration, calculate the quantity of analyte, and store the data in a computer, all without human intervention.

15.7   Coulometric Precipitation Titrations

The most common precipitant generated coulometrically is Ag$^+$ ion. (One of the uses for this technique is for the clinical determination of chloride ion in blood.) The Ag$^+$ can be produced by making the working electrode (Figure 15.6) from heavy silver wire and producing Ag$^+$ ion from it by the anodic reaction

$$Ag \rightarrow Ag^+ + e^- \tag{15.20}$$

The secondary analytical reaction of the Ag$^+$ ion can involve species such as halides

$$Ag^+ + X^- \rightarrow AgX(s) \tag{15.21}$$

where X$^-$ is Cl$^-$, Br$^-$, or I$^-$. It can also precipitate mercaptans, RSH, such as ethyl mercaptan, C$_2$H$_5$SH.

$$Ag^+ + HSR \rightarrow AgSR(s) + H^+ \tag{15.22}$$

Mercury(I) ion is a precipitant that can be generated at a mercury metal anode by the half-reaction

$$2Hg(l) \rightarrow Hg_2^{2+} + 2e^- \tag{15.23}$$

The reagent thus generated can undergo secondary analytical reactions with the halides

$$Hg_2^{2+} + 2X^- \rightarrow Hg_2X_2(s) \tag{15.24}$$

For the coulometric titration of Cl$^-$ with Hg$_2^{2+}$, it is useful to make the medium up to 70% by volume methanol to decrease the solubility of Hg$_2$Cl$_2$. In a noncomplexing medium, some Hg^{2+} may be generated along with Hg$_2^{2+}$ in the electrolytic oxidation of mercury metal. Why this happens is apparent from the closeness of the E^0 values for the following half-reactions:

$$Hg_2^{2+} + 2e^- \rightleftarrows 2Hg(l) \qquad E^0 = 0.546 \text{ V vs. SCE} \qquad (15.25)$$

$$2Hg^{2+} + 2e^- \rightleftarrows Hg_2^{2+} \qquad E^0 = 0.678 \text{ V vs. SCE} \qquad (15.26)$$

$$Hg^{2+} + 2e^- \rightleftarrows Hg(l) \qquad E^0 = 0.612 \text{ V vs. SCE} \qquad (15.27)$$

Fortunately this makes no difference because one Hg^{2+} ion reacts with two Cl^- ions to form $HgCl_2$, just as one Hg_2^{2+} ion forms $Hg_2Cl_2(s)$ with two Cl^- ions. Since the production of one Hg^{2+} ion involves the loss of two electrons, as does the production of one Hg_2^{2+} ion, the stoichiometry of the determination is not affected by the production of some Hg^{2+}.

15.8   Coulometric versus Volumetric Titrations

From the preceding discussion of constant-current coulometric titrations, it is seen that they use the same titration reactions as those employed for volumetric oxidation-reduction, neutralization, and precipitation titrations. Where they can be employed, the major advantages of coulometric titrations over volumetric titrations are the following:

1. No need for a chemical primary standard.
2. No need to prepare and standardize standard solutions.
3. Highly reactive titrant reagents that would rapidly decompose if prepared as standard solutions can be generated coulometrically. These include bromine, chlorine, and titanium(III) ion.
4. Superior technique when very small quantities of analyte are to be determined because a low current can be employed; corresponding volumetric determinations require small volumes of dilute titrant, both of which cause difficulties.
5. Identical electronics and very similar electrolysis cells can be employed for neutralization, precipitation, complexometric, and oxidation-reduction titrations.
6. Especially amenable to automation, including automatic sampling devices, and data acquisition and analysis of data.

Although variations in current and error in the measurement of current and time are mentioned in the older literature as sources of error in coulometric titrations, these errors are insignificant with modern electronics. End point error (see Section 5.10) with both visual and electrochemical means of end point detection is common to both volumetric and coulometric titrations. The lack of attainment of 100% current efficiency is a problem with coulometric titrations that is not shared by volumetric titrations. Nevertheless, current efficiencies of 99.5% are often reported for coulometric titrations, and even 99.9% is attained in a number of procedures. Corrections can be made for less than 100% current efficiency, if the deviation from 100% is known and not greater than 2 or 3%.

Overall, in those applications where it works, coulometric titration is comparable to volumetric titration in precision and accuracy. In many cases, particularly those involving a large number of routine determinations, coulometry is superior.

Programmed Summary of Chapter 15

The major terms and concepts introduced in this chapter are contained in this summary in a programmed format. To derive the most benefit from the summary, you should fill in the blanks for each question and then check the answers at the end of the book to see if your choices were correct.

The basic distinction between coulometry and electrogravimetric analysis is that with coulometry the (1) _____ is measured, whereas with electrogravimetric analysis the (2) _____ of a deposit formed by an (3) _____ reaction is measured. Electrogravimetric and coulometric methods are conceptually very simple because they depend upon measurement of the two fundamental quantities of (4) _____. Electrogravimetric procedures are most commonly used for the determination of (5) _____. These are normally deposited from media in which they are (6) _____ such as NH_3 or CN^-; the quality of the deposit is normally favored by low (7) _____. A side-reaction that tends to give poor quality metal deposits is the evolution of (8) _____, an effect eliminated by addition of a (9) _____. An element whose electrogravimetric determination does not involve deposition of a metal is (10) _____, which is deposited on the anode as (11) _____. Electrolysis at a stirred mercury cathode is commonly used to remove (12) _____. Constant potential coulometry depends upon regulation of the (13) _____ to give the desired electrolysis reaction, and measurement of the (14) _____ to enable calculation of the quantity of analyte. This measurement basically involves integration of a (15) _____. In a constant-current coulometric titration, much of the titration reaction occurs away from the working electrode by reaction of a (16) _____ species present at a relatively high concentration. A reaction by such a species occurring away from the working electrode is called a (17) _____. In a conventional volumetric titration the buret stopcock is opened and closed several times near the end point to allow the titration reaction to come to equilibrium and to establish the end point titrant volume. The analogous operation in constant-current coulometry is (18) _____. In constant-current coulometry the counter (auxiliary) electrode must be (19) _____ to prevent the reverse of the titration reaction from occurring. The most common precipitant generated coulometrically is (20) _____, produced by the half-reaction (21) _____. A major advantage of coulometry over standard volumetric titration in respect to primary standards and standard solutions is that with coulometry (22) _____. A problem in coulometric titrations that does not occur in volumetric titrations is the lack of attainment of (23) _____.

Questions

1. Compare electrogravimetry, constant-potential coulometry, and constant-current coulometry with respect to quantities measured exactly, regulation of current, and regulation of potential.

2. In Section 15.2, the degree of separation of Pb^{2+} and Cd^{2+} possible with controlled-potential electrodeposition was calculated. What assumption do you think was made regarding the nature or composition of the working electrode surface at the point of incipient Cd^{2+} deposition? In what respect might this assumption be erroneous in regard to the Nernst equation?

3. Which metal is commonly determined electrogravimetrically by deposition of a compound of it at the anode?

4. What factors usually contribute to a good quality metal deposit in an electrogravimetric determination?

5. For what purpose is a cathode depolarizer employed in electrogravimetric analysis? What is a common cathode depolarizer?

6. Why is a mercury cathode particularly advantageous for the electrolytic removal of contaminant metal ions from solution? (It may be helpful to consider some of the advantages of the mercury electrode in polarography).

7. In a constant-potential coulometric determination, is the current normally taken down to 0? If not, what is the value to which it is taken? What correction is normally made in calculating the charge, C?

8. In a volumetric determination the hypothetical metal ion Mt^+ is oxidized to Mt^{2+} by a hypothetical oxidant Me^{3+}, which is reduced to Me^+. Outline the conditions under which Mt^+ might be determined by coulometric titration; include mention of the relative values of pertinent E^0's.

9. Why must the solution in contact with the auxiliary or counter electrode be physically isolated from the analyte solution in a constant-current coulometric titration?

10. What is the definition of a secondary analytical reaction in a coulometric titration?

11. What is the most common precipitant generated coulometrically? What is the electrode used to generate the precipitant?

12. What is the "primary standard" employed in coulometric titrations?

13. What electrochemical means of end point detection can be employed in coulometric titrations?

14. What visual means of end point detection can be employed in coulometric titrations?

Problems

1. This current-time curve was obtained during the controlled-potential coulometric determination of Cd^{2+} at a mercury cathode. The volume of sample was 50.0 mL. Calculate the milliequivalents (meq), millimoles, and concentration of Cd(II) in the sample. Subtract what appears to be the background current in the calculation.

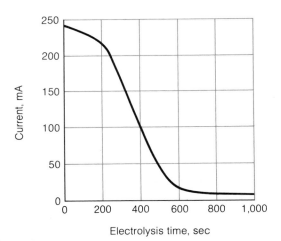

Electrolysis time, sec

2. How many coulombs correspond to 1 meq?
3. Given for the hypothetical metals Me and Mt,

$$Me^{2+} + 2e^- \rightleftharpoons Me \qquad E^0_{Me}$$

$$Mt^+ + e^- \rightleftharpoons Mt \qquad E^0_{Mt} \qquad E^0_{Me} > E^0_{Mt}$$

It is desired to separate Me^{2+} from Mt in a solution in which the concentration of each is 1.00×10^{-2} M using controlled-potential coulometry. In order to get a 99.9% separation, how much greater (more positive) must E^0_{Me} be than E^0_{Mt}?

4. For an electrogravimetric determination a current of 1.37 A was used with a working electrode of 11.25-cm^2 area. What was the current density at the working electrode?

5. For the constant-potential coulometric determination described in Problem 1, the initial current density at the mercury cathode was 0.050 A/cm^2. Estimate the current density after half the cadmium had been reduced (this is not the same as half of the total electrolysis time).

6. A solution for electrogravimetric analysis was made 1.00 M in nitric acid to act as a cathode depolarizer. If a current of 0.500 A was employed for 30 minutes, and 92.0% of the current at the cathode was employed for the reduction of NO_3^-, what was the concentration of this ion at the end of the electrolysis, given a total solution volume of 250 mL?

7. Allyl alcohol, $CH_2\!\!=\!\!CHCH_2OH$ was determined by constant-current coulometry with Br_2 electrogenerated from Br^- followed by addition of Br_2 across the double bond. Assuming 100% current and reaction efficiency, calculate the number of millimoles of allyl alcohol if a current of 0.250 A was used for a total titration time of 11.32 min to the end point.

8. A quantity of solid organic acid weighing 1.037 g was dissolved and titrated at a constant current of 0.500 A by electrogenerated OH^- ion. The titration took 13.67 min. What was the equivalent weight of the acid?

9. A deposit of 0.337 g of PbO_2, fw = 239.2, was formed at the anode in the electrogravimetric determination of Pb(II) from 250 mL of a plating-bath solution. What was the concentration of Pb in the plating-bath solution in grams per liter?

16 Separations in Analytical Chemistry

LEARNING GOALS

1. Overview of separation processes.
2. Single-stage and multistage separations.
3. Various processes for separations between bulk phases, including gaseous diffusion, liquid-liquid extraction, gas-liquid extraction, crystallization, sublimation, and distillation.
4. Separations that make use of thin-layer phases.
5. How precipitation can be used in separations through control of acidity, sulfide formation, and organic precipitants.
6. How organic solvents can be used with water for liquid-liquid extraction employing both batch and continuous procedures.
7. The partition coefficient in calculating percentages of analyte extracted in liquid-liquid extraction.
8. What is meant by countercurrent fractionation.
9. Some common examples of liquid-liquid extraction.
10. Some practical aspects of solvent extraction.
11. Ion exchange separations, ion exchange equilibria, and types of ion exchange resins.

16.1   Importance of Separations

Purpose of Separations in Chemical Analysis

Both qualitative and quantitative chemical analysis are largely matters of **separation** of the analyte either into a pure form or into a medium free of interferences for the method of determination employed. For example, Zn^{2+} ion can be relatively difficult to titrate with EDTA in the presence of Cu^{2+} because the two EDTA chelates are of about the same stability. However, separation of interfering copper by electrodeposition on a mercury cathode (see Figure 15.4) can be employed to remove the interfering ion and enable the titration of Zn^{2+} with EDTA to be carried out. Since there are interferences with the majority of analytical methods, more often than not some sort of separation must be carried out as part of the sample analysis. Several important analytical methods are, themselves, separations. Both liquid chromatography (Chapter 18) and gas chromatography (Chapter 19) are separation methods. With these techniques, the time required for an analyte to go through the chromatographic column can be used to identify it, and the response of the detector at the end of the column for each analyte separated provides a quantitative measure of the analyte. The mass spectrometer, which separates analyte ions produced in a vacuum on the basis of the mass-to-charge ratio of the ionized analyte molecule and ionic fragments of it, is a separation technique that is now one of the most commonly employed methods for the analysis of complex organic samples, such as biological materials or hazardous waste substances.

Separations and Masking

This chapter considers means of physically separating analytes and interferences. It should be recalled that another approach to dealing with interferences is to immobilize them in some way such that they do not interfere with the determination of the analyte. The most common way that this is done is through *masking*, which is most often accomplished by adding a complexing agent to "tie up" an interfering metal ion. A typical case is the addition of citrate as a chelating agent to bind to interfering iron(III) in the potentiometric determination of fluoride (see Section 13.11 for a discussion of fluoride ion–selective electrode response).

16.2   Separation Processes

Distillation

Conceptually, the simplest separation process is that of two components of a mixture. An example of such a separation would be the evaporation of a solvent from a solution to leave a pure solid solute. Another very common example is **distillation**, in which two (or more) liquids of different volatilities are separated by boiling and condensation, with the more volatile component collected in the pure or enriched form in the vapor condensate.

Distribution between Phases

Separations in chemical analysis, however, generally involve the distribution of one or more species between two different **phases**, which might consist of two immiscible liquids, a solid and a liquid, a solid and a gas, or a gas and a liquid. The precipitation of $BaSO_4$ by the addition of a $BaCl_2$ solution to a sulfate solution is an example of a separation based upon distribution between a solid and a liquid. Virtually all the $BaSO_4$ goes into the solid form, whereas the chloride ions and the cations associated with the soluble sulfate stay in solution.

Single-Stage and Multistage Separations

In favorable cases a separation can be accomplished with only one equilibration between the two phases involved, as happens when $BaSO_4$ precipitates from solution. Such a separation is a **single-stage process**. In many cases only a partial separation is attained by one stage. For example, suppose an organic compound dissolved in water is to be separated from inorganic salts in the same aqueous solution by extraction into dichloromethane, CH_2Cl_2. If only part of the organic compound transfers to the dichloromethane each time the two liquids are contacted, several such equilibrations employing fresh dichloromethane each time can result in removal of essentially all the organic compound. This is an example of a **multistage fractionation** or extraction. Separation of two species becomes much more difficult when both are soluble in the two phases employed and when their distribution between the two phases is almost the same for each equilibrium. Under these circumstances a multistage separation is required.

16.3   Separations between Bulk Phases

Analytical chemical separations can be divided into the two general categories of those that occur between two **bulk phases**—that is, gas, liquid, solid—and those that occur between a bulk phase and a thin layer, discussed in the

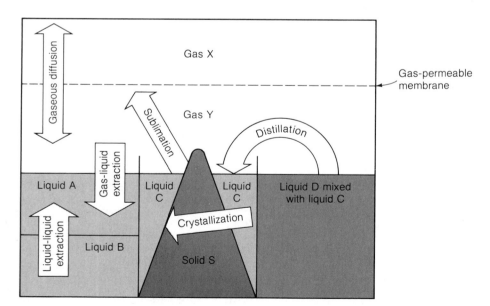

FIGURE 16.1   Possible separations between bulk phases consisting of gases, liquids, or solids

following section. Figure 16.1 summarizes interchange between bulk phases. **Gaseous diffusion** is the process by which continual molecular motion of gas molecules enables them to migrate; gases of different molecular weight can be separated by their different rates of movement through gas-permeable membranes. **Liquid–liquid extraction** occurs when a solute transfers from a liquid to another one in which it is more soluble. **Gas–liquid extraction** is the dissolution of a gas into a liquid. **Crystallization**, shown here for "S" dissolved in liquid C, is the process of solid formation from a solute in a solution. **Sublimation**, used to purify some of the solid reagents employed in chemical analysis, is the direct formation of vapor from a solid. Distillation, shown in Figure 16.1 for the separation of more volatile liquid C from less volatile liquid D, takes place when a liquid is evaporated and then condensed to a liquid form. It is widely used for purifying solvents required in some analytical applications.

16.4 Separations on Thin-Layer Phases

The Nature of Thin-Layer Phases

Modern separation practices, particularly the burgeoning area of chromatography (see Chapter 17), make use of **thin-layer phases** consisting of the surface layer of solids or thin layers of liquids adsorbed onto solid particles. As shown in Figure 16.2, there are two basic kinds of thin layers. One consists of the surface of a solid material that holds molecules in essentially fixed sites on the surface, a phenomenon called **adsorption**. In the other case shown, a low-volatility liquid is held onto the surface of a solid support material, and molecules dissolve in this liquid, where they are free to move around in the liquid layer by diffusion processes. The thin layers are typically of the order of a nanometer thick. The solid particles that support the thin-layer surfaces are designated the **stationary phase**. The stationary phase is surrounded by a gas or liquid called the **mobile phase**, so called because it is free to move relative to the stationary phase. The species that undergoes exchange between the mobile phase and the thin layers on the stationary phase consists of molecules or ions and is called the **solute**.

Chromatographic Separations

As discussed in detail in Chapter 17, different kinds of solutes in a mobile phase can be separated from each other by causing the mobile phase to move over the stationary phase held in a column. Those solutes that are held more strongly in the thin layers on the stationary phase in the column emerge from the column later than those solutes held less strongly. This forms the basis of a *chromatographic separation*. When the mobile phase is a gas and the solute is adsorbed on the surface of a solid particle, the separation process is called *gas-solid*

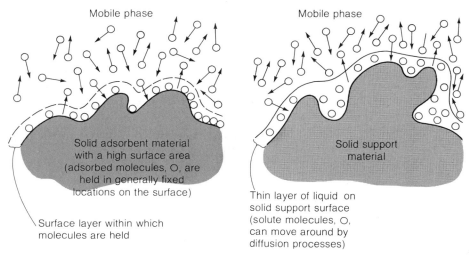

FIGURE 16.2   The two major types of thin-layer phases. On the left is a portion of a particle, the surface of which holds and exchanges molecules in a thin layer, with an outer boundary designated by a dashed line. On the right a thin layer of liquid capable of dissolving molecules is shown bound to the surface of a solid support material. The left illustration shows adsorption, the right, absorption by the thin-layer liquid phase. Motion of the molecules in the mobile phase is shown by arrows. A molecule entering one of the thin layers is shown as ⊘ , and one leaving is shown ⊘

chromatography (GSC), but when the solute is free to diffuse in a liquid that coats the solid support, the technique is called *gas-liquid chromatography (GLC).* When the mobile phase is a liquid that flows over solid particles coated with a layer of liquid immiscible with the mobile-phase liquid, the chromatographic technique is called *liquid-liquid chromatography.* Chromatography with a liquid mobile phase and a solid phase that adsorbs solutes is *adsorption chromatography* or, when the solutes are ionic, *ion exchange chromatography.*

16.5   Separation by Precipitation

Considerations in Precipitation Separations

Precipitation separation involves the preferential formation of a solid from solutes in solution without the removal from solution of solutes that the analyst wishes to keep in the dissolved state. This would seem to be a straightforward process where the feasibility of separations could be determined from differences in solubility or solubility product (see Chapter 7). In practice, *coprecipitation* of normally soluble salts (see Section 8.5) and *inclusion* and *occlusion* of impurities (see Section 8.6) tend to give incomplete separations. Other problems, such as *peptization* of colloidal precipitates (Section 8.5) may occur. Generally the approaches used in the formation and isolation of precipitates for separation purposes are those used in gravimetric determinations.

Precipitation of Oxides and Hydroxides

Hydroxide precipitation by acidity control is the most general means of separating solutes from water with precipitation. In a hot, concentrated nitric acid medium, oxides of Nb(V), Sb(V), Si(IV), Sn(IV), Ta(V), and W(V) precipitate from solution leaving most other metal ions dissolved. In a pH 10 NH_3–NH_4Cl buffer, Al(III), Cr(III), and Fe(III) precipitate as hydrated oxides or hydroxides, whereas the alkali and alkaline earth metal ions do not. Because of the formation of soluble complex ions with NH_3, Cu(II), Co(II), Mn(II), Ni(II), and Zn(II) stay in solution in this buffer medium. A buffer of acetic acid–ammonium acetate results in the precipitation of Al(III), Cr(III), and Fe(III) as hydroxides, leaving most doubly charged metal cations in solution. Because of the extreme insolubility of the hydroxides of Al(III), Cr(III), and Fe(III), these ions precipitate even in this relatively acidic medium. An oxidizing medium of NaOH–Na_2O_2 precipitates most +2 metal cations, Fe(III), and rare earths, but leaves Al(III), Cr(VI), U(VI), V(V), and Zn(II) in solution because of their formation of soluble hydroxy complexes and oxy anions (for example, CrO_4^{2-}).

Precipitation of Sulfides

Next to formation of hydroxides, oxy anions, and hydrated metal oxides, **sulfide formation** is the most common means of precipitating metal ions from solution. Precipitation is controlled with pH because of the weakly acidic nature of H_2S and HS^-.

$$H_2S \rightleftarrows H^+ + HS^- \qquad pK_{a1} = 7.2 \tag{16.1}$$

$$HS^- \rightleftarrows H^+ + S^{2-} \qquad pK_{a2} = 14.9 \tag{16.2}$$

Sulfide is generated readily from *homogeneous solution* (Section 8.7) by the thioacetamide

$$\underset{\displaystyle \overset{\|}{S}}{H_3C-C-NH_2} + H_2O \rightarrow \underset{\displaystyle \overset{\|}{O}}{H_3C-C-NH_2} + H_2S(aq) \tag{16.3}$$

Practically all metal ions, except the alkali metals and alkaline earths, form sulfide precipitates. These precipitates have a large range of solubility, which favors separations.

The major solution conditions for sulfide precipitation are (1) 3 M HCl, (2) 0.3 M HCl, (3) pH 6 acetate buffer, and (4) pH 9 NH_3–$(NH_4)_2S$ buffer. The first three media listed are also saturated with H_2S gas. The least soluble of the sulfides are those of Hg(II), Cu(II), and Ag(I), which precipitate in all the media just listed. Ions of As(III), As(V), Sb(III), and Sb(V) precipitate in media 1 to 3, but not in medium 4. Ions of Bi(III), Cd(II), Pb(II), and Sn(II) precipitate with sulfide in media 2 to 4, but not in the first medium. Tin(IV) forms sulfide precipitates in media 2 and 3 but remains soluble in 1 and 4. Iron(II) and Mn(II) form sulfide precipitates in pH 9 NH_3–$(NH_4)_2S$ buffer, but not in media 1 to 3.

Separations with Organic Precipitants

The production of precipitates with organic precipitants was discussed in Section 8.10. As noted previously, the precipitants are normally relatively high molecular weight chelating agents. For general use, 8-hydroxyquinoline has been widely employed; it precipitates a number of metal ions but a degree of selectivity can be achieved by regulation of pH and the use of masking agents. Dimethylglyoxime is a virtually specific precipitant for Ni(II) and is therefore useful for separations involving this metal.

$$H_3C—C{=}N—OH$$
$$H_3C—C{=}N—OH$$

8-Hydroxyquinoline   Dimethylglyoxime

Separation of Analytes by Coprecipitation

In some cases a precipitate serves as **collector** to deliberately coprecipitate traces of an analyte that would otherwise be soluble. Examples are coprecipitation of small quantities of Zn(II) and Pb(II) with CuS and collection of microgram quantities of titanium with gelatinous $Al(OH)_3$ precipitate. The collector is chosen so that it does not interfere with the subsequent determination of the analyte.

16.6   Liquid-Liquid Extraction

Transfer of Solute between Solvents

Liquid-liquid extraction, or **solvent extraction**, is the process whereby a solute is transferred from one liquid phase to another that is immiscible with the first (Figure 16.3). Most commonly, solutes are transferred from an aqueous phase to an organic solvent. Since organic solvents that are not miscible with

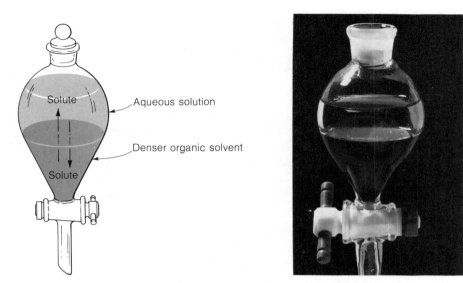

FIGURE 16.3 **Left:** Distribution of a solute between an aqueous solution and a denser organic solvent in a separatory funnel. This is the basis of solvent extraction. **Right:** Separatory funnel, showing two immiscible liquid phases, that could be used for solvent–solvent extraction of an analyte.

water are not generally good solvents for ionic materials, a neutral species is usually extracted (an illustration of the solubility rule that "like dissolves like"). Ionic solutes can be extracted as **ion pairs** with no net charge, or metal ions can be reacted with charged organic chelating agents, **organic extractants**, that neutralize the charge of the cation and give it an "organic-like" nature so that it is much more soluble in the organic solvent. Both of these approaches are discussed later in this chapter.

Solvents Used for Extractions

Diethyl ether and hydrocarbons such as benzene are common organic **extractant solvents** less dense than water; those denser than water are organochlorine compounds, such as chloroform, dichloromethane, and carbon tetrachloride. *(Care must be exercised in handling solvents for solvent extraction. Benzene is suspect as a possible cause of leukemia, and toluene,* ⬡—CH₃*, is now commonly substituted for it. Prolonged heavy exposure to carbon tetrachloride causes liver damage.)* Although limited, the mutual miscibility of water and most organic solvents is such that volumes of liquids after equilibration may differ from those of the original volumes. Corrections can be made for this if the effect is severe. Furthermore, because of the mutual presence of the liquids in solution in each other, the relative solubilities of a solute in the two liquids in contact may differ from the relative solubilities of the solute in the separate liquids.

Solvent extractions can be performed by placing a solution containing the solute to be extracted in a separatory funnel, adding the extracting solvent, mixing, settling, and separating the phases by draining off the bottom liquid. (As with many apparently simple procedures there are complications and precautions. A very volatile solvent such as dichloromethane builds a high vapor pressure in the funnel, and the vapor must be vented by inverting the funnel and opening the stopcock. Emulsions consisting of colloidal suspensions of one of the solvents in the other may form, hindering separation.)

Distribution of Solute between Two Solvents

The distribution of a solute between two solvents is governed by the **distribution law**, which is expressed by a **partition coefficient** or **distribution coefficient, K:**

$$K = \frac{[A_{org}]}{[A_{aq}]} \tag{16.4}$$

In this equation A represents the formula of the extracted species, and the subscripts org and aq designate whether it is in organic or aqueous solution. The constant K describes the equilibrium distribution of the species A between the two phases

$$A_{aq} \rightleftarrows A_{org} \tag{16.5}$$

It does not account for further reactions of the neutral species, such as the dimerization of A in the organic solvent

$$A_{org} + A_{org} \rightleftarrows (A_2)_{org} \tag{16.6}$$

or acid–base reactions in water

$$A_{aq} + H^+ \rightleftarrows HA^+ \tag{16.7}$$

Like solubility and precipitation equilibria, the preceding is an example of heterogeneous equilibria involving the distribution of species between two phases.

For solute and solvent systems that follow Equation 16.4 for the partition coefficient, the amount of solute removed in one extraction can be calculated rather simply. Suppose that V_{org} mL of organic solvent are used to extract solute A from V_{aq} mL of water. If the *total* millimoles of A available is designated as mmol A_t, the millimoles of A in the organic solvent after equilibration as mmol A_{org}, and the millimoles of A in the aqueous phase after equilibration as mmol A_{aq}, the concentrations of A in the two phases after equilibration are given by,

$$[A_{aq}] = \frac{\text{mmol } A_{aq}}{V_{aq}} \tag{16.8}$$

$$[A_{org}] = \frac{\text{mmol } A_{org}}{V_{org}} = \frac{\text{mmol } A_t - \text{mmol } A_{aq}}{V_{org}} \tag{16.9}$$

Recalling the formula for the partition coefficient

$$K = \frac{[A_{org}]}{[A_{aq}]} \tag{16.4}$$

the formula for the fraction of A remaining in the aqueous phase after the first extraction, q_1, is developed as follows:

$$q_1 = \frac{\text{mmol } A_{aq}}{\text{mmol } A_t} = \frac{\text{mmol } A_{aq}}{\text{mmol } A_{aq} + \text{mmol } A_{org}} = \frac{[A_{aq}] V_{aq}}{[A_{aq}] V_{aq} + [A_{org}] V_{org}} \tag{16.10}$$

Dividing through by $[A_{aq}]$ yields

$$q_1 = \frac{V_{aq}}{V_{aq} + ([A_{org}]/[A_{aq}]) V_{org}} = \frac{V_{aq}}{V_{aq} + K V_{org}} \tag{16.11}$$

Using the same approach, the fraction of solute remaining after a second extraction with V_{org} mL of a fresh organic solvent is

$$q_2 = \frac{V_{aq}}{V_{aq} + K V_{org}} q_1 = \left(\frac{V_{aq}}{V_{aq} + K V_{org}} \right)^2 \tag{16.12}$$

The fraction of solute remaining in the aqueous phase after n identical extractions with organic solvent is

$$q_n = \left(\frac{V_{aq}}{V_{aq} + K V_{org}} \right)^n \tag{16.13}$$

If the subscripts 0 and n are taken to denote concentrations initially and after n extractions, respectively, the concentration of solute remaining in the aqueous phase after n extractions is

$$[A_{aq}]_n = \left(\frac{V_{aq}}{V_{aq} + K V_{org}} \right)^n [A_{aq}]_0 \tag{16.14}$$

Effect of Number of Extractions

It is instructive to use the equations just derived to calculate the fraction of solute remaining in the aqueous phase after several extractions with a relatively small volume of solvent compared to a lesser number of extractions with a larger volume of solvent. Assume that 100.0 mL of an aqueous solution initially 0.0100 M in solute A is to be extracted into a total volume of 500 mL of chloroform. The value of the distribution coefficient K is 2.35. Compare two series of extractions, each using a total of 500 mL of chloroform, one with two 250-mL extractions, the other with ten 50-mL extractions. The data are given in Table 16.1. For the first case, employing 50-mL portions, each extraction leaves the following fraction in aqueous solution of A that was dissolved in water at the beginning of the extraction:

$$\frac{V_{aq}}{V_{aq} + KV_{org}} = \frac{100}{100 + 2.35 \times 50} = 0.460 \tag{16.15}$$

The corresponding fraction for a 250-mL extraction is 0.145. Figure 16.4 shows a plot of fraction remaining versus total volume of chloroform

TABLE 16.1   Extraction of solute A into chloroform from 100.0 mL water

Extraction number, n	50-mL extractions		250-mL extractions	
	q_n	Total chloroform used, mL	q_n	Total chloroform used, mL
1	0.460	50	0.145	250
2	0.211	100	0.021	500
3	0.097	150		
4	0.045	200		
5	0.020	250		
6	0.009	300		
7	0.004	350		
8	0.002	400		
9	0.0009	450		
10	0.0004	500		

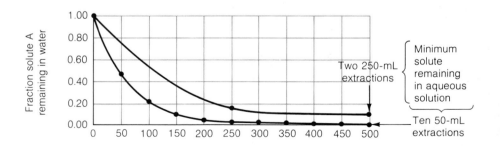

FIGURE 16.4   Fraction of solute A remaining in water versus total volume of chloroform extractant used; from data in Table 16.1.

extractant used. It is seen that ten extractions with 50 mL each remove essentially all the solute, whereas about 2% remains after two extractions with 250 mL each. Somewhat more impressive is the fact that two 250-mL extractions leave about 50 times as much solute in the water as ten 50-mL extractions. Furthermore, examination of the data in Table 16.1 reveals that five extractions with a total of only 250 mL of organic solvent reduces the solute level in water to the same level (2% of that originally present) as do two 250-mL extractions with a total of 500 mL of organic solvent.

The simple **batch extraction** procedure just discussed can be used to separate solutes in water where there is a favorable distribution coefficient (that is, 5 to 10) for extraction into an organic solvent and the other has an extraction coefficient of less than about 0.1. In such cases about five extractions with fresh solvent are sufficient to bring about a satisfactory extraction.

Continuous Extraction

Continuous extraction, sometimes called **exhaustive extraction**, can be used to extract solutes with extremely low distribution coefficients. An example of the apparatus employed is shown in Figure 16.5. In this case the organic solvent must be one that is less dense than water, such as diethyl ether. The solvent is continually distilled from its flask and condenses such that it flows into a long tube immersed in water. A sufficient "head" of solvent in this tube forces the solvent out into the aquatic solution where it extracts the desired solute from the water. Continuous recirculation of the organic solvent by redistillation enables extraction of solutes with quite low partition coefficients.

Countercurrent fractionation involves the separation of two solutes with only slightly different partition coefficients by a series of individual extractions in which fresh solvent of the two different media involved is introduced in each step of the extraction. The two solvent phases move in opposite directions as shown schematically in Figure 16.6. With each move, each tube of aqueous medium is equilibrated with the tube containing organic solvent opposite it. A tube of pure organic solvent and one of pure aqueous medium join the group of tubes undergoing equilibration with each move. The net result is that the solute with the greater aqueous affinity is found predominantly in the first three or four tubes of aqueous medium (left) and the solute with the greater organic affinity is found in the first three or four tubes of organic solvent (right) after equilibration is complete.

Countercurrent fractionation is no longer widely used for analytical separations. However, the theory that explains its operation describes some basic aspects of chromatographic technique and theory, an extremely important area discussed in Chapters 17 through 19. On an industrial scale, furthermore, countercurrent extraction is used extensively to purify large quantities of chemicals. In industrial processes a long vertical column is

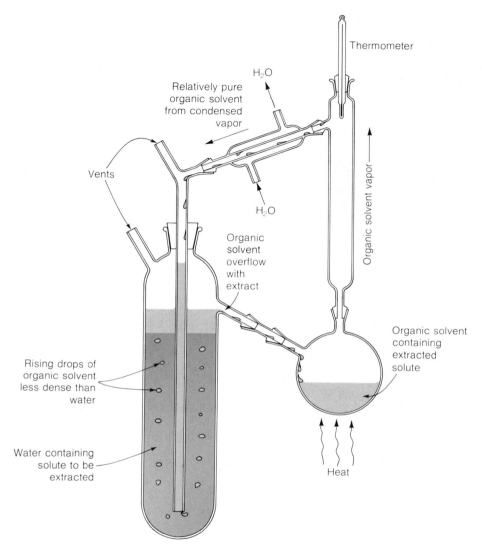

FIGURE 16.5   Device for continuous extraction of solute from water by a solvent less dense than water

commonly used and the denser of two immiscible solvents is introduced at the top and the less dense solvent at the bottom. Solutes are exchanged between the solvents as they pass through the column in opposite directions under gravitational forces.

Organic solvent less dense than water→

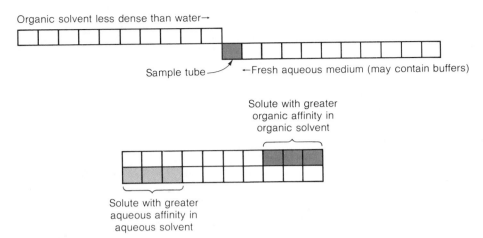

Sample tube → ←Fresh aqueous medium (may contain buffers)

Solute with greater
organic affinity in
organic solvent

Solute with greater
aqueous affinity in
aqueous solvent

FIGURE 16.6 Countercurrent fractionation of two solutes, one (right) more soluble in organic solvent, the other (left) more soluble in water. The tubes before equilibration are shown at the top where, except for the sample tube, they contain either pure organic solvent or pure aqueous medium. The tubes after equilibration are shown at the bottom of the drawing.

16.7   Some Examples of Liquid-Liquid Extraction

Organic Extractants

It is well known and readily understood that hydrated inorganic salts tend to stay in aqueous solution because of the charge of the ions and their hydrophilic (water-attracting) nature, whereas organic molecules without substantial hydrophilic groups (carboxylate, sulfonate, hydroxyl) are relatively more soluble in organic solvents, such as toluene and dichloromethane. Metal ions are strongly hydrophilic, and their extraction into organic solvents requires neutralization of the metal ion charge and "disguising" their nature, usually by binding to organic molecules. This is shown by the binding of Cu(II) by one of the more common organic extractants, *acetylacetone*:

$$2 \quad \begin{array}{c} H_3C-C-OH \\ \| \\ H-C \\ \backslash \\ H_3C-C=O \end{array} + Cu(H_2O)_6^{2+} \rightarrow \quad [\text{copper acetylacetonate complex}] \quad + 2H^+ + 6H_2O$$

(16.16)

Acetylacetone (enol form)   Organic-soluble copper acetylacetonate

Other commonly used metal extractants include cupferron, diphenylthiocarbazone (dithizone), and 8-hydroxyquinoline (oxine):

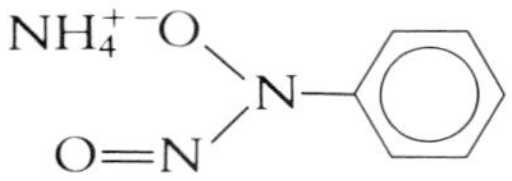

Ammonium salt of *N*-nitroso-
N-phenylhydroxylamine
(cupferron)

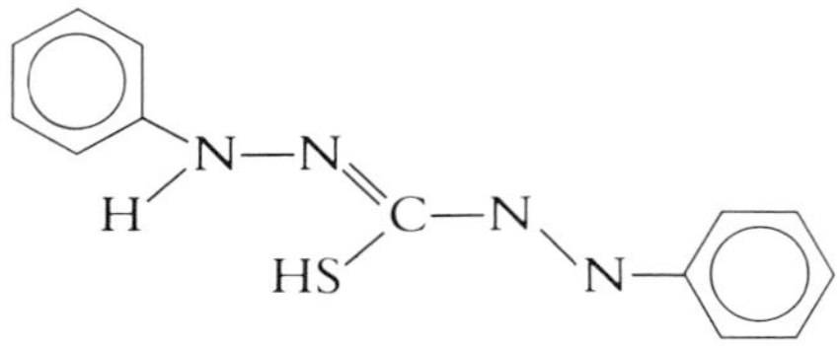

Diphenylthiocarbazone
(dithizone)

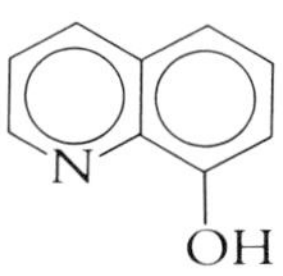

8-hydroxyquinoline
(oxine)

Extraction of Organically Bound Metals

Extractable organically bound metal ions are normally *metal chelates* with relatively large organic ligands (see Chapter 10). So that the chelates will extract into organic solvents, the chelating agents chosen are those that form neutral coordination compounds with the analyte metal ions.

In addition to metal chelates, another common type of extractable species consists of neutral **ion-association complexes**. The simplest of these are *ion pairs* composed of a positive and a negative ion. In organic solvents with low dielectric constants *ion clusters* occur consisting of electrically neutral aggregates of a number of ions. One of the better known ion pairs in chemical analysis is $H^+FeCl_4^-$, which is extractable into diethyl ether from an aquatic medium of 6 M HCl. This extraction is employed to remove most of the iron from a dissolved iron ore or steel sample prior to the determination of traces of other metals present, such as Al, Cr, Ni, or Ti.

The tetrabutylammonium cation, $(n\text{-}C_4H_9)_4N^+$, and the tetraphenylarsonium cation, $(C_6H_5)_4As^+$, form extractable ion pairs with a number of anions. These large, bulky, organic-like cations do not have a primary hydration sphere consisting of water molecules bonded directly to the cation. They form pairs with large, singly charged anions, such as MnO_4^-, ReO_4^-, or TcO_4^-. The resulting ion pair is not hydrogen-bonded appreciably to water molecules, and in fact disrupts the strongly hydrogen-bonded matrix of the

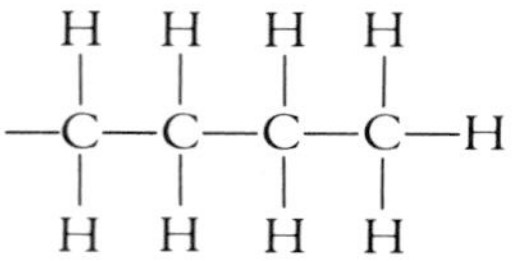

The *n*-butyl group,
n-C$_4$H$_9$

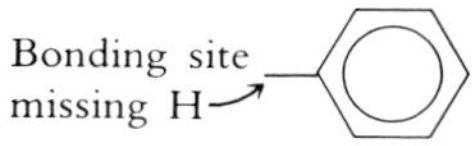

Bonding site
missing H

The phenyl group

FIGURE 16.7   Disruption of hydrogen-bonded matrix of water molecules by a nonhydrogen bonding ion pair; dashed lines represent hydrogen bonds

water solvent (Figure 16.7). Thus there is a strong tendency for the neutral ion pair to be "ejected" into an organic solvent.

Some chelated metal ions can be extracted as ion pairs when the ligand is a bulky organic group. For example, iron(II) perchlorate can be extracted into chloroform when the iron(II) is bound as the 1,10-phenanthroline complex, $Fe(phen)_3^{2+}$. Since the extracted species consists of three ions (see below), it is called an *ion aggregate*, a term used when more than two ions associate.

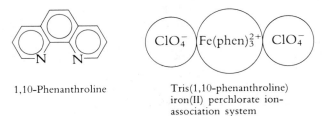

1,10-Phenanthroline

Tris(1,10-phenanthroline) iron(II) perchlorate ion-association system

Solvents That Coordinate with Metal Ions

Some solvents, including some alcohols, ketones, esters, and ethers, themselves have the ability to coordinate with metal ions to form species extractable into the solvent. The resulting large size of the metal ion and the resemblance of its surface to the solvent medium contribute to the extractability of salts, such as chlorides and nitrates. The esters of orthophosphoric acid are especially good for the extraction of inorganic salts. The best known of these is tri-*n*-butyl phosphate (TBP) (above), used for extractions in the laboratory and on an industrial scale. In the latter case it is widely used for the extraction of uranyl nitrate, $UO_2(NO_3)_2$, from uranium ore solutions and from fission products. The effectiveness of this solvent in the extraction of cations is largely due to its sterically (structurally) available electron-donor oxygen atom.

Formulas and Uses of Common Extractants

The formulas of some of the more common extractants were given on page 461. Some others that have been found to be useful are presented in Table 16.2. Many of these act as *colorimetric reagents* in that they form intensely colored complexes in organic solvent media. The absorption of light at appropriate wavelengths can be used to measure the concentrations of complexes (see Spectrophotometry, Chapter 20). The selectivity of extractants for metal ions is often increased by using EDTA (see Chapter 10) as a masking agent to keep interfering metals in the aqueous phase.

Some Examples of Extraction Procedures

Solvent extraction is an exacting science — some would call it an art — demanding very careful control of conditions and extraction parameters. It is beyond the scope of this chapter to discuss these in detail here. However, several typical procedures will be outlined to show some of the general approaches used.

Iron(III) at levels of several parts per million can be extracted into a chloroform solution of 1% 8-hydroxyquinoline (see formula page 458). The

TABLE 16.2 Commonly used extractants (extraction reagents)

Name	Formula	Organic solvents used	Analytes for which the extractant is used
Thenoyltrifluoroacetone (TTA)		Used as a 0.1–0.5 M solution in benzene or toluene	Many metals, used in separating metals in the lanthanide and actinide groups
Dimethylglyoxime	$H_3C—C≡N—OH$ $\|$ $H_3C—C≡N—OH$	Chloroform	Nickel in pH 4–12 tartrate or pH 7–12 citrate
1-Nitroso-2-naphthol		Chloroform	Co(II) in acidic medium, Fe(II) in base
Toluene-3,4-dithiol (dithiol)		Chloroform or pentyl acetate	Mo(VI), W(VI), and Re(VI) from acidic solution
Tri-n-octylphosphine oxide (TOPO)	$(n\text{-}C_8H_{17})_3PO$	Used as a 0.1 M solution in cyclohexane	From 1 M HCl: Fe(III), Mo(VI), Sn(IV), $H_2Cr_2O_7$, $ZrCl_4$, $UO_2(NO_3)_2$
Sodium diethyldithiocarbamate	$(C_2H_5)_2N—\overset{\overset{\textstyle S}{\|}}{C}—S^-Na^+$	Carbon tetrachloride, chloroform, butyl acetate	Particularly useful for Cu(II), which can be extracted in the presence of iron(III)

extraction works over a wide range of pH from 2 to 10. At pH 2 to 2.5, however, potentially interfering aluminum, cerium(III), cobalt, and nickel do not affect the determination. The level of iron in the sample is determined by the light absorbed at 470 nm by the 8-hydroxyquinoline complex.

Lead can be determined by its reaction with diphenylthiocarbazone (dithizone). A chloroform-extractable lead chelate forms at pH 9.5 where the dithizone is a monoprotic weak acid. A very dilute 0.005% solution of dithizone in chloroform is employed. The absorption of 510-nm light by the lead-dithizone complex extracted into chloroform is measured to allow calculation of the concentration of lead analyte.

Dimethylglyoxime forms a chloroform-extractable chelate, bis(dimethylglyoximato)nickel(II) (structure following) with dimethylglyoxime. The red chelate is formed in a pH 7 to 12 aqueous medium containing citrate by the addition of an aqueous solution of dimethylglyoxime. The chelate is extracted into chloroform and measured by light absorbed at 366 nm. Cobalt forms a brown water–soluble complex with dimethylglyoxime, but it is not extracted by chloroform.

Other examples of liquid–liquid extraction procedures for the determination of metals extracted into chloroform include uranium(VI) as the 8-hydroxyquinolate, copper(II) as the 2,9-dimethyl-1,10-phenanthroline (neocuproine) chelate, and beryllium as the acetylacetonate complex.

16.8   Practical Aspects of Solvent Extraction

As noted in Section 16.6, liquid–liquid extraction can be either a batch extraction or a continuous extraction. In cases where the desired material has a favorable distribution coefficient, the batch extraction is favored because it is simple and can be carried out with a separatory funnel (Figure 16.3). Mechanically, batch extraction with a separatory funnel consists of simply adding the two liquids to the funnel, shaking, allowing the two liquids to settle and separate, and draining the lower layer through the stopcock. The extraction can be repeated several times to completely remove the desired constituent. The operation is facilitated with a solvent denser than water, such as chloroform, because the aqueous phase can be left in the funnel after each extraction. With a less dense solvent, such as ether, both layers must be removed separately from the funnel and the aquatic layer placed back in the device. Separatory funnels have been developed in which the upper layer is pushed out through a stopcock in the top of the funnel by mercury draining into the bottom of the device; these are seldom worth the trouble in routine applications. In some cases, such as the extraction of organics from "dirty" samples (coal-coking byproduct water is such a sample), emulsions may form in which the liquid layers do not separate well. This may necessitate centrifugation to separate the two phases. Water droplets entrapped in the organic extract can be removed by filtration through dry filter paper, followed by washing the paper with fresh organic solvent.

Stripping is employed to remove the extracted material from the organic solvent in cases where the solvent interferes with the determination. For a nonvolatile solute this can be accomplished by heating the organic solution and drawing the solvent off under a vacuum (vacuum distillation). In other cases chemicals can be added that precipitate the analyte or cause it to go back into an aqueous phase.

Backwashing is employed to remove impurities from the organic solvent back into aqueous solution, leaving the desired substance in the organic solvent. This is essentially a method of analyte purification.

16.9   Separation by Ion Exchange

The Ion Exchange Process

Ion exchange is a process in which ions from solution bind to functional groups of opposite charge on a solid material (or immiscible liquid), releasing ions of like charge. This is illustrated by the reaction

$$H^{+-}\{Cat(solid)\} + Na^+(aq) \rightleftharpoons Na^{+-}\{Cat(solid)\} + H^+ \qquad (16.17)$$

Solid cation   Solid cation
exchanger with   exchanger with
bound H^+ ion   bound Na^+ ion

and the reaction

$$OH^{-+}\{An(solid)\} + Cl^-(aq) \rightleftharpoons Cl^{-+}\{An(solid)\} + OH^-(aq) \qquad (16.18)$$

Solid anion   Solid anion
exchanger with   exchanger with
bound OH^- ion   bound Cl^- ion

Types of Solid Ion Exchangers

Solid ion exchangers consist of polymers with a high molecular weight to which are bound many negatively charged functional groups (cation exchangers) or positively charged functional groups (anion exchangers). Typical structures of ion exchangers are shown in Figures 16.8 and 16.9. The cation exchanger shown in Figure 16.8 is called a **strongly acidic cation exchanger**, because in the H^+ form it contains the $-SO_3^-H^+$ group, which is a strong acid (releases H^+ readily). When the functional group binding the cation is the $-CO_2^-$ group, the exchange resin is called a **weakly acidic cation exchanger**, because the $-CO_2H$ group is a weak acid. Figure 16.9 shows a **strongly basic anion exchanger** in which the functional group is a quaternary ammonium group, $-N^+(CH_3)_3$. In the hydroxide form, $-N^+(CH_3)_3OH^-$, the hydroxide ion is readily released, which is what makes this a *strongly basic* anion exchanger.

Weakly basic anion exchangers have functional groups of the type $-N(CH_3)_2$ or $-NHCH_3$. In neutral water these groups are not charged—

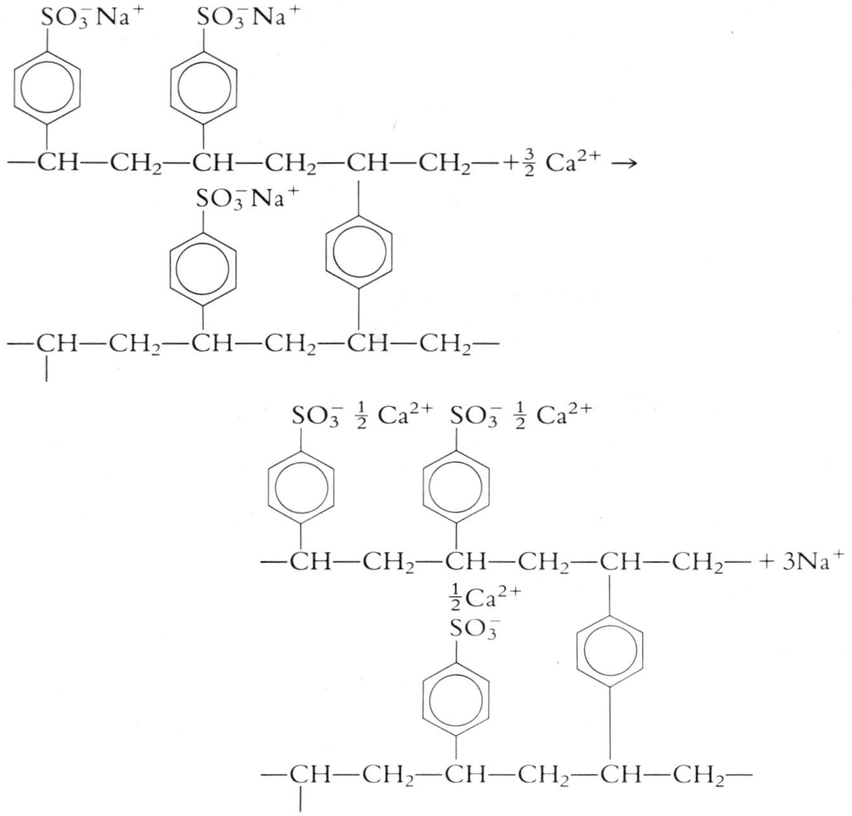

$$SO_3^- Na^+ \quad SO_3^- Na^+$$

$$-CH-CH_2-CH-CH_2-CH-CH_2-+\tfrac{3}{2}\,Ca^{2+} \rightarrow$$

$$SO_3^- Na^+$$

$$-CH-CH_2-CH-CH_2-CH-CH_2-$$

$$SO_3^- \tfrac{1}{2} Ca^{2+} \quad SO_3^- \tfrac{1}{2} Ca^{2+}$$

$$-CH-CH_2-CH-CH_2-CH-CH_2- + 3Na^+$$

$$\tfrac{1}{2}Ca^{2+}$$

$$SO_3^-$$

$$-CH-CH_2-CH-CH_2-CH-CH_2-$$

FIGURE 16.8   Strongly acidic cation exchange resin showing partial structure of the resin polymer. The reaction shows the exchange of Ca^{2+} in water for Na^+ on the resin, a process used in water softening.

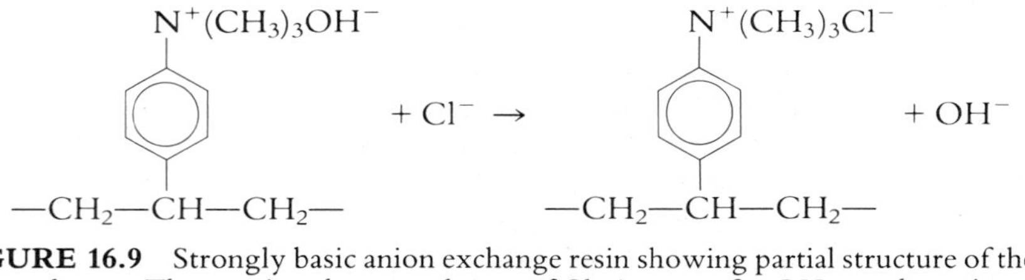

$$N^+(CH_3)_3OH^- \qquad\qquad\qquad N^+(CH_3)_3Cl^-$$

$$+ Cl^- \rightarrow \qquad\qquad\qquad\qquad + OH^-$$

$$-CH_2-CH-CH_2- \qquad\qquad\qquad -CH_2-CH-CH_2-$$

FIGURE 16.9   Strongly basic anion exchange resin showing partial structure of the resin polymer. The reaction shows exchange of Cl^- in water for OH^- on the resin.

that is, the equilibrium of the reaction

$$H_2O + \{H_3C-\underset{\underset{H}{|}}{\overset{\overset{H}{|}}{N}}-An\,(solid)\} \rightleftarrows OH^- {}^+\{H_3C-\underset{\underset{H}{|}}{\overset{\overset{H}{|}}{N}}-An\,(solid)\} \qquad (16.19)$$

lies strongly to the left. However, the addition of a strong acid, such as HClO$_4$, pushes the reaction strongly to the right

$$HClO_4 + \{H_3C-\underset{\underset{H}{|}}{N}-An\,(solid)\} \rightleftarrows ClO_4^- {}^+\{H_3C-\underset{\underset{H}{|}}{\overset{\overset{H}{|}}{N}}-An\,(solid)\}$$

$$(16.20)$$

In this kind of acidic solution the resin can function to exchange other anions, such as Cl$^-$

$$Cl^- + ClO_4^- {}^+\{H_3C-\underset{\underset{H}{|}}{\overset{\overset{H}{|}}{N}}-An\,(solid)\} \rightleftarrows Cl^- {}^+\{H_3C-\underset{\underset{H}{|}}{\overset{\overset{H}{|}}{N}}-An\,(solid)\}$$

$$(16.21)$$

The addition of strong base (NaOH) readily regenerates weakly basic resins that are in the salt form with uncharged functional groups.

A specialty form of resin with an exceptionally strong affinity for chelatable metal ions is the **chelating resin** containing the chelating iminodiacetate group (Chelex 100, Bio Rad Laboratories). It is normally used in the sodium form, which is shown chelating copper(II) ion in water.

Chelating ion exchange resin in the sodium form; unshared pair of electrons available for binding metal ion shown on N

Cu^{2+} ion chelated to chelating ion exchange resin

$$(16.22)$$

16.10   Ion Exchange Equilibria

As with many other phenomena used in analytical chemistry, ion exchange equilibria are governed by the *law of mass action*. For example, the equilibrium constant of the reaction

$$H^{+-}\{Cat(solid)\} + Na^{+}(aq) \rightleftharpoons Na^{+-}\{Cat(solid)\} + H^{+} \qquad (16.17)$$

Solid cation   Solid cation
exchanger with   exchanger with
bound H^{+} ion   bound Na^{+} ion

Concentrations: $[H^{+}(res)]$ $[Na^{+}(aq)]$ $[Na^{+}(res)]$ $[H^{+}(aq)]$

is given by the expression

$$K = \frac{[Na^{+}(res)][H^{+}(aq)]}{[Na^{+}(aq)][H^{+}(res)]} \qquad (16.23)$$

Although $[Na^{+}(res)]$ and $[H^{+}(res)]$ refer to concentrations of the ions on a solid, they vary from 0 to some maximum value. For singly charged ions, the maximum value is the **ion exchange capacity** of the resin, typically in units of equivalents per 100 g. For a doubly charged ion the maximum concentration on the resin in moles of ion per 100 g of resin is half the ion exchange capacity; for triply charged ions, one-third. In cases where ions of unequal charge are exchanged—for example, exchange of Ca^{2+} in solution for Na^{+} on the resin, the equilibrium constant takes a form such as

$$K = \frac{[Na^{+}(aq)]^{2}[Ca^{2+}(res)]}{[Na^{+}(res)]^{2}[Ca^{2+}(aq)]} \qquad (16.24)$$

where the concentrations are raised to their appropriate powers.

Ion exchange equilibrium calculations are simplified when, as is often the case, one ion predominates both in solution and on the resin. For example, suppose that Reaction 16.17 is carried out in an acidic medium, such that H^{+} is predominant in both aquatic and resin phases. In that case the K expression for that reaction (16.23) rearranges to

$$\frac{[Na^{+}(res)]}{[Na^{+}(aq)]} = K\underbrace{\frac{[H^{+}(res)]}{[H^{+}(aq)]}}_{\text{Constant ratio}} = K_{d} \qquad (16.25)$$

This yields a distribution coefficient K_{d} analogous to that shown earlier in the chapter for extraction equilibrium between an aqueous and a nonaqueous solvent phase.

16.11   **Uses of Ion Exchange Resins**

The most important applications of ion exchange resins are in liquid chromatography, discussed in Chapter 18. However, there are some useful batch ion exchange processes in which the ion exchange resin is contacted with individual quantities of solution, much as is done in a batch liquid-liquid extraction (Section 16.6). Some of these applications are discussed here.

One application of batch ion exchange processes is *isolation and concentration of analyte*. Chelating resins (Section 16.9) are especially effective in removing trace quantities of chelatable metals from large volumes of water, such as natural water samples. The metals can subsequently be eluted from the resin with a small volume of acid and determined by techniques such as atomic absorption spectrophotometry (see Chapter 22).

Interfering ions, particularly those with a charge opposite in sign from that of the analyte, can often be removed by a batch ion exchange process. For example, orthophosphate species ($H_2PO_4^-$, HPO_4^{2-}) tend to form interfering phosphate precipitates with Ba^{2+} or Ca^{2+} in the gravimetric determination of these ions as the sulfate or oxalate, respectively. The interfering orthophosphate can be removed by exchange for noninterfering Cl^- ion by anion exchange with a chloride-form anion exchange resin. Similarly, some cations, most commonly Al(III) and Fe(III), coprecipitate with barium sulfate in the gravimetric determination of sulfate as $BaSO_4(s)$. The interfering cations are readily removed by passing the sample through a sodium-form cation exchange resin, which does not hold the sulfate at all.

The *total salt content* of a sample can be determined by contacting the sample solution with a cation exchange resin in the H^+ form. Each milliequivalent of cation releases 1 meq of H^+ — for example, 1 mmol of Na^+ releases 1 meq of H^+; 1 mmol of Ca^{2+} (2 meq of Ca^{2+}) releases 2 meq of H^+. The salt content of the sample in milliequivalents is then simply determined by titrating the H^+ released with a base titrant. The same thing can be done by equilibrating the sample with an anion exchange resin in the OH^- form and titrating the OH^- released with a standard acid.

Programmed Summary of Chapter 16

The major terms and concepts introduced in this chapter are contained in this summary in a programmed format. To derive the most benefit from the summary, you should fill in the blanks for each question and then check the answers at the end of the book to see if your choices were correct.

Separations in chemical analysis generally involve the distribution of one or more species between two different (1) _____. Complete separation in a case where a single separation is only partial requires a (2) _____.

A broad category of separations involves distribution of a species between two bulk phases; the alternative is distribution between one bulk phase and a (3) _____. The two major kinds of thin layers consist of (4) _____. A thin layer used for separations is held on a (5) _____ phase, whereas the solute species are distributed between the thin layer and a (6) _____ phase. Although differences in solubility may be favorable for a separation by precipitation, some problems that may arise in such a separation are (7) _____ of normally soluble salts, (8) _____ of impurities, and (9) _____ of colloidal precipitates. Three major ways of separating metal ions by precipitation are control of acidity to precipitate (10) _____, formation of inorganic (11) _____ salts, and separations with usually high-molecular-weight (12) _____ precipitants. Metals extracted into organic solvents must be in an uncharged form, usually as uncharged (13) _____ or combined with (14) _____ that neutralize the charge of the cation and form with it an organic-like species. The mathematical formulation of the distribution law for an uncharged species A soluble in both water and an organic solvent is (15) _____. A process by which an extracting solvent is continually distilled and recirculated through a water solution of the solute to be extracted is called a (16) _____ extraction. An extraction process in which portions of two immiscible solvents are moved in opposite directions relative to each other is called (17) _____. Acetylacetone and 8-hydroxyquinoline are both examples of (18) _____. Two examples of large cations without a primary hydration sphere that form extractable ion pairs with large, singly charged anions are (19) _____. The best known example of a solvent which, itself, combines with and extracts metal ions is (20) _____. Dimethylglyoxime is most commonly used for the extraction of (21) _____. In solvent extraction, stripping is used to (22) _____ and backwashing is employed to (23) _____. A process that involves equilibration between ions in solution and a material in another phase containing functional groups capable of binding the ions is (24) _____. The functional group $-N(CH_2CO_2^-)_2$ is found on (25) _____ ion exchange resins. The maximum number of equivalents of ion that an ion exchange resin can hold per unit weight is called the resin's (26) _____.

Questions

1. Name two analytical techniques that are by nature separation processes.
2. Complete: Separations for chemical analysis generally involve the distribution of two or more substances between two different _____.
3. Match:
   a. Sublimation
   b. Liquid-liquid extraction
   c. Gaseous diffusion

d. Distillation

e. Crystallization

1. Separation of $^{235}UF_6(g)$ from $^{238}UF_6(g)$
2. Gravimetric determination of Ba^{2+}
3. Physical purification of I_2, which does not commonly form a liquid phase
4. Separation of low-boiling dichloromethane and higher-boiling carbon tetrachloride
5. Removal of phenol from water into dichloromethane

4. What distinguishes the two major types of thin-layer phases used to adsorb solutes?

5. What phenomena tend to complicate separation by precipitation?

6. Why do some metals precipitate as sulfides from a 3 M HCl medium but not from a 0.3 M HCl medium?

7. What can be said about the charge of a species normally extracted into an organic solvent?

8. What would be wrong with expressing the distribution coefficient as the ratio of total solute in the organic phase to total solute in the aqueous phase?

9. Which will give the greatest extraction into an organic solvent, one extraction with V_{org} mL of solvent or five extractions with ⅕ V_{org} mL of organic solvent?

10. What is meant by a continuous extraction?

11. What is introduced at each step in a countercurrent fractionation?

12. Why, in a simple sense, do salts tend not to be extracted from water into organic solvents?

13. Match:

a. "Phen"
b. Dithizone
c. Cupferron
d. Oxine

1. Ammonium salt of *N*-nitroso-*N*-phenylhydroxylamine
2. 8-Hydroxyquinoline
3. Diphenyldithiocarbazone
4.

14. What are two large, organic-like cations used to form extractable ion pairs?

15. What is the mechanism of the solvent tri-*n*-butyl phosphate in extracting metal salts?

16. Commonly used extractants bond to inorganic salts (usually metal cations), enabling their extraction into solvents such as chloroform. What additional analytical function do many of these extractants have?

17. What are some of the complications that may be encountered in the practice of solvent extraction employing a separatory funnel?

18. What is the definition of ion exchange?

19. Match the following applying to functional groups on ion exchangers:

a. —CO_2H
b. —$N^+(CH_3)_3$
c. —SO_3^-
d. —$NHCH_3$

1. Strongly acidic
2. Strongly basic
3. Weakly basic
4. Weakly acidic

20. How does the maximum molar concentration of ions on an ion exchange resin differ with ion charge?

21. When can the equilibrium constant describing equilibrium with an ion exchange resin be regarded as a distribution coefficient?

22. What are some uses of batch ion exchange processes in chemical analysis?

Problems

1. Organic compound Y has a distribution coefficient K of 0.75 for its distribution between dichloromethane and water. Calculate the fraction of Y remaining in 100.0 mL of aqueous solution after one, two, and three extractions with 50.0 mL of the dichloromethane and a final extraction with 150 mL of dichloromethane.

2. Three extractions of compound Y from 75.0 mL of water with dichloromethane removed 70.0% of Y from the water. The value of K was 0.75. Each of the three extractions was accomplished with an equal volume of dichloromethane. What volume of dichloromethane was employed each time?

3. A 0.3207-g steel sample was dissolved in acid, appropriately treated, and diluted to 100.0 mL. Nickel as the dimethylglyoximate was extracted from the aqueous sample into 50.0 mL of chloroform. The procedure was such that 96.0% of the nickel was extracted into the chloroform. Measurement of light absorbed at 366 nm revealed 4.58 $\mu g/mL$ of Mn in the chloroform extract. What was the percentage of Ni in the steel sample?

4. A mixture of exactly 50% NaCl (fw = 58.44) and 50% Na_2SO_4 (fw = 142.0) was dissolved in about 100 mL of water and equilibrated with strong base ion exchange resin in the hydroxide form. The OH^- liberated required 36.47 mL of 0.00971 M HCl for titration. What weight of the salt mixture was dissolved in the water originally?

5. A solute X was present in 100 mL of water at a concentration of 2.75×10^{-2} mol/L. The water was extracted with two 50-mL portions of dichloromethane. The second extraction removed 5.00×10^{-4} mol of X from the water. What is the value (or possible values) of the distribution coefficient K?

6. The species HB is extractable into chloroform with a distribution coefficient $K = [HB_{org}]/[HB_{aq}] = 11.7$. In water the nonextractable species H_2B^+ and B^- also form. The H_2B^+ can be regarded as a diprotic acid ionizing in water to yield HB and B^-, and it has a K_{a1} of 2.33×10^{-4} and a K_{a2} of 4.07×10^{-9}. Calculate the pH at which the maximum number of millimoles of B species is extracted and calculate the value of the fraction

$$\frac{[HB_{org}]}{[H_2B_{aq}^+] + [HB_{aq}] + [B_{aq}^-]}$$

at the pH of maximum extraction. (This problem may require a little review of aqueous acid-base equilibrium.)

7. Water hardness caused by Ca^{2+} and Mg^{2+} hardness ions was measured by equilibrating a 100-mL sample of the water with a chelating ion exchange resin in the sodium form and measuring the sodium released by atomic absorption spectrometric analysis (see Chapter 22). The water contained initially 2.61×10^{-4} M Na^+ and after equilibration with the chelating ion exchange resin the concentration of Na^+ was 1.37×10^{-3} M. What was the concentration of hardness ions in the water sample in millimoles per milliliter?

8. A 20.0-g (dry-weight basis) portion of a cation exchange resin in the H^+ ion form containing 17.3 milliequivalents of H^+ per 100 g resin was equilibrated with 250 mL of water containing a concentration of Na^+ ion of 2.46×10^{-3} mmol/mL. After equilibration the water contained 1.87×10^{-3} mmol/mL Na^+. What was the pH of the water after equilibration? What was the value of the equilibrium constant K (Equation 16.23), with aqueous concentrations expressed in millimoles per milliliter and concentrations of species on the ion exchange resin in equivalents per 100 g?

Principles of Chromatography

1. What is meant by chromatography.
2. Distinction between adsorption chromatography and partition chromatography.
3. Distinction between gas chromatography and liquid chromatography.
4. What is meant by the column chromatographic techniques of ion exchange chromatography, molecular exclusion chromatography, affinity chromatography, and capillary chromatography.
5. Paper chromatography and thin–layer chromatography involving separations on planar surfaces.
6. Important aspects of chromatographic columns, including column packing, solid support material, eluent, and solute.
7. The function and nature of chromatographic detectors.
8. Description of chromatographic peaks.
9. General characteristics of chromatograms.
10. Separation of chromatographic bands (peaks) and zone broadening.
11. Plate theory and the terms and concepts used to describe this theory.
12. Rate theory and the terms and concepts used to describe this theory.
13. Qualitative and quantitative aspects of peak resolution.
14. Causes of and remedies for irregular peak shapes.
15. Major ways in which chromatography is used for chemical analysis.

17.1   Definition and History of Chromatography

What Is Chromatography?

Chromatography is the separation of two or more substances by virtue of their different affinities toward a *stationary phase* and a *mobile phase* that moves in relation to the stationary phase. Figure 17.1 illustrates the classic picture of chromatography by the separation of two solutes on a column packed with an adsorbent solid, through which flows a solvent. Solute Y has a relatively greater affinity for the stationary phase material than does X in comparison to their solubilities in the mobile phase. As a result solute Y is washed down the column more slowly by the solvent and emerges after X, resulting in the separation of X and Y.

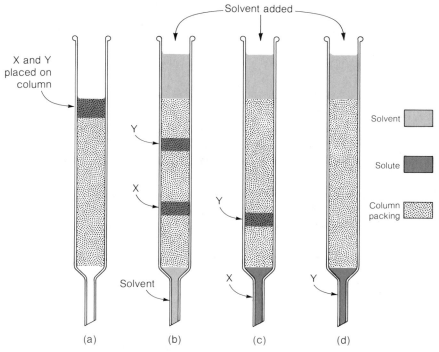

FIGURE 17.1   Illustration of chromatographic separation: (a) mixture of solutes X and Y added to top of column; (b) after solvent has flowed through the column for some time, X and Y are separated on the column; (c) X elutes from the column; (d) Y elutes from the column to complete the separation

History of Chromatography

The first research paper on chromatography was published by Mikhail Tswett, a biochemist on the faculty of the Veterinary School of Warsaw, Poland. Tswett used a column packing (stationary phase) of calcium carbonate and a petroleum ether (low-boiling hydrocarbon) mobile phase to separate plant pigments into their component colors. Since the separated pigments appeared as different-colored bands on the column, Tswett used the word *chromatos*—Greek for *color*—to describe the technique. Though he published a number of papers on the topic, World War I terminated his promising career and chromatography was virtually forgotten until revived by biochemists in 1931.

The chromatographic technique just described and illustrated in Figure 17.1 is **adsorption chromatography**, in which the solute adheres to the surface—that is, is adsorbed by a solid. Better chromatographic separations are often obtained when the stationary phase is a liquid immobilized on a solid and the mobile phase is a liquid (sometimes water) immiscible with the adsorbed liquid phase. This is called **liquid-liquid partition chromatography** and was first described by two biochemists, A. J. P. Martin and R. L. M. Synge in 1942. Remarkably little note was taken of this advance for about a decade, although the two investigators described the principles of chromatography and suggested that an adsorbed liquid phase and a gaseous mobile phase should be an excellent way of separating compounds in the vapor state. In 1952, Martin and A.T. James demonstrated **gas-liquid partition chromatography**. The technique immediately caught on and within four or five years was in routine use in laboratories throughout the world. Its development continues at an appreciable pace even today. In 1954 Martin and Synge shared the Nobel Prize in chemistry for their pioneering work in chromatography.

Despite its enormous potential, liquid-liquid partition chromatography did not begin its tremendous growth period until about 1970. This was due to the lack of suitable detectors for the chromatographic effluents, and even more because of the fact that efficient separations require small particles in the column, which in turn demand high mobile phase pressures for an appreciable mobile phase flow rate; these are experimental conditions that were hard to attain. The development of suitable high-pressure pumps and injection systems, column packings that did not collapse under pressure, and refractive index and ultraviolet light-absorbing detectors has resulted in the spectacular growth of **high performance liquid chromatography, HPLC,** known at an earlier stage of development as high-pressure liquid chromatography.

Chromatography is by far the leading analytical separation technique. Considering all its variations, chromatography is the most widely applied method of chemical analysis. It is capable of separating organic chemical isomers (see Section 1.6), which have only minute differences in structure between them. Some of the chromatographic detectors used can detect extremely small quantities of analytes. Other detectors, particularly the mass spectrometer used with a gas chromatograph, are capable of identifying any one of hundreds of different compounds by computer matching. (This is accomplished in a mass spectrometer by comparing the mass spectra obtained experimentally with those stored in computer memory.)

17.2   Types of Chromatography

Diverse Nature of Chromatography

There are many varieties of chromatography, of which several major ones are illustrated in Figure 17.2. This figure is based upon differences in the mobile phases and the solid surfaces or layers with which solutes in these mobile phases interact. In a sense, chromatography is the continuous extraction

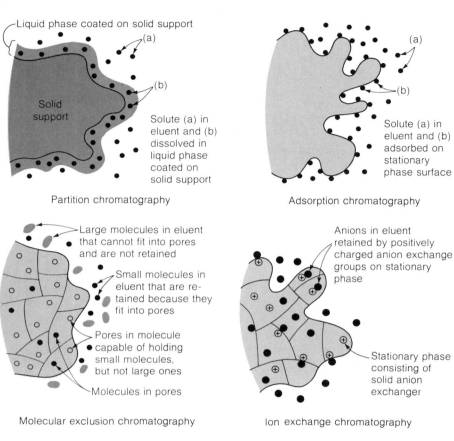

Liquid phase coated on solid support

(a)

(b)

Solid support

Solute (a) in eluent and (b) dissolved in liquid phase coated on solid support

Partition chromatography

(a)

(b)

Solute (a) in eluent and (b) adsorbed on stationary phase surface

Adsorption chromatography

Large molecules in eluent that cannot fit into pores and are not retained

Small molecules in eluent that are retained because they fit into pores

Pores in molecule capable of holding small molecules, but not large ones

Molecules in pores

Molecular exclusion chromatography

Anions in eluent retained by positively charged anion exchange groups on stationary phase

Stationary phase consisting of solid anion exchanger

Ion exchange chromatography

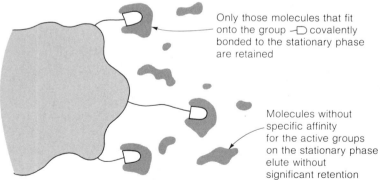

Only those molecules that fit onto the group ⊸ covalently bonded to the stationary phase are retained

Molecules without specific affinity for the active groups on the stationary phase elute without significant retention

Affinity chromatography

FIGURE 17.2   The major types of chromatography classified according to differences in the solid surfaces or layers on solid surfaces with which solutes interact

version of countercurrent fractionation (see Section 16.6 and Figure 16.6). In place of discrete extractions, however, chromatography involves continual exchange and equilibration between mobile and stationary phases. However, a major mathematical model of chromatography (Section 17.5) treats it as a series of equilibrations involving a large number of column "plates."

Adsorption Chromatography

The form of chromatography first described—mechanically simple, though theoretically complex—is *adsorption chromatography*, in which a solute carried by a liquid or gas mobile phase is adsorbed and released by a solid surface on particles packed in a column. One major factor that accounts for the separation of the solutes is their different affinities for the solid surfaces.

Partition Chromatography

In general, much better separations are obtained when the solid surface is coated with a liquid, or a "liquid-like" phase chemically bonded to the solid support. In this kind of separation, solutes are separated by virtue of their differences in solubility in the stationary liquid phase. In traversing the column, each solute molecule spends part of its time in the immobile liquid phase and the mobile liquid or gas phase. This means that the solutes are *partitioned* between the immobile liquid phase and the mobile phase, so that the technique is called **partition chromatography**.

Ion Exchange Chromatography

A type of chromatography that involves very specific interaction between a stationary solid phase and a mobile liquid phase is **ion exchange chromatography** (see Ion Exchange, Section 16.9) in which the passage of ions through a column is slowed by cation exchange or anion exchange with, for example, $-SO_3^-$ and $-N^+(CH_3)_3$ groups, respectively. The stationary phase in this type of chromatography is an ion exchange resin.

Molecular Exclusion Chromatography

A stationary phase consisting of a gel with pores within a rather narrow size range for each type of gel can be used to separate molecules on the basis of size. Molecules too large to fit into the pores at all bypass the pores in the gel and go right through the column, whereas those that fit into the pores are retarded by entering the pores to various degrees. Two types of separations are accomplished, one in which all molecules above a certain size emerge from the column in a group, and the other in which smaller molecules are separated from each other on the basis of molecular size. Either liquid or gas mobile phases can be employed. The general term for this approach is called **molecular exclusion chromatography**, also known as **gel permeation** or **gel filtration** chromatography.

Affinity Chromatography

A relatively new approach to chromatography offering tremendous selectivity is that based upon strong, selective interactions between solute molecules and organic molecules chemically bonded to the stationary phase support. Since the separation is based upon a strong *affinity* between the stationary and solute molecules, this kind of chromatography is termed **affinity chromatography**. It is especially useful in biochemical separations where, for example, the active component of the stationary phase might be an antibody to a specific protein, which retains that protein from a mixture of proteins passed through the column. The retained protein can be eluted from the column by a change in the mobile phase, such as a change in pH or addition of a salt solution.

Gas Chromatography

There are two major types of chromatography in which a gas is the mobile phase. These are **gas-solid chromatography, GSC**, with an adsorptive solid stationary phase held in a column, and the more common and versatile **gas-liquid chromatography, GLC**, in which the stationary phase consists of liquid held by, or bonded to the pores of, a granular solid. Alternatively for GLC, the stationary phase may be coated on the walls of a narrow-bore column, which is the basis of the highly useful and rapidly growing technique of **capillary column chromatography**.

Planar Chromatography

With the exception of capillary column chromatography, all the chromatographic techniques discussed above, with either liquid or gas mobile phases, involve **packed columns** in which the stationary phase consists of small solid particles or a gel and is packed inside a chromatographic column. Two very widely used chromatographic techniques are based upon separations with a flat, planar stationary phase. One of these, **paper chromatography**, has either paper or a liquid held in the pores of paper as a stationary phase. **Thin-layer chromatography** uses a layer of solid bonded to a flat surface; sometimes a layer of liquid immiscible with the mobile phase and held on a flat surface is used in place of the solid material.

17.3   The Language of Chromatography

Chromatographic Terms

In order to understand chromatography, it is important to know a number of terms applied to it. Some of these are defined here, others later in the chapter as the concepts that they describe are introduced. It will be helpful to refer to Figures 17.1 and 17.2 in discussing these terms.

As shown in Figure 17.1, most chromatographic separations occur in a **chromatographic column**. These range all the way from short, thick glass columns, such as may be made from a laboratory 50-mL buret, to *capillary columns* for gas chromatography that are made of polymer-coated flexible silica tubing up to 50 meters long, and only a few micrometers in inside diameter. The granular solid inside a packed column is the **column packing**; when, as is often the case, it is coated with a liquid or chemically bonded to a thin layer of material that acts to retain the solutes being separated, the column packing is called the **solid support**. The mobile phase, such as the solvent added as shown in Figure 17.1, is termed an **eluent**; the process of passing eluent through the column to bring the solutes through is called **elution**, and the fluid issuing from the end of the column is called the **eluate**. At the end of the column a **detector** shows when a solute emerges from the column and provides a measure of the quantity of solute.

The Chromatogram

A recording of electronically amplified detector response as a function of time and/or volume of eluent through the column is a **chromatogram**, as shown in Figure 17.3. Here the solutes appear on the chromatogram as bell-shaped "rounded triangular" **chromatographic peaks**. Ideally these peaks are shaped like a normal probability distribution curve (see Chapter 3, Section 3.7

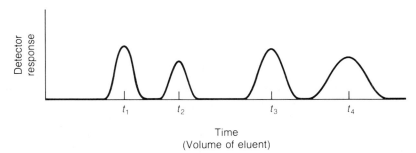

FIGURE 17.3   A chromatogram showing detector response as a function of time or volume of eluent. Individual solutes emerging with the eluate are displayed as chromatographic peaks.

and Figure 3.4). For some common detectors, the area of a peak due to a specific compound is proportional to the quantity of that compound, which allows for a quantitative analysis. The time that it takes the peak to emerge from the column is called the **retention time**. This time is measured from the point at which the solute starts moving through the column (injection of solute or beginning of eluent flow) to the highest value of the detector response for the peak. The retention times from the first peak to the last peak to elute as shown in Figure 17.3 are labeled t_1 through t_4. If eluate volume is substituted for time (the two are linearly related for a constant mobile phase flow rate), a **retention volume** can be assigned to a peak. This is the volume of eluate issuing from the column from the instant of peak injection to the time of maximum detector response as the peak elutes. Retention times and volumes differ with chromatographic conditions and even with individual columns constructed from the same materials and packings; therefore, it is usually more meaningful to express **relative retention times** and **relative retention volumes** in which comparison is made with a standard solute introduced with the sample. The **retention ratio** R is defined as

$$R = \frac{\text{Time for solvent to go through column}}{\text{Time for solute to go through column}} \tag{17.1}$$

17.4   Basics of Chromatographic Theory

Distribution between Phases

A chromatographic separation is a dynamic process in which each solute spends part of the time in the mobile phase and part of the time in the stationary phase (see swimmer analogy in Figure 17.4). The process can be viewed as a large number of equilibrations of solute between mobile and stationary phases. These are described quantitatively by the **partition coefficient** K

$$K = \frac{C_s}{C_m} \tag{17.2}$$

where C_s and C_m are the solute concentrations in the stationary and mobile phases, respectively. Under conditions of constant K, C_m and C_s are related linearly, giving rise to the term **linear elution chromatography**.

Linear elution chromatography views the solute as undergoing a series of equilibrations between the solid phase and the solvent mobile phase. Whenever the solute is in the gas or liquid mobile phase, it is carried away from the point of sample injection toward the point of sample elution, *downstream*, so to speak. *The rate at which the solute moves through the column is proportional to the time it spends in the mobile phase relative to the time spent in the stationary phase.* It is

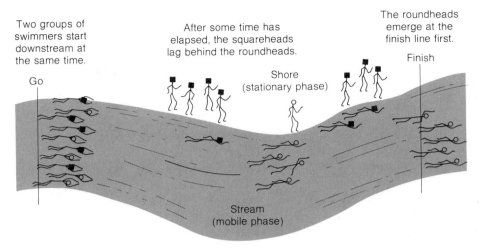

FIGURE 17.4 The swimmer analogy to chromatographic separation. The squareheads like the shore relatively better and spend more time on it, where they are restrained from moving downstream by the rules of the race (solutes do not migrate significantly in the stationary phase). The roundheads, who prefer the water (mobile phase), move downstream faster and "elute" first at the finish line.

readily seen that the time spent in the mobile phase is small (and the rate slow) for large values of K, and vice versa. The net result of this is, as shown in Figure 17.1(b), the separation on the column of solutes into distinct **bands** (areas on the column containing each solute). If elution is continued long enough, both bands emerge at different times with the column eluate.

Separation and Broadening of Solute Bands

As illustrated for the separation of two solutes, X and Y, in Figure 17.5, two major things happen to the solute bands. The first of these is that the bands *separate*, which is of course the goal of any chromatographic process. The second is that the bands become broader. This is called **zone broadening**; it is an unavoidable, but undesirable aspect of chromatographic separations. It is apparent that lengthening the column increases the degree of separation. However, this also increases zone broadening, sometimes to the extent that the peak becomes indistinguishable from the background.

Plate Theory and Rate Theory

Chromatographic theory must explain both separations and peak broadening. The older *plate theory* may be used to explain much of the basic observation of chromatographic behavior, and it is outlined in the following section. However, plate theory is unable to explain some of the important subtleties of

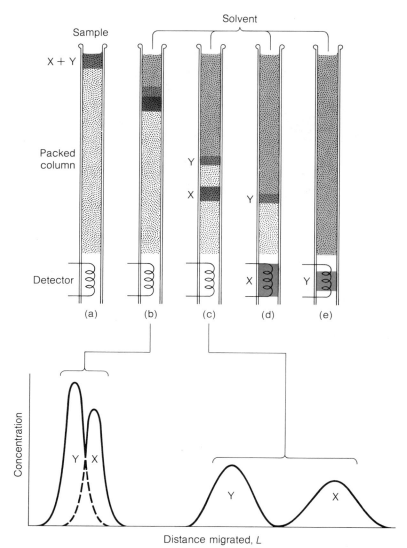

FIGURE 17.5 Chromatography of solutes X and Y showing separation of the peaks and zone broadening of each peak. Chromatograms corresponding to stages of separation b and c are shown in the lower part of the figure.

zone broadening; these are explained by the *rate theory* or *kinetic theory*, as outlined in Section 17.6.

17.5   Chromatographic Plate Theory

Basis of Plate Theory

The discussion of chromatographic theory requires consideration of several features of chromatographic peaks. An ideal peak has a bell-shaped Gaussian configuration (normal error curve, see Section 3.7), as shown in Figure 17.6. The **plate theory** of Martin and Synge (first published in 1941) views a chromatographic column as being divided into a number of adjacent segments called **theoretical plates** analogous to individual steps in a *countercurrent separation* (Section 16.6). In each of these hypothetical plates, complete equilibration of solutes between stationary and mobile phases is envisaged to occur. These equilibrations are seen as occurring sequentially from the first to the last plate in the column. Greater separation efficiency occurs with a greater **number of theoretical plates, N**. The length of each small column segment

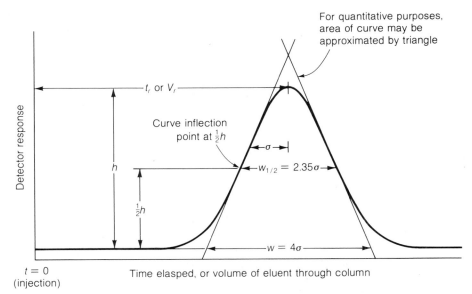

FIGURE 17.6 Major features of an idealized chromatographic peak in a Gaussian configuration. The parameters shown are t_r, retention time; V_r, retention volume; h, peak height; $w_{1/2}$, width at ½ h; and w, width of the peak. It is seen that w is measured by extrapolating the essentially linear sides of the peak. The symbol σ is the standard deviation used to plot the Gaussian curve.

corresponding to a theoretical plate is called the **height equivalent of a theoretical plate, H** (or **HETP**). The parameters H and N are related to the **length L of packing** in the column by the equation

$$L = NH \tag{17.3}$$

The value of N can be determined from the bandwidth w and the retention time t_r or retention volume V_r of a chromatographic peak (see Figure 17.6). The formula for N is

$$N = \frac{t_r^2}{\sigma^2} = \frac{16t_r^2}{w^2} = \frac{5.54t_r^2}{w_{1/2}^2} \tag{17.4}$$

where all the terms are defined in Figure 17.6. Note that $w_{1/2}$ *is not* $1/2w$, but is *the width of the peak at* $1/2h$. It should be emphasized that a theoretical plate is a useful, but abstract, idea and that N for the same column may differ with the mobile phase and even with different solutes.

The value of N obviously increases with the length of the column. Therefore the best way to express column separation efficiencies is by the height equivalent to a theoretical plate, H. This is expressed as

$$H = \frac{L}{N} \tag{17.5}$$

Substituting the value of N from Equation 17.4 yields the following:

$$H = \frac{\sigma^2 L}{t_r^2} \tag{17.6}$$

Deficiencies of Plate Theory

The plate theory has a number of deficiencies that have led to its replacement by the rate theory. Among these deficiencies are several assumptions that fail in some cases. The assumption of a constant partition coefficient, $K = C_s/C_m$, with varying solute concentration tends to be invalid in adsorption chromatography, because of crowding of available adsorption sites on the solid adsorbent. The plate theory assumes rapid equilibrium of the solute between the mobile and stationary phases compared to the rate of mobile phase movement; the failure of this assumption at faster mobile phase flow rates leads to less efficiency as manifested by lower N values. Diffusion distances of solutes may be significant relative to H, so that plate-to-plate diffusion occurs that results in peak spreading not accounted for by the plate theory; this phenomenon is called **longitudinal diffusion**. The important effect of mobile phase velocity is neglected in plate theory. The dimensions of the phases are also not considered in plate theory, although one would intuitively deduce that these dimensions are important.

17.6   Chromatographic Rate Theory

Basis of Rate Theory

The **rate theory of chromatography** explains chromatographic separations by taking into account dynamic processes that occur during the separation, including longitudinal diffusion, and eddy diffusion. Thus rate theory accounts for bands broader than those predicted by plate theory. Rate theory considers the effect on band shape of rate of elution, irregular and variable paths followed by solutes going through the tortuous pathways formed by particles packed in the column, and the diffusion of solute forward and backward along the length of the column. The broadening of chromatographic bands, **chromatographic zone broadening**, is illustrated in Figure 17.7.

The van Deemter Equation

The effects just listed are taken into account by the **van Deemter equation** for plate height H

$$H = Av^{1/3} + \frac{B}{v} + Cv \tag{17.7}$$

where v is the rate of mobile phase movement through the column and A, B, and C are empirical constants for a specific column and mobile phase. The B/v term in the van Deemter equation arises from the fact that solute is continually diffusing to the leading and trailing edges of the solute zone. Although the concentration always remains higher in the center of the zone, the zone continues to spread out both before and following mid-peak due to longitudinal diffusion. The term $Av^{1/3}$ in the van Deemter equation results from irregular, random paths taken by solute molecules traversing the stationary phase particles in a column and by **eddy diffusion** caused by solute species swirling around in trapped masses (pools) of mobile phase. The Cv term accounts for the fact that there is not instantaneous equilibration of the solute between mobile and stationary phases on the column. This term increases with increasing v because as the mobile phase moves faster, the likelihood increases of a molecule trailing the solute zone as the zone is swept through the column (loss of equilibrium).

The efficiency of a chromatographic separation depends upon the value of N—that is, the number of theoretical plates in the column. The most efficient separations are achieved at minimum H, the height equivalent to a theoretical plate. As shown by the van Deemter relationship, H is a function of the mobile phase flow rate v. Plotting the contributions of each of the factors in the van Deemter equation to H versus the mobile phase flow rate v yields a plot of the

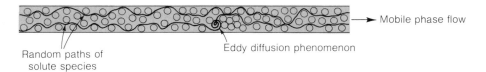

Random paths of
solute species   Eddy diffusion phenomenon

(a)   Solute species take random paths and may be trapped in eddies in the column; the term
      $Av^{1/3}$ accounts for these phenomena in the van Deemter equation

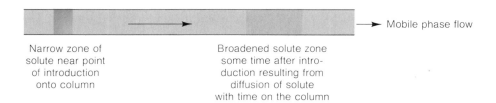

Narrow zone of   Broadened solute zone
solute near point   some time after intro-
of introduction   duction resulting from
onto column   diffusion of solute
                                        with time on the column

(b)   Illustration of longitudinal diffusion, term B/v in the van Deemter equation

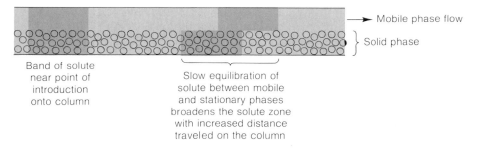

Band of solute
near point of   Slow equilibration of
introduction   solute between mobile
onto column   and stationary phases
                                  broadens the solute zone
                                  with increased distance
                                  traveled on the column

(c)   Slow solute equilibration between stationary and mobile phases increases bandwidth, as
      accounted for in the Cv term of the van Deemter equation

FIGURE 17.7   Phenomena causing chromatographic zone broadening

type shown in Figure 17.8. It is seen that initially the term B/v predominates, so that at low flow rate velocities H is undesirably large. This is because the solute zone moves through the column so slowly that there is a lot of time for the solute to undergo longitudinal diffusion that contributes to high values of H. As B/v drops toward insignificant values, Cv becomes predominant, so that as the time decreases for the solute to equilibrate with the stationary phase with increasing v, the value of H increases. The contribution of $Av^{1/3}$ to H is almost constant throughout the normal range of mobile phase flow rates employed; in some forms of the van Deemter equation, this term is replaced by just a constant, A.

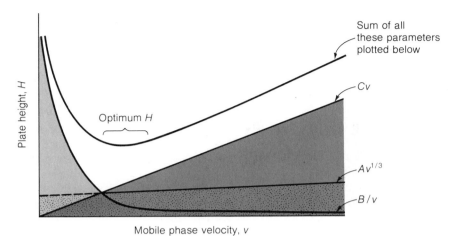

FIGURE 17.8 Plot of van Deemter equation parameters

It can be seen from Figure 17.8 that the lowest range of H occurs in a relatively flat portion of the plot. This provides a comfortable range of mobile phase velocities for acceptable values of H.

Reduced Plate Height and Velocity

Different solutes and different columns can be compared by considering the average stationary phase particle diameter d and the diffusion coefficient D of the solute in the mobile phase. If these two parameters are known, the **reduced plate height** h

$$h = \frac{H}{d} \tag{17.8}$$

and the **reduced velocity v**

$$\mathbf{v} = \frac{vd}{D} \tag{17.9}$$

may be defined. Examination of Equation 17.8 shows that h expresses plate height in multiples of stationary phase particle diameter; for example if H equals twice the particle diameter, h is 2.

The form of the van Deemter equation in terms of the parameters just defined is

$$h = A'\mathbf{v}^{1/3} + \frac{B'}{\mathbf{v}} + C'\mathbf{v} \tag{17.10}$$

where A', B', and C' are constants analogous to those in the form of the van Deemter equation presented in Equation 17.7. In the van Deemter equation, $A'\mathbf{v}^{1/3}$ causes a slow increase in h with increasing $\mathbf{v}$, $C'\mathbf{v}$ results in a larger increase in h with increasing $\mathbf{v}$, and $B'/\mathbf{v}$ decreases with increasing $\mathbf{v}$. A plot of $\log h$ versus $\log \mathbf{v}$ yields a minimum value of h, which is the optimum condition for chromatographic separation. Examination of Equation 17.8 shows that small particle size favors higher column efficiency. Furthermore, the optimal flow rate increases with decreasing particle diameter. However, very small particles require very high pressures on the mobile phase to achieve optimum flow rates, an especially difficult situation in liquid chromatography. The values of the diffusion coefficient (D) of the solute in the mobile phase are much greater in gases than in liquids, so that the optimum gas chromatographic flow rates are considerably higher than in liquid chromatography.

17.7   Solute Migration Rate

Retention time of a solute t_r is measured from the instant that a sample is injected onto a column. The average time that it takes a mobile phase molecule to traverse the column from beginning to end is the same time that it would take a peak from a totally unretained solute to emerge from the column, and is designated as t_m. It is useful to define the *average rates of migration* of solute $\bar{v}$ and of mobile phase u as follows:

$$\bar{v} = \frac{L}{t_r} \tag{17.11}$$

$$u = \frac{L}{t_m} \tag{17.12}$$

(Recall that L is the length of the packing in the column.) Another parameter useful in discussing solute migration rates is the *fraction of time that is spent by the solute in the mobile phase*. This is a function of the partition coefficient of the solute K (larger with increased affinity for the stationary phase), the volume of the stationary phase V_s, and the volume of the mobile phase V_m, as follows:

$$\begin{array}{l}\text{Fraction of time that solute} \\ \text{is in the mobile phase}\end{array} = \frac{V_m}{V_m + KV_s} \tag{17.13}$$

This fraction can be used to express the following relationship:

$$\bar{v} = u\,\frac{V_m}{V_m + KV_s} \tag{17.14}$$

Examination of the preceding equation shows that if K is very large, the rate of movement of solute is very slow compared to that of the mobile phase.

The **capacity factor** k'

$$k' = \frac{KV_s}{V_m} \tag{17.15}$$

is a measure of how much solute can be handled by a column without overloading. From this equation and Equations 17.11, 17.12, and 17.14, the following can be derived:

$$\bar{v} = u \, \frac{1}{1 + k'} \tag{17.16}$$

$$\frac{L}{t_r} = \frac{L}{t_m}\left(\frac{1}{1 + k'}\right) \tag{17.17}$$

$$k' = \frac{t_r - t_m}{t_m} \tag{17.18}$$

17.8   Resolution of Chromatographic Peaks

Degree of Peak Overlap

In an ideal chromatogram of a mixture of solutes, each solute is manifested by a separate peak distinct from the others and not overlapping with any other. Two such peaks of equal size are shown in Figure 17.9. The degree of separation of two peaks is called the **resolution**. Sometimes chromatographic peaks are not well separated. Three different degrees of peak overlap are shown in Figure 17.10. The lower plots illustrate ideal Gaussian peaks that overlap as shown in the shaded areas. The detector response for each of the two individual peaks is added in each of the three cases shown to give the composite chromatograms in the upper plots. In Figure 17.10(a) the degree of overlap is small, so the upper composite plot clearly shows two peaks. Two

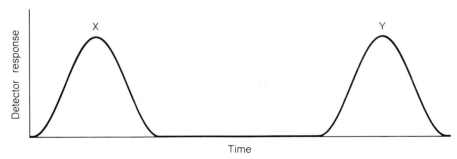

FIGURE 17.9   Two well-resolved chromatographic peaks of solutes X and Y

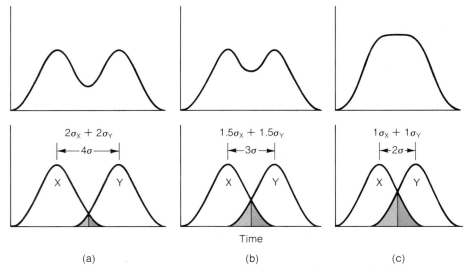

FIGURE 17.10   Identical chromatographic peaks from solutes X and Y showing various degrees of overlap. The upper trace in each case shows the actual detector response; the lower trace shows the two component peaks making up each trace. Overlap is shown by shading. (a) Resolution = 1.00, 2.3% peak overlap; (b) resolution = 0.75, 6.5% peak overlap; (c) resolution = 0.50, 16% peak overlap

peaks are still apparent in Figure 17.10(b), but the degree of overlap is clearly higher. Figure. 17.10(c) shows so much overlap that the upper plot consists of one somewhat flattened peak without an exact Gaussian shape.

Quantitative Treatment of Resolution

Figure 17.10 provides a qualitative picture of resolution, but it can be treated mathematically as well. Consider the resolution of two Gaussian peaks of identical height and width, one from solute X emerging from the column before the other peak from solute Y. Where Δt_r and ΔV_r represent peak separation (measured between the two peak maxima) in units of time and eluent volume, respectively, the mathematical expression of resolution is

$$\text{Resolution} = \frac{\Delta t_r}{w} = \frac{\Delta V_r}{w} \tag{17.19}$$

As shown in Figure 17.6, w is the peak width in units of time or volume, and is equal to 4σ, where σ is the standard deviation corresponding to the Gaussian peak. Resolution is a function of the number of theoretical plates in a column N, α is the ratio of partition coefficients for the two solutes, and f_Y is the fraction of time that solute Y is in the stationary phase. The value of α is given by

$$\alpha = \frac{K_Y}{K_X} \qquad (17.20)$$

where the K's are the partition coefficients of the two solutes. Where V_s is the stationary phase volume and V_m is the mobile phase volume in the column, f_Y is calculated by the following:

$$f_Y = \frac{K_Y V_s}{V_m + K_Y V_s} \qquad (17.21)$$

Using the parameters defined above, resolution is expressed mathematically by the following equation:

$$\text{Resolution} = \frac{\sqrt{N}}{4}\left(\frac{\alpha - 1}{\alpha}\right) f_Y \qquad (17.22)$$

The value of N increases linearly with column length, but resolution increases as a function of $\sqrt{N}$. Therefore, doubling the resolution on a column requires increasing its length by a factor of 4.

17.9   Irregular Peak Shapes

Peak Tailing

So far the discussion has dealt with ideal Gaussian-shaped chromatographic peaks. In practice, peaks are often skewed, with **peak tailing** (Figure 17.11) the most common problem. These results can be explained by solutes that behave in a nonideal manner, or the application of experimental parameters that result in such behavior. A common cause of skewed peaks is a deviation from the linear elution chromatography condition. Recall from Section 17.4 that this condition is characterized by a constant partition coefficient K

$$K = \frac{C_s}{C_m} \qquad (17.2)$$

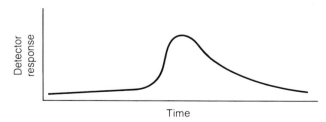

FIGURE 17.11   Illustration of peak tailing

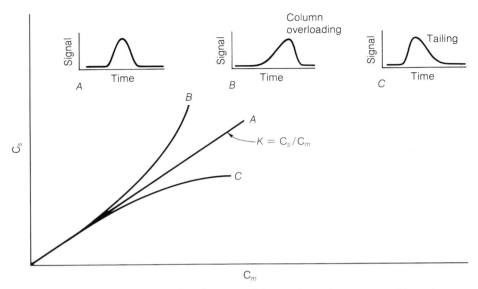

FIGURE 17.12   Three types of isotherms and the resultant chromatographic peaks

where C_s and C_m are the solute concentrations in the stationary and mobile phases, respectively (see Section 17.4). For ideal chromatographic behavior, a plot of C_s versus C_m at constant temperature (such a plot is commonly called an *isotherm*) is a straight line as shown in the middle plot in Figure 17.12.

Tailing commonly is caused by the presence of a fraction of particularly strong binding sites on the stationary phase. This results in the retention of small quantities of solute to a higher degree than larger quantities, causing the downward bend of the lower isotherm in Figure 17.12. The selective higher retention of a portion of the solute composing the peak causes this portion to lag behind the rest of the peak to a greater degree than would occur by normal longitudinal diffusion; the resulting effect on peak shape is a gradual decline of the trailing edge of the peak compared to the abrupt rise of the leading edge.

Column Overloading

The opposite of a chromatogram with a tailing edge is one with a leading edge, also shown in Figure 17.12. These result from **column overloading**, in which the stationary phase has insufficient capacity to retain all the solute in the solute band. Excessive mobile phase flow can also result in a leading edge. The result of this—or of column overloading—is that some of the solute "passes up" the stationary phase to create a peak showing a gradual increase in solute preceding the peak maximum, with an abrupt decline following the max-

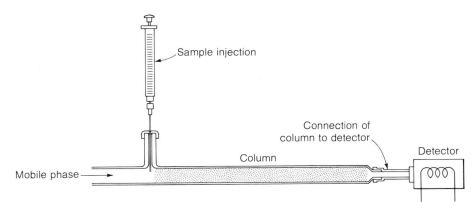

FIGURE 17.13 Schematic drawing of a chromatograph showing potential sources of peak spreading other than the column

imum. The isotherm in this case shows a greater-than-linear increase of C_s with increasing C_m because the solute condenses on the stationary phase with increasing concentration, creating a stationary phase surface that is composed of solute material; the result is increased stationary phase attraction for the solute because "like dissolves like."

Precolumn and Postcolumn Effects

Some less-than-ideal peaks result from phenomena that occur pre- and postcolumn. Spreading results from the fact that analyte cannot be introduced onto the column as an infinitesimally thin band, but starts out as a band of finite width even before any on-column peak spreading occurs. The proper design of injection chambers or devices is therefore important in reducing peak spreading, as is the injection technique itself. Further spreading occurs in any dead space between the column and the detector or within the detector, as shown in Figure 17.13. This can be reduced by connecting the outlet end of the column as closely as possible to the detector—that is, with minimum dead volume.

17.10   Chemical Analysis by Chromatography

Specific approaches to chemical analysis by chromatography are discussed in Chapters 18 and 19. Chromatography's greatest attribute is its ability to *separate* compounds—in especially favorable cases even those that differ only in minute details of structure. Chromatography is capable of *qualitative analysis*. This capability is enhanced by compound-selective detectors, especially the mass spectrometer, which can identify compounds specifically. In addition, chromatography can be used for *quantitative analysis*, an application that depends strongly upon the detector.

Qualitative Chromatographic Analysis

With the use of a universal detector—that is, one that responds about equally well to whatever solute elutes from the column—the major qualitative analysis parameter of chromatography is retention time t_r. This parameter can be employed to distinguish among a limited group of compounds when their relative retention times are known. Variations in the order of t_r with different mobile and stationary phases can also be used for qualitative analysis. The use of selective detectors greatly enhances the qualitative analysis capabilities of chromatography. Detectors employed may range from those that respond preferentially to a class of compounds (such as electron-capture detectors for gas chromatographically separated chlorinated hydrocarbons) to the compound-specific mass spectrometer. The roles played by detectors in combination with chromatography for qualitative analysis are discussed further in the next two chapters.

Quantitative Chromatographic Analysis

Under carefully controlled conditions, both the height and area of an analyte peak vary linearly with the quantity of solute producing the peak. Therefore, comparison of peak height or area for an unknown quantity of analyte with that for a known quantity of it run under identical conditions enables a quantitative determination of analyte. In order to obtain good quantitative data, the manner and rate of sample injection must be carefully controlled, along with eluent flow rate and composition and column temperature. Furthermore, the column must not be overloaded.

For most chromatographic peaks, peak area measurement yields the most accurate quantitative results because areas are generally independent of peak broadening, which may differ between samples and standards. Various methods have been employed to measure peak areas, including mechanical integration on the recorder, use of a planimeter, weight of paper composing the peak cut out from its plot on chart paper, and use of the product of peak height and width. In modern practice electronic integration with computer calculation of peak area and other pertinent parameters is employed.

For the quantitative determination of a number of peaks, one or more mixtures containing known quantities of the desired analytes may be run and the peak areas compared between the knowns and the unknowns. The greatest source of error in this approach consists of variations in sample volumes introduced. These can be minimized by special sample valves designed to introduce a specific volume of sample onto the chromatographic column. Analytes can be determined to within 1% by the use of **internal standards** added to the sample in known quantities. These standards consist of com-

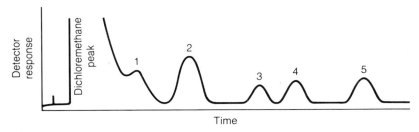

FIGURE 17.14   A gas-liquid chromatogram of an organic hazardous waste sample extracted with highly volatile dichloromethane solvent, showing the solvent peak off-scale and analyte peaks 1 to 5

pounds similar to the analyte to be determined, producing a chromatographic peak close to, but distinct from, the analyte peak. This technique is especially useful for eliminating errors arising from variations in injection rate or volume. Of course it is necessary to know the relative detector response—the **detector response factor**—of the analyte and standard.

In gas chromatography it is sometimes possible to elute all the components of a sample consisting of a mixture of gases or a mixture of volatile liquids. The peak areas of each component can be measured, corrected for detector response factor, and the percentage of each component of the mixture calculated. This is called the **area normalization method**. It is impractical when there is a large amount of solvent present relative to the analytes dissolved in the solvent because the solvent overloads the column, cannot be measured quantitatively, and thus makes impossible the measurement of the absolute percentage of analyte in the sample (see Figure 17.14).

Programmed Summary of Chapter 17

The major terms and concepts introduced in this chapter are contained in this summary in a programmed format. To derive the most benefit from the summary, you should fill in the blanks for each question and then check the answers at the end of the book to see if your choices are correct.

Chromatography is defined as the separation of two or more substances because of their differences in affinity (1) _____. Five different types of chromatography classified according to differences in the solid surfaces or layers on solid surfaces with which solutes interact are (2) _____ chromatography. The two major types of chromatographic separations performed on a planar solid surface are (3) _____ chromatography. For column chromatography, in addition to having the column packed with particles constituting the stationary phase, the stationary phase may be on (4) _____, a variation called (5) _____. In referring to the process by which the mobile phase is carried through the column, the mobile phase fluid is called (6) _____, its passage through the column is a process called (7) _____, and fluid emerging from the column is called (8) _____. A chromatogram is plotted such that the plot shows the emergence of each solute from the column; each solute is detected by the detector and shown as a (9) _____ on the chromatogram. The time that it takes each of these to emerge from the column is called the (10) _____ and the volume of eluent that passes through the column during this time is called the (11) _____. As solute passes through a column, bands arising from different solutes normally separate, and each band undergoes (12) _____, which should be minimized in order to enable detection of a particular solute. The older and simpler of the two major theories of chromatography is the (13) _____. This theory divides the column into a number of hypothetical (14) _____, and the length of the column corresponding to each of these is called (15) _____. A more sophisticated chromatographic theory is the (16) _____. This theory takes into account (17) _____ processes that occur during separation, including (18) _____ and (19) _____ diffusion. Account is taken of these factors by the (20) _____ equation. The degree of peak separation is called (21) _____. If two equal-size peaks of Gaussian shape are separated by exactly 2σ, the degree of overlap is (22) _____ percent. Where N is the number of theoretical plates on a column, the degree of peak resolution increases as a function of (23) _____. The two most common irregularities in peak shape are (24) _____ and peaks with leading edges, the latter resulting from (25) _____. The chromatographic detector (primarily for gas chromatography) with the highest compound specificity is the (26) _____. The two aspects of chromatography most commonly employed for qualitative chromatographic analysis are (27) _____ of peaks and type of (28) _____ on the instrument. The characteristic of a chromatographic peak that gives the greatest accuracy for quantitative analysis is (29) _____. In some cases the percentages of chromatographically separated species can be determined by the (30) _____ method, which involves measuring peak areas and correcting for differences in detector response, followed by calculation of percentage composition.

Questions

1. What materials were separated in the first published accounts of what has come to be known as chromatography?
2. Define how each of the four words in "gas–liquid partition chromatography" describes this particular technique.

3. Match the following:

   a. Ion exchange chromatography
   b. Affinity chromatography
   c. Partition chromatography
   d. Molecular exclusion chromatography

e. Adsorption chromatography

1. One kind of molecule interacts very specifically with chemical groups on the stationary phase
2. Different degrees of attraction of different molecules to the surface of a solid stationary phase
3. May involve groups such as $-SO_3^-$ and $-N^+(CH_3)_3$ on the column
4. Depends upon dimensions of molecules
5. Conceptually much like a series of solvent extractions

4. Which type of chromatography is most promising for the separation of a single protein from a mixture of proteins?
5. What is a chromatogram in very general terms?
6. What is retention ratio, and why is it more exact than either retention time or retention volume?
7. What is the essential condition for linear elution chromatography?
8. What is the term given to distinct regions occupied by individual solutes on a chromatographic column?
9. What is the basis of the plate theory of chromatography?

10. What are some deficiencies in plate theory and some important factors not considered in its development?
11. What is the term in the van Deemter equation that accounts for irregular, random paths taken by solute molecules traversing the stationary phase particles in a column?
12. What chromatographic parameter is to be minimized for most efficient chromatographic separations?
13. What characteristic of columns that separate solutes especially well demands that most modern liquid chromatographic separations [high-performance liquid chromatography (HPLC)] employ high pressures on the mobile phase?
14. How is the fraction of time spent by the solute in the mobile phase related to the volume of the mobile phase, the volume of the stationary phase, and the partition coefficient?
15. How is peak resolution defined qualitatively and quantitatively?
16. What effect does doubling column length have upon resolution?
17. What is chromatography's greatest analytical attribute?

Problems

1. For the plot of a chromatographic peak shown here, each division on the time axis represents 5 seconds and the plot begins at $t > 0$. Calculate w, $w_{1/2}$, and σ.
2. The reduced plate height h is 4.0 for a column in which the average stationary particle size diameter in 25 μm and the column is 1.150 m long. What are the values of height equivalent to a theoretical plate H and the number of theoretical plates N?

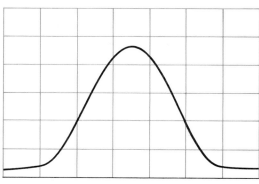

Time

3. Calculate the fraction of time that a solute is in the mobile phase in a column where $V_m/V_s = 0.728$ and $K = 3.485$.

4. Peak X is an internal standard for the measurement of the quantity of Y in a chromatographic sample. The detector gives 1.28 times the response to X that it gives for an equal mass of Y. The quantity of X producing peak X is 1.80 μg. What is the quantity of Y producing peak Y?

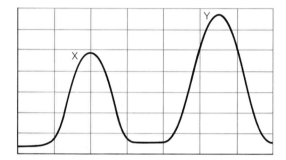

5. For a chromatographic peak, t_r is 103 s and the width of the peak at $\frac{1}{2}h$ is 5.6 s. From this information calculate the number of theoretical plates on the column.

6. If the column in Problem 5 was 3 m long, what was the value of H?

7. For two solutes X and Y separated chromatographically on a column with 1,180 theoretical plates, the values of partition coefficients were $K_Y = 1.07$ and $K_X = 1.23$. The stationary phase volume of the column was 11.23 mL and the mobile phase volume was 7.93 mL. Calculate the resolution of the two peaks.

8. What is the resolution of two Gaussian peaks of identical width (3.27 s) and height eluting at 67.3 s and 74.9 s, respectively?

9. A sample containing five organic gases, A, B, C, D, and E, was separated into its components by gas chromatography. During the sampling process the mixture of gases was contaminated by air, and a total of six well-resolved peaks was obtained (the air peak eluting at the retention time of the column). The peak areas were calculated by computer; these areas are given in the table below along with the detector response factor for each gas. From the data given, calculate the weight percentages of gases A through E, compensating in the calculation for the air impurity.

Gas	Area of peak, units	Detector response factor*
Air	911	0.111
A	20,213	1.000
B	13,721	0.873
C	31,419	1.138
D	27,286	1.649
E	9,815	1.129

* Response factor of A normalized to exactly 1; detector gives 0.873 times the response for B as it does for an equal weight of A.

10. If the peak shown in Problem 1 had a retention time of 2 min, 39 s on a column 2 m long, what was the height equivalent to a theoretical plate?

Liquid Chromatography

L E A R N I N G   G O A L S

1. Distinction between column and planar liquid chromatography.
2. Distinction between the major planar chromatographic techniques of thin-layer and paper chromatography.
3. Principles and applications of thin-layer chromatography.
4. Principles and applications of paper chromatography.
5. Conventional nonpressurized column chromatography.
6. Use of gradient elution in liquid chromatography in place of isocratic elution.
7. Principles of ion exchange chromatography.
8. Principles and uses of molecular exclusion chromatography.
9. What is meant by affinity chromatography.
10. Advantages and uses of high-performance liquid chromatography, HPLC.
11. Distinction between normal-phase chromatography and reverse-phase chromatography.
12. Principles and uses of ion chromatography.

18.1   Types of Liquid Chromatography

Definition and Types of Liquid Chromatography

All of the earlier chromatographic work from the turn of the century until the development of gas chromatography in the early 1950s employed a liquid mobile phase. Gas chromatography turned out to be such an ideal analytical technique and so theoretically interesting that it eclipsed liquid chromatography for about two decades. With the development of high-performance liquid chromatography (HPLC) (see Section 18.8) in the early 1970s, liquid chromatography has re-emerged as a technique of great utility and interest.

Liquid chromatography is the general term applied to separation of solutes based upon their differing affinities for a *solid* stationary phase and a *liquid* mobile phase. One of the most vivid such separations is represented in Figure 18.1, showing the separation of carotenes from a petroleum ether extract of leaves by elution down a column packed with magnesia (MgO) using a petroleum ether mobile phase. Here the different colored components of the leaf extract appear as distinct bands visible through the glass column wall.

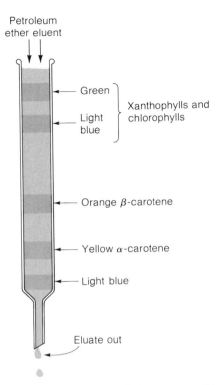

Petroleum
ether eluent

Green ⎤
        ⎬ Xanthophylls and
Light ⎦ chlorophylls
blue

Orange β-carotene

Yellow α-carotene

Light blue

Eluate out

FIGURE 18.1   Liquid chromatographic separation of a petroleum ether extract of leaves on a magnesia-packed column with petroleum ether as the mobile phase

The two broadest divisions of liquid chromatography are those between **column chromatography**, in which the solid support is packed into a cylindrical column, and **planar chromatography**, in which the solid phase is in the form of a coating on a planar surface, such as a sheet of paper. As discussed later in this chapter, the most widely used variation of column liquid chromatography is HPLC. Furthermore, there are several major variations of planar chromatography. Each of these will be discussed separately.

18.2   Thin-Layer Chromatography

Planar Chromatography

There are two general types of planar chromatography in which the solid support consists of a thin sheet of material. One of these is paper chromatography, which is discussed in the following section. The other type is *thin-layer chromatography*, in which the stationary phase consists of an adherent coating applied to a plane surface, such as a microscope slide.

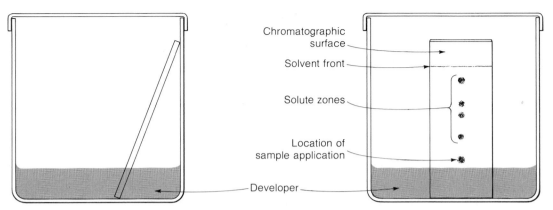

FIGURE 18.2 End view and side view of plates for thin-layer chromatography

Basics of Thin-Layer Chromatography

Figure 18.2 shows the basic features of a simple, typical setup for thin-layer chromatography. The plate upon which the solid phase is coated is held at an angle inside a container holding a small quantity of the mobile phase liquid. The container is covered to maintain an atmosphere saturated with solvent vapor, which serves to prevent evaporation of the mobile phase from the plate surface. The process of separating analytes on a thin-layer plate is called **development**, and the liquid eluent is called the **developer**. In the example shown in Figure 18.2, the developer travels upward through the layer of solid phase material, drawn by small pores in the solid layer by a phenomenon known as **capillary action**. This is *ascending flow; descending flow* is also possible if the liquid is introduced to the top of the plate. The sample is placed (spotted) just above the surface of the developer liquid, and various solutes in the sample migrate at different rates behind the leading edge of the developer, called the **solvent front**. The solvent front is usually allowed to reach near the top of the plate before development is stopped. After development, the various solutes are located at different distances from the point of sample application in spots called **solute zones**. Usually some reagent must be applied to make these solute zones visible. Both solutions of sulfuric acid and iodine produce dark spots with organic solutes. Ninhydrin reacts with amino acids to produce spots ranging from purple to pink. Spots of fluorescent

Ninhydrin

materials (see Section 21.8) fluoresce (emit visible light) under ultraviolet light; alternatively, a fluorescent material can be incorporated with the stationary phase so that nonfluorescent solute zones are revealed as dark areas where the solute has covered the fluorescing substance. A quantitative determination of a solute can be made by measuring its area or using a device known as a scanning densitometer to measure light fluorescing or reflecting from the spot. The spot can even be scraped from the plate and the quantity of solute measured by some physical or chemical method.

Solid Phases for Thin-Layer Chromatography

The solid phase employed in thin-layer chromatography is normally applied as an approximately 0.25-mm-thick layer of material coated on a glass plate. One of the major solid phases employed is *silica*—also known as *silica gel* and *silicic acid*—a hydrated form of SiO_2 produced by precipitation from an acidic silicate solution followed by washing and drying. This material has a vast surface area of around 500 m^2/g. The active sites on silica that react with solutes are Si—OH groups about 0.5 nm apart. The other solid phase commonly used for thin-layer chromatography is alumina, consisting of various forms of Al_2O_3. Both alumina and silica are *adsorbents*, meaning that solutes are attracted to the solid surface. Neither adheres very well to the glass plates upon which they are commonly coated so a **binder** is usually added to hold the solid adsorbent together and cause it to adhere to the glass. The most common inorganic binder is plaster of Paris, $CaSO_4 \cdot \frac{1}{2}H_2O$, added as 5 to 20% of the solid phase. When water is added, plaster of Paris hydrates to form hard $CaSO_4 \cdot 2H_2O$. Starch makes an excellent binder, although it reacts with the most common reagents used to make solute zones visible. The particles composing the adsorbent layer are very small, usually less than 15 μm in diameter.

The most important characteristic of a solid adsorbent used in chromatography is the ability to retain solutes, known as the solid's **activity**. Activity is dependent upon the mobile phase used and varies greatly with the means of preparing and treating the solid phase. In general, the greatest activity of silica is obtained by heating it to 150 to 200°C to drive off water, which otherwise masks the active Si—OH groups. Heating to higher temperatures results in loss of activity because of dehydration

$$2Si\text{—}OH \xrightarrow[\text{heat}]{} Si\text{—}O\text{—}Si + H_2O(g) \tag{18.1}$$

<small>Surface sites</small>

and resultant loss of active Si—OH sites. The activity of alumina, however, increases with the temperature to which it has been heated, up to about 1000°C.

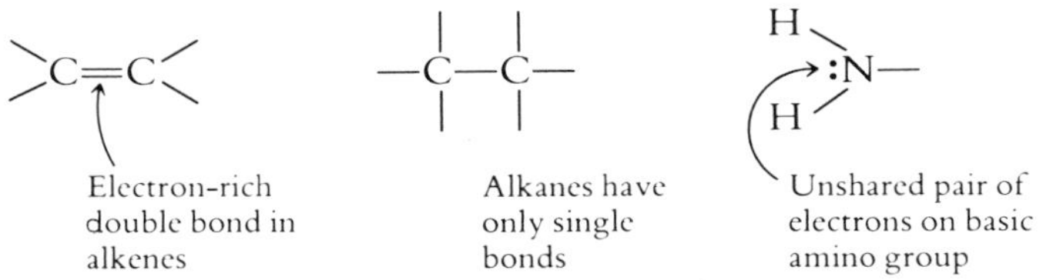

| Electron-rich double bond in alkenes | Alkanes have only single bonds | Unshared pair of electrons on basic amino group |

FIGURE 18.3 Silica tends to hold solutes by hydrogen bonding to electron-rich groups, thus having a relatively strong attraction for alkenes and bases.

The ability of silica to retain solutes containing oxygen or nitrogen atoms is primarily through hydrogen bonding

$$\text{Si—OH—} \underset{\text{Hydrogen bond}}{\text{—}} \text{(Solute)}$$

where the solute is the electron donor. Electron-rich alkenes are thus more strongly retained on silica than are alkanes (see Section 1.6). Bases, which are electron donors by definition, are especially strongly retained (Figure 18.3).

Alumina appears to have three mechanisms for retaining solutes. Electron-donor Lewis bases are attracted to the strongly positive field around the Al^{3+} ion. Acidic solutes apparently are retained by attraction to basic O^{2-} ion sites on the alumina surface. Aromatic compounds appear to be attracted by a third type of interaction, possibly **charge-transfer interaction** involving transfer of π electrons from the aromatic ring to the Al^{3+} site.

The Mobile Phase in Thin-Layer Chromatography

The mobile phase solvent plays a role in solute retention by competing with the solutes for active sites on the solid phase adsorbent. If the solvent has a relatively low attraction to the adsorbent, solutes will tend to adhere to the latter and move only slowly relative to the mobile phase. If the mobile phase solvent is strongly attracted to the adsorbent, the solvent displaces solutes from the solid phase, and the solutes tend to move at about the same rate as the solvent.

Retardation Value

A thin-layer chromatographic separation is normally allowed to proceed until the solvent front (leading edge of the mobile phase) has nearly reached the end of the thin-layer plate. The individual solute zones are then detected by means previously described. The degree of migration is expressed by the **retardation**

value, R_f, defined as,

$$R_f = \frac{\text{distance traveled by solute}}{\text{distance traveled by mobile phase solvent front}} \qquad (18.2)$$

The retardation factor is about 0.87 times the retention ratio R, defined in Chapter 17 as

$$R = \frac{\text{time for solvent to go through column}}{\text{time for solute to go through column}} \qquad (18.3)$$

This is because the solvent front moves faster than does the bulk of the solvent due to capillary attraction for the solvent by the dry solid phase ahead of the solvent front. In calculating retardation values, the distance traveled by the solute is measured from the starting point to the middle of the solute zone, where the solute zone appears as a spot made visible by treatment with a suitable reagent.

Applications of Thin-Layer Chromatography

Thin-layer chromatography has numerous applications as a simple, inexpensive separation method. It is used to determine the purity of synthetic products, for natural product separations (for example, amino acids), and in a variety of other applications. It is also useful to determine the feasibility of more sophisticated separations involving column chromatography.

18.3   Paper Chromatography

Paper as a Stationary Phase

Planar chromatography can be carried out on sheets of paper serving the function of a solid phase. The paper employed as a stationary phase is carefully purified and manufactured to close tolerances of thickness and porosity. The paper normally consists of cellulose fibers containing appreciable amounts of water. This water can act as the active stationary phase, such that paper chromatography can function as a liquid-liquid partition separation, with solute distributed between the water in the paper and the organic solvent mobile phase. It is possible to displace the water with paraffin oil or silicone oil sorbed to the paper. This gives a **reverse phase** stationary medium in which water or some other polar solvent functions as a mobile phase. Some commercial papers contain an adsorbent, permitting adsorption chromatography; others contain an ion exchange resin, making ion exchange paper chromatography possible. The remarkable simplicity of paper chromatography has led to its widespread use since it was introduced in 1944 by A.J.P. Martin and co-workers.

Modified Cellulose

Cellulose paper that has been treated chemically to change its properties is called *modified cellulose*. The most common modifications consist of the introduction of carboxymethyl, diethylaminoethyl, and phosphate groups onto the cellulose.

$$
\begin{matrix}
\text{O} & & & \text{O} \\
\| & & & \| \\
\text{—C—OCH}_3 & & \text{—N}^+(\text{C}_2\text{H}_5)_3 & & \text{—O—P—OH} \\
& & & & | \\
& & & & \text{OH}
\end{matrix}
$$

Carboxymethyl   Diethylaminoethyl   Phosphate

These enable ion exchange separations to be carried out on a planar medium. Each phosphate group bound to cellulose has both a strongly acidic and very weakly acidic hydrogen. It functions as a rather selective separator for various forms of metal ions. The carboxymethyl group acts above pH 4 to 5 as a weakly acidic cation exchanger. Diethylaminoethyl groups bonded to cellulose impart the properties of a *strong base ion exchanger* (see Section 16.9). A major advantage of modified cellulose is that most of the functional groups are located at the surface of the solid, thus enabling rapid, close contact with solutes. In addition to being available in sheet form, modified cellulose is also marketed as a powdered column packing.

Techniques of Planar Paper Chromatography

The techniques of planar paper chromatography are much like those described in the preceding section for thin-layer chromatography. It is essentially a microanalytical technique employing samples not exceeding 100 μg. As with thin-layer chromatography, the solute zones (spots) can be made visible by spraying with a suitable reagent; if the solute is a metal, a metallochromic indicator (see Section 10.9) can be employed. The spot can also be cut out, leached from the paper, and analyte in it determined by standard analytical procedures.

    Two-dimensional paper chromatography (Figure 18.4) can be used to enhance the separation capabilities of planar paper chromatography. The first step in two-dimensional paper chromatography consists of first developing with one solvent having the capability of resolving some of the solutes, but usually not all. Next the paper is eluted at a right angle to the first solvent flow using a solvent with different properties that can resolve solutes not separated by the first solvent. As a result, the zones are located in a scattered pattern around the paper, rather than in a straight line. If the solvents are well chosen, a much improved separation is obtained. The same basic technique can also be applied to thin-layer chromatography.

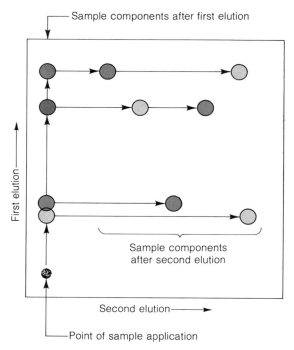

FIGURE 18.4   Two-dimensional paper chromatography showing the use of two different solvents moving at right angles to each other at different times to resolve overlapping solute zones

18.4   Column Chromatography

Nature and Types of Column Chromatography

As noted in Section 18.1, column liquid chromatography is a chromatographic technique utilizing a solid phase packed into a cylindrical column and employing a liquid mobile phase. It is the oldest version of chromatography in the form that makes use of relatively large column-packing particles through which the mobile phase flows by gravity, as well as the newest version of chromatography when employed as HPLC (see Section 18.8). Column chromatography used to be carried out in generally large columns 50 cm to 5 m or more in length and 1 to 50 cm in diameter. Packed with particles of 150 to 200 μm in diameter, these columns sustained gravity flow rates of the order of tenths of a milliliter per minute. "Force-feeding" such columns by introducing the mobile phase with a pressure pump or increasing the flow by vacuum tends to increase plate heights (see Section 17.5) and is thus impractical. Despite these disadvantages, classic column chromatography still finds applications, particularly in separating large quantities of readily separable material. Such separations are sometimes said to be on a **preparative scale**

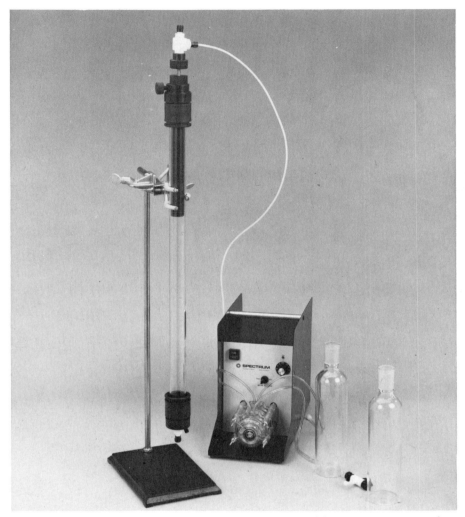

FIGURE 18.5 A glass chromatographic column equipped for temperature control and flow regulation. Temperature is maintained by constant-temperature water circulating through the water jacket, and a flow adapter at the top controls eluent flow rate.

as a step in chemical synthesis. Column separations in large columns with relatively large column-packing materials employing a mobile phase at low pressure can be referred to as **classical column chromatography**. Figure 18.5 illustrates a moderately sophisticated column for classical column chromatography.

Stationary Phases

Silica and alumina (see Section 18.2) are the most common solid phases employed for classical column chromatography. These solids have a relatively high attraction for polar solutes and may retain very polar solutes too strongly. For such solutes, powdered cellulose can be used as a column packing. Other solid phases used include magnesia ($MgO \cdot xH_2O$), florisil (coprecipitated silica and magnesium oxide), and activated charcoal.

Eluents

A variety of liquid mobile phases are employed in classical column chromatography. The choice of eluting solvent is crucial, because the retention of solute by the column is more dependent upon the displacement of solute from the solid phase by the solvent than it is upon the ability of the solid phase to attract solute. The ability of a solvent to displace solute depends upon the solvent's **eluent strength, ε^0**, a relative measure of the energy with which the solvent is adsorbed to the solid phase, where the value for n-pentane is set at zero. Some typical values of eluent strength are given in Table 18.1; such an order of ε^0 values is called an **elutropic series**.

Isocratic and Gradient Elution

Isocratic elution is the term applied when the composition of the mobile phase does not change as a chromatogram is run. Improved column chromatographic separations can be obtained by employing a **gradient elution** process

TABLE 18.1   Eluent strengths of typical chromatographic solvents*

Solvent	ε^0
Fluoroalkanes	−0.25
n-Pentane	0.00
Carbon tetrachloride	0.18
Toluene	0.29
Dichloromethane	0.42
Acetone	0.56
Acetonitrile	0.65
Methanol	0.95
Acetic acid	Very high

* Eluent strengths for solvents on an alumina column

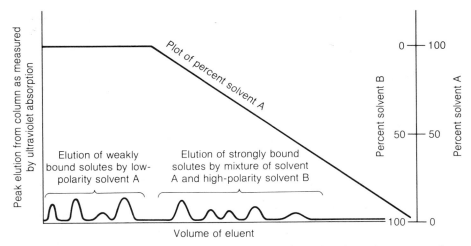

FIGURE 18.6 Gradient elution of solutes from a column by a low-eluent-strength solvent A, and a mixture of A with a high-eluent-strength solvent B. At the end of the chromatogram, the eluent is 100% B.

in which solvents of increasing eluent strength are employed to displace solutes that are progressively bound more strongly to the solid phase (Figure 18.6). Solvent compositions can be changed in steps or linearly from a solvent with a low eluent strength (low polarity) to a high-polarity solvent with a high eluent strength. It should be noted that the introduction of just a small quantity of high-polarity solvent into one with low polarity results in a disproportionate increase in eluent strength.

Peak Detection

The detection of peaks from a column effluent in a classical column chromatographic separation is most commonly accomplished by the absorption of ultraviolet light by the solute compounds as the effluent traverses the beam of light in a flow-through cell. These and other detector systems are discussed in more detail in Section 18.8 on HPLC.

Separation of Copper(II) and Nickel(II)

A typical column chromatographic separation is that of copper from nickel on a cellulose column. The mixture of copper(II) and nickel(II) is contained in 8 M hydrochloric acid. After the sample is introduced onto the column packed with powdered cellulose, the copper is eluted as an orange-brown solution with acetone containing 2 mL of concentrated HCl per 100 mL of acetone. Water is then added to the eluate, the eluate is neutralized with base, the acetone is evaporated, and copper(II) determined complexometrically by EDTA titration. Nickel is eluted with aqueous 1 M HCl. Again, acetone is removed from the eluate solution by heating, the solution is neutralized, and the nickel content of this eluate is determined by titration with EDTA.

18.5   Ion Exchange Chromatography

The ion exchange process, batch separation by ion exchange, and the nature of ion exchange resins were discussed in Section 16.9. Ion exchange resins make excellent solid phases for the column chromatographic separation of ionic solutes. In addition to ion exchangers in which the resin consists of conventional styrene–divinylbenzene of methacrylic acid–divinylbenzene copolymers, ion exchange groups may be attached to several gel-type solid phase materials employed for chromatography. In addition to cellulose, these include dextran (Sephadex, a glucose polymer cross-linked by glycerin), as well as polyacrylamide and the polysaccharide, agarose. These polymers have large pore sizes and a low density of ion-exchanging functional groups, so they can be used to separate relatively heavy molecules (molecular weight exceeding 500).

Several inorganic materials may act as ion exchangers in column chromatography. Solid arsenate, molybdate, phosphate, and tungstate salts of Sn, Th, Ti, W, and Zr function as cation exchangers. Hydrous oxides of these same metals exchange both cations and anions. Cadmium sulfide, CdS, is particularly effective in exchanging transition metal cations; some other metal sulfides also exchange transition metal cations. Mineral zeolites are also used for ion exchangers.

As discussed in Section 16.10, the *selectivity* of an ion exchange resin is based upon an equilibrium reaction of the type

$$Na^{+-}Cat(solid) + K^+(aq) \rightleftarrows K^{+-}Cat(solid) + Na^+(aq) \tag{18.4}$$

and is expressed in terms of the selectivity coefficient K

$$K = \frac{[K^+(res)][Na^+(aq)]}{[Na^+(res)][K^+(aq)]} \tag{18.5}$$

where *res* and *aq* denote species on the resin and in aqueous solution, respectively. This equilibrium constant expresses the *relative selectivity* of the resin for K^+ over Na^+.

There is a wide range of choices for ion exchanger solid phases to be used for chromatographic separations. Normally the first choice is made between ion exchange resins (these are usually styrene-divinylbenzene or methacrylic acid–divinylbenzene polymer types) for lower-weight ions and ion exchanging cellulose or agarose, dextran, or polyacrylamide gels for large species, such as proteins or nuclear acids.

Ion exchange resins used for most classical column chromatographic separations normally are in the form of 100 to 400 mesh-size spheres (higher mesh size denotes smaller particles; the higher the mesh size, the slower the separation because of slow flow rates through the column). Strong acid ion exchangers (with the $—SO_3^-$ functional group) and strong base ion exchangers (with the $—N^+R_4$ functional group*) can be employed over a wide concentration range. A weak acid ion exchanger cannot be used for separations below pH 4 because of protonation of the $—CO_2^-$ group, and removal of H^+ from the $—N^+R_3H$ group prevents use of weak base ion exchangers in high-pH media.

Ion Exchange Separations

Ion exchange resins are useful for the chromatographic separation of metal ions. A number of metal ions are held to approximately the same degree by resins; in such a case, variations in the mobile phase composition in respect to ionic strength, pH, or other parameters often results in a good separation. Complexing agents, such as the halide ions, often enable a separation to be achieved. Consider three hypothetical divalent metal ions, A^{2+}, B^{2+}, and C^{2+}, all held to essentially the same degree by a cation exchange resin. Of these ions, B(II) forms the BCl_4^{2-} complex at low Cl^- concentration, ACl_4^{2-} is formed at high Cl^- concentration, and C(II) does not form chloro complexes to a significant degree. After introduction onto the column, B(II) could be eluted in 0.1 M HCl, since the BCl_4^{2-} *anion* is not retained on a *cation* exchange column, ACl_4^{2-} might be eluted with 6 M HCl, and C^{2+} could be eluted last with a solution of 1 M $NaNO_3$.

Mobile phase compositions may be changed by *gradient elution*, as well as the stepwise process mentioned above. A slowly changing (shallow) gradient assists in separating poorly resolved peaks. A rapidly changing (steep) gradient is employed to speed the elution of strongly retained, well-resolved peaks.

One of the first major accomplishments of ion exchange chromatography was the separation of the chemically almost identical rare earth ions (Ce^{3+}, Er^{3+}, Eu^{3+}, Gd^{3+}, La^{3+}, Lu^{3+}, Tb^{3+}, Tm^{3+} Yb^{3+}). Although these ions are held to essentially the same degree by a cation exchange resin, addition of citrate, which forms chelates of varying stabilities with these ions, enables their separation.

---

*R is an alkyl group (see Section 1.6), typically the methyl group, $—CH_3$.

18.6   Principles of Molecular Exclusion Chromatography

Molecular exclusion chromatography separates molecules on the basis of their size, and to a certain extent their shape, which may be different for molecules of similar weights but different structures. The technique is also called *gel-permeation* or *gel-filtration chromatography*. It is especially useful for the separation of protein and carbohydrate molecules of biochemical interest. It is also employed on samples of synthetic polymers.

The general principle of molecular exclusion chromatography is shown in Figure 18.7. The solid phase has pores in a well-defined size range; molecules less than a certain size can enter these pores, larger molecules cannot. Therefore, large molecules spend all of their time in the mobile phase and elute rapidly, whereas smaller molecules spend a large portion of time in the stationary phase.

The affinity of a solute for the stationary phase is expressed in terms of the partition coefficient K (see Section 17.4) as follows:

$$K = \frac{\text{solute concentration in stationary phase}}{\text{solute concentration in mobile phase}} \qquad (18.6)$$

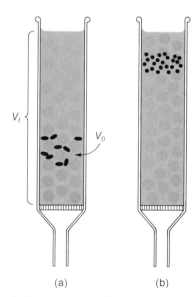

(a)   (b)

FIGURE 18.7   Illustration of molecular exclusion chromatography: (a) Molecules too large to enter the stationary phase pores elute within the void volume of the solvent, V_0. (b) A larger volume of solvent is required for elution of molecules small enough to penetrate the pores because of the time spent by these molecules in the stationary phase.

The retention volume of a peak V_r (see Section 17.5), is given in terms of volume of mobile phase and volume of stationary phase (V_m and V_s, respectively; see Section 17.7) by the following:

$$V_r = V_m + KV_s \qquad (18.7)$$

In molecular exclusion chromatography, V_m is denoted as the *void volume* V_0, and includes only that volume of the solvent *outside* the pore spaces contained by the matrix of porous beads. The preceding equation may be rearranged to express K as

$$K = \frac{V_r - V_m}{V_s} = \frac{V_r - V_0}{V_s} \qquad (18.8)$$

Because so much of the "solid" phase is composed of pores containing solvent, V_s is almost equal to the volume of the mobile phase liquid contained in the pores of the porous beads making up the column packing; V_s *would be* equal to $V_t - V_0$ *if* the bead matrix took up no volume at all (V_t is the volume of that segment of the column into which the beads are packed and is equal simply to $\pi r^2 \times$ length). However, the volume of solid phase occupied by mobile phase liquid in pores is proportional to $V_t - V_0$ because of the constant fraction of the gel occupied by the liquid.

The effective attraction of the solid phase for a solute (that is, the ability of the solid phase to "trap" the solute) is expressed by the "K average," K_{av}, given by

$$K_{av} = \frac{V_r - V_0}{V_t - V_0} \qquad (18.9)$$

Molecules too large to penetrate the pores of the gel traverse the column within its void volume of eluent so that $V_r = V_0$ and $K_{av} = 0$. For small molecules that enter the pores of the gel without restriction, but which are not attracted (adsorbed) by the matrix surface, V_r is almost equal to V_t and K_{av} is very close to 1. Intermediate-size molecules that partially penetrate the pores have values of K_{av} between 0 and 1. These three different situations are represented in Figure 18.8.

The value of V_0 is equal to V_r of a compound composed of molecules that are not adsorbed by the solid matrix and are too large to enter the pores. The most commonly used such compound is a blue dye of molecular weight 2 million called Blue Dextran 2000. Values of K_{av} exceeding 1, greater than described by the theory outlined above, occur when the solute is adsorbed by the solid matrix. This happens most readily with aromatic molecules.

The gels used as solid phases in molecular exclusion chromatography are composed either of dextran (Sephadex), a glucose polymer cross-linked by glycerin or an acrylamide polymer cross-linked by N,N'-methylenebisacrylamide (Bio-Gel P). These gels are classified according to their molecular weight fractionation range based upon spherical (globular) molecules. Some large molecules, such as polysaccharides, do not fit into

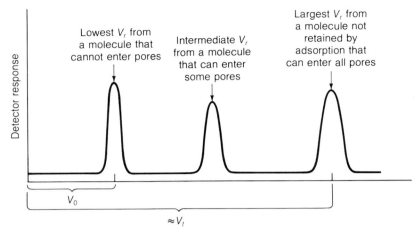

FIGURE 18.8   Retention volumes of three peaks separated by molecular exclusion chromatography showing extremes of molecules that cannot enter the pores at all and those with complete access, as well as an intermediate case in which there is partial access to pores

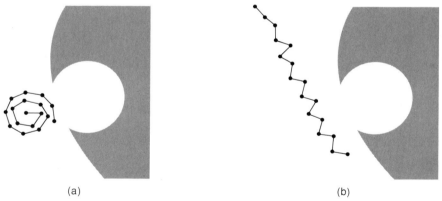

FIGURE 18.9   The effect of molecular shape upon access to solid phase pores in molecular exclusion chromatography. (a) A globular molecule of roughly spherical shape will fit into a much smaller pore than (b) will a linear molecule of the same molecular weight.

pores of gels in their molecular weight range because they are elongated (Figure 18.9), and a larger–pore-size gel must be used for such molecules.

For gel-filtration media, some typical molecular weight fractionation ranges based upon molecular weights of globular proteins are 0 to 700 (Sephadex G-10), 1000 to 6000 (Bio-Gel P-6), 10,000 to 1,500,000 (Sephacryl S-300), and 1,000,000 to 150,000,000 (Bio-Gel A-150 m). Gel-filtration media are available to cover all molecular weight ranges up to 150,000,000. Gels are available in several particle sizes; smaller particles give better resolution but may have unacceptably slow eluent flow rates when packed in relatively long columns.

18.7   Affinity Chromatography

Principles of Affinity Chromatography

Affinity chromatography functions by virtue of the specific interaction between a molecule attached to the stationary phase via a covalent bond and a solute in the mobile phase (Figure 18.10). This solute is retained specifically when the sample is eluted through the column. A mobile phase of different composition is then employed to displace the specifically sorbed solute.

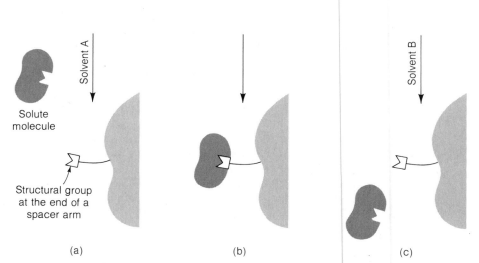

FIGURE 18.10 Affinity chromatography showing (a) solute introduced onto column in solvent A, (b) solute attached to specific binding group on column, and (c) solute eluted from column with solvent B. The binding group on the column is located at the end of a chain of carbon atoms called a spacer arm to avoid steric hindrance by the resin surface that could prevent the solute molecule from reaching the binding group.

Applications of Affinity Chromatography

To date affinity chromatography has found its greatest use in biochemistry. A typical application is the separation of molecules having coplanar *cis*-diol groups

$$
\begin{array}{cc}
\text{H} & \text{H} \\
\text{O} & \text{O} \\
| & | \\
-\text{C}-\text{C}- \\
| & |
\end{array}
$$

using a solid phase having a phenylboronic acid group bonded via a long chain to a resin of Affi-Gel 601:

chain constituting the spacer arm

The long chain at the end of which the phenylboronic acid group is located functions as a spacer arm (see Figure 18.10).

18.8   High-Performance Liquid Chromatography

Distinguishing Features of HPLC

High-performance liquid chromatography (HPLC) is a form of column chromatography with a liquid mobile phase employing much smaller particles than conventional column chromatography and using high pressures to force the mobile phase through the column against the high back-pressures developed when very small particle column packings are employed. HPLC yields much better peak resolution than does classical column chromatography. With small column packing particles of the order of 5 to 10 μm in diameter, all solute molecules follow much more uniform paths in traversing the column. This reduces the degree of eddy diffusion, making the first term in the van Deemter equation (Equation 17.7, Section 17.6) smaller and reducing plate height H. Furthermore, the thicknesses of mobile and stationary phase segments are reduced with smaller particles (that is, any particular portion of the stationary phase is very close to mobile phase and vice versa), decreasing the time of equilibration for solutes between phases. The favorable effect of lower particle diameters is illustrated in Figure 18.11. It is readily seen that smaller-diameter particles yield sharper peaks than do the larger-diameter particles. As a result, the *detection limit* (minimum quantity of solute detectable)

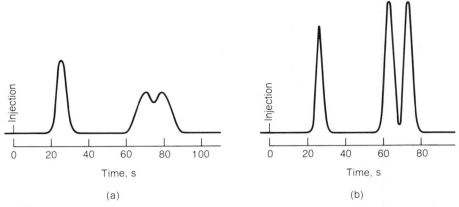

FIGURE 18.11   HPLC chromatographic separations of three peaks with a column packed with 15-μm-diameter particles (a) and 5-μm-diameter particles (b)

is improved; so is peak resolution as shown by the virtually complete resolution of the second and third peaks in the second chromatogram.

HPLC Instrumentation and Operation

Although the advantages of HPLC just outlined had been known for two decades before the technique achieved major application in chemical analysis, it was first necessary to solve mechanical problems inherent to the use of the high pressures required and to develop column-packing materials that could withstand these high pressures of several thousand pounds per square inch (1 psi $\cong$ 0.0067 MPa). A schematic diagram of an HPLC instrument is shown in Figure 18.12. Each of the major components is discussed separately below.

The **solvent system**, which need not be an integral part of the chromatograph, provides for storage, filtration, and degassing of solvents. It is important to remove gases such as N_2 and O_2 from the solvents because they may form small bubbles in the column, causing band spreading and erratic detector performance postcolumn. Gas removal is accomplished by distillation, heating and stirring, or by application of a vacuum. If only one solvent composition is employed, the separation is called an *isocratic elution*. The use of two or more solvents with changing composition during the separation is termed *gradient elution*. This often provides much greater separation efficiency as discussed for classical column chromatography in Section 18.4 (see Figure 18.6). With a modern chromatograph, solvents are mixed in a mixing vessel according to preprogrammed computerized directions. Most commonly, solvent composition is changed by volume in either a linear or exponential

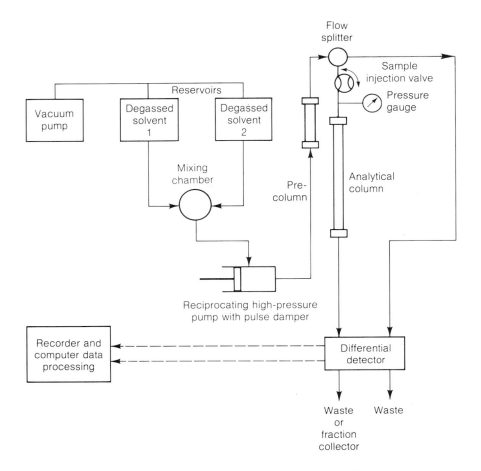

FIGURE 18.12 Above: Schematic diagram of an instrument for HPLC separations. **Below:** System for HPLC analysis.

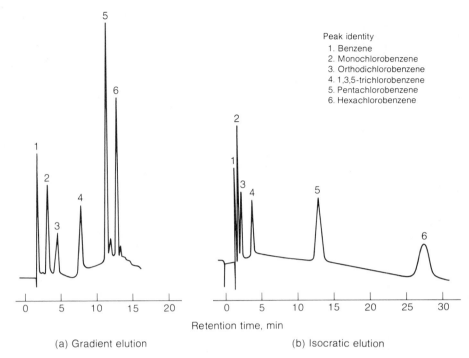

FIGURE 18.13   Comparison of (a) gradient elution with (b) isocratic elution in the HPLC separation of chlorinated benzene compounds

manner. The advantages of gradient elution are illustrated in Figure 18.13. In the example illustrated, the total separation time was shortened by gradient elution, while retaining the good resolution of the earlier peaks. Note how much sharper the last two peaks are as they elute earlier in the gradient elution mode.

Most HPLC pumps are now of the reciprocating type with a damped pulse (fluctuation in pressure due to reciprocating motion) and are capable of delivering solvent at about 0.5 to 5 mL/min and a pressure of at least 1000 psi (6.7 MPa). The maximum pressure delivered typically is of the order of 5000 psi (33.3 MPa).

Precolumns packed with a larger-particle-size version of the material contained in the analytical column can be used in liquid–liquid HPLC. Because of the relatively large particle size, the pressure drop across the precolumn is comparatively small. The mobile phase going through the precolumn becomes saturated with the liquid with which the stationary phase particles are coated, so that this liquid is not stripped from the analytical column. Also, impurities in the solvent are removed by the precolumn, preventing contamination of the analytical column.

Sample injection is normally accomplished in HPLC with a rotary sample valve. The sample is injected into a loop at low pressure, then swept onto the column by diverting the mobile phase flow through the loop.

Columns employed in HPLC are usually straight precision-bore stainless steel tubing with an inside diameter of 2 to 5 mm ranging in length from as short as 10 cm to over 150 cm. Column packing materials are of three major types as shown in Figure 18.14. **Microporous particles**, most commonly of silica gel, but including alumina and Celite (diatomaceous earth), are used at diameters of 5 to 10 μm. These packings can be used for adsorption chromatography in which the solutes are attracted to the solid. Or the packing can be coated and impregnated with a liquid immiscible with the mobile phase for partition chromatography. Silica gel contains water, so that water is a common stationary phase material on a silica gel support. Other stationary phase materials commonly sorbed to silica gel include aliphatic alcohols, glycols, or nitromethane.

Another form of column packing consists of **pellicular particles**. These are beads, normally of glass, about 40 to 50 μm in diameter and covered with an approximately 2 to 3-μm-thick layer of active stationary phase material. In addition to silica gel or alumina, this material may consist of ion exchange resin or some other substance capable of binding solutes. Pellicular particles have only about one-tenth the capacity of porous silica gel, alumina, or Celite. Therefore, it is easier to overload a pellicular column with sample solute. However, greater separation efficiency is achieved with pellicular column

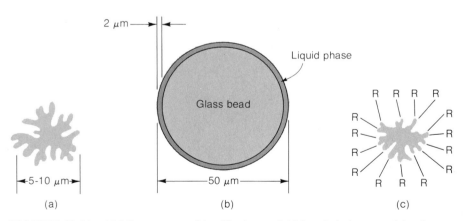

FIGURE 18.14   (a) Microporous, (b) pellicular, and (c) bonded phase particles for HPLC column packing

packing because the layer of active material on the column packing is so thin that very rapid exchange and equilibria of solutes between mobile and stationary phases are obtained.

The most generally satisfactory solid phase packing materials now commonly available for HPLC consist of **bonded phase particles**, in which a polar or nonpolar material is *covalently* bonded to microporous silica particles. A typical polar phase is the $—CH_2CH_2CH_2CN$ group bonded to the silica by an $OSi—$ bond; a common nonpolar phase is the $—(CH_2)_{17}CH_3$ group. These bonded phase materials behave like liquids that are attached to the column packing so that they cannot bleed off (slow loss by leaching) during the course of chromatographic separations. Separations in which a polar phase is bonded to the support are called **normal-phase chromatography**. This requires a mobile phase less polar than the bonded phase, and the degree of displacement of solutes from the column is increased by increasing the solvent polarity. **Reverse-phase chromatography** is conducted with a nonpolar phase bonded to the solid support and use of a polar solvent mobile phase. Reverse phase chromatography has become the more important of the two techniques. It is not nearly so much afflicted by peak tailing (see Figure 17.11) common with polar packings and, since it uses a polar solvent such as water for the mobile phase, it is less affected by the presence of polar impurities in the mobile phase solvent.

HPLC Detectors

A need that has been only slowly and partially met in HPLC is that for sensitive, universally applicable detectors. (Such detectors, available early in the development of gas chromatography, were largely responsible for the phenomenal growth of this analytical tool during its early years.) Figure 18.15 shows a detector for HPLC based upon the absorption of ultraviolet radiation. A rather simple UV detector can be employed using ultraviolet radiation from a mercury source at a wavelength of 254 nm or from fluorescence of a solid fluorescent material excited by the 254-nm mercury line and emitting at 280 nm. The 254- or 280-nm lines can be selected by a simple light filter and provide some selectivity for solutes. The detector is quite sensitive for a number of organic compounds because many functional groups absorb at these wavelengths. Spectrophotometers capable of measuring the absorption of any wavelength of light from about 200 to about 1000 nm (see Chapters 20 and 21) can be used as somewhat more selective detectors. The best attainable detection limit with such detectors is around 0.5 μg solute per liter of sample. Somewhat more sensitive in most favorable cases is fluorometry (see Chapter 21) in which light emitted by a molecule that has absorbed ultraviolet radiation is measured. The absorption of infrared radiation is a relatively selective, but insensitive mode of detecting HPLC peaks.

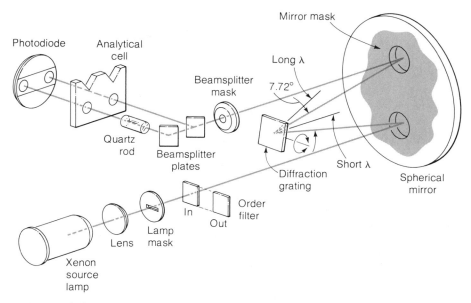

Mirror mask

Photodiode   Analytical
cell

Beamsplitter
mask

Long λ

7.72°

Quartz
rod

Beamsplitter
plates

Diffraction
grating

Short λ

Spherical
mirror

In

Out

Order
filter

Lamp
mask

Lens

Xenon
source
lamp

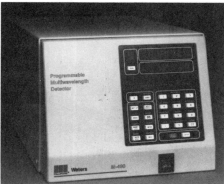

FIGURE 18.15   Above: Ultraviolet detector for HPLC. **Below:** Programmable multiwavelength detector for HPLC peaks.

Changes in the refractive index of mobile phase as a result of the presence of solute peaks is widely employed for HPLC detection, a technique which provides a virtually universal detector. These detectors are not useful in gradient elution HPLC, however. Conductometric and polarographic electrochemical means of detection have been used for HPLC. The most powerful tool for the specific identification of compounds is the mass spectrometer used as an HPLC detector. The elimination of solvent from the mobile phase is one of the most important problems with the use of the mass spectrometric detector.

18.9   Ion Chromatography

The Nature of Ion Chromatography

The determination of low levels of a number of ions in solution has long been a problem for analytical chemists. This is particularly true of some of the common anions, such as HPO_4^{2-}, NO_3^-, and SO_4^{2-}. Ions at low concentrations can be separated by HPLC using ion exchanger stationary phases (see Sections 18.5 and 18.8) with a relatively concentrated solution of a non-analyte salt as an eluent. A major problem has been the detection of analyte anions at levels of parts per million in the presence of a considerably higher concentration of eluent electrolyte. Ions are readily detected by their electrical conductivity in solution, but a large concentration of non-analyte salt in the eluent completely masks conductivity changes that otherwise would be seen during the elution of analyte ion peaks.

Around 1975 the problem of detection outlined above was solved, giving rise to **ion chromatography**, an analytical technique that has now been extended to a wide variety of cations, anions, and neutral species that can be placed in an ionizable form.[1] The solution to the detection problem is the use of a **suppressor** that converts the eluent electrolyte to a weak electrolyte. The way in which this can be done is illustrated by the following example.

Example of Ion Chromatographic Separation

Consider the separation of low concentrations of analyte nitrate and sulfate salts using sodium carbonate solution as an eluent. An anion exchanger is used as a stationary phase. It has a greater affinity for SO_4^{2-} ion than for NO_3^- ion, so nitrate elutes before sulfate. A solution of sodium carbonate is used as an eluent; the CO_3^{2-} serves to displace both sulfate and nitrate from the stationary phase anion exchanger column. The suppressor, located between the analytical column and the detector, consists of a source of H^+ ion. A strong cation exchanger in the H^+ ion form can serve this purpose. The reaction with sodium carbonate in solution is

$$2Na^+(aq) + CO_3^{2-}(aq) + 2\{H^{+-}Cat(solid)\} \rightarrow CO_2(aq) + H_2O + 2\{Na^{+-}Cat(solid)\} \qquad (18.10)$$

From this reaction it is seen that the suppressor converts ionic, electrically conducting Na_2CO_3 in solution to non-ionic, non-conducting dissolved CO_2 [a small fraction of which is in equilibrium with the very poorly conducting weak electrolyte carbonic acid, $H_2CO_3(aq)$]. Furthermore, the sodium ions eluting with the analyte nitrate and sulfate anions are replaced by H^+ ions by

---

[1] For an informative review article on ion chromatography see Hamish Small, "Modern Inorganic Chromatography," *Analytical Chemistry*, 55 (1983), 235A–242A.

the following exchange reactions:

$$Na^+(aq) + NO_3^-(aq) + H^{+-}Cat(solid) \rightarrow H^+(aq) + NO_3^-(aq) + Na^{+-}Cat(solid) \tag{18.11}$$

$$2Na^+(aq) + SO_4^{2-} + 2\{H^{+-}Cat(solid)\} \rightarrow 2H^+(aq) + SO_4^{2-} + 2\{H^{+-}Cat(solid)\} \tag{18.12}$$

The H^+ ions are much more conductive than Na^+ ions in solution and they co-elute with the NO_3^- and SO_4^{2-} analyte ions. This adds greatly to the conductivity and, therefore, detector response arising from the analyte ions.

Although conductivity is the most common means of detecting ions separated by ion chromatography, other methods can be used for appropriate analyte species. These methods include amperometry (see Chapter 14) and ultraviolet, visible, and fluorescence spectrophotometry (see Chapter 21).

Applications of Ion Chromatography

The applications of ion chromatography are growing at a rapid rate.[2] Ion chromatographic methods have been developed for most of the common anions including arsenate, arsenite, borate, carbonate, chlorate, chlorite, cyanide, the halides, hypochlorite, hypophosphite, nitrate, nitrite, phosphate, phosphite, pyrophosphate, selenate, selenite, sulfate, sulfite, sulfide, trimetaphosphate, and tripolyphosphate. Ionic metals amenable to ion chromatographic determination include the alkali metals, the alkaline earths, aluminum, cadmium, chromium, cobalt, copper, iron(II), iron(III), lead, nickel, and tin. Among the ionized or ionizable organic chemicals determined by ion chromatography are alcohols, organic acids, amines, and cationic and anionic surfactants. Amino acids constitute a very important class of biochemical compounds that can be determined by ion chromatography. Carbohydrates can also be determined by this method.

Programmed Summary of Chapter 18

The major terms and concepts introduced in this chapter are contained in this summary in a programmed format. To derive the most benefit from the summary, you should fill in the blanks for each question and then check the answers at the end of the book to see if your choices are correct.

The two major divisions of chromatography based upon mobile phase are (1) _____ chromatography. The two major divisions of chromatography based upon the geometrical configuration of the stationary phase are (2) _____ chromatography. Thin-layer chromatography has a solid phase that is coated onto (3) _____. The various solutes are carried by

[2] Gregory O. Franklin, "Development and Applications of Ion Chromatography," *American Laboratory*, June, 1985, 66–79.

solvent along this solid phase behind an advancing edge of solvent called (4) _____. The final location of each portion of solute is called (5) _____. Some chemicals that cause these to become visible are (6) _____. The two most common adsorbents for thin-layer chromatography are (7) _____ held as a thin adherent coating by the presence of a (8) _____. The degree of migration of a spot in thin-layer chromatography is expressed by a parameter called the (9) _____. The material inside of paper that commonly interacts with a solute, thereby functioning as a stationary phase in paper chromatography is (10) _____. This liquid held by the paper may be displaced and replaced by (11) _____ or (12) _____, which gives a (13) _____ stationary medium. A variation of paper chromatography that makes consecutive use of two different solvents moving at right angles to each other is called (14) _____. The stationary phases most commonly employed in classical column chromatography consist of (15) _____. The eluent strength is a measure of the ability of a solvent to (16) _____. The active groups in the solid phase in ion exchange chromatography are (17) _____. In ion exchange the selectivity coefficient K expresses the relative affinities of the stationary phase for (18) _____. The functional group in strong acid ion exchangers is the (19) _____ group and in strong base ion exchangers it is the (20) _____ group. Molecular exclusion chromatography separates solutes by virtue of differences in their (21) _____; in this application the active entities on the stationary phase consist of (22) _____. The function of Blue Dextran in molecular exclusion chromatography is the determination of the (23) _____ of the column. The basis of affinity chromatography is specific interaction between a solute molecule and a (24) _____ on the stationary phase. High-performance liquid chromatography involves column packing that consists of particles that are (25) _____ than those in conventional classical column chromatography, and therefore requires much higher (26) _____ on the mobile phase. The use of only one solvent composition for a liquid chromatographic separation is known as (27) _____ elution, whereas a variable composition of solvent during the separation is called (28) _____. HPLC separations using a nonpolar stationary phase and a polar solvent mobile phase are classified as (29) _____. Many anions at parts per million levels may be determined by (30) _____.

Questions

1. What are the two broadest divisions of chromatography based upon the geometrical configuration (shape) of the stationary phase?

2. What is the distinction between paper and thin-layer chromatography?

3. In planar chromatography, what is meant by solvent front and by solute zone?

4. What is wrong with saying that the solvent front in planar chromatography moves at the same rate as the average solvent molecule?

5. What is the primary mechanism by which silica retains solutes?

6. What is meant by the activity of a solid phase?

7. What effect does the nature of the mobile phase and its interaction with the solid phase have upon the rate of solute migration?

8. What would be the definition (hard to measure physically) of the retardation value R_f in terms of *times* of solute and solvent movement?

9. How is a reverse phase medium made in paper chromatography?

10. What kinds of chromatographic separations are made feasible by treating cellulose to produce modified cellulose?

11. What is the principle of two-dimensional paper chromatography?

12. In what sense is column chromatography "one of the newest forms of chromatography?"

13. Why does HPLC require high pressures?

14. What is meant by gradient elution in liquid chromatography?

15. What is the result of adding a small quantity of a high-polarity solvent to a low-polarity solvent insofar as the solvent's behavior as a mobile phase is concerned?

16. For what types of solutes are gel-type ion exchangers used?

17. Why cannot a weak acid cation exchanger be used below a pH of approximately 4?

18. For what class of elements has ion exchange chromatography been especially successful?

19. Upon what molecular property is molecular exclusion chromatography based?

20. For what purpose is the dye Blue Dextran used?

Problems

1. Calculate as closely as possible the values of R_f for each of the four solute zones shown in Figure 18.2.

2. Suppose in Figure 18.2 that there was another solute that moved just enough less than the fastest-moving solute zone shown to be resolved from it. What should be the R_f for the slower-moving solute zone? You have not been given a mathematical formula for calculating the resolution of these solute zones, so try to use reasoning from visual examination of the zones to arrive at an answer.

3. Assume silica gel with a surface area of $500 \text{ m}^2/\text{cm}^3$. How thick would a layer of silica gel be if 1 cm^3 of it were spread out as a layer of 500 m^2 area?

4. What is the value of V_t in a molecular exclusion column of diameter 3.88 cm and length 50 cm?

5. In the column described in the preceding example, V_r for Blue Dextran was found to be 550 cm^3. What would be the value of V_r for a solute with $K_{av} = 0.582$?

6. The value of V_r for another solute separated by molecular exclusion in the system described in the preceding two examples was 625 mL. What was the value of K_{av}? What does such a value mean?

19 | Gas Chromatography

19.1 | Introduction

Gas Chromatography and Gas Chromatograms

First described in the literature in the early 1950s, gas chromatography is a remarkably sensitive and selective method for the qualitative and quantitative determination of substances that are stable in the vapor phase. **Gas chromatography** is based on the phenomenon that occurs when a mixture of volatile materials is transported by a carrier gas eluent through a column containing an absorbing material coated on a solid material (or, less commonly, an adsorbing solid material); each volatile component is partitioned between the stationary phase and the carrier gas. The length of time required for a volatile analyte to traverse the column depends upon the degree to which it is retained by the stationary phase. Solutes that are retained to even slightly different degrees will emerge from the column at different times. The time at which the analyte emerges, and its quantity, can be measured by suitable detectors. As shown in Figure 19.1, a recorder trace of the detector response appears as peaks of different sizes depending upon the quantity of material and the type and sensitivity of the detector. In modern practice, the information displayed by a gas chromatogram, such as that shown in Figure 19.1, is stored by a computer, which calculates parameters such as peak area, percentage of total sample represented by each peak, retention time, and other pertinent information.

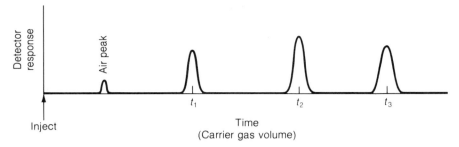

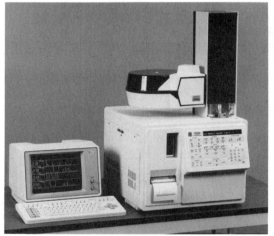

FIGURE 19.1   Above: Gas chromatogram showing the separation of three solutes. **Below:** Varian's 3500 High Resolution Gas Chromatograph with the Model 8035 AutoSampler, and the 600 Series Chromatography Data System.

Gas-Liquid and Gas-Solid Chromatography

Gas chromatography is broadly divided into two major categories. The most common of these is gas–liquid chromatography (GLC), in which the stationary phase is coated with a liquid phase into which the solutes are partitioned. Less commonly used gas-solid chromatography (GSC) is the form in which an adsorbent solid is used as the active stationary phase.

Components of a Gas Chromatograph

The essential components of a gas chromatograph are shown schematically in Figure 19.2. The **carrier gas** is a gas constituting the mobile phase. It must be very pure, and generally consists of argon, helium, hydrogen, or nitrogen. The sample is injected as a single compact plug into the carrier gas stream immediately ahead of the column entrance. If the sample is liquid, it is essential to heat the injection chamber to vaporize the liquid quickly, otherwise tailing

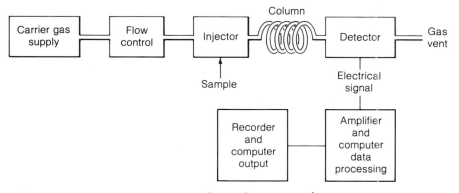

FIGURE 19.2   Major components of a gas chromatograph

occurs, which results in spreading of the peak. If the sample is heated too much, pyrolysis (thermal decomposition) may occur and produce erroneous results. The *column*, through which the carrier gas and analyte species flow and in which separations occur, is the key component of the gas chromatograph. In packed-column gas chromatography, the prevalent form until the early 1980s, the column consists of a metal or glass tube packed with a high-surface-area granular solid coated with a low-volatility liquid. Superior results are usually obtained with a *capillary column* consisting of a very long, small-diameter quartz tube with the stationary liquid phase coated on the inside of the column. A column is generally chosen for its ability to separate compounds in a group of compounds from each other—that is, its selectivity.

The component that primarily determines the sensitivity of gas chromatographic analysis and, for some classes of compounds, the selectivity as well, is the *detector*, a device which shows when an analyte peak is emerging from a column and responds to the quantity of analyte in the peak. The existence of the *thermal conductivity detector*, which sensitively reflects small changes in the conduction of heat by gases passing over it, was largely responsible for the rapid early development of gas chromatography. The electrical signal response of a detector is amplified electronically, then read out on a recorder and/or stored and displayed by a computer.

19.2   Carrier Gas

The carrier gas is usually supplied from tanks of pressurized gas; sometimes nitrogen gas evaporated from liquid N_2 stored for cryogenic applications is used in large chromatographic installations. The gas must be chemically

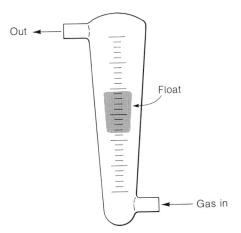

FIGURE 19.3   Rotameter for measuring gas flow rates

unreactive toward the sample and chromatograph components, which is a major reason for the use of argon, helium, hydrogen, and nitrogen. For use with a thermal detector, the thermal conductivity of the gas is a factor. Very pure commercial carrier gas is often further purified by passing it over granules of a solid adsorbent material called a **molecular sieve**.

The flow rate of the carrier gas is carefully regulated and measured by gauges, pressure regulators, and flowmeters. Desired flow rates in the range of 25 to 50 mL/min are usually obtained at column inlet pressures of 10 to 50 psi (pounds per square inch above the pressure of the surrounding atmosphere, 1 psi = 0.0067 MPa). Flow rate can be measured by a **rotameter** (Figure 19.3) placed before the column. Sometimes a **soap-bubble meter** is used at the end of the column. The effluent gas pushes a film of water containing soap or detergent through a graduated tube, and flow rate is calculated from the time required for the film to traverse the distances between the graduations in the tube.

19.3   Sample Injection

Apparatus for Sample Injection

Figure 19.4 shows two means of introducing a sample into a gas chromatograph. A gas-sampling valve can be used for this purpose, especially when the gas is introduced from a pressurized container or a vacuum line. More

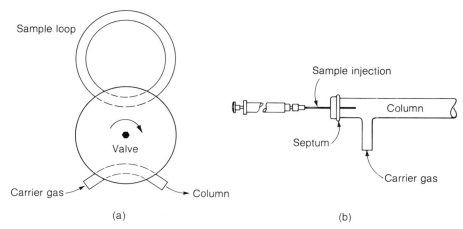

FIGURE 19.4 Gas chromatographic sample injection by (a) gas-sampling valve and (b) sample injection through a septum

commonly the sample is injected into the carrier gas through a rubber or silicone septum at the head of the column with a hypodermic syringe. In the case of liquid samples, the injection system is heated and provides the additional service of vaporizing the sample. Although in most applications injector port temperatures high enough to cause decomposition must be avoided, **pyrolysis gas chromatography** makes use of special heated sample introduction ports to produce thermal decomposition products from nonvolatile materials. Measurement of the decomposition products enables determination of the nature and quantity of the nonvolatile analyte.

Sample Injection Conditions

The volume of liquid samples used with packed columns is normally in the range of a few tenths to 10 μL; such samples can be injected with reasonable accuracy by hand or with automated sample injection devices. However, capillary columns accept only much smaller samples of around 10^{-3} μL of liquid. This requires the use of a **sample splitter**, in which only a fraction of the sample injected actually passes over the column.

The injection port is normally maintained at a temperature about 50° C higher than the boiling point of the least-volatile component to evaporate the sample and compensate for heat loss during vaporization. Lining the injection port with glass helps prevent the metal-catalyzed pyrolysis of heat-sensitive sample liquids. An even better system enables injection of the sample directly onto the column packing.

Columns for Gas Chromatography

Packed and Capillary Columns

Used from the beginning of gas chromatography, *packed columns* composed of metal or glass tubes commonly have dimensions of 1 to 8 mm internal diameter and lengths of 1 to 10 m. As implied by the name, packed columns are packed with a granular solid support material, upon which is normally coated a stationary liquid phase that is the active stationary phase material for the gas chromatographic separation. Such a column typically has 300 to 3000 theoretical plates per meter; the best packed columns may have more than 20,000 theoretical plates. *Capillary columns* are only 0.3 to 0.5 mm in internal diameter. They are commonly 10 to 50 m long, but may exceed 100 m. The bore of a capillary column is coated with an approximately 1-μm layer of liquid phase. These columns are much more efficient than packed columns, with the number of theoretical plates ranging up to several hundred thousand. As noted in the preceding section, however, capillary columns have only very limited capacities, less than 0.01 μL of liquid analyte.

Capillary columns are in the form of coils to accommodate their great length in a reasonable space. Packed columns are either coiled or in the form of single or multiple U-tubes, so that the column can be placed in a heated oven, as required for many chromatographic separations.

Solid Support Material

The solid support used in packed columns usually consists of small particles of firebrick or kieselguhr, the silica skeletons of diatomaceous algae. The solid support may consist of glass beads. Particles of Teflon, a fluorocarbon polymer, are also used. The solid support should be made up of small spherical particles of uniform size. These should be resistant to mechanical breakdown and have a high specific surface area, not less than 1 m^2/g. Furthermore, the solid support should have as little interaction as possible with the solute. Solid supports consisting of kieselguhr are sold commercially as Celite, Celetom, Chromosorb W, and Embacel. Those consisting of firebrick are sold as C 22, Chromosorb P, and Sterchamol.

Liquid Phase

The liquid phase coated on the solid support is the key component of the column stationary phase in GLC. Such a liquid should have a low volatility, with a boiling temperature at least 200° C higher than the maximum temperature to which the column will be subjected. In addition, the liquid phase should be chemically unreactive. Solvent characteristics of the liquid phase must differ at least slightly for each of the solutes so that a separation of the analytes is obtained.

Many chemical compounds have been used as liquid phases for GLC. Some of these are listed in Table 19.1. In general, the choice of liquid phase for a particular type of analyte is determined experimentally; the simple rule that "like dissolves like" is used for guidance in selecting the liquid phase.

Loading

Loading is a term used to express the percentage of the stationary phase composed of the liquid phase. Extremes of loading cover the range of 1 to 30%; at the lower level the support surface may lack complete coverage, and at the higher end pools of liquid phase may be present, decreasing column efficiency. For most applications loadings of 2 to 10% are used. Column capacity for solute, retention times, and degree of separation of peaks are all increased by higher loading; lower loading yields smaller values of height equivalent to a theoretical plate, H, and more rapid elution.

Liquid Phase in Capillary Columns

In the case of capillary columns, more properly called **open bore columns** or **open tubular columns**, the liquid phase is applied as a coating on the inside wall of the column. Thus there is no granular solid support inside the column bore. This yields two major advantages. It eliminates band broadening caused by carrier gas flow around the particles of column packing, and the column has a much lower resistance to the flow of mobile phase. The net result is that columns up to 100 m long and containing as many as 1 million theoretical plates can be used for extremely efficient and fast separations. An example of the superb separations obtained with a capillary column is illustrated in Figure 19.5.

Coating Solid Support and Packing Columns

The solid supports used in GLC may be coated with liquid phase and packed into the column in the laboratory. The liquid phase is dissolved in a volatile solvent and mixed with screened and sized support material; then the volatile solvent is evaporated to leave the liquid phase coated on the solid particles. The amount of liquid phase used should produce a particle coating 5 to 10 μm thick. After preparation, the solid phase should appear dry and be non-agglomerating (the particles do not stick together). Column packing material is poured into a straight column that is shaken to ensure uniform packing. After packing, the column is coiled or bent so that it will fit into a temperature-regulating oven. Prepacked columns are readily available from laboratory supply houses.

TABLE 19.1   Liquid phases commonly employed in gas chromatographic columns

Compound structure	Types of solutes best retained*	Maximum operating temperature, °C
1 Squalane	Essentially nonpolar	125
SE-30	Essentially nonpolar	350
Dinonyl phthalate	Somewhat polar	175
OV-17 (methylphenyl silicone)	Somewhat polar	325

Name	Structure	Polarity	Temp.
Zonyl E-7	$H\text{-}(CF_2)_n\text{-}CH_2\text{-}O\text{-}C$ ... $C\text{-}O\text{-}CH_2\text{-}(CF_2)_n\text{-}H$; $H\text{-}(CF_2)_n\text{-}CH_2\text{-}O\text{-}C$ ... $C\text{-}O\text{-}CH_2\text{-}(CF_2)_n\text{-}H$	Polar	200
XE-60	$H_3C\text{-}Si\text{-}CH_2\text{-}CH_2\text{-}C\equiv N$ with $O\text{-}Si(CH_3)_3$	Polar	275
Carbowax 20M	$HO\text{-}(CH_2\text{-}CH_2)_n\text{-}H$	Very polar	210
Versamid	$HO\left[C\text{-}R\text{-}C\text{-}NH\text{-}R'\text{-}NH\right]_n H$ (each C with $=O$)	Very polar	275
Tetrahydroxyethylenediamine	$HO\text{-}CH_2\text{-}H_2C$, $HO\text{-}CH_2\text{-}H_2C$ $N\text{-}CH_2\text{-}CH_2\text{-}N$ $CH_2\text{-}CH_2\text{-}OH$, $CH_2\text{-}CH_2\text{-}OH$	Very polar	150 150

* Essentially nonpolar compounds include alkanes, alkenes, aromatic hydrocarbons, halocarbons, sulfides, mercaptans, CS_2; somewhat polar compounds include aldehydes, ketones, esters, ethers, tertiary amines, nitriles without $\alpha\text{-}H$, nitro compounds without $\alpha\text{-}H$; polar compounds include alcohols, carboxylic acids, phenols, primary and secondary amines, oximes, nitriles with $\alpha\text{-}H$, nitro compounds with $\alpha\text{-}H$; very polar compounds include amino alcohols, polyhydroxy alcohols, hydroxy acids, polyprotic acids, polyphenols.

† This structure represents a hydrocarbon chain in which the vertical lines stand for methyl, CH_3, groups.

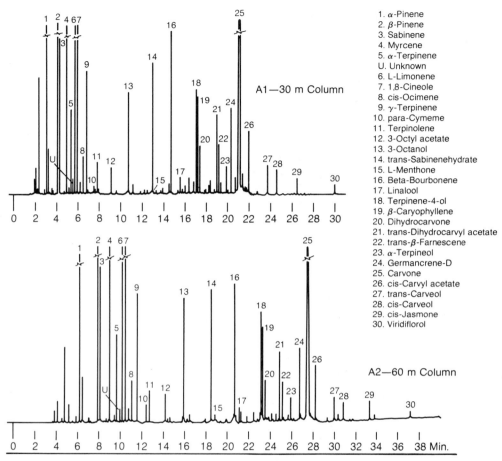

1. α-Pinene
2. β-Pinene
3. Sabinene
4. Myrcene
5. α-Terpinene
U. Unknown
6. L-Limonene
7. 1,8-Cineole
8. cis-Ocimene
9. γ-Terpinene
10. para-Cymeme
11. Terpinolene
12. 3-Octyl acetate
13. 3-Octanol
14. trans-Sabinenehydrate
15. L-Menthone
16. Beta-Bourbonene
17. Linalool
18. Terpinene-4-ol
19. β-Caryophyllene
20. Dihydrocarvone
21. trans-Dihydrocarvyl acetate
22. trans-β-Farnescene
23. α-Terpineol
24. Germancrene-D
25. Carvone
26. cis-Carvyl acetate
27. trans-Carveol
28. cis-Carveol
29. cis-Jasmone
30. Viridiflorol

SUPELCOWAX 10 fused silica capillary column, 30 m × 0.25 mm ID (Figure A1) or
60 m × 0.25 mm ID (Figure A2), 0.25 μm d$_f$. Col. Temp.: 4 min. at 75°C, then to 200°C at 4°C/min.
and hold 5 min., Linear Velocity: 25 cm/sec., He, set at 160°C, Inj. & Det. Temp.: 250°C, Det.:
FID. Sens.: 1 × 10^{-11} AFS. Sample: 0.2 μl neat oil, Split Ratio: 100:1.

FIGURE 19.5 Capillary column gas chromatographic separation of native
spearmint oil components on Supelcowax 10 capillary columns (From *The Supelco
Reporter*, March, 1984, Supelco, Inc., Bellefonte, Pa.)

Column Temperatures and Temperature Programming

Columns are usually housed in an oven because accurate temperature regulation is essential for reproducible results and because the temperature is often
raised during the course of a separation (temperature programming). If the
analyte compounds all boil within a relatively narrow range, the column

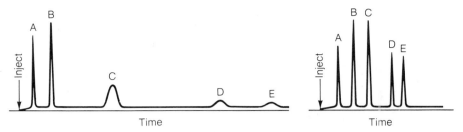

FIGURE 19.6 Effect of temperature programming. On the left is a gas chromatogram of a mixture run at a constant temperature (isothermally) at 50° C; on the right is the same mixture run with a column temperature programmed to increase at 10° C/min.

should be maintained at a temperature slightly above the average boiling point of the sample constituents. For a broad range of analyte boiling temperatures, **temperature programming** (analogous to gradient elution), in which the column temperature is increased in a continuous or stepwise fashion (see Figure 19.6), is employed. The choice of column temperature is usually a trade-off between better resolution at lower temperatures and faster separations at higher temperatures.

19.5   Detectors for Gas Chromatography

Purpose of the Detector

The purpose of a *gas chromatographic detector* is to respond to the passage of solute with the eluent from a column. Detectors should have the following characteristics to the maximum possible degree: (1) high sensitivity, (2) linear response, (3) rapid response, and (4) uniform response to different compounds in those classes of compounds for which the detector is used. No detector meets these criteria completely. In addition to detectors that respond to most analytes, there are detectors that respond to specific classes of compounds or to specific elements in compounds. Examples of these are electron-capture detectors for organohalides or atomic absorption detectors for volatile organometallic compounds. The ultimate in detector selectivity is the mass spectrometer, which produces a mass spectrum that specifically identifies a compound.

Thermal Detectors

Thermal conductivity detectors are of interest because of their simplicity, sensitivity, stability, longevity, and reliability. Since the earliest days of gas chromatography, they have remained the standard gas chromatographic detector.

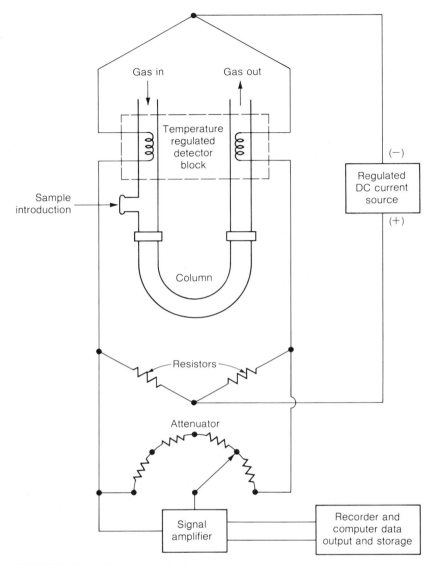

FIGURE 19.7   Schematic of a thermal conductivity detector and associated electronics

This type of detector responds to the degree to which a gas conducts heat, which in turn is affected by low levels of analyte in the eluent. Figure 19.7 is a schematic of a thermal conductivity detector. The elements of these detectors may consist of a fine wire of platinum or tungsten. The wire is heated electrically, and its temperature changes as a solute in the carrier gas stream passes over it. This slight change in temperature is manifested by a minuscule change in resistance that is measured very sensitively by a Wheatstone bridge circuit and amplified by the associated electronic circuitry. Alternatively, a **thermistor** consisting of a semiconducting material can be used in place of the metal wires. The use of thermal conductivity sensors in the gas stream, one ahead of the injector and one at the column exit, enables compensation for variations in parameters such as gas flow rates and pressures and fluctuations in electrical power.

Hydrogen and helium are the gases of choice for use with thermal conductivity detectors because their thermal conductivities are six to ten times those of most common organic compounds. As a result, a small quantity of an organic compound in a gas stream causes a measurable change in thermal conductivity.

Flame Ionization Detector

The **flame ionization detector, FID**, works on the principle that ionic species produced when organic compounds are burned in a hydrogen and air flame conduct electricity through the flame. These species are produced when CH radicals that are formed as intermediates in the combustion process react with O atoms in the flame

$$CH + O \rightarrow CHO^+ + e^- \tag{19.1}$$

to form electrically conducting CHO^+ ions. Figure 19.8 shows a flame ionization detector. The burner serves as an anode, and the positive ions are attracted to a cathode above the flame. As long as no organic compounds are in the carrier gas eluent, the current is almost zero. However, a measurable current flows when as little as 10^{-12} g of carbon in organic compounds flows through the flame per second. For alkanes the FID is about 1000 times as sensitive as the thermal conductivity detector. Furthermore, the detector response is linear to solute mass over about seven orders of magnitude.

The main requirement of a carrier gas used with an FID is that it be free of organic compounds. This permits the use of N_2 as a carrier gas instead of much more expensive He that is usually used with thermal conductivity detectors. The high sensitivity, linearity of response, stable background (baseline), and ability to be used with inexpensive N_2 carrier gas have made the FID the detector of choice for hydrocarbons.

Electron-Capture Detector

The **electron–capture detector, ECD**, is based upon the principle that some organic compounds, most commonly organohalides, take up electrons in an ionized gas, changing a current flowing through the gas. With this detector, the carrier gas, usually N_2, is passed over material containing a radioactive element that emits electrons (a β-emitter, see Chapter 23). The β-emitters used

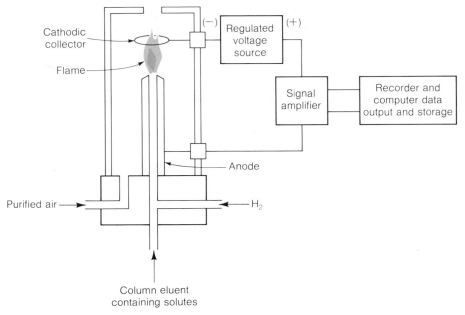

FIGURE 19.8   A flame ionization detector (FID)

for this purpose are usually nickel-63 or tritium (hydrogen-3) adsorbed on platinum or titanium foil. Electrons from the β-emitter ionize the carrier gas, releasing more electrons. This makes the carrier gas electrically conducting. When a solute capable of picking up electrons is present in the carrier gas, the number of free electrons is decreased and the current is therefore decreased.

An important property of the ECD is its insensitivity to hydrocarbons so that ECD-detectable compounds can be measured in a high-hydrocarbon background. In addition, an ECD does not respond to amines or alcohols. ECDs have proved to be especially sensitive for the detection of chlorinated hydrocarbons, such as those in pesticides or chlorinated hydrocarbon solvents. The detector is nondestructive, so that sample peaks can be collected after detection, or the eluent can be run through a second detector, such as a flame ionization detector, in series or in parallel with the ECD.

19.6   The Nature of Gas Chromatograms

Retention Time

Figure 19.9 shows a typical gas chromatogram. The time required for a solute to emerge from a column may be expressed by the *retention time*, t_r. This can be corrected for the time required for the carrier gas to go through the column,

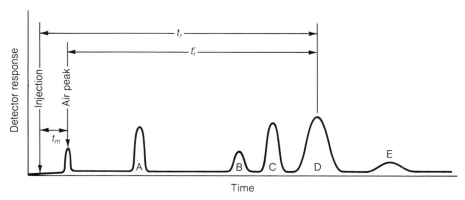

FIGURE 19.9   Gas chromatogram illustrating retention parameters

t_m, to give the *adjusted retention time*, t'_r, equal to $t_r - t_m$. The value of t_m is readily determined with a thermal conductivity detector by injection of a small quantity of air with the sample, which shows up as a small peak of unretained air, as shown in Figure 19.9. Separation of a solute from other solutes is favored by a high **capacity factor, k'**, defined as

$$k' = \frac{t_r - t_m}{t_m} \tag{19.2}$$

It is readily visualized that the capacity factor is a measure of the attraction of the stationary phase for a solute. It is a function of the partition coefficient K (see Section 17.7), according to the equation

$$k' = \frac{K V_s}{V_m} \tag{19.3}$$

where V_s and V_m are the volumes of the stationary phase and mobile phase, respectively. An acceptable range of k' is generally from 0.1 to 50. Column failure is often manifested by changes in capacity factor.

Kovats Index

Relative retention times can be expressed on a logarithmic scale by the **Kovats index, I**. The Kovats index is commonly calculated for an unknown peak in terms of retention times for peaks with known numbers of carbon atoms, n. For example, suppose that peak B in Figure 19.9 is for an alkane with n_B carbon atoms and an adjusted retention time of t'_r (n_B), peak D is for an alkane with n_D carbon atoms and an adjusted retention time of t'_r (n_D), and peak C is for an unknown compound with an adjusted retention time of t'_r (unknown). As would be the case for any common column, n_B is less than n_D

(the alkane with fewer carbon atoms elutes first). The Kovats index is the following:

$$I = 100\left[n_B + \frac{\log t_r' \text{ (unknown)} - \log t_r' \text{ } (n_B)}{\log t_r' \text{ } (n_D) - \log t_r' \text{ } (n_B)}\right] \qquad (19.4)$$

For linear alkanes, I is 100 times the number of carbon atoms. This is apparent when the "unknown" is the same as B in Equation 19.4, in which case I becomes simply $100n_B$. In a homologous series of compounds differing only in the number of CH_2 groups in the compound, $\log t_r'$ tends to vary in a linear manner with n. Comparison of I values for an unknown compound with known compounds on several different columns can be used to identify the unknown by making use of tabulated values of I.

19.7   Qualitative Determinations by Gas Chromatography

Chromatograms such as those shown in Figure 19.5 illustrate the superb abilities of gas chromatography to separate compounds that are stable in the vapor phase. Gas chromatography is excellent for showing the purity of compounds, where the presence of contaminants is revealed by chromatographic peaks.

Retention time data are the most obvious means of qualitative analysis of a compound by gas chromatography. The extremely careful control of chromatographic parameters required for reproducible retention times limits the usefulness of retention time data for qualitative analysis. The best way to perform qualitative gas chromatographic analysis is through the use of a known (authentic) sample of the suspected substance. When this authentic sample is introduced into the sample being analyzed, the height of the peak of the suspected compound should increase and no new peaks should appear. Such evidence is particularly convincing when it occurs on different columns and at different temperatures. This type of analysis with an authentic sample is called **cochromatography**.

19.8   Quantitative Determinations by Gas Chromatography

Quantitative measurements of gas chromatographic analytes can be performed using peak area, which is proportional to the quantity of a compound responsible for a peak. With very sharp peaks, peak height can be substituted for peak area. In modern practice peak area is calculated by a computer interfaced with the gas chromatograph. Other means of calculating peak area include mechanical integrators attached to the recorder, measurement of the peak area with a mechanical planimeter, and even cutting out the peaks from chart paper and weighing them on an analytical balance.

Different compounds give different gas chromatographic detector responses. The response of a thermal conductivity detector varies with the thermal conductivity of the solute. The response of an FID varies with the number of carbon atoms in the compound. Compensation for such differences can be made by the measurement of an empirical *detector response factor, F*. This factor enables comparison of the area of a peak due to analyte with that of a peak arising from an **internal standard** compound added in known quantity to the sample. The analyte concentration can be calculated from

$$\text{Analyte concentration} = \frac{\text{analyte peak area}}{\text{standard peak area}} \times \text{standard concentration} \times F \tag{19.5}$$

In performing quantitative gas chromatographic analysis, it is essential to be aware of the limitations of the detector response and to stay within the region of linear detector response.

If a quantitative gas chromatographic determination is performed on just one or two compounds, *standard addition* can be used. This consists of running two chromatograms, one on the sample and a second on the sample to which has been added a known concentration of the analyte. In favorable cases the height of the analyte peak is proportional to the peak area and, with linear detector response, to analyte concentration. This leads to the relationship

$$\frac{\text{Peak height, unknown}}{\text{Peak height, unknown plus spike}} = \frac{\text{concentration, unknown}}{\text{concentration, unknown plus spike}} \tag{19.6}$$

where the spike refers to the added analyte. For example, suppose that the following data were obtained for the determination of phenol:

Peak height, unknown = 37.5 chart divisions (arbitrary divisions on a recorder chart)

Peak height, unknown plus spike = 52.1 chart divisions

C_s = concentration of added phenol spiked into sample = 50.0 μg/L
C_u = unknown concentration of phenol, μg/L

Substituting these data into Equation 19.6 and solving yields the following:

$$C_u = (C_u + C_s) \frac{\text{peak height, unknown}}{\text{peak height, unknown plus spike}} \tag{19.7}$$

$$C_u = (C_u + 50.0 \ \mu\text{g/L}) \frac{37.5}{52.1}$$

$$C_u = 128 \ \mu\text{g/L}$$

Programmed Summary of Chapter 19

The major terms and concepts introduced in this chapter are contained in this summary in a programmed format. To derive the most benefit from the summary, you should fill in the blanks for each question and then check the answers at the end of the book to see if your choices are correct.

The two major divisions of gas chromatography based upon the nature of the stationary phase are (1) _____ and (2) _____. Of these, the one generally giving the best chromatograms is (3) _____. Starting with the controlled-flow carrier gas supply, the major components of a gas chromatograph are (4) _____. The most common carrier gases are (5) _____. A rotameter is used to (6) _____. The two major types of columns used in gas chromatography are (7) _____. In gas-liquid chromatography using a packed column the (8) _____ is coated onto the (9) _____ material. The two major types of injector systems are (10) _____. A sample splitter is employed to prevent overloading a (11) _____ column. The operation in gas chromatography analogous to gradient elution in liquid chromatography is (12) _____. The type of gas chromatographic detector most sensitive for hydrocarbons is the (13) _____. The type of detector selective for organohalide compounds is the (14) _____. The adjusted retention time is adjusted for (15) _____. An authentic sample is used in qualitative gas chromatographic determinations to (16) _____. For quantitative gas chromatographic analysis, a comparison is often made between the peak area of the analyte to that of (17) _____.

Questions

1. What is the distinction between GLC and GSC?
2. What is the primary criterion for column choice?
3. What gas chromatograph component determines the sensitivity of gas chromatographic analysis? What other parameter is sometimes determined by this component?
4. What is the primary use for a molecular sieve in gas chromatography?
5. What is the purpose of a sample splitter

on a gas chromatograph?
6. For what purpose is granular firebrick or kieselguhr used in gas chromatography?
7. What are the major desirable characteristics of the liquid phase used in GLC?
8. What problems may be encountered at low liquid phase loading on a column? What problems may arise from excessively high loading?
9. What are two major advantages of capillary columns?
10. What is the effect of increasing column

temperature upon resolution and speed of separation respectively?

11. Upon what basis are solutes usually divided in determining the most suitable liquids for stationary phases?

12. What are the main criteria for a gas chromatographic detector to be used for a wide range of compounds?

13. Match

    a. Thermal conductivity detector
    b. Flame ionization detector
    c. Electron-capture detector
    d. Mass spectrometer

    1. Uses a β-emitter
    2. Electrical conductivity increases as solute passes through
    3. Gives the specific identity of analyte
    4. Particularly good for hydrocarbons

14. What gas chromatographic detector should be chosen for the determination of small quantities of chlorinated hydrocarbon contaminants in gasoline?

15. For what quantity is the adjusted retention time adjusted?

16. What is the best way to perform qualitative analysis with a gas chromatograph?

17. Under what circumstances may peak height be substituted for peak area in quantitative gas chromatographic analysis?

18. Where S represents a standard and U is an unknown compound in gas chromatographic analysis, C_S and C_U are their respective concentrations and A_S and A_U are their respective peak areas, give a formula by which the response factor F can be calculated.

Problems

1. Assume that a gas chromatograph equipped with a capillary column is injected with a 1×10^{-3} μL sample of liquid dichloromethane solvent (molecular weight = 84.9, density = 1.34 g/mL, boiling point = $40°$C). What length of a 0.5-mm-diameter capillary column would the dichloromethane vapor produced occupy at $50°$ C and 1 atm pressure?

2. For a particular solute separated by gas chromatography, the solute peak was detected at 5 min, 12 s after injection, and the air peak came at 23 s. What was the capacity factor for the solute?

3. For a solute on a particular column the partition coefficient K is 8.3 and the ratio V_s/V_m is 11.2. What is the value of the capacity factor k'?

4. Calculate the inside surface area of a capillary column with an inside diameter of

0.5 mm and a length of 12.0 m.

5. Injection of 8.0 μg of X and 8.0 μg of Y into a gas chromatograph resulted in peaks with areas of 123 mm^2 for X and 141 mm^2 for Y. What is the response factor F of the detector for X relative to Y?

6. A 50.0-mL sample of a wastewater sample was acidified and phenol extracted from it with a 50.0-mL portion of dichloromethane. A 10-μL portion of the dichloromethane extract was injected into a gas chromatograph, giving a peak height of 53.7 (arbitrary chart units) for phenol. Another 50.0-mL portion of the sample was spiked with 10.0 μg of pure phenol, acidified, and extracted with 50.0 mL of fresh dichloromethane. Injection of a 10-μL portion of this extract gave a phenol peak height of 97.1. What was the concentration of phenol in the original sample,

assuming peak height to be linear with phenol concentration in the dichloromethane extract?

7. Carrier gas flow in a gas chromatograph was measured by a soap-bubble meter. A soap film moved from the 2.50-mL mark to the 23.00-mL mark over a period of 34.7 s. What was the carrier gas flow rate in mL/min?

8. Carrier gas flows through a 0.3-mm inside diameter, 20-m-long capillary column at a rate of 2.15 mL/min. What is the time required for the carrier gas to go through the column, t_m?

An Introduction to Spectrophotometric Methods of Analysis

L E A R N I N G   G O A L S

1. The nature of electromagnetic radiation.
2. The role of electronic ground states and excited states of atoms and molecules in the absorption and emission of electromagnetic radiation.
3. Basic components of apparatus for spectrophotometric measurements.
4. How transmittance, absorbance, and absorptivity are related in Beer's law pertaining to absorption spectrophotometry.
5. What is meant by an absorption spectrum.
6. Description of a light band.
7. Qualitative analysis by spectrophotometry.
8. Quantitative analysis by spectrophotometry.

20.1 Introduction

Anyone who has worked with chemicals and chemical solutions in the laboratory knows that color and intensities of colors are important chemical characteristics. A small quantity of permanganate ion, MnO_4^-, in solution will give the solution a light pink color. A more concentrated permanganate solution has an intense purple color, suggesting a quantitative relationship between the apparent intensity of the color (a function of light absorbed at certain wavelengths) and the concentration of the permanganate in the solution. Most chemistry students have had the experience of heating sodium chloride salt on a wire in a flame and observing the intense yellow flame emitted by the hot sodium atoms. Again, an analytical application is suggested.

Both the absorption and emission of visible light can be used for qualitative and quantitative analysis. Light is a form of **electromagnetic radiation**, which also includes gamma rays, X-rays, ultraviolet radiation, infrared radiation, microwaves, and radio waves. All of these have analytical applications; in this text we will emphasize ultraviolet, visible, and infrared radiation, whose applications are most accessible to the ordinary chemical laboratory. Analytical methods involving ultraviolet, visible, and infrared electromagnetic radiation are classified as **spectrophotometric methods of analysis,** the topic that is introduced in this chapter.

20.2   **Spectrophotometry and Electromagnetic Radiation**

Wave and Particle Nature of Electromagnetic Radiation

Electromagnetic radiation is a form of energy with a dual particle and wave nature that propagates through a vacuum at a speed of 2.99792458 $\times 10^8$ m s^{-1}. This very high number is commonly known as "the speed of light"; light, itself, is a form of electromagnetic radiation.

The wave nature of electromagnetic radiation is illustrated in Figure 20.1. Electromagnetic radiation is commonly characterized by a **wavelength**, λ, which is the distance "crest-to-crest" or "node-to-node" of the wave as shown in Figure 20.1. The number of waves (cycles) per second is the **frequency**, ν, in units of reciprocal seconds, s^{-1}. One s^{-1} is called a **hertz** (Hz); for example, when ν is 3×10^3 s^{-1}, the frequency is 3 kilohertz (KHz).

The product of wavelength and frequency is

$$\lambda\nu = c \tag{20.1}$$

where c is the **speed of the electromagnetic radiation** (speed of light). The speed of electromagnetic radiation passing through a material (such as light through glass) is never greater than that through a vacuum, given above as 2.99792458 $\times 10^8$ m s^{-1}, commonly rounded to 3.00×10^8 m s^{-1}. The speed of electromagnetic radiation in a material that is transparent to it is given by

$$\text{Speed of electromagnetic radiation} = n \times 3.00 \times 10^8 \tag{20.2}$$

where n is the **refractive index** of the material. The refractive index depends upon the wavelength, as well as the material through which the electromagnetic radiation passes.

In addition to its wave nature, electromagnetic radiation may be viewed as small particles, or "packets of energy," called **photons**. A photon has an energy E of

$$E = h\nu \tag{20.3}$$

where h is **Planck's constant**. The value of h is 6.626176×10^{-34} J-s, where J stands for the joule, the SI unit of energy equal to 0.23889 cal.

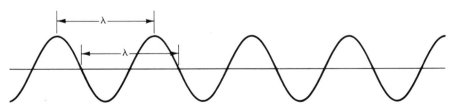

FIGURE 20.1   Representation of the wave nature of electromagnetic radiation, where λ is the length of an entire wave

It is often convenient to refer to electromagnetic radiation in terms of **wave number**, $\bar{\nu}$, which is the following:

$$\bar{\nu} = \frac{1}{\lambda} \tag{20.4}$$

The SI unit for $\bar{\nu}$ is m^{-1}, but the unit most commonly found in the chemical literature is cm^{-1}. One can also express the energy associated with electromagnetic radiation as

$$E = hc\bar{\nu} = h\,\frac{c}{\lambda} \tag{20.5}$$

The Electromagnetic Spectrum

As noted in the preceding section, electromagnetic radiation occurs over a very wide range of wavelengths and associated energies. These compose the **electromagnetic spectrum** shown in Figure 20.2. For analytical applications, all parts of the spectrum from gamma rays (measured in neutron activation analysis, Section 23.7) through radio waves [nuclear magnetic resonance (NMR)] are used. For much routine chemical analysis, the visible region of light, encompassing only a very narrow wavelength range, is the part of the spectrum most commonly used. Since analysis in this region

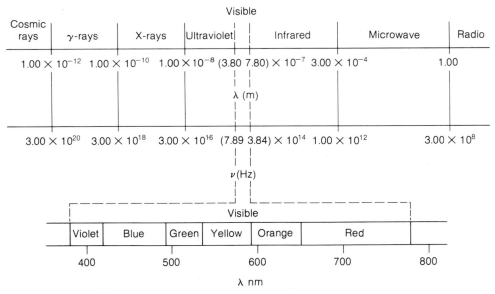

FIGURE 20.2 The electromagnetic spectrum

involves colored solutions and colored visible light, it is sometimes called **colorimetric analysis**. In this text, the absorption and emission of electromagnetic radiation in the ultraviolet, visible, and infrared regions of the spectrum are emphasized.

20.3   The Emission and Absorption of Light

Excitation of Atoms and Molecules

The absorption of a photon of light by an atom or molecule increases the internal energy of that body. This can be seen in the example of a common atmospheric reaction

$$NO_2 \times h\nu \text{ (ultraviolet or short-wavelength visible light)} \rightarrow NO_2^* \qquad (20.6)$$

in which a molecule of NO_2 in the atmosphere absorbs light of energy $h\nu$ to yield a molecule of higher energy, NO_2^*, in an **excited state** (designated with an asterisk, *): An excited state may also be reached by the input of energy from other sources, most commonly heat.

Electronically excited molecules and atoms are produced when molecules or atoms absorb electromagnetic radiation in the ultraviolet or visible region of the spectrum. A molecule or atom may possess several excited states, but ultraviolet or visible radiation generally excites a species only to one of the lower-lying excited energy levels.

The nature of the electronically excited state can be understood by considering the disposition of electrons in an atom or molecule. The electrons are contained in **orbitals**, each of which may be populated by no more than two electrons. Most molecules have an even number of electrons, usually two per orbital, having paired (opposite) spins, as represented by the arrows in Figure 20.3. Absorption of light may promote an electron (an outer electron in

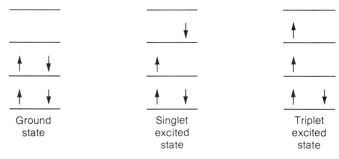

FIGURE 20.3   Representation of electrons in orbitals in the ground state and in excited singlet and triplet states

an atom, or usually a bonding electron in a molecule) to a vacant orbital of higher energy. In cases where the electron is promoted from a ground state in which it had occupied an orbital with another electron, a **singlet excited state** results when the two electrons retain their opposite spins and a **triplet excited state** when they do not.

Emission of Light

Emission of light may occur when a molecule or atom goes back to the ground state from an excited state. This is one of the means by which the species can lose excess energy acquired by excitation.

Effects of Energy of Vibration and Energy of Rotation

As discussed in Chapter 22, gaseous atoms exhibit the simplest light absorption and emission phenomena because their excited states consist only of well-defined electronic energy levels. For molecules and ions, a number of **vibrational energy levels** may be superimposed upon each electronic energy level and various **rotational energy levels** are superimposed on the vibrational energy levels. These make the absorption or emission of light much more complicated than is the case with isolated atoms in the gas state.

Figure 20.4 shows three different electronic states of a molecule. It is seen that each of these has a number of vibrational energy levels, each of which in turn may have a number of rotational energy levels, as shown by the inset.

Processes Following Absorption of a Photon by a Molecule

Figure 20.4 summarizes the major phenomena that can occur when a molecule absorbs visible or ultraviolet radiation. The absorption of a light photon (path $h\nu_A$) can raise the molecule from the ground electronic state S_0 to the first excited singlet state S_1. The molecule in the excited singlet state may lose vibrational energy (path 2). From the lowest vibrational level of S_1, one of three pathways may be followed: (1) emission of a photon of light ($h\nu_F$) of a lower energy (longer wavelength) than that absorbed, a phenomenon called **fluorescence**; (2) a radiationless transition from S_1 to a high vibrational energy level of S_0 (path X), a process called **internal conversion**; and (3) **intersystem crossing** to a high vibrational energy level of the lowest triplet excited state T_1. From the lowest vibrational energy level of the triplet excited state, energy may be lost by radiationless intersystem crossing to S_0 (path Z) or by emission of a photon $h\nu_P$. The latter process is called **phosphorescence**. Phosphorescence is a relatively slow process in that the molecule may stay in the T_1 state for from 10^{-4} up to 100 s prior to phosphorescence emission. Fluorescence, however, occurs within 10^{-8} to 10^{-4} s after the absorption of the radiation that excites the molecule ($h\nu_A$). Fluorescence is widely used in chemical analysis (of

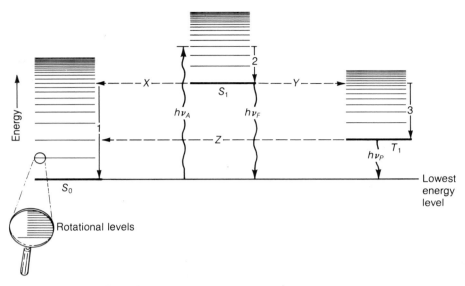

FIGURE 20.4   Three electronic states of a molecule, showing different vibrational and rotational energy levels in each. In this illustration, S_0 is the ground electronic state, S_1 is the lowest excited singlet state, and T_1 is the lowest excited triplet state. Transitions involving absorption or emission of visible or ultraviolet radiation are $h\nu_A$ for absorption of light, $h\nu_F$ for emission of light by fluorescence (emission from an excited singlet state formed by absorption of a photon), and $h\nu_P$ for emission of light by phosphoresence (delayed emission from an excited triplet state). Paths 1, 2, and 3 denote loss of energy within an electronic state not involving emission of ultraviolet or visible radiation. Paths X, Y, and Z are radiationless transitions between electronic states.

vitamins for example); phosphorescence is also used analytically, though much less so than fluorescence.

20.4   Absorption Spectrophotometric Terminology

Measurement of Absorption

In the broadest sense, **spectrophotometric analysis** makes use of selected wavelengths of electromagnetic radiation for the qualitative or quantitative determination of sample analytes. The way that this is done for the absorption of light is best understood in reference to Figure 20.5, which shows a spectrophotometer used to measure the absorption of ultraviolet, visible, or infrared radiation.

To follow the light path through a spectrophotometer, one begins with the **light source**. This source may be as simple as a small tungsten bulb, it

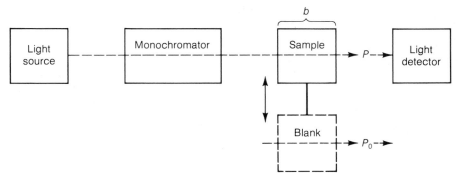

FIGURE 20.5 Block diagram of apparatus for spectrophotometric measurements of absorbed ultraviolet, visible, or infrared radiation (for infrared measurements the monochromator is after the sample)

may be a gas-filled tube in which excited gas molecules emit light, or it may be an electrically heated element that emits infrared radiation.

In most spectrophotometers, light from the source is focused into a beam of light that passes through a **monochromator**. In this device a *grating, prism,* or *filter* selects a **band** of radiation (see Figure 20.9) with a narrow wavelength span; since this is a single color of light in the visible region, it is said to be **monochromatic light**. The beam of monochromatic light emerging from the monochromator has a certain **radiant power**, defined as the light energy per unit time in the cross-sectional area of the beam measured.

When the absorption of light by an analyte is to be measured, the monochromatic light beam is first passed through a **blank** containing all the reagents except the sample. The light traversing the blank is assigned a radiant power of P_0. The light beam passing through the sample containing light-absorbing analyte has a radiant power of P, where $P \leq P_0$. The **path length** of the light beam path through both the blank and the sample is designated b. The values of P and P_0 are measured by a suitable **light detector**.

Terms Applying to the Absorption of Monochromatic Light

The measurement of the absorption of monochromatic light is known as **absorption spectrophotometry**. There are several key terms applying to such measurements, as defined below.

The radiant power of light through a blank and through a sample has just been defined. The **transmittance, T,** is given by

$$T = \frac{P}{P_0}$$

(20.7)

The **percent transmittance, $\% T$**, is

$$\% T = \frac{P}{P_0} \times 100 \tag{20.8}$$

If no light is absorbed by the sample relative to the blank, $\% T$ is 100%; if all the light is absorbed, $\% T$ is 0. Therefore, $\% T$ ranges from 0 to 100. It is desirable to have a parameter that varies in direct proportion to the concentration of light-absorbing substance, C. Such a parameter is **absorbance, A**

$$A = \log \frac{P_0}{P} = \log \frac{100}{\% T} \tag{20.9}$$

EXAMPLE

If $\% T = 53.2\%$, what is the value of A?

$$A = \log \frac{100}{\% T} = \log \frac{100}{53.2} = 0.274$$

In the older literature and in some disciplines, even today, absorbance is termed *optical density*. Absorbance varies from 0 when $\% T$ is 100 to ∞ when $\% T = 0$. The key relationship in absorption spectrophotometry is,

$$A = abC \tag{20.10}$$

This is the Lambert–Beer law, known commonly as **Beer's law**. It is easy to remember Beer's law expressed as $A = abC$! The term a is the **absorptivity**; it is characteristic of an absorbing species at a specified wavelength. Thus, at a particular wavelength and constant b, a plot of A versus C is a straight line when Beer's law is obeyed. In the older literature the absorptivity is called the *extinction coefficient*. When a molar concentration M is used, the absorptivity is given the symbol ε and is called **molar absorptivity**. Using molar concentration and molar absorptivity, Beer's law is expressed as

$$A = \varepsilon b M \tag{20.11}$$

EXAMPLE

What are the units of ε when b is in centimeters?

$$\varepsilon = \frac{A}{bM} = \frac{1}{\text{cm} \times \text{mol/L}} = \text{L/mol-cm}$$

Absorption Spectrum

A plot of absorbance A for a substance versus wavelength is called an **absorption spectrum**. Figure 20.6 shows a typical absorption spectrum for a

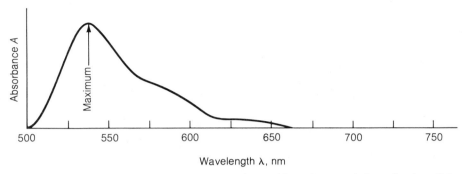

FIGURE 20.6 An absorption spectrum in the visible region consisting of a plot of *A* versus *λ* for a solution of a molecule or ion at constant concentration and constant light path length *b*.

molecule or ion in solution. Absorption spectra in the visible or ultraviolet region for species in solution tend to be rather broad, as is the case for the example shown in Figure 20.6. This is because of energy transfer between solute and solvent molecules and because of the numerous vibrational levels in the ground and excited states of the solute analyte molecules; the net result is a wide range of energies (wavelengths) over which light may be absorbed.

For quantitative analysis, a **maximum**, or **peak**, in the absorption spectrum is usually chosen for the analytical wavelength. This is for maximum sensitivity, because at a peak the analyte gives the highest absorbance per unit concentration. Furthermore, on a spectrum maximum, small changes in wavelength do not cause substantial changes in absorbance, as is the case on a steeply rising or falling portion of the spectrum.

20.5 Instrumentation for Absorption Spectrophotometry

Simple Single-Beam Spectrophotometer

Figure 20.7 shows a simple spectrophotometer, the Bausch and Lomb Spectronic 20. The elegant design of this instrument has remained basically the same for more than three decades. It consists of a light source (lamp), monochromator (grating), sample holder, and a detector (phototube). Associated with these components are optical components consisting of lenses, slits, and a light control; these make up an optical system that guides and focuses the light beam and regulates its intensity. The wavelength of light passed through the sample from the monochromator is adjusted by rotating the grating with a wavelength cam.

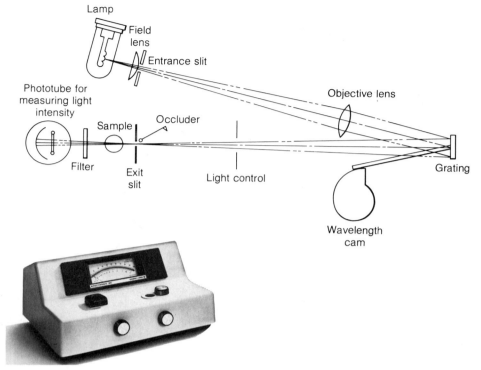

FIGURE 20.7   Above and left: Bausch and Lomb Spectronic 20 single-beam spectrophotometer. (Courtesy Bausch and Lomb, Analytical Systems Division, Rochester, N.Y.)

    The measurement of absorbance with a Spectronic 20 can be described with reference to the components just discussed. The first step is to select a wavelength by rotating the wavelength selector knob, attached to the wavelength cam, to the desired value of the wavelength expressed on a dial as nanometers. With nothing in the sample compartment, an occluder blocks out the light beam to the phototube. The 0% T calibration knob is used at this point to adjust the % T reading to zero on the meter. The blank containing all the reagents except the analyte is next placed in the sample holder; the blank is contained in a **cuvette**, which in its simplest form consists of a test tube of known, uniform diameter (constant path length b). With the cuvette containing the blank in the sample holder, the transmittance is 100% and the meter is set at 100% T with the light control knob. Finally the light-absorbing sample is placed in the same cuvette, or one of identical dimensions, and inserted in the sample holder. The meter then reads % T directly; it is also calibrated in units of absorbance A.

Spectrophotometer Components

Having described the general construction and operation of a spectro-photometer for the measurement of light absorbance, it is now appropriate to consider each of the major components in more detail. The simplest, least expensive source for absorption spectrophotometry is the *tungsten lamp*, in which light is emitted from a tungsten wire filament heated by electrical resistance to about 2900 K. The tungsten lamp can serve as a source from 320 to 2500 nm; this region encompasses the visible region (400 to 800 nm), as well as the longer-wavelength ultraviolet and shorter-wavelength infrared (near ultraviolet and near infrared, respectively).

For an ultraviolet source in the 160 to 375 nm region, a *deuterium arc lamp* is used. In the electrical discharge of this lamp, deuterium gas (D_2, the hydrogen-2 isotope) is dissociated to electronically excited deuterium atoms

$$D_2 \xrightarrow[\text{discharge}]{\text{electrical}} D^* + D^* \tag{20.12}$$

The excited atoms thus formed revert to the ground state and recombine to form deuterium molecules, emitting a continuum ultraviolet radiation in the process.

Infrared radiation is manifested as radiant heat, so an infrared source consists of a heated element. Most commonly this is a *globar* composed of a silicon carbide rod heated by electrical resistance to 1500 to 2000 K.

Monochromators serve the function of spreading **polychromatic** (white) light into its monochromatic (single-wavelength) components. In modern practice the **diffraction grating** is the most commonly used device for that purpose. Figure 20.8 shows a diffraction grating. Light viewed at an angle ϕ

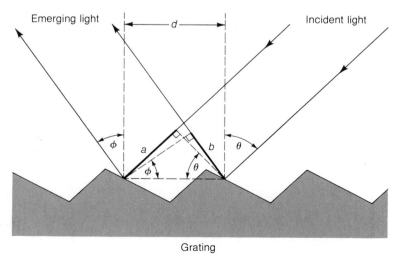

FIGURE 20.8   Diffraction grating component of a monochromator system

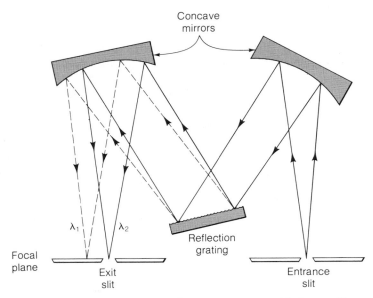

FIGURE 20.9   Grating monochromator system with Czerny–Turner optics

from the line perpendicular to the grating is *constructively reinforced* at wavelengths that are unit multiples n of a wavelength λ. Wavelengths other than these undergo destructive interference and are not seen at the angle ϕ. Thus the light diffracted from a grating at a particular angle has wavelengths of

$$n\lambda = d(\sin\theta - \sin\phi) \qquad (20.13)$$

This gives bands with maximum intensities at λ, 2λ, 3λ, and so on. The undesired bands can be filtered out of the light beam with filters. More than just a grating is required to compose a monochromator system. Figure 20.9 shows a monochromator system with associated slits and mirrors.

The light emerging from a monochromator is in the form of a **band** as shown in Figure 20.10. This band encompasses a narrow range of wavelengths with maximum intensity at a **nominal wavelength** λ_{nom}. The range of wavelength covered by the band is the **bandwidth**; a narrow bandwidth is accomplished with narrow slits and high-quality monochromator components. A narrow bandwidth enables distinction of fine features (for example, closely adjacent peaks) in an absorption spectrum. The ability of a spectrophotometer to distinguish fine features in a spectrum through the use of narrow-bandwidth light and high-quality optics is known as **resolution**. When measurements of absorbance are made for quantitative purposes at a wavelength corresponding to a broad absorption peak, a broader band is preferred because of its higher intensity.

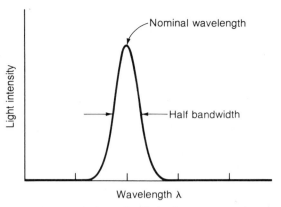

FIGURE 20.10 Band of light from a monochromator

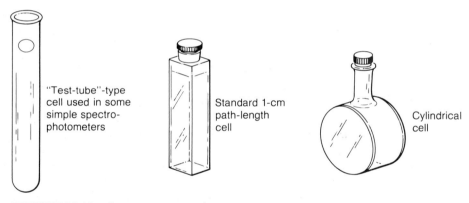

FIGURE 20.11 Common spectrophotometer cuvettes

Samples for spectrophotometric measurement are held in cuvettes or **cells** (Figure 20.11). These come in a variety of shapes and path lengths. Most cells are designed to hold liquids, some to hold gases. Some cells are designed so that solutions can flow through them, enabling absorbance measurements to be made in a flow system. Cells for use in the visible region may have glass windows, whereas those for ultraviolet measurements have quartz (SiO_2) windows, because glass absorbs ultraviolet radiation. Infrared cells generally have windows of infrared-transmitting NaCl; KBr and AgCl are also used.

Spectrophotometric detectors produce an electrical response to the absorption of photons. A number of different kinds of detectors have been developed for various purposes. The simplest of these to understand—the one

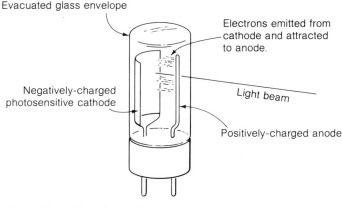

Evacuated glass envelope

Electrons emitted from cathode and attracted to anode.

Negatively-charged photosensitive cathode

Light beam

Positively-charged anode

FIGURE 20.12   Phototube

used in simple spectrophotometers—is the **phototube** (Figure 20.12). The key component of a phototube is a *photosensitive cathode* coated with a thin layer of semiconductive material such as Cs_3Sb or K_2CsSb containing alkali metals. Photons striking the photosensitive cathode cause emission of electrons from the cathode. These are attracted to the anode. This flow of electrons is a small current that is electronically amplified to give an electrical signal that is proportional to light intensity on the photocathode.

A much more sensitive device than the simple phototube is a **photomultiplier**, in which the emitted electrons strike a series of electrodes called *dynodes* held at successively more positive electrical potentials. Each electron striking a dynode results in the emission of several electrons, so that the absorption of one photon of incident radiation can result in the collection of as many as a billion electrons at the anode, an enormous amplification of the signal.

Having discussed the individual components of a spectrophotometer, mention should be made of spectrophotometer designs more sophisticated than that shown in Figure 20.8. An obvious shortcoming of a single-beam instrument is the necessity to interchange sample and reference cuvettes in the sample holder with each measurement. This makes the acquisition of absorbance measurements at different wavelengths for construction of an absorption spectrum extremely cumbersome. Drift in source intensity and detector response between the times of sample and reference measurement can lead to erroneous absorbance values. These disadvantages are overcome with a **double-beam spectrophotometer** as shown in Figure 20.13. Such an instrument has a **chopper** that sends the light beam alternatively through a reference cuvette and through a sample cuvette. Electronically, the instrument continuously compares the ratio of P_0 to P, the radiant power through the references and sample cuvettes, respectively. Furthermore, a *scanning mono-*

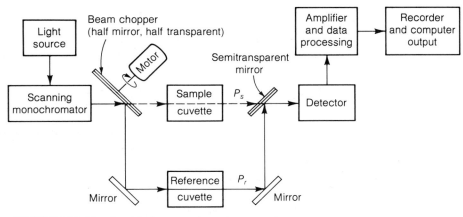

FIGURE 20.13 Double-beam spectrophotometer

chromator can be employed to scan the wavelength of light through the sample and reference. Therefore, an absorption spectrum can be taken by the instrument within a few minutes.

A *baseline spectrum* obtained by scanning the desired wavelength region with blank solution in both cuvettes should be a straight line at $A = 0$. The absorbance is readily set to zero with the zero-adjust knob on the spectrophotometer. Some modern computer-controlled instruments automatically straighten the baseline and adjust it to zero from a scan of the blank solution.

Spectrophotometer Instrument Error

Although spectrophotometers will give readings of absorbance ranging from 0 to 2 or 3, minimum **instrument error** is encountered with A in the range of approximately 0.2 to 0.8. At very high absorbance values, the intensity of light through the sample cuvette is very low and difficult to measure; small errors in P give relatively large errors in the ratio P_0/P from which A is calculated. At very low absorbance values, the difference between P and P_0 is very small and difficult to distinguish, again introducing error.

20.6   Emission Spectroscopy

The measurement of ultraviolet or visible radiation emitted by an analyte is widely used in chemical analysis. **Emission** occurs when an excited species loses its excess energy by returning to the ground electronic state (see

Figure 20.4), releasing energy in the form of a photon as part of the process. When the emitting species is an atom, the process is **atomic emission**. This is a comparatively simple process and only photons of a few specific energies corresponding to differences in electronic energy levels are emitted.

The processes by which molecules or ions emit light are fluorescence and phosphorescence. These two phenomena are discussed in Section 20.3, where it is noted that fluorescence is the much more widely used for chemical analysis.

Instrumentation used for emission spectroscopy is similar to that shown schematically in Figure 20.5. Indeed, an atomic absorption spectrophotometer that measures absorption of light by atoms is usually equipped for atomic emission analysis, and ultraviolet-visible spectrophotometers can be equipped for fluorescence measurement. In atomic emission spectroscopy, the light source consists of gaseous analyte atoms heated in a flame, electrical discharge, or plasma of high-temperature ionized gas. The monochromator is set to select the desired emission wavelength, and the radiant power measured by the light detector and amplifier gives the analyte concentration. In fluorescence measurements (almost always of ions and molecules in solution, but possible for gaseous atoms as well), an intense ultraviolet light source is used to excite fluorescent species in the sample, and the fluorescent light intensity is measured by a detector set at a right angle to the excitation source. A monochromator is required between the sample and the detector to eliminate intense scattered light of the excitation source, and a monochromator may also be placed between the light source and the sample. In simple laboratory units these are inexpensive filters.

20.7   Qualitative Analysis by Spectrophotometry

Absorption Spectrophotometry for Qualitative Analysis

The shape of a spectrum can be used for a qualitative determination of the compound responsible for the spectrum. Visible and ultraviolet spectra of solutions are usually too featureless for compound identification, but infrared spectra, such as the one shown in Figure 20.14, often show a number of peaks characteristic of the compound, which enables compound identification by exactly matching known spectra of compounds. The intricate nature of infrared spectra is a consequence of the fact that infrared radiation excites various vibrational transitions in the molecules; these transitions occur at rather distinct energies (wavelengths). More important than compound identification is the ability of infrared to show peaks resulting from the excitation of vibrational transitions in specific groups on a molecule. For example, the absorption of infrared radiation at a particular wavelength is due to C—H stretching vibration; a band at 1.97 μm results from a combination of N—H bending and stretching vibrations; and C=O, C=C, and C=N bonds show characteristic absorption bands.

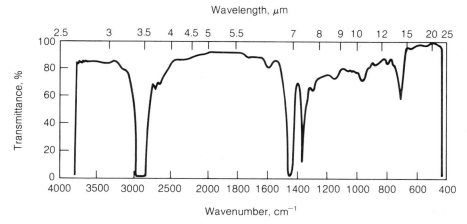

FIGURE 20.14   Infrared spectrum of 10W–30 motor oil. (Courtesy of Exxon Research and Engineering Co.)

Atomic Spectra for Qualitative Determinations

In contrast to molecular spectra in the visible and ultraviolet region, atomic spectra are excellent for qualitative analysis. This is because of the very narrow bandwidths of atomic absorption and emission lines that are observed for atoms in the gas phase. For example, when a solution containing compounds involving several elements is injected into the very hot plasma ("flame" composed of ionized gas) of an inductively coupled plasma atomic emission spectrometer (Section 22.4), the atoms of the various elements become excited by the heat of the plasma and emit sharp lines like those shown in Figure 20.15. The wavelengths of the several lines emitted by an element are

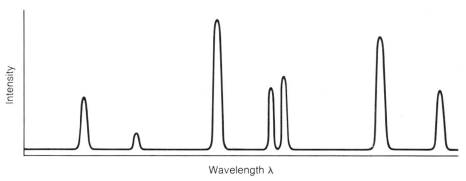

FIGURE 20.15   Inductively coupled plasma atomic emission spectrum showing atomic emission peaks from several elements

characteristic of the element and the intensity (peak height) is proportional to concentration. Therefore, both a qualitative and a quantitative analysis may be obtained with this technique. Simultaneous determinations of up to 30 elements can be made using a monochromator with an exit slit for each of the elements set to allow passage of a band of light characteristic of each element. For multielement determinations each element requires a separate photo-multiplier detector and amplifier.

20.8   Quantitative Analysis by Spectrophotometry

Both the absorption and emission of light can be used for quantitative determinations of compounds and elements. Details of such determinations are discussed in Chapters 21 and 22.

Quantitative analysis by absorption spectrophotometry is normally based upon a *Beer's law plot* as shown in Figure 20.16. Such a plot is prepared using known concentrations of analyte and measuring the absorbances of the analyte as a function of known values of C. If Beer's law

$$A = abC \tag{20.10}$$

is obeyed, a linear plot is obtained. The determination of the concentration of an unknown simply involves the measurement of the absorbance A of the unknown and reading its concentration from the plot. Even if Beer's law is not followed, as shown by the curved plots in Figure 20.16, it is still possible to measure A of the unknown and read its concentration from the calibration curve. Depending upon the analyte, the type of analysis, and the matrix in which the analyte is found, various interferences can occur. Compensation can be made for such interferences by using as a blank a solution containing everything but the analyte. Additional details on the practical aspects of

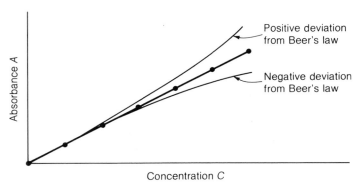

FIGURE 20.16   Plot of absorbance versus concentration showing a linear plot when Beer's law is followed, as well as positive and negative deviations from Beer's law

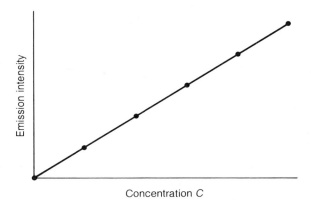

FIGURE 20.17   Plot of emission intensity versus concentration for an emission band of an element determined by inductively coupled plasma atomic emission spectroscopy. Unlike absorbance, atomic emission values using this technique may be linear and accurate over several orders of magnitude of concentration.

quantitative determinations by absorption spectrophotometry are given in Chapters 21 and 22.

The emission of light can also be used for quantitative determinations. Figure 20.17 is a plot of emission intensity versus concentration for an element determined quantitatively by inductively coupled plasma atomic emission spectrometry. Measurement of the emission intensity from an element of unknown concentration followed by a reading from the calibration plot enables calculation of the unknown concentration.

With modern computerized instrumentation, both absorption and emission calibration curves can be stored by computer and the concentration of an unknown calculated by the computer from the unknown's absorbance or emission intensity reading. Sophisticated computer software can do calculations that compensate for various interferences, thus greatly extending the utility of quantitative spectrophotometric analysis.

Programmed Summary of Chapter 20

The major terms and concepts introduced in this chapter are contained in this summary in a programmed format. To derive the most benefit from the summary, you should fill in the blanks for each question and then check the answers at the end of the book to see if your choices were correct.

A form of energy with a dual particle and wave nature that propagates through a vacuum at a speed of 3.00×10^8 m s^{-1} is called (1) _____. The distance "crest-to-crest" of a wave of electromagnetic radiation is its (2)

_____ and the number of waves (cycles) per second is the (3) _____. The ratio of the speed of light through a material to that in a vacuum is called the (4) _____ of the material. In addition to its wave nature, electromagnetic radiation may be regarded as small packets of energy called (5) _____, representative of the particle nature of the radiation. In the formula $E = h\nu$, E is (6) _____ and h is (7) _____. In the reaction M + energy → M*, the asterisk denotes an (8) _____. Absorption of photons from the ultraviolet or visible region of the spectrum results in the production of (9) _____ excited atoms or molecules. Fluorescence is the emission of a photon from an excited (10) _____ state; phosphorescence is emission of a photon from an excited (11) _____ state. Of the two, the one that occurs most quickly after absorption of a photon causing excitation is (12) _____. In terms of radiant power through a blank, P_0, and radiant power through an equal path length of sample, P, the transmittance is defined as (13) _____. In terms of percent transmittance (% T), the absorbance is $A =$ (14) _____. The mathematical expression of Beer's law is (15) _____. A plot of absorbance versus wavelength for an absorbing substance is called an (16) _____, upon which the (17) _____ is usually chosen as the analytical wavelength. The major components of a spectrophotometer for absorption spectrophotometric measurements are (18) _____. The source for the visible region is commonly a (19) _____. The most common device used as a monochromator is a (20) _____. The basic purpose of a monochromator is to isolate a band of light that encompasses a narrow (21) _____. The peak of such a band of light is called the (22) _____, whereas the wavelength range that it covers is called the (23) _____. The containers in which liquid samples are held for spectrophotometric measurement are called (24) _____. A phototube for the measurement of light consists of a cathode covered with a material that, when struck by light, (25) _____ that are attracted to an anode, resulting in a measurable (26) _____. A more sensitive version of the phototube is the (27) _____, which contains a series of electrodes called (28) _____. Drift in source intensity and detector response, as well as the inconvenience of having to alternate sample and blank cuvettes, are overcome by use of a (29) _____ spectrophotometer. The range of absorbance values giving the most accurate results for analytical measurements is (30) _____. In inductively coupled plasma atomic emission spectroscopy, accurate emission values that are linear with analyte concentration may occur over a range of as much as (31) _____. In atomic emission spectroscopy, the light source consists of heated (32) _____. Qualitative infrared analysis is often possible because of the (33) _____ nature of many infrared spectra. Qualitative analysis by atomic emission is possible because the atomic emission spectrum of a typical element (usually a metal) typically consists of (34) _____. Quantitative analysis by absorption spectrophotometry is normally based upon a (35) _____ plot.

Questions

1. What is the dual nature of electromagnetic radiation?
2. Why is light represented by $h\nu$?
3. The emission of light by an atom, such as an atom of Ca, can result from heating the atom in a flame. Represent the processes involved by two reactions similar to Reaction 20.6.
4. What is the definition of radiant power?
5. What are transmittance T and absorbance A in terms of the values of radiant power P and P_0?
6. In order for a plot of A versus C to be a straight line, Beer's law must be obeyed. What other two parameters must remain constant for a linear plot?
7. In terms of the Beer's law equation, what is the relationship between a and ε?
8. Ultraviolet radiation shining on a particular rock causes various minerals in the rock to glow with visible light. After the ultraviolet source is turned off, some minerals cease glowing immediately, whereas others continue to glow for a while. What two phenomena are illustrated?
9. With a single-beam spectrophotometer, what should be the intensity of light on the detector used to set 0% T and 100% T, respectively?
10. What is resolution in spectrophotometric terms?
11. What is the principle of a phototube?
12. What is the major distinguishing feature of a double-beam spectrophotometer compared to a single-beam instrument?
13. Why are infrared spectra particularly useful for qualitative analysis?
14. How is positive deviation from Beer's law manifested in a plot of A versus C?

Problems

1. What is the frequency in hertz of 500-nm visible light (in a vacuum)?
2. What is the energy in joules of a mole (6.02×10^{23}) of photons of 500-nm light?
3. A phototube produces a current that is proportional to the radiant power of light impinging on it. When the light source is cut off, it has a small background current from stray light and electronic noise that must be subtracted from any current reading obtained when light is impinging on the phototube. Suppose this background current is 86 μA for a particular phototube. With a blank solution in the spectrophotometer cuvette, the current is 930 μA; and with analyte solution in the cuvette, the current is 537 μA. Calculate % T and A.
4. In absorption spectrophotometry, minimum instrument error occurs in a range of absorbance values from about 0.2 to about 0.8. To what % T values do these correspond?
5. For a light-absorbing analyte obeying Beer's law, an instrument reading of 33.7 % T was obtained for a 15.0 mg/L solution of analyte and 62.4% T for a solution of unknown concentration at the same wavelength and same light path length b. What was the concentration of the unknown?
6. What is the energy in joules in a mole of photons from monochromatic ultraviolet radiation for which $\lambda = 371.2$ nm?
7. Derive a conversion factor to change absorptivity, a, to molar absorptivity, ε,

for a compound of molecular weight 239.3, for which a was measured with C expressed in units of mg/L.

8. Consider a compound with the absorption spectrum shown in Figure 20.6. If the % T was 92.3% at 600 nm for a 2.40 × 10^{-3} M solution of the compound, estimate the molar absorptivity of the compound at the wavelength of maximum absorbance shown for the spectrum in Figure 20.6.

9. What is the difference in % T for a change of 0.01 A at an A value of 0.6, which is in the optimum analytical range of A? What is the corresponding difference in % T for $A = 1.5$? What conclusions can you draw from these two values?

21

Solution and Molecular Spectrophotometric Analysis

L E A R N I N G   G O A L S

1. How examination of absorption spectra can be used as a guide for quantitative absorption spectrophotometric analytical techniques.
2. The role of a Beer's law plot in quantitative absorption spectrophotometric analyses.
3. Conversion of analytes to light-absorbing species by various means.
4. The role of light-absorbing, extractable metal chelates in the determination of metal ions.
5. How mixtures of analytes can be determined by spectrophotometric measurements.
6. Significance of isosbestic points.
7. Essentials of spectrophotometric titrations.
8. Equilibrium constant determination by spectrophotometry.
9. Basics of luminescence analysis, including fluorescence and chemiluminescence.
10. Distinction between fluorescence emission and excitation spectra.

21.1   Quantitative Spectrophotometric Analysis

Advantages of Absorption Spectrophotometric Determinations

The absorption of electromagnetic radiation in the range of 200 to 1000 nanometers (nm) forms the basis of a large number of chemical determinations. This range constitutes the near-ultraviolet (200–400 nm), visible (400–800 nm), the near-infrared (800–1000 nm) regions of the spectrum. For simplicity we will refer to these techniques under the heading of **absorption spectrophotometry**, and to the radiation as light, realizing that in some cases it is near-ultraviolet or near-infrared radiation. Since so many analyses have been developed for the absorption of visible light, the terms **colorimetric analysis** and **colorimetry** are also encountered in the analytical chemical literature.

Quantitative absorption spectrophotometry is applicable to a wide range of substances. Some analytes themselves absorb light, enabling their direct measurement. More frequently, an analyte can be converted to a species that absorbs light as a measure of analyte concentration. Good detection limits are attained for many analytes in absorption spectrophotometry. Thus, molar concentrations of analytes at 10^{-5} to 10^{-4} M levels are commonly determined, and detection limits as low as about 1×10^{-7} M have even been attained. Concentration values accurate to within 1 to 3% are routine with quantitative absorption spectrophotometry, and accuracies of 0.1 to 0.3% are attainable with special care. A degree of selectivity is afforded by the different wavelengths at which various species absorb light, and even more selectivity is introduced by the chemical reagents used to develop color from analytes, solvent extraction of absorbing species, use of masking agents to eliminate the effects of interfering species, and other chemical means. In general, absorption spectrophotometry is not very expensive, with some instruments costing as little as a few hundred dollars. Many procedures are carried out with relative ease by technician-level personnel, and generally the techniques are comparatively convenient.

Solution Emission Analysis

Emission analysis can be applied to species in solution. The most useful such application is fluorescence (sometimes called molecular fluorescence to distinguish it from atomic fluorescence, which is also possible), in which ultraviolet radiation is used to excite emission at a longer wavelength. Fluorescence is very sensitive and selective for some analytes. In specialized cases, analytes may react with reagents to produce electronically excited species that emit light. Since excitation occurs via a chemical reaction, this technique is called **chemiluminescence**. This phenomenon has some limited, but important and growing, applications in chemical analysis.

| 21.2 | **Direct Spectrophotometric Determinations** |

Direct Measurement of Inorganic Species

Some inorganic species have high enough molar absorptivities at their absorption spectra peaks to enable their direct determination by absorption spectrophotometry. Two of the best examples are chromium(VI) and manganese(VII) in acidic media; their absorption spectra are shown in Figure 21.1.

Chromium(VI) can be measured spectrophotometrically by absorbance at 440 nm in 1 M H_2SO_4–0.7 M H_3PO_4. Under these conditions absorbance at 440 nm is little affected by shifting equilibria among chromium(VI) species so that Beer's law is followed over a chromium concentration range of 2×10^{-4} to 5×10^{-3} M.

Permanganate is readily measured spectrophotometrically at 545 nm. If the manganese is not originally in the permanganate form, potassium persulfate, $K_2S_2O_8$, can be employed to oxidize it.

$$2Mn^{2+} + 5S_2O_8^{2-} + 8H_2O \rightarrow 2MnO_4^- + 10SO_4^{2-} + 16H^+ \qquad (21.1)$$

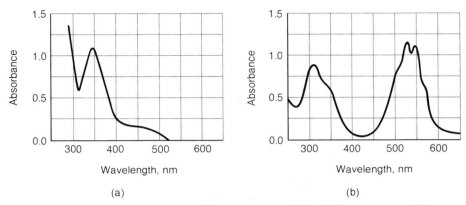

FIGURE 21.1 Absorption spectra of (a) chromium(VI) and (b) manganese(VII) at 5.00×10^{-3} M concentrations in $1\ M\ H_2SO_4$–$0.7\ M\ H_3PO_4$ medium; absorbance measured in a 1-cm cell. The manganese is present as MnO_4^-. Several chromium species may be present, including $Cr_2O_7^{2-}$, $HCr_2O_7^-$, $HCrO_4^-$, and H_2CrO_4; because of these, chromium(VI) solutions do not obey Beer's law exactly over very wide concentration ranges.

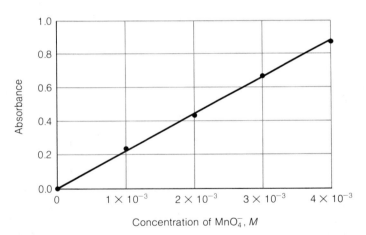

FIGURE 21.2 Plot of absorbance in a 1-cm cuvette at 545 nm versus concentration of MnO_4^-, M; calibration curve for the spectrophotometric determination of permanganate

Potassium persulfate can also be added to permanganate solutions to prevent the reduction of MnO_4^- ion. Figure 21.2 shows a calibration curve prepared for the spectrophotometric determination of permanganate. In this case the plot is a straight line showing that Beer's law is being followed.

Determination of Mn

As an example of the kind of analysis that would utilize the calibration curve in Figure 21.2, consider the determination of Mn in a 1.238-g sample of steel dissolved in acid, the manganese in the steel oxidized by persulfate (Reaction 21.1), and the solution diluted to 100 mL. If this solution had a value of A of 0.427 in a 1-cm cell, what was the percentage of Mn in the steel? From the calibration curve the molar concentration of permanganate in the final solution is

$$[MnO_4^-] = 2.0 \times 10^{-3} \ M \tag{21.2}$$

The molar concentration of Mn is likewise $2.0 \times 10^{-3} \ M$. In units of grams per liter, the concentration of Mn is

$$C_{Mn} = 2.0 \times 10^{-3} \ \text{mol/L} \times 54.94 \ \text{g/mol} = 0.110 \ \text{g/L} \tag{21.3}$$

Since the solution was diluted to 0.100 L (100 mL), the total mass of Mn from the sample was

$$\text{Mass of Mn} = 0.100 \ \text{L} \times 0.110 \ \text{g/L} = 1.10 \times 10^{-2} \ \text{g} \tag{21.4}$$

The percentage of Mn in the steel sample is

$$\% \, Mn = \frac{1.10 \times 10^{-2} \ \text{g}}{1.238 \ \text{g}} \times 100 = 0.89 \tag{21.5}$$

21.3   Conversion of Analytes to Absorbing Species

Spectrophotometric Determination of Copper(II) as the EDTA Chelate

The number of analytes that absorb light well enough for direct spectrophotometric determination is quite limited. It is much more common to bind the analyte to a reagent to form a light-absorbing species, or to have the analyte undergo a chemical reaction resulting in the formation of a light-absorbing species. The most straightforward example of a color-forming reaction is the complexation or chelation of a metal ion with a reagent that forms a colored metal chelate or complex. A good example of this is the intensely colored chelate formed by Cu^{2+} and the strong chelating agent EDTA (see Figure 10.3). The reaction for the formation of the EDTA chelate of Cu^{2+} is

$$Cu^{2+} + H_x Y^{x-4} \rightarrow CuY^{2-} + xH^+ \tag{21.6}$$

where $H_x Y^{x-4}$ is a partially protonated form of the EDTA anion. Figure 21.3 clearly illustrates the feasibility of measuring CuY^{2-} spectrophotometrically. The absorption spectra in Figure 21.3 suggest that copper(II) might be determined spectrophotometrically as CuY^{2-} at 260 nm where there is a plateau (an essentially level region) on the CuY^{2-} absorption curve and minimal absorbance from either unchelated EDTA or $Cu^{2+}(aq)$. Experimental studies have shown that copper(II) can be determined experimentally in a

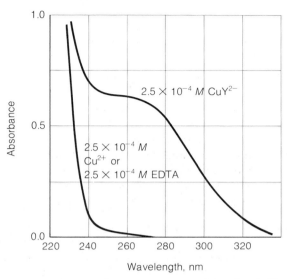

FIGURE 21.3 Absorption spectra of CuY^{2-}, Cu^{2+}, and EDTA in a 1-cm cuvette with water as a blank. The latter two species have an essentially identical "absorption edge" at about 240 nm in the ultraviolet region; above this wavelength absorption is minimal for both EDTA and Cu^{2+}, although the latter gets its blue color in aqueous solution from weak absorption in the visible region above 400 nm (not shown).

concentration range of 5×10^{-5} to 5×10^{-4} M in an approximately five-fold excess of EDTA (to ensure that all the copper(II) is in the CuY^{2-} form) by measurement of A at 260 nm. The same concentration of EDTA in the absence of Cu(II) may be used as a blank. Many samples with a number of constituents may contain ultraviolet-absorbing species that can cause interferences. If this is the case, and the interference is not too severe, the blank may consist of sample solution *without* EDTA. If this approach is used, at lower concentrations of copper(II) analyte there may be some slight interference as a result of absorption by unchelated EDTA. Such interference is minimized by using a lower excess of EDTA at low copper concentrations. For relatively high copper concentrations in solutions with high EDTA excess, absorbance can be measured at 320 nm. It is seen from Figure 21.3 that EDTA does not absorb at 320 nm, and ultraviolet-absorbing interfering species are less likely to absorb at this wavelength than at 260 nm. However, the absorptivity of CuY^{2-} at 320 nm is sufficient to allow measurement of copper(II) down to about 2×10^{-4} M.

Diphenylthiocarbazone as a Colorimetric Reagent

Diphenylthiocarbazone, commonly called dithizone, forms neutral, low-polarity complexes with metal ions that are readily extracted into organic

$$\underset{\text{Thione form}}{\underset{\displaystyle \text{S}=\text{C}}{\overset{\displaystyle \overset{\text{H}}{\underset{}{\text{N}}}—\overset{\text{H}}{\underset{}{\text{N}}}—\bigcirc}{\underset{\text{N}=\text{N}—\bigcirc}{}}}} \rightleftarrows \underset{\text{Thiol form}}{\underset{\displaystyle \text{HS}—\text{C}}{\overset{\displaystyle \text{N}—\overset{\text{H}}{\underset{}{\text{N}}}—\bigcirc}{\underset{\text{N}=\text{N}—\bigcirc}{}}}} \qquad (21.7)$$

solvents, such as carbon tetrachloride and chloroform (see Section 16.7). Some of the metal complexes are intensely colored, and their concentrations can be measured by absorption spectrophotometry. Lead is the most common example of a metal determined by the absorption of light by its dithizone complex. Denoting un-ionized dithizone as HDz, the reaction with lead is

$$2\text{HDz} + \text{Pb}^{2+} \rightarrow 2\text{H}^+ + \text{Pb}(\text{Dz})_2 \qquad (21.8)$$

in which the lead complex, $\text{Pb}(\text{Dz})_2$, is formed. The absorption spectra of HDz and $\text{Pb}(\text{Dz})_2$ dissolved in carbon tetrachloride are shown in Figure 21.4. It is seen from the spectra that absorbance of uncomplexed HDz interferes with that of the $\text{Pb}(\text{Dz})_2$ complex. Fortunately, $\text{Pb}(\text{Dz})_2$ can be quantitatively extracted from water into carbon tetrachloride at pH 11.0, where essentially all the uncomplexed dithizone is in the form of Dz^- ion, which stays behind in the water. Following extraction, the lead dithizonate can be measured spectrophotometrically in chloroform solution from its absorbance at 520 nm.

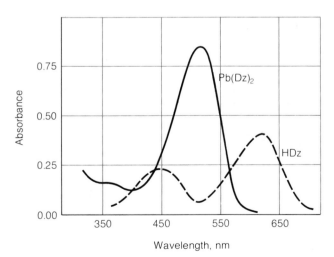

FIGURE 21.4   Absorption spectra of dithizone (HDz) and the lead–dithizone complex ($\text{Pb}(\text{Dz})_2$) in carbon tetrachloride

Organic Chelating Agents for Spectrophotometric Metal Determination

A number of organic chelating agents have been found to be useful for the spectrophotometric determination of metal ions. Some of these are listed in Table 21.1.

21.4 Analysis of Mixtures

Absorption Spectra of Mixtures

Figure 21.5 shows the absorption spectra in carbon tetrachloride of (A) 1.00×10^{-5} M mercury(II) dithizonate, $Hg(Dz)_2$; (B) 1.00×10^{-5} M copper(II) dithizonate, $Cu(Dz)_2$; and (C) a solution that is 1.00×10^{-5} M in both $Hg(Dz)_2$ and $Cu(Dz)_2$. The concentrations of each of these chelates can be determined in a mixture of the two by absorbance measurements at two different wavelengths, given Beer's law behavior by each constituent.

Additivity of Absorbances

For subsequent discussion, it should be noted at this point that the absorbances of solutes can be added to give the total absorbance of the solution, if there is no interaction among the solutes. This means, for example, that the total absorbance at a specified wavelength and specified path length b of a mixture of solutes X, Y, and Z is given by

$$A_{total} = A_X + A_Y + A_Z \qquad (21.9)$$

where A_X, A_Y, and A_Z are the absorbances that each of the designated solutes would have at the same concentration in the absence of the other two solutes.

Mixture of Copper(II) and Mercury(II) Dithizonates

To illustrate the spectrophotometric determination of a mixture of solutes, consider the dithizonates of copper(II) and mercury(II), the spectra of which are shown in Figure 21.5. Calculation of the concentrations of these two analytes can be made with a knowledge of their molar absorptivities at the two different wavelengths where each exhibits an absorption maximum—that is, 490 nm for $Hg(Dz)_2$ and 545 nm for $Cu(Dz)_2$.

The molar absorptivity of $Hg(Dz)_2$ can be calculated at 490 nm with the knowledge that $A = 0.545$ in a 1.00-cm cell at a concentration of 1.00×10^{-5} mol/L. Substitution into the Beer's law relationship yields

$$\varepsilon_X^{490} = \frac{A}{bM} = \frac{0.545}{1.00 \text{ cm} \times 1.00 \times 10^{-5} \text{ mol/L}}$$

$$= 5.45 \times 10^4 \text{ L/mol-cm} \qquad (21.10)$$

TABLE 21.1   Organic chelating agents used for the spectrophotometric determination of metal ions

Name and structure of reagent	Analytes	Typical determinations
Phenylfluorone	Ge, Sn, Co, Fe, In, Nb, Ni, Ti, Zr, Mo(VI)	Reaction to form 2:1 chelate with Ge(IV) in 0.5 M HCl, extraction with benzyl alcohol, measurement of A at 505 nm
Acetylacetone	Be, Fe(III), Mo(VI), UO_2^{2+}, V(III), V(V)	Formation of 2:1 Be chelate at pH 7 in the presence of EDTA and citrate, extraction into chloroform, A at 295 nm
Eriochrome Black T	Cd, Co(II), Mg, Zn, rare earths	Formation of 2:1 chelate with Cd at pH 6 in presence of 1,10-phenanthroline, extraction with chloroform, A at 522 nm
Pyridylazonaphthol (PAN)	Cd, Co(III), Cu(II), Hf, Mn(II), Ni, Os(VIII), Zn, Tl(III), Zr	Formation of 2:1 chelate with Hf at pH 4.0 in 40% methanol, A at 545 nm in the same medium
8-Hydroxyquinoline (oxine)	Al, Co(II), Cr(III), Ga, Hf, In, Mn(II), Rh, Mo(VI), Nb(V), Pd(II), Ru(IV), Th, Tl(III), U(VI), V(III)	Formation of 3:1 chelate with Al at pH 4.5–11.3 in aqueous tartrate–EDTA–H_2O_2 medium, A at 388 nm in the same medium
Sodium diethyldithiocarbamate	Bi, Co(III), Cu(II), Fe(II), Fe(III), Ni, Pd(II), Pt(IV), Sb(III), Te(IV), UO_2^{2+}	Formation of 3:1 chelate with Bi, A at 366 nm in carbon tetrachloride extract

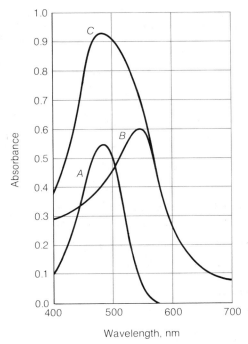

FIGURE 21.5 Plot of absorbance in a 1-cm cell versus wavelength for the following solutions in carbon tetrachloride: solution A, 1.00×10^{-5} M mercury(II) dithizonate, $Hg(Dz)_2$; solution B, 1.00×10^{-5} M copper(II) dithizonate, $Cu(Dz)_2$; and solution C, 1.00×10^{-5} M in each of these chelates

where the subscript X designates $Hg(Dz)_2$ and subsequently the subscript Y will designate $Cu(Dz)_2$. The three other molar absorptivities needed are those of $Cu(Dz)_2$ at 490 nm (ε_Y^{490}) and those of $Hg(Dz)_2$ and $Cu(Dz)_2$ at the absorbance maximum of $Cu(Dz)_2$, 545 nm. The molar absorptivities of $Hg(Dz)_2$ and $Cu(Dz)_2$ at 545 nm are designated as ε_X^{545} and ε_Y^{545}, respectively. All four molar absorptivities and the data from Figure 21.5 required to calculate them are given in Table 21.2 on page 576.

As an example of the calculation of concentrations of $Hg(Dz)_2$ and $Cu(Dz)_2$ by the measurement of absorbance values at 490 and 545 nm, consider a mixture of these two chelates in carbon tetrachloride having absorbance values of 0.825 at 490 nm and 0.460 at 545 nm in 1.00-cm cuvettes ($b = 1.00$ cm). The two equations to be solved are

$$A(490 \text{ nm}) = \varepsilon_X^{490}b[Hg(Dz)_2] + \varepsilon_Y^{490}b[Cu(Dz)_2] \tag{21.11}$$

$$A(545 \text{ nm}) = \varepsilon_X^{545}b[Hg(Dz)_2] + \varepsilon_Y^{545}b[Cu(Dz)_2] \tag{21.12}$$

TABLE 21.2   Molar absorptivities of Hg(Dz)$_2$ and Cu(Dz)$_2$ at 490 and 545 nm as calculated from values read from Figure 21.4. Subscript X refers to Hg(Dz)$_2$ and subscript Y refers to Cu(Dz)$_2$. Absorbance values were taken for $1.00 \times 10^{-5}\,M$ solutions in 1.00-cm cuvettes.

Wavelength, nm	Absorbance values		Molar absorptivities, L/mol-cm			
	A_X	A_Y	ε_X^{490}	ε_Y^{490}	ε_X^{545}	ε_Y^{545}
490	0.545	0.400	5.45×10^4	4.00×10^4	—	—
545	0.095	0.600	—	—	9.5×10^3	6.00×10^4

Substituting known values of absorbances and molar absorptivities into these equations gives the following two equations with two unknowns:

$$0.825 = 5.45 \times 10^4[\text{Hg(Dz)}_2] + 4.00 \times 10^4[\text{Cu(Dz)}_2] \qquad (21.13)$$

$$0.460 = 9.5 \times 10^3[\text{Hg(Dz)}_2] + 6.00 \times 10^4[\text{Cu(Dz)}_2] \qquad (21.14)$$

Solution of these equations gives the following concentrations:

$$[\text{Hg(Dz)}_2] = 1.076 \times 10^{-5}\ M \qquad (21.15)$$

$$[\text{Cu(Dz)}_2] = 5.97 \times 10^{-6}\ M \qquad (21.16)$$

EXAMPLE

A mixture of Hg(Dz)$_2$ and Cu(Dz)$_2$ in carbon tetrachloride has absorbance values of 0.542 and 0.325 at 490 and 545 nm, respectively, in a 1-cm cell. What are the values of [Hg(Dz)$_2$] and [Cu(Dz)$_2$]?

Solve

$$0.542 = 5.45 \times 10^4[\text{Hg(Dz)}_2] + 4.00 \times 10^4[\text{Cu(Dz)}_2]$$
$$0.325 = 9.5 \times 10^3[\text{Hg(Dz)}_2] + 6.00 \times 10^4[\text{Cu(Dz)}_2]$$
$$[\text{Hg(Dz)}_2] = 6.75 \times 10^{-6}\ M \qquad [\text{Cu(Dz)}_2] = 4.35 \times 10^{-6}\ M$$

21.5   Isosbestic Points

Spectra of an Indicator at Different pH Values

A series of spectra taken of solutions containing the same total concentration (analytical concentration) of a solute that can undergo a conversion between two species on a 1:1 basis will cross at a common absorbance value called the **isosbestic point**. The most common example of this is an acid–base indicator for which the color, and therefore the spectra, changes through loss or gain of

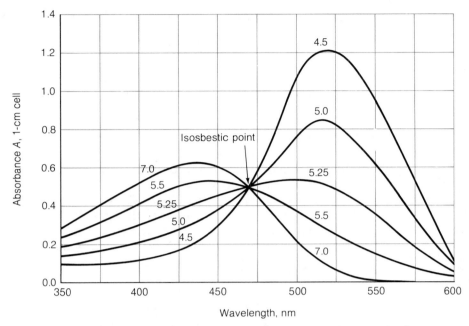

FIGURE 21.6 Absorption spectra of 5.0×10^{-4} M methyl red in a 1-cm cell at pH values shown

H^+ ion. Figure 21.6 shows the spectra of methyl red indicator at different pH values. The two methyl red species are the red form, HIn

and the yellow form, In^-, produced by the loss of the ionizable H

$$HIn \rightleftharpoons H^+ + In^- \qquad pK_a \cong 5.1 \tag{21.17}$$

   Red   Yellow

(Below pH 2.5, H^+ adds to the $—CO_2^-$ group to produce a third species, H_2In^+.) As seen from the spectra, the red HIn species has an absorption maximum at 525 nm, and the yellow In^- form has a maximum at 435 nm. The absorption spectrum of HIn and that of In^- at the same concentration must cross, and this intersection point occurs at a wavelength at which their absorptivities are the same. Furthermore, if the analytical concentration of the indicator—that is, the sum of [HIn] and [In$^-$]—is maintained at a constant value and the pH varied in the vicinity of pK_a where [HIn] and [In$^-$] are of comparable magnitude, the resulting spectra will all intersect at the wavelength where the absorptivity of HIn is equal to that of In$^-$. Such spectra and their intersection point are shown in Figure 21.6. The intersection point is the isosbestic point, which for the case of methyl red occurs at 465 nm.

Significance of Isosbestic Point

The existence of an isosbestic point when a parameter such as pH, oxidation-reduction potential E (see Chapter 11), or ligand concentration (see Chapter 10) is varied is indicative of a 1:1 transition between two species such as that shown for methyl red in Equation 21.17. When an isosbestic point is not observed, it is likely that more than two species are involved. Such is the case, for example, in the formation of complex ions with closely spaced formation constants. Consider a metal ion M^{2+} that forms complexes with the ligand L with formulas ML^{2+} and ML_2^{2+}

$$M^{2+} + L \rightleftarrows ML^{2+} \qquad K_1 = \frac{[ML^{2+}]}{[M^{2+}][L]} \qquad (21.18)$$

$$ML^{2+} + L \rightleftarrows ML_2^{2+} \qquad K_2 = \frac{[ML_2^{2+}]}{[ML^{2+}][L]} \qquad (21.19)$$

Suppose a series of solutions is prepared in which the analytical concentration of the metal [M(II)] is kept constant and the analytical concentration of ligand C_L is varied from 0 up to the value of [M(II)]—that is, a ratio of ligand to metal ranging from 0 to 1. The spectra of these solutions would not show an isosbestic point if K_2 is almost as large as K_1 because as the 1:1 ligand-metal ratio is approached, an appreciable amount of ML_2^{2+} will form, giving the *three* species M^{2+}, ML^{2+}, and ML_2^{2+} at significant concentrations.

21.6   Spectrophotometric Titrations

Nature of Spectrophotometric Titrations

The absorption of light by analyte, titrant, or product can be used to follow the course of a titration. Consider the generalized titration reaction

$$A + T \rightarrow P \qquad (21.20)$$

where A, T, and P are analyte, titrant, and product, respectively. Suppose that the titration is followed spectrophotometrically at 500 nm in a 1.00-cm cell and that the molar absorptivity of A is 0 (no light absorption), that of T is

TABLE 21.3 Spectrophotometric titration data

Volume added titrant, mL	[P], mmol/mL	Absorbance from P*	[T], excess mmol/mL	Absorbance from T*	Total absorbance
0	0	0	0	0	0
0.25	4.98×10^{-5}	0.100	0	0	0.100
0.50	9.90×10^{-5}	0.200	0	0	0.200
0.75	1.48×10^{-4}	0.299	0	0	0.299
1.00	1.96×10^{-4}	0.399	0	0	0.399
1.25	2.44×10^{-4}	0.499	0	0	0.499
1.50	2.91×10^{-4}	0.599	0	0	0.599
1.75	3.38×10^{-4}	0.698	0	0	0.698
2.00	3.85×10^{-4}	0.798	0	0	0.798
2.25	4.31×10^{-4}	0.898	0	0	0.898
2.50	4.76×10^{-4}	0.998	0	0	0.998
2.75	4.74×10^{-4}	0.998	4.74×10^{-5}	0.011	1.009
3.00	4.72×10^{-4}	0.998	9.43×10^{-5}	0.023	1.021
3.25	4.70×10^{-4}	0.998	1.41×10^{-4}	0.034	1.032
3.50	4.67×10^{-4}	0.998	1.87×10^{-4}	0.045	1.043

* Corrected for dilution by the factor (50.0 mL + mL titrant added)/50.0 mL

225, and that of P is 1995. Assume the titration of 50.0 mL of a solution containing $5.00 \times 10^{-4} \, M$ A with a titrant solution of $0.01000 \, M$ T. (A simple way to follow such a titration is to remove a portion of the solution after each addition of titrant, measure absorbance, and return the solution from the cuvette to the titration vessel.) The data to be expected from this titration are given in Table 21.3. Note that the absorbance values are corrected for the slight dilution of the titrant solution. For example, at 2.00 mL added titrant, the absorbance is due to product P and has an uncorrected value of

$$A = 1.995 \times 10^3 \text{ mL/mmol-cm} \times 1.00 \text{ cm} \times 3.85 \times 10^{-4} \text{ mmol/mL}$$
$$= 0.767 \tag{21.21}$$

Corrected for dilution, the absorbance is

$$A = 0.767 \, \frac{50.00 \text{ mL} + 2.00 \text{ mL}}{50.00 \text{ mL}} = 0.798 \tag{21.22}$$

Spectrophotometric Titration Plots

Figure 21.7 is a plot of absorbance versus volume of titrant. The intersection of the two linear portions of the plot at 2.50 mL of added titrant denotes the end point volume.

Spectrophotometric titrations are useful for reactions with relatively low equilibrium constants, such that the titration reaction is not complete at

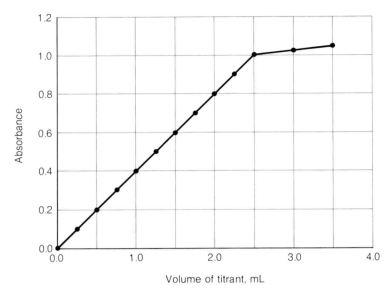

FIGURE 21.7   Titration of 50.0 mL of analyte A with 0.01000 M titrant T, reacting according to Reaction 21.20. This is a plot of absorbance versus volume of titrant at a wavelength of 500 nm, where the product absorbs strongly and the titrant absorbs weakly.

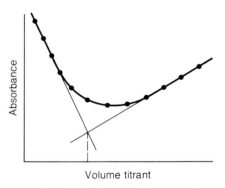

FIGURE 21.8   Spectrophotometric titration plot for an incomplete reaction

equilibrium at the end point, but is complete in a large excess of analyte or titrant. The curve for such a titration is shown in Figure 21.8, where both the analyte and the titrant absorb light at the wavelength used, but the product does not. The titration plot is rounded at the end point, which is determined by extrapolation of the linear segments.

21.7   Spectrophotometric Determination of Equilibrium Constants

The Role of Spectrophotometry in Equilibrium Constant Determination

Absorbance measurements provide an excellent means of measuring the concentrations of species involved in equilibrium reactions. In some cases, such as complexes of transition metal ions (see Chapter 10), the central metal ion, the several complexes that it forms, and even the ligand may absorb light and have different, but overlapping spectra. By varying the ligand and metal ion concentrations, a large number of spectra of the different species present can be obtained, and analysis of absorbance values at various wavelengths and different concentrations can serve to identify the species present and the absolute or relative values of their equilibrium constants. In many cases other types of measurements are employed to measure concentrations of non-absorbing species. For example, if the ligand L^- is the conjugate base of the weak acid HL, the latter species can be added in excess and the concentration of L^- regulated by careful control of pH measured by a glass electrode.

Mole Ratio Plots and Metal Complex Formation Constants

Obviously, the spectrophotometric determination of formation constants can be a very complicated and sophisticated process requiring computer analysis of the numbers measured. Such a treatment is beyond the scope of this book; however, a simple case will be discussed to show the general principle involved. Consider complexation of the metal ion Mt^{2+} with a ligand L. Suppose that a series of solutions is prepared in which the analytical concentration of the metal, [Mt(II)], is held constant at 1.00×10^{-3} M, and the analytical concentration of the ligand, C_L, is varied from 0 to 3.00×10^{-3} M. Furthermore, suppose that neither solutions of pure Mt^{2+} nor pure L absorb light in the 400- to 800-nm visible region of the spectrum, but that the solutions of their mixtures described in the preceding sentence show an increased intensity of color with spectra of similar shape showing an absorption maximum at 525 nm. If the absorbances are measured at 525 nm in a 1.00-cm cell, the data given in Table 21.4 are obtained. These data are plotted as absorbance versus $C_L/[M(II)]$ in Figure 21.9. This is called a **mole ratio plot**. In this case it is a curve with approximately linear segments at high and low values of C_L. Extrapolation of these two linear segments to their intersection point shows that they intersect at a ligand-metal ratio of 1:1, indicating the formation of the MtL^{2+} complex. If this complex had a very high formation constant such that it was almost completely formed at the point where $C_L/[M(II)]$ is exactly 1, the data points would lie on the two straight lines; instead they lie on a curve, characteristic of incomplete formation of the complex when ligand and metal are present at the stoichiometric ratio. If the molar absorptivity of the complex were known, it would be possible to calculate $[MtL^{2+}]$ at any point on the plot

TABLE 21.4   Absorbance at 525 nm in a 1.00-cm cell for solutions that are 1.00×10^{-3} M in Mt(II) and varying values of the analytical concentration of ligand L, C_L

C_L, mmol/mL	$C_L/[Mt(II)]$	Absorbance
0.200×10^{-3}	0.200	0.145
0.400×10^{-3}	0.400	0.283
0.600×10^{-3}	0.600	0.410
0.800×10^{-3}	0.800	0.517
1.000×10^{-3}	1.000	0.600
1.200×10^{-3}	1.200	0.656
1.400×10^{-3}	1.400	0.692
2.00×10^{-3}	2.000	0.742
3.00×10^{-3}	3.00	0.769

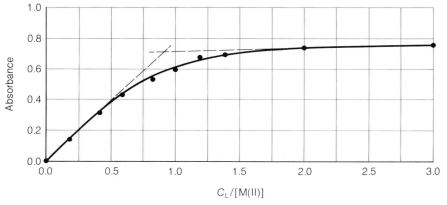

FIGURE 21.9   Mole ratio plot for the data given in Table 21.4

and calculate the formation constant of this complex from the following relationships:

$$K = \frac{[MtL^{2+}]}{[Mt^{2+}][L]} \qquad \text{Formation constant expression} \qquad (21.23)$$

$$[MtL^{2+}] = \frac{A}{\varepsilon} \qquad (21.24)$$

(A is the absorbance in a 1-cm cell and ε is the molar absorptivity of the complex.)

$$[Mt^{2+}] = 1.000 \times 10^{-3} - [MtL^{2+}] = 1.000 \times 10^{-3} - \frac{A}{\varepsilon} \qquad (21.25)$$

$$[L] = C_L - [MtL^{2+}] = C_L - \frac{A}{\varepsilon} \qquad (21.26)$$

(In the above, $[Mt^{2+}]$ is the concentration of uncomplexed metal ion and $[L]$ is the concentration of *uncomplexed* ligand.)

$$K = \frac{A/\varepsilon}{(1.000 \times 10^{-3} - A/\varepsilon)(C_L - A/\varepsilon)} \tag{21.27}$$

With a correct value of ε, the substitution of C_L and the corresponding value of A into Equation 21.27 should give the same value of K for the curved regions of the mole ratio plot, and this should be the correct value of K. Figure 21.9 shows the limiting value of A to lie around 0.7 or 0.8 in an excess of ligand, where essentially all the M(II) is in the form of MtL^{2+}—that is, $[MtL^{2+}] = 1.000 \times 10^{-3}$ M under these conditions. Trial and error show that use of 0.800 for the limiting value of A such that

$$\varepsilon = \frac{\text{limiting value of } A}{\text{maximum value of } [MtL^{2+}]} = \frac{0.800}{1.000 \times 10^{-3}} = 800 \tag{21.28}$$

gives a consistent set of K values with an average of $K = 1.20 \times 10^4$ when the appropriate numbers are substituted into Equation 21.27.

EXAMPLE

For a problem of the type just discussed, use a K of 1.000×10^8 and prepare a mole ratio plot like that in Figure 21.9.

Because of the high value of the formation constant the plot will appear as shown with a very sharp break at a mole ratio of $1:1$.

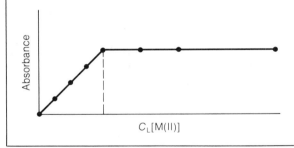

21.8   The Emission of Light from Species in Solution

Luminescence and Chemiluminescence

So far the discussion of the spectrophotometric determination of analytes has centered around the absorption of light. The light energy absorbed is denoted by pathway A in Figure 20.6. This represents light energy absorbed in raising a molecule to an excited state. In many cases it is useful to look at the light emitted by a molecule in an excited state, a phenomenon called **luminescence**. Luminescence is placed in several categories, depending upon the mode of exciting the analyte species. A standard determination of NO uses a chemical reaction to produce excited NO_2 molecules

$$NO \text{ (air or gas stream)} + O_3 \rightarrow NO_2^* + O_2 \qquad (21.29)$$

followed by measurement of the light emitted when the excited molecule, NO_2^*, reverts to the ground state:

$$NO_2^* \rightarrow NO_2 + h\nu \text{(energy of emitted photon)} \qquad (21.30)$$

Since the excited molecule emitting the light measured was produced by a chemical reaction, this phenomenon is called **chemiluminescence**.

Fluorescence

Most common molecular and ionic species in solution measured by luminescence are excited by the absorption of a photon, followed by essentially immediate emission of a photon of lower energy. This overall process is represented by the following pathways in Figure 20.4 (page 550): absorption of a photon $(h\nu_A)$, vibrational transitions (2), and emission of a photon $(h\nu_F)$. The overall process is called **fluorescence**, and in favorable cases the fluorescent emission is employed for chemical analysis.

Figure 21.10 illustrates the general configuration of a spectrofluorometer in which light of a specific wavelength can be used to excite fluorescence of a species in solution. The fluorescent light emitted, of wavelength λ_{em}, is observed at right angles to the excitation light of wavelength λ_{ex}. The

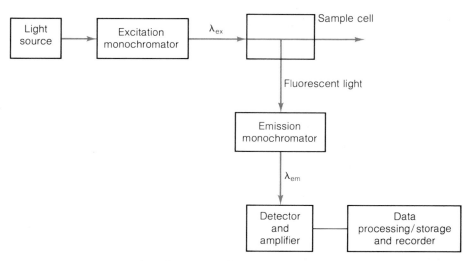

FIGURE 21.10 General configuration of a spectrofluorometer for exciting fluorescence with light of λ_{ex} and measuring its intensity at λ_{em}.

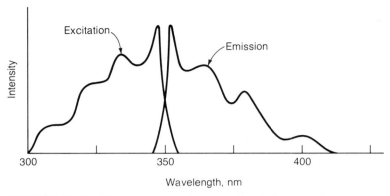

Intensity

Excitation

Emission

300   350   400

Wavelength, nm

FIGURE 21.11   Fluorescence excitation and emission spectra

fluorescent light has lower energy so that $\lambda_{em} > \lambda_{ex}$. Since fluorescence emission is viewed at right angles to the exciting radiation, and at a different wavelength, fluorescence is in a sense the detection of light against a dark background. This contrasts to absorbance spectrometry (Section 20.4), which involves the measurement of differences of the intensity of light through a sample and through a blank. Therefore, strongly fluorescing materials can be detected at lower concentrations measured over a wider range than analytes measured by absorbance.

With a spectrofluorometer it is possible to excite emission at a specific value of λ_{ex} and scan the intensity of light emitted at various values of λ_{em}. The resulting spectrum is an **emission spectrum**. Scanning λ_{ex} at constant λ_{em} yields an **excitation spectrum**. These spectra enable finding the most efficient wavelengths to excite fluorescent molecules and to observe emission from excited molecules. Excitation and emission spectra are plotted in Figure 21.11.

The basic equation for measuring the concentration of a fluorescent species is

$$I = kP_0 C \qquad\qquad (21.31)$$

where I is the intensity of fluorescent light measured, k is a constant for a particular system, P_0 is the radiant power of the radiation exciting fluorescence, and C is the concentration of the analyte species. This equation applies in systems where the fluorescent light is not absorbed appreciably. In favorable cases I may vary linearly with C over several orders of magnitude, so that the useful analytical range of fluorescence is greater than that of absorbance measurements. Furthermore, within limits, I can be increased by increasing P_0, thus increasing the intensity of fluorescence analysis.

Fluorescence can be used to determine a wide range of substances, including proteins, riboflavin (vitamin B_2), and drugs. In addition, non-fluorescent species can often be determined by the attachment of fluorescing chemical groups.

Programmed Summary of Chapter 21

The major terms and concepts introduced in this chapter are contained in this summary in a programmed format. To derive the most benefit from the summary, you should fill in the blanks for each question and then check the answers at the end of the book to see if your choices were correct.

Although chromium(VI) absorbs light, its direct spectrophotometric determination is complicated by the existence of (1) _____. Inorganic manganese can be determined in solution in the form of (2) _____. Copper(II) can be determined spectrophotometrically using EDTA as a reagent by measuring the absorbance of the species (3) _____ at a wavelength of (4) _____ in the presence of excess EDTA. The characteristics of metal chelates formed by diphenylthiocarbazone and similar chelating agents that favor the extraction of the chelates into organic solvents are (5) _____ and (6) _____. A ketone (refer to Table 1.2 for the ketone functional group) that is used as an organic chelating agent in the spectrophotometric determination of metal ions is (7) _____. In order to determine the concentrations of two different species in solution by measurement of absorbance at two different wavelengths several criteria must be satisfied. The two species must not (8) _____ in solution. Furthermore, the absorbances of the two species must be (9) _____ and the species must have substantially different (10) _____ at the two wavelengths employed. An isosbestic point occurs when two different species in equilibrium at a constant total concentration have the same (11) _____ at a specific wavelength. A spectrophotometric titration involves following the absorbance of (12) _____ as a function of volume of added titrant. Spectrophotometric titrations are especially useful for titration reactions with low (13) _____. A common means for determining spectrophotometrically the number of ligands bound to a metal ion in a metal complex and the formation constant of the complex makes use of a plot called a (14) _____. If the formation constant is relatively low, the plot (answer to 14) is (15) _____ at the point where ligand and metal are present at the stoichiometric ratio of ligand to metal, whereas if the equilibrium constant is quite high, the plot shows a (16) _____ at this point. The general term for the emission of light by a species losing energy from an excited state is (17) _____. If excitation has occurred via a chemical reaction, the emission of light is termed (18) _____, an example of which is the emission of light from (19) _____ formed by the gas phase reaction of (20) _____. If a molecule is excited by a photon with subsequent essentially immediate emission of light of a longer wavelength than that of the exciting radiation, the phenomenon is called (21) _____. With this phenomenon an excitation spectrum is a plot of (22) _____, whereas an emission spectrum is a plot of (23) _____.

Questions

1. What are the lowest detection limits to be expected in absorption spectrophotometry and what is the highest degree of relative accuracy normally encountered?

2. What two common inorganic species can be determined by direct spectrophotometry?

3. What is the most common or straightforward method of producing a colored species from a metal analyte?

4. What is an excellent extractant and color-forming reagent for the determination of metals?

5. In what kind of situation is an isosbestic point most commonly encountered?

6. Match the type of spectrophotometric titration described with the appearance of the accompanying titration curves.
   a. High equilibrium constant reaction, only titrant colored
   b. High constant, only product colored
   c. Low constant, product strongly colored, analyte weakly colored
   d. High constant, only analyte and titrant colored

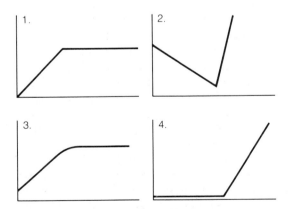

7. What is a mole ratio plot?

8. How does chemiluminescence vary from fluorescence?

9. What is the distinction between an emission spectrum and an excitation spectrum in fluorescence?

10. What is the fundamental equation for fluorescence analysis and what is the meaning of each term?

11. What is a major reason that chromium (VI) solutions obey Beer's law less well than manganese(VII) solutions?

12. For what purpose is persulfate used in spectrophotometric permanganate determinations?

13. From examination of the data contained in Table 21.1, what are three properties common to most organic reagents (or their compounds with metals) used for the spectrophotometric determination of metals?

14. For a mixture of solutes X, Y, and Z, absorbing light at a particular wavelength, express A_{total} in terms of a_X, a_Y, a_Z, C_X, C_Y, and C_Z (see Equation 21.9). Assume all measurements were made in a 1-cm cell.

15. Under what conditions is an isosbestic point obtained for several solutions of the same indicator at different pH values? What does the existence of an isosbestic point reveal about an indicator? (You might want to consider ionizable H and the values of K_a's.) What does the existence of an isosbestic point in spectra from a series of solutions having a constant metal concentration but variable ligand concentrations tell about the metal complex species formed?

16. Corrections for dilution were made in the spectrophotometric titration data given in Table 21.3. What would be the effect of omitting such corrections if a relatively dilute titrant were used, for example, such that the end point volume of the titration mixture was around twice the

initial volume? What would be the effect if the titrant were relatively quite concentrated?

17. Answer this question if you have studied amperometric titrations previously. In terms of degree of completion of titration reaction, what advantage is shared by amperometric and spectrophotometric titrations?

18. Consider the mole ratio plot shown in the example on page 583. Explain why you think that the exact value of a formation

constant could or could not be calculated from such a plot. What about a maximum or minimum value of a formation constant?

19. What two conditions of fluorescence analysis provide the rationale for the statement that "fluorescence is in a sense the detection of light against a dark background"?

20. What are the meanings of the terms "near–ultraviolet" and "near–infrared"?

Problems

1. Calculate the slope of the calibration plot in Figure 21.2 and give its significance.

2. A 1.225-g copper ore sample was dissolved in acid and diluted to 100 mL. A 25.0-mL aliquot of the solution was mixed with 5.00 mL of 5.0×10^{-3} M EDTA, adjusted to pH 10.0 and diluted to 50.0 mL. The absorbance of this solution at 260 nm was 0.450; 5.00×10^{-4} M EDTA at pH 10.0 was employed as a blank, and 1-cm cells were used. Using Figure 21.3 and assuming Beer's law behavior, calculate the percentage of copper in the ore.

3. Analyte X has an absorption maximum at 390 nm with $\varepsilon_X^{390} = 7.25 \times 10^3$. Analyte Y has an absorption maximum at 510 nm with a molar absorptivity of $\varepsilon_Y^{510} = 3.86 \times 10^3$. The spectra overlap such that $\varepsilon_Y^{390} = 975$ and $\varepsilon_X^{510} = 346$. A mixture of X and Y showed absorbance values of 0.558 at 390 nm and 0.684 at 510 nm measured in 1-cm cells. What were the concentrations of X and Y?

4. Manganese in manganese nodules (rich sources of Mn on the ocean floor) is to be determined spectrophotometrically by procedures outlined in Section 21.2 with

reference to the calibration curve in Figure 21.2. If the nodules are 5 to 9% Mn, outline a spectrophotometric procedure for determining the manganese. Include the appropriate dilution factors assuming that a 2-cm path length cell is used.

5. The conjugate base A^- of a weak acid HA has an absorption maximum at 375 nm in the near–ultraviolet; this is a wavelength where HA does not absorb at all. When the analytical concentration (a term you should look up, if you do not know its exact meaning) of HA is 1.00×10^{-3} mol/L, the HA is 89.5% dissociated and the absorbance (1-cm cell) is 0.0325. Plot absorbance as a function of analytical concentration of HA from 1.00×10^{-4} to 0.100 mol/L, choosing as many points as needed to make a good plot, and explain any apparent deviations from Beer's law.

6. The spectrophotometric determination of copper as the EDTA chelate, CuY^{2-}, was explained in Section 21.3, and spectrophotometric titrations were explained in Section 21.6. Outline conditions under which EDTA in food processing wastewater at about 5×10^{-4} M concentra-

tion might be determined by spectrophotometric titration with Cu^{2+} titrant. What approach and conditions might be used for the spectrophotometric titration of an EDTA waste solution that is about 5×10^{-3} M in EDTA, if the analyte solution is not to be diluted?

7. Using the information given in Section 21.3 and the spectra in Figure 21.4 as a basis, outline conditions under which lead might be spectrophotometrically titrated with dithizone, HDz. What wavelength would be used? What about extraction steps? What would be the approximate range of lead concentration that could be determined? What is the expected shape of the titration curve? After considering all of these factors, what can you say about the feasibility of the titration?

8. The metal ion Mt^{2+} forms a 2:1 complex ion, MtZ_2^{2+}, with the ligand Z. This is the only complex formed between Mt^{2+} and Z. At its 380 nm peak, where neither Mt^{2+} nor Z absorbs, the complex ion has a molar absorptivity ϵ of 1.037×10^3 mol/L-cm. A solution in which the analytical concentration of Mt(II) is 1.200×10^{-3} mol/L and that of

Z is 2.400×10^{-3} mol/L has a value of A of 0.700 at 380 nm in a 1-cm cell. Calculate the overall formation constant β_2 (see Chapter 10) of MtZ_2^{2+}.

9. A spectrophotometric titration was carried out at 465 nm where both the analyte and titrant absorb well, but the product of the titration reaction does not. An initial volume of 50.0 mL of analyte was titrated. The equilibrium constant of the titration reaction is such that the reaction is incomplete at the equivalence point. The following titration data values are given:

mL titrant	A (not corrected for dilution)
0	0.617
10.0	0.425
50.0	0.725
60.0	1.131

From the information given, calculate the end point volume.

10. From the information presented in Figure 21.3, estimate the molar absorptivity of the EDTA chelate of copper(II), CuY^{2-}, at 260 nm.

22 | Atomic Spectrophotometric Analysis

LEARNING GOALS

1. Distinction between atomic absorption and atomic emission.
2. Basis of the high element-selectivity of atomic spectroscopic techniques.
3. Components of an atomic absorption spectrophotometer and the importance of the hollow cathode lamp source.
4. Distinction between flame atomic absorption and flameless atomic absorption.
5. Atomization and atomizers in atomic spectroscopy.
6. Electrothermal excitation in emission spectroscopy.
7. Advantages of emission spectroscopy with electrothermal excitation for qualitative analysis of small samples.
8. Principles and major applications of flame atomic emission spectroscopy.
9. Spectroscopic properties of elements and flame conditions that must be considered in flame atomic emission.
10. Principles, applications, and advantages of inductively coupled plasma atomic emission spectroscopy.
11. Causes and effects of line broadening.
12. The meaning of spectral interferences and matrix spectral interferences.

22.1   Introduction

The Nature of Atomic Spectroscopy

Atomic spectroscopy consists of the observation of the emission or absorption of visible or ultraviolet radiation (which we will refer to as light) by elemental atoms in the gas state. Atomic spectroscopy is most applicable to metals, which, with the notable exception of mercury, are not present at significant levels as gaseous atoms at normal temperatures. Usually, by the time a sample is prepared for atomic spectroscopic analysis, the analyte is present as compounds, particularly ionic compounds, in solution. Therefore, it is necessary to get the analyte element from the chemically combined state in solution to the atomic state in the gas phase. This is done by exposing a small portion of the sample to a high temperature in one of several devices. Such a device may simply be a flame into which the aqueous sample solution is aspirated. It may be a plasma of ionized noble gas heated by microwave or radiofrequency radiation (see Figure 20.2). An electrothermal source may be used consisting of an arc or spark between two conducting electrodes of graphite or metal. Analyte from a very small volume of sample may be placed in the gas phase by heating the sample inside a cylinder consisting of graphite heated by its resistance to the passage of a high electrical current. Whatever device is used, it must vaporize any solvent in which the analyte is dissolved, vaporize the analyte element or its compounds, and provide an atmosphere in which the element being determined is converted to free atoms. For atomic emission, the free analyte atoms must also become electronically excited so that they emit light.

Excitation and De-excitation of Atoms

Atomic spectroscopy is based upon the excitation of ground electronic state atoms to an excited state and their subsequent reversion to the ground state (de-excitation). These phenomena were discussed in a general sense in Chapter 20, Section 20.3. Figure 22.1 illustrates the basic phenomena involved in atomic spectroscopy. An atom in an electronic ground state, E_0, may reach an electronically excited state, E^*, by the absorption of energy. If this energy is in the form of a photon of light with a **resonance energy**, ΔE, exactly equal to the difference in energy between E^* and E_0, the phenomeon is called **atomic absorption**. For atomic absorption, the determination of an element consists of measuring the decrease in intensity of light resulting from its passage

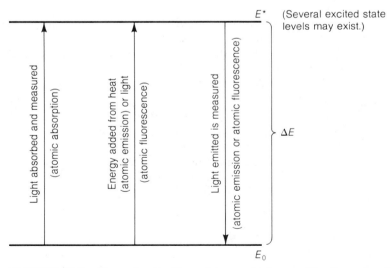

FIGURE 22.1 Ground state, E_0, and one excited state, E^*, of an atom. The difference in energies between these two states is ΔE. This energy difference is equal to the energy of a photon required to excite an atom from E_0 to E^* and is equal to the energy of the photon emitted when the reverse occurs.

through a cloud of analyte atoms. If heat energy excites the analyte atoms, and the light is measured that is emitted when the atom reverts from an excited state, E^*, to the ground state, E_0, the process is called **atomic emission**. Excitation of the atom by light, and measurement of the light emitted as the atoms return to the ground state is called **atomic fluorescence**. Both atomic emission and atomic absorption are widely used analytical techniques; atomic fluorescence is not commonly applied to routine analysis.

Advantages of Atomic Spectroscopy

The greatest advantage of atomic spectroscopy is its selectivity for individual elements. This is because of the very specific values of ΔE between the ground and excited states of atoms of various elements. This may be understood by considering the relationships,

$$\Delta E = h\nu \tag{22.1}$$

and

$$\lambda = \frac{c}{\nu} \tag{22.2}$$

where h is Planck's constant, ν is the frequency of light, λ is the wavelength of light, and c is the velocity of light (see Chapter 20, Section 20.2). From these equations it may be concluded that if ΔE has a definite value with virtually no spread, then the corresponding λ will also have a specific value. A value of λ for absorption or emission is called a line. If light is both emitted and absorbed at the same wavelength, the line is called a **resonance line**. Atomic lines are very narrow, with bandwidths (see Section 20.5 and Figure 20.10) typically less than 0.01 nm. Such narrow lines are very desirable for analytical purposes because there is little chance for overlap of lines with nominal wavelengths (see Figure 20.10) that are very close together; this minimizes interference in both absorption and emission. Furthermore, the concentration of emission intensity over a very narrow region gives a sharp, intense peak that increases sensitivity for the analyte element. The narrow atomic line bandwidths are due to the fact that the analyte atoms are in the gas phase, and there are no solution effects. Because there may be several—sometimes numerous—values of E^* for a particular element, atoms of the element may emit several prominent lines as shown in Figure 22.2. In many cases, therefore, the spectroscopist has several choices of emission or absorption wavelengths.

In addition to its selectivity for specific elements, other advantages of atomic spectroscopy include relative simplicity, detection limits as low as a few parts per billion, and suitability for routine and automated analysis. Atomic spectroscopy is free of many chemical interferences, although in some cases chemical and spectral interferences (see Section 22.8) can be severe.

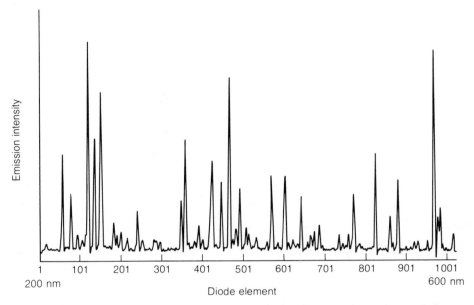

FIGURE 22.2   Emission lines from titanium produced in an inductively coupled plasma atomic emission spectrometer (see Section 22.5). The light intensities are measured on a series of light-sensitive diode elements located to receive progressively longer wavelengths of the emission spectrum from the instrument monochromator over a wavelength range of approximately 200 to 600 nm. This illustration shows the very narrow line bandwidths of atomic emission spectra.

22.2   Atomic Absorption Spectroscopy

Importance of Atomic Absorption

Entering commercial production in the mid–1960's, **atomic absorption spectroscopy** (AAS, or simply AA) rapidly became the method of choice for most commonly determined metals. The basic principle of this technique is the absorption of monochromatic light by a cloud of atoms of analyte metal; AAS is not suitable for the direct determination of most nonmetals. The atomic absorption source produces intense electromagnetic radiation with a wavelength exactly the same as that absorbed by the atoms, resulting in extremely high selectivity.

Atomic Absorption Spectrophotometer Components

The basic components of an atomic absorption instrument are shown in Figure 22.3. The key to the success of AAS is the **hollow cathode lamp**,

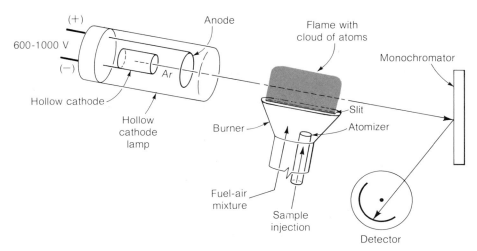

FIGURE 22.3   The major components of an atomic absorption spectrophotometer

which is the source of the light that is absorbed. This device consists of a glass tube with a quartz (SiO_2) end window to allow for the passage of ultraviolet radiation; it contains a noble gas (primarily argon) and two electrodes with a high potential applied between them. The cathode is the component that makes the hollow cathode lamp work for AAS. It is a hollow cylinder, or cup, the inside of which is coated with one or more analyte metals. In operation, a high potential of 600 to 1000 V is applied between the cathode and the anode. This results in a current of several milliamperes and produces positively charged noble gas ions (Ar^+) that impinge upon the negatively charged hollow cathode with a very high energy. Bombardment of the hollow cathode surface with energetic positive ions causes atoms of the analyte metal to be "sputtered" from the surface. Some of these atoms are electronically excited by the energy imparted by collision with the Ar^+ ions and emit electromagnetic radiation in the ultraviolet or visible region with lines characteristic of the metal atoms knocked from the cathode surface. This radiation is guided by the appropriate optics through a flame into which the sample is aspirated. In the flame, most metallic compounds are decomposed, and the metal is reduced to the elemental state, forming a cloud of atoms in the ground electronic state. These absorb a fraction of the radiation produced by the same kind of metal atoms in the hollow cathode lamp. The fraction of radiation absorbed increases with the concentration of the analyte in the sample according to the Beer's law relationship (Equation 20.10). From the flame, the attenuated light beam goes to a monochromator to eliminate extraneous light produced by the flame, finally passing to a photomultiplier tube detector (see Section 20.5). The detector signal is amplified and either recorded on a strip-chart recorder or processed by a computer data analysis system.

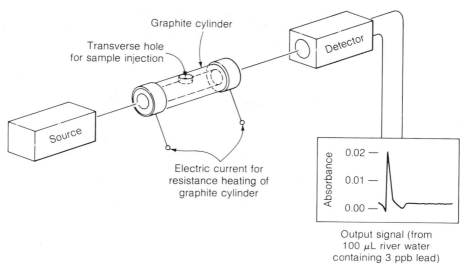

FIGURE 22.4 Graphite furnace for atomic absorption analysis and typical output signal.

Flameless Atomic Absorption

The flame in the system just described is the most common **atomizer**, or device that converts chemically bound atoms of analyte in ions or molecules in solution to gas-phase analyte atoms in the light beam. There are other types of atomizers that for some elements offer much lower detection limits. The most common of these is the graphite-rod furnace shown in Figure 22.4. It consists of a hollow graphite cylinder oriented so that the light beam passes lengthwise through it. A small sample (up to 100 μL) is inserted in the tube through a small hole drilled transversely in the side of the graphite tube. After the sample is injected, the tube is heated in several steps by passing electrical current through it. First the tube is heated mildly to around 125° C for about 30 seconds to drive off water. Next the current through the graphite tube is increased for about 60 seconds to yield a higher temperature of the order of 1000° C, which destroys organic matter in the sample; if this is not done, a cloud of smoke produced inside the furnace in the subsequent step diminishes the light beam and constitutes an interference. Finally the tube is heated for about 10 seconds to around 2700° C to atomize the analyte. This step creates a cloud of analyte atoms in the tube for a brief period resulting in a peak-shaped signal like the one shown in Figure 22.4.

As shown in Table 22.1 the major advantage of the graphite furnace atomizer over conventional flame techniques is much lower detection limits for some metals. In some cases the graphite furnace provides detection limits as much as 1000 times lower than those of conventional flame techniques.

TABLE 22.1   Comparison of detection limits among conventional atomic absorption, graphite furnace atomic absorption, and inductively coupled plasma optical emission spectroscopy

	Detection limit, mg/L*		
	Atomic absorption		Plasma emission
Element	Graphite furnace	Conventional flame	
As	0.003	0.1	0.006
Sb	0.001	0.1	0.004
Hg	0.002	0.5	0.03
Cd	0.000003	0.001	0.0001
Pb	0.0002	0.001	0.001
Be	0.00002	0.002	0.00002
Mn	0.00002	0.002	0.00003
Fe	0.0001	0.005	0.0004
Sn	0.025	0.02	0.025
B	—	0.7	0.04
Cu	0.00002	0.002	0.0004
Cr	0.0001	0.003	0.0002

* These **detection limits** are defined as the concentration of metal (mg/L) producing absorption equivalent to twice the noise level (background fluctuation) at zero absorption. Another term used to describe spectrophotometric performance is **sensitivity**, which is the minimum concentration of the analyte that will result in absorption of 1% of the incident light.

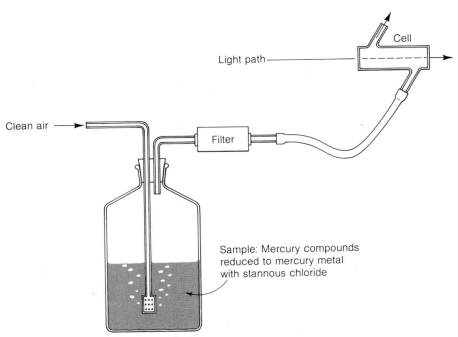

FIGURE 22.5   Flameless atomic absorption analyzer for mercury

Fortunately, the graphite furnace provides much lower detection limits for several environmentally significant heavy metals, particularly lead and cadmium. Another advantage of the graphite furnace in some cases is the very small quantity of sample required.

Although it is the most volatile elemental metal, mercury does not have a favorable flame AAS detection limit and is difficult to determine with the use of a graphite furnace atomizer. Around 1970, a special technique was developed for the flameless AAS determination of mercury. This approach involves the room-temperature reduction of mercury to the elemental state by stannous chloride in solution followed by sweeping the mercury into an absorption cell with air. Nanogram (10^{-9} g) quantities of mercury can be determined by measuring absorption of ultraviolet radiation at the 253.7 nm mercury line. A diagram of a flameless atomic absorption mercury analyzer is shown in Figure 22.5.

Interferences with Atomic Absorption

Although in favorable cases AAS is virtually an element-specific technique, it does suffer from some spectral and chemical interferences. These are discussed for both atomic absorption and emission in Section 22.8.

22.3   Emission Spectroscopy

Electrothermal Excitation

Emission spectroscopy is the name commonly given to the atomic emission in which analyte atoms are electrothermally excited by heating in an electrical discharge between two electrodes, usually made of high-purity graphite. Emission spectroscopy can reveal the presence of as little as 0.001% or less of most metals in a solid sample of only a few milligrams and has comparably low detection limits for metals in solution. It can also detect several nonmetals, such as arsenic, boron, phosphorus, and silicon. It is an excellent qualitative technique for many elements and can yield quantitative measurements to within about ±50% in routine practice. Much better quantitative results are obtained for some elements when special techniques are used.

Components of an Emission Spectrograph

Figure 22.6 shows the basic components of an emission spectrograph. The sample is held in a depression at the end of an electrode, above which is another electrode. Current in the form of a direct current or alternating current arc or of a spark flows between the electrodes, vaporizing the analyte and exciting the analyte atoms. These atoms emit lines of characteristic

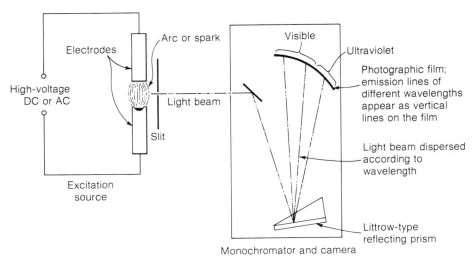

FIGURE 22.6 An emission spectrograph equipped for detection with photographic film

FIGURE 22.7 A strip of exposed and developed photographic film from an emission spectrograph showing atomic emission lines

wavelengths in the ultraviolet and visible regions of the spectrum. A beam of the emitted light passes through the appropriate optical system consisting of slits, lenses, and mirrors to a monochromator where the beam is dispersed according to wavelength in the visible and ultraviolet regions of the spectrum. The beam intensity is recorded as a function of wavelength on a strip of photographic film (Figure 22.7). On this film, atomic emission lines appear as dark lines. The intensity of a particular line is a function of the element from which it originated, which specific line of the element it is, the concentration of the element in the sample, and exposure time.

Continuous Spectra and Band Spectra

In addition to lines, two other kinds of emissions may be recorded on the film in an emission spectrograph. One of these consists of **continuous spectra** lacking any well-defined lines. These spectra result from incandescent (hot, glowing) solids. **Band spectra** consist of a series of lines that become progressively closer together until they reach a limit called the head of the band. Band spectra are usually emitted by excited molecules or molecular fragments.

Chemical Analysis by Emission Spectroscopy

Qualitative analysis is accomplished in emission spectroscopy from the wavelengths of emitted lines characteristic of various elements. Wavelengths are commonly determined by comparison to iron lines. Iron is ideal for this purpose because it has an abundance of lines and appears as a contaminant in virtually any sample.

Quantitative emission spectrographic determinations with film make use of a **densitometer**, which reads the intensity of a darkened line on the photographic film by its absorption of light from a beam passed over the line. **Internal standards** of elements of known concentration are added to the sample to provide a quantitative comparison. These are elements other than the analyte that have emission lines close to the emission lines of the analyte element. Comparison of the intensities of the known and unknown emission lines enables a quantitative comparison to be made.

Direct readers consisting of photomultiplier tubes set at locations on the monochromator corresponding to analyte lines may be used for detection in place of photographic films. These are especially useful for routine quality control determinations such as those required in the metals industries.

22.4   Flame Atomic Emission Spectroscopy

Excitation of Emission in a Flame

Flame atomic emission spectroscopy, henceforth referred to simply as **flame emission**, consists of monitoring the intensity of atomic emission lines from analyte atoms in a flame. Flame emission has been used for the determination of low levels of more than 40 metals, and is superior to atomic absorption for many of these. Flame emission is the method of choice for the biologically important metals calcium, potassium, sodium, and lithium.

In modern practice the source for flame emission consists of a **laminar flow burner** in which sample, fuel, and oxidant are thoroughly mixed before reaching a slotted burner head. The sample is introduced through a capillary tube and **nebulized** (converted to fine aerosol droplets) in a flow of air or other oxidant. The fuel (usually natural gas or acetylene) is likewise added near the point of sample nebulization. The fuel/oxidant/sample aerosol mixture passes around a set of baffles that cause all but the smallest sample droplets to settle from the gas stream in the mixing chamber and drain to waste. The fuel-oxidant mixture containing the analyte aerosol burns above a thin,

5- to 10-cm-long slot in the burner head and emission is observed from the flame at an angle perpendicular to the flame. The same burner and spectrophotometer are often used for both flame emission and atomic absorption. For the latter the light beam from the hollow cathode lamp (see Section 22.2) traverses the length of the long narrow flame (see Figure 22.3), which affords a longer absorption path length.

Effect of Type of Flame on Emission

The nature of the flame is a key aspect of the practice of flame emission analysis. The two most important variables are flame temperature and the relative proportions of fuel and oxidant; a rich flame has somewhat higher than the stoichiometric proportion of fuel, whereas a lean flame has a relatively higher concentration of oxidant. Normally an approximately stoichiometric flame works best. However, interference due to the formation of stable metal oxides in the flame can be largely avoided by a rich flame.

Low-temperature flames are readily achieved with a mixture of natural gas and air. These flames are best for those elements that readily form atoms. Such elements include analytically important alkali metals and cadmium, copper, lead, and zinc. Calcium, barium, and the other alkaline earths have a tendency to form refractory oxides that are best decomposed in a relatively hot acetylene/air flame. Several elements, including aluminum, beryllium, and the rare earths form particularly stable oxides and require flames fueled with acetylene and employing either oxygen or nitrous oxide as oxidants.

Spectrochemical Properties of Elements

Aside from forming compounds, such as oxides, in flames, various elements differ in their **spectrochemical properties**. One of these is the population in the flame of free atoms formed by the following overall process:

$$X(\text{chemically bound}) + \text{heat of flame} \rightarrow X(g) \tag{22.3}$$

Analyte atom X, as an ion in solution, or otherwise chemically bound	Free atom of X in the gas phase capable of excitation and emission

Another crucial spectrochemical property is the tendency of the atom to reach an excited state, X^*, as shown below:

$$X(g) + \text{energy from flame} \rightarrow X^*(g) \tag{22.4}$$

Ionization of the atom in the flame

$$X(g) + \text{energy from flame} \rightarrow X^+(g) + e^- \tag{22.5}$$

is to be avoided because it detracts from the number of free atoms in the flame. Although ions can be excited to emit light, their spectra are different from those of the free atoms.

Maximization of the population of free ground-state atoms, $X(g)$, and of excited atoms, X^* (g) (always a small fraction of the ground-state atoms), is desirable in atomic emission, whereas the formation of ions should be avoided. In addition to the temperature and fuel variables discussed above, these populations depend upon **flame zone** viewed. The flame zone is a function of the height of the part of the flame above the burner slit. Progressively higher zones are the primary reaction zone, interconal zone, and secondary reaction zone. The hottest part of the flame is that just above the primary reaction zone.

22.5   Inductively Coupled Plasma Atomic Emission Spectroscopy

The Plasma Excitation Source

Inductively coupled plasma atomic emission spectroscopy, referred to here as **plasma emission**, and commonly known by the abbreviation ICP, is an atomic emission technique in which the "flame" consists of an ionized gas, or **plasma**, heated by the action of a 4–50 megahertz, 2–5 kilowatt radiofrequency signal. The plasma gas, normally argon, reaches temperatures of 10,000 K, and in the sample region from which atomic emission is actually viewed is about 7,000 K, about twice that of the hottest chemical flames (acetylene/nitrous oxide at 3,200 K). The plasma source is shown in Figure 22.8.

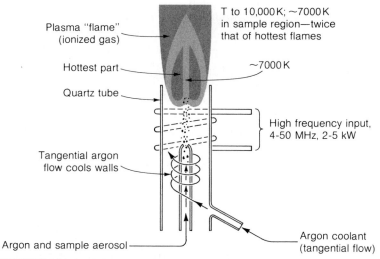

FIGURE 22.8 Schematic diagram showing inductively coupled plasma, used for atomic emission spectroscopy.

Advantages of Plasma Emission

Plasma has a number of advantages as an atomic emission source. Emission is an exponential function of absolute temperature; therefore, the very hot plasma results in lower detection limits for many elements compared to chemical flames. Furthermore, a number of elements that cannot be determined by other atomic spectroscopic techniques are readily determined by plasma emission. These include the environmentally important metalloids, arsenic, boron, and selenium. Unlike chemical flames, analyte reactions with the noble gas in the plasma are virtually nonexistent. Furthermore, the very high temperatures of the plasma prevent formation of chemically bonded analyte species. As a result, the kinds of chemical reactions and interactions common to many flame atomic emission sources are virtually absent from inductively coupled plasmas.

Perhaps the greatest advantage of plasma emission is that it is the best available **multielement analyzer** with capabilities for the measurement of up to 30 elements in one sample on a single analytical run. (As of this writing, an inductively coupled plasma atomizer interfaced with a mass spectrometer detector shows some promise of becoming an even more powerful multielement analyzer with detection limits superior even to those of furnace AAS, discussed in Section 22.2.) Multielement plasma-emission measurements may be made sequentially by having the spectrophotometer scan to observe a series of emission lines for the elements in question. Since only one line is being observed at a time, only one photomultiplier tube and associated amplifier circuitry is required. A relatively long time and, consequentially, a relatively large sample are required. A faster method of multielement analysis with plasma emission is simultaneous analysis, in which the spectrophotometer is set to display a number of lines from a corresponding number of elements, each of which is detected by a separate photomultiplier tube and associated electronics. Both sequential and simultaneous plasma emission analysis require computer processing of the vast amount of data generated.

Plasma Atomic Emission for Multielement Analysis

There are several reasons why the plasma is successful as a multielement technique, whereas atomic absorption and flame emission are usually used for only one element per run. As mentioned, the high temperature of the plasma causes the greatest possible number of elements to become atomized and emit analytically useful electromagnetic radiation. The plasma is a uniform emission source, so that unlike chemical flames, conditions need not (and usually cannot) be varied for individual elements. The plasma has little tendency toward self-absorption, in which ground-state atoms in the outer layers of the source absorb light emitted by excited atoms nearer the center of the source (see Section 22.7). Therefore, the plasma tends to give a linear response to element concentration over several orders of magnitude of analyte concentration, an obvious requirement where dilution of the sample because of elements present at a relatively high concentration would lower the concentrations of other elements below detection limits and result in loss of the multielement capability.

The last column of Table 22.1 gives detection limits of several elements by plasma emission. These tend to fall between those of conventional flame and graphite furnace atomic absorption.

22.6 Line Broadening

The Nature of Line Broadening

As discussed earlier, the single factor that makes atomic spectroscopy so element-specific and relatively free of interferences is the existence of very narrow atomic lines of specific wavelengths for various elements distributed around the ultraviolet and visible regions of the spectrum. In fact, if it were possible to observe emission from stationary atoms at low pressures, the natural line width would be only about 10^{-4} Å (angstroms). However, in all experimental situations atoms are in motion and are usually surrounded by other atoms and molecules, resulting in line widths of 0.01 to 0.05 Å; these are still very narrow. The fact that observed lines are always broader than the natural line width is given the name of **line broadening**.

Doppler Broadening

A major contributor to line broadening is **Doppler broadening** due to the rapid motion of gas-phase atoms, even at room temperature, and particularly at the high temperatures of a flame or plasma. When Doppler broadening occurs, light from an atom moving rapidly away from the observer has a very slightly lower frequency (greater wavelength, or "red shift") compared to a stationary atom. Similarly, light from an atom moving toward the observer at a great velocity has a somewhat higher frequency (shorter wavelength, or "blue shift") than light from a stationary atom. Doppler broadening is observed for absorbing as well as emitting atoms. An atom moving rapidly toward a light source will "see" light of a higher frequency than does an atom that is either stationary or moving perpendicular to the source. For an absorbing atom moving away from the source, the opposite is true.

Pressure Broadening

Although the ground-state energy of an isolated atom is a very specific value, it can be changed slightly by interactions with other atoms. This occurs as a consequence of collisions between rapidly moving atoms, which, of course, are very frequent in the gas state. The consequent slight perturbation in the ground state energy thus results in broadening; the phenomenon is called **pressure broadening**.

22.7   Self-Absorption

Causes of Self-Absorption

Resonance lines are those that are both emitted and absorbed by an atom. Both phenomena may occur in a flame. Emission is most abundant from the high population of excited atoms, X^* (still only about 0.01% of the total), in the center of the flame. In the cooler surrounding regions, ground state atoms will absorb some of the emitted resonance radiation. This phenomenon is called **self-absorption**. Self-absorption increases with analyte concentration as shown in Figure 22.9. The net result is a negative deviation from linearity at higher concentrations. In extreme cases of self-absorption, the emission intensity may actually decrease with increasing analyte concentration. Such an extreme case is called **self-reversal**.

Under conditions that lead to self-reversal, the emitting atoms are hotter than the absorbing atoms. As a result of Doppler broadening (see Section 22.6), the emission line is wider than the absorption line, with the result that light is absorbed primarily from the middle of the emission line. This decreases the height of the emission peak, and in extreme cases may even put an indentation in the top of it or cause it to appear as two lines.

Absence of Self-Absorption in Plasma Sources

The configuration of the inductively coupled plasma source is such that all of the emitting atoms are in a region of uniform high temperature inside the

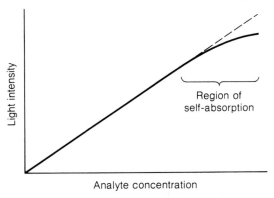

FIGURE 22.9   Atomic emission plot showing self-absorption

plasma "flame," and are surrounded by a sheath of non-absorbing argon atoms free of analyte atoms. The result is that the plasma is free of self-absorption effects. As noted in Section 22.5, the lack of self-absorption enables plasma emission measurements to be made over several orders of magnitude of analyte concentration, which is essential for plasma emission's application to multielement determinations.

22.8   Spectral Interferences

Causes of Spectral Interference

Spectral interferences in atomic spectroscopy occur from emission or absorption from species other than analyte atoms at the wavelength used for analyte measurement. The narrowness of atomic lines tends to decrease spectral interferences, a fundamental advantage of atomic spectroscopy in both the absorption and emission modes. Because of the extremely narrow lines from the hollow cathode light source, spectral interference due to overlap of atomic lines is not a problem in atomic absorption. In atomic emission, however, the spectrophotometer is depended upon to isolate the band of the spectrum in which the analytical line is viewed. Since even the best monochromators have a bandwidth exceeding atomic line widths, spectral interference from close-lying atomic lines can be a problem. The problem is more severe in plasma emission because of the multitude of lines excited by the exteme heat of the plasma. With both flame and plasma emission, however, spectral interference due to adjacent atomic lines is a relatively rare and manageable interference.

In flames, molecular species may exhibit broad band emission or absorption, thus constituting spectral interference. Combustion products in the flame may scatter light from the analyte emitted in the flame in atomic emission; the result is a negative interference. In atomic absorption scattering of light from the hollow cathode source appears as false absorption by the analyte, yielding a positive interference. When these interferences are due to the flame, compensation may be made by measuring the light intensity at the analytical wavelength in the absence of analyte.

Matrix Spectral Interferences

Matrix spectral interferences result from constituents in the sample matrix itself, and compensation is much more difficult than for flame interferences. For example, when calcium is present in a barium sample matrix (which is usually the case because calcium is the more abundant of the two alkaline earth metals), a CaOH species may form in a flame. This species has a band absorption in the 540–560 nm region, which overlaps the 553.6 nm barium analytical line. This interference in atomic absorption can be overcome by using nitrous oxide in place of air as the oxidant for acetylene fuel, producing a hotter flame that prevents CaOH formation.

Programmed Summary of Chapter 22

The major terms and concepts introduced in this chapter are contained in this summary in a programmed format. To derive the most benefit from the summary, you should fill in the blanks for each question, then check the answers at the end of the book to see if your choices are correct.

Atomic spectroscopy consists of the observation of the (1) _____ or (2) _____ of light by (3) _____ in the gas phase. Atomic absorption is the absorption by ground-state atoms in the (4) _____ phase of light having a (5) _____, exactly equal to the difference between E_0 and E^*. Atomic emission consists of the measurement of light emitted by (6) _____ gas-state atoms. If the excitation has occurred through absorption of a photon, the phenomenon is called (7) _____. The selectivity of atomic absorption and emission for various elements is due to the very narrow (8) _____, which occur for atoms that are in the (9) _____ phase, and the selectivity is also due to the (10) _____ wavelengths of the lines for each element. A major means of obtaining element specificity with atomic absorption is by the use of the (11) _____ lamp source, which emits light from excited atoms of (12) _____ as the analyte element. For classical atomic absorption the (13) _____ that converts analyte to gas-phase atoms consists of a (14) _____. An alternative to this device that gives much lower detection limits for some elements with a much smaller sample size is (15) _____ atomic absorption, in which the atomizer consists of a cylinder made of (16) _____ that is heated by (17) _____. Mercury compounds in solution can be determined by atomic absorption by first (18) _____, then getting the elemental mercury into an absorption cell by (19) _____. Electrothermal excitation of emission occurs by means of (20) _____. It is most useful for the (21) _____ of a number of elements at very low levels in a small sample. The spectrophotometric method of choice for the determination of lithium, sodium, potassium, and calcium is (22) _____. The two most important variables in the atomizer of this device are (23) _____ and (24) _____. Three important spectrochemical properties for analytes determined by this technique are (25) _____, (26) _____, and avoidance of the formation of (27) _____. The thermal source ("flame") in inductively coupled plasma atomic emission spectroscopy consists of (28) _____, heated by the action of a (29) _____. The temperature of the portion of this source from which atomic emission is normally observed is about (30) _____. Some advantages of the plasma as an atomic emission source are (31) _____. The broadening of atomic lines occurs by two mechanisms. One of these is called (32) _____, and is due to the (33) _____ of the atoms. The other is called (34) _____, and is due to interactions between atoms resulting from (35) _____ of the atoms. The phenomenon of self-absorption occurs in a flame when ground-state atoms in the (36) _____ regions of the flame absorb light emitted by atoms in the (37) _____ region of the flame. In a plot of light intensity from a line as a function of analyte element concentration, self-absorption is manifested as a (38) _____. The extreme case of self-absorption is (39) _____. In atomic spectroscopy, interferences from emission or absorption at the analytical wavelength by species other than analyte atoms is called (40) _____. Spectral interferences that result from interactions of chemical species in a flame are called (41) _____ spectral interferences.

Questions

1. Match the following:

   a. Flame atomic emission
   b. Atomic absorption
   c. Plasma atomic emission
   d. Atomic fluorescence
   1. Radiofrequency signal as the ultimate source of energy to produce excited atoms
   2. Light as a source of energy to excite atoms from which emission is observed
   3. Chemical energy to excite atoms
   4. Normally uses a hollow cathode lamp

2. Why are atomic absorption and atomic emission so virtually element-selective?
3. What may be said about the practical aspects of atomic spectroscopy in areas such as detection limits, applicability, and ease of use?
4. Why is the hollow cathode lamp so useful in atomic absorption? Describe it briefly.
5. What is an atomizer in atomic spectroscopy?
6. Suppose an analyst freeze-dried some fish tissue, ground and homogenized it, and weighed out a small portion to be placed directly and quantitatively in a graphite tube furnace. What major spectroscopic interference might be expected?

7. What is the major advantage of the graphite furnace atomizer over conventional flame techniques?
8. What is the function of an electrical arc or spark in emission spectroscopy?
9. Identify the three types of spectra illustrated by the sketch of a photographic film exposed in an emission spectrograph.

10. What are some important metals for which flame emission is the analytical method of choice?
11. What is the distinction between nebulization and atomization in atomic spectroscopy?
12. How may interference due to the formation of stable metal oxides in the flame be avoided in flame emission?
13. What are some of the advantages of inductively coupled plasma atomic emission spectroscopy?
14. What are the two major contributing factors to line broadening in atomic spectroscopy?

15. The light intensity from flame emission of an analyte read 371 (arbitrary units) at 250 ppm, 400 at 300 ppm, and 368 at 350 ppm. What pehnomenon in flame emission do these figures most likely illustrate?

Problems

1. A 200-μg/L standard lead solution gave an absorbance reading of 0.177 on an atomic absorption instrument with a flame atomizer and automatic correction for background. A 50-mL groundwater sample, acidified and diluted quantitatively to 100 mL, gave an absorbance reading of 0.023. What was the concentration of lead in the groundwater? Assume Beer's law is followed.

2. Standard addition techniques are often employed in atomic spectroscopy to partially compensate for interferences. Four different solutions were prepared for the standard addition determination by graphite furnace atomic absorption of cadmium in interstitial water "squeezed" from lake sediment. Each of these solutions contained 5.00 mL of the water sample, 0.5 mL of concentated HNO_3, and a small measured quantity of added cadmium standard; each was diluted quantitatively to 10.00 mL. The concentrations after dilution of the *added* cadmium in μg/L and the corresponding peak absorbance values (in parentheses) were the following: 0 (0.0125), 1.5 (0.0202), 3 (0.0280), 4.5 (0.0354). Does it appear that a Beer's law relationship is being followed? If so, what is the concentration of the cadmium in the unknown?

3. A flame emission instrument was calibrated with "element X" at the following values of μg/L of X and emission intensities (arbitrary units) in parentheses: 0 (0), 10 (4.00), 20 (7.99), 30 (11.98), 40 (15.00), 50 (17.98). Unaware of the calibration curve, an analyst used the 20-μg/L standard, got a reading of 8.00 for emission, and a reading of 18.00 on an unknown sample of X. The analyst assumed a completely linear response and calculated an unknown concentration of 45.0 μg/L. What phenomenon does the calibration curve show? What is the percentage error in the analyst's reported result?

4. Assume that all of the lead atoms in a 10-μL aliquot of 5.0 μg/L lead standard solution were vaporized in a graphite furnace flameless atomic absorption atomizer. How many lead atoms would be in the cloud of gaseous lead atoms?

5. In Section 22.2 it is noted that "nanogram quantities" of mercury can be determined by flameless mercury atomic absorption. How many atoms are in a nanogram of mercury?

6. Mercury may be determined by the absorption by atoms of the 253.7 nm mercury line. To what value of ΔE does a λ of 253.7 nm correspond?

7. As a footnote to Table 22.1 a simple definition is given for detection limit for atomic spectroscopy. The signal shown represents

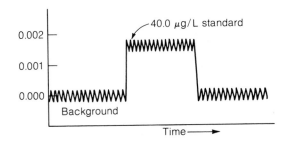

the background (no analyte) and the absorbance of a 40.0 $\mu g/L$ standard of a metal, M, determined by conventional flame atomic absorption analysis. The standard is rather close to the detection limit. From the plot shown, estimate the detection limit for the metal, M.

8. In Section 22.3 it is stated that emission spectroscopy with electrothermal excitation can reveal the presence of as little as 0.001% or less of most metals in a solid sample of only a few milligrams. To what mass of contaminant metal does 0.001% correspond in a solid sample of 5.0 mg?

Nuclear and Radiochemical Analysis

1. Types of radioactivity, including alpha particles, beta particles, positrons, gamma rays; the meaning of electron capture.
2. What is meant by the numbers and asterisk in a notation such as $^{236}_{92}U^*$
3. General configuration and meaning of a gamma-ray spectrum.
4. Activity as applied to radionuclides.
5. Decay rate equation and the meaning of half-life.
6. Decay schemes.
7. Use of ionization chambers for the detection of ionizing radiation.
8. Scintillation counting, liquid scintillation counting, and solid semi-conductor detectors.
9. The statistical aspects of radioactivity measurements.
10. What is meant by neutron activation analysis and the role of the n-gamma reaction in producing radionuclides for activation analysis.
11. Roles of reaction cross sections, neutron flux, and internal standards in neutron activation analysis.
12. Uses of radiolabeled compounds.
13. What is meant by isotope dilution analysis.
14. Principles and applications of radioimmunoassay.

23.1   Introduction

Radionuclides

The term **nuclear and radiochemical analysis** covers a number of analytical techniques that involve the use, production, and measurement of **radioactive nuclei**—that is, nuclei that evolve ionizing radiation in the form of gamma rays (high-energy electromagnetic radiation, see Figure 23.1) or charged particles. **Radionuclides**, as radioactive nuclei are commonly called, are sometimes themselves analytes, as is the case when radioactive contaminants are measured to determine the safety of drinking water. Since an individual event resulting from the decay of a radionuclide can be detected, radionuclides can be measured in very small quantities in favorable cases. Furthermore, radionuclides of particular elements can often be detected with high specificity. These two properties make the use of radionuclides in chemical analysis very advantageous in some cases.

Neutron Activation Analysis and Radiochemistry

For chemical analysis per se the most widely applied technique is **neutron activation analysis** (Section 23.7), in which stable nuclei are made radioactive by exposure to neutrons and the radioactivity measured. The nature of the radioactivity produced provides a qualitative elemental analysis and the intensity of the radioactive emissions yields a quantitative analysis. In favorable cases neutron activation analysis can be accomplished with no chemical processing of the neutron-activated sample, and is called **instrumental neutron activation analysis**. In other cases the sample made radioactive by exposure to neutrons is subjected to chemical separations and processing, known in general as **radiochemistry** or **radiochemical separations**.

Radioactive Tracers

The other major use of radionuclides in analytical chemistry is as **tracers**, taking advantage of the detectability of radionuclides to follow radioactive elements, or compounds labeled with radionuclides of an element, through analytical schemes and separations. **Isotope dilution analysis** involves the addition of radiolabeled compounds to samples in which the quantitative isolation of the analyte is not possible. Measurement of the activity before and after isolation enables quantitative measurement of the analyte. This chapter deals with the production, measurement, and uses of radionuclides in analytical chemistry.

23.2   **Types of Radioactivity**

Radioactive Decay

As mentioned in the preceding section, radionuclides can emit both electromagnetic radiation and charged particles. In so doing, the radionuclides reach a state of stability—sometimes after several emission steps—a process known as **radioactive decay**. This section addresses the types of particles and electromagnetic radiation given off by radionuclides as they decay.

Alpha Emission

Alpha emission is the ejection from a nucleus of an energetic particle with a charge of $+2$ and a mass number of 4 (each particle has a mass of approximately 4 atomic mass units, u). Such a particle is, of course, a ^4_2He nucleus. The emission of charged particles from a nucleus can be represented by a

nuclear reaction such as

$$^{226}_{88}\text{Ra} \rightarrow ^{4}_{2}\alpha + ^{222}_{86}\text{Rn} \tag{23.1}$$

showing the decay of radioactive radium-226 to radon-222. In an equation of this type, the superscript **mass numbers** (denoted A) and the subscript **nuclear charges** (Z) must balance on both sides of the equation. Because of their relatively huge mass, alpha particles do not travel far in matter but cause an enormous amount of ionization throughout their path. Therefore, they are particularly damaging in the body and their determination at very low levels is very important in workplace atmospheres and in drinking water.

Beta Emission

Beta particles consist of energetic electrons emitted from nuclei. For example, the decay of tritium, a radioactive isotope of hydrogen commonly used to label organic compounds, occurs by the following nuclear reaction:

$$^{3}_{1}\text{H} \rightarrow ^{3}_{2}\text{He} + ^{0}_{-1}\beta \tag{23.2}$$

Note that the mass of the beta particle is essentially zero compared to the much heavier parent and product nuclei, and that algebraic addition of nuclear charge gives the same total nuclear charge on either side of the equation. Another important example of a beta-emitting element, commonly called a **beta emitter**, is that of radioactive carbon-14:

$$^{14}_{6}\text{C} \rightarrow ^{14}_{7}\text{N} + ^{0}_{-1}\beta \tag{23.3}$$

The carbon-14 radionuclide is even more commonly used to label organic compounds than is tritium.

Because of the much lower mass and lower charge of beta particles compared to alpha particles, beta particles have a much greater ability to penetrate matter than do alpha particles. However, some beta emissions are very weak. For example, the beta particles emitted by tritium and carbon-14, which are especially important in labeling compounds, are particularly low in energy, and must be detected by special techniques, most commonly liquid scintillation counting (see Section 23.5).

Positrons

Positrons are positively charged particles of very low mass that may be viewed as positive electrons. They arise in the nucleus from conversion of a nuclear proton into a neutron and are given the symbol $^{0}_{1}e$. A typical radioactive decay reaction exhibiting positron emission is

$$^{122}_{53}\text{I} \rightarrow ^{122}_{52}\text{Te} + ^{0}_{1}e \tag{23.4}$$

A positron is an **antiparticle** that undergoes mutual annihilation upon contact with a negatively charged electron; energy is released from the annihilation in the form of photons of gamma radiation, described later in this section.

Electron Capture

Electron capture (ec), otherwise known as **K capture**, occurs when a nucleus captures an electron from the K or L shell of an atom. The net effect, like positron emission, is to reduce the charge of the nucleus by 1 and leave the mass number unchanged. This is shown by the nuclear reaction

$$_{-1}^{0}e + {}_{26}^{55}Fe \longrightarrow {}_{25}^{55}Mn \tag{23.5}$$

Electron capture is detected by measuring X-rays that are emitted when an electron falls into a vacant K or L orbital location left by the capture of the electron.

Gamma Radiation

Gamma radiation is a very short wavelength, high-energy electromagnetic radiation associated with nuclear processes. The emission of a gamma photon from a nucleus is a means of dissipating excess nuclear energy but does not cause a change of charge or a measurable change of mass in the nucleus. Nuclear reactions often leave a nucleus in a state with excess energy known as an energized, or excited, state, designated by an asterisk, *. An example of gamma emission from an excited nuclear state is

$$_{92}^{236}U^* \longrightarrow {}_{92}^{236}U + \gamma \tag{23.6}$$

Note that the nucleus in the excited state loses energy by emission of gamma radiation, but the product nucleus has the same mass number and nuclear charge as the reactant excited nucleus. The excited nucleus, itself, must be formed by a nuclear reaction. In the case of $_{92}^{236}U^*$ the precursor nuclear reaction consists of emission of an alpha particle from a radioactive plutonium isotope:

$$_{94}^{240}Pu \longrightarrow {}_{92}^{236}U^* + {}_{2}^{4}\alpha \tag{23.7}$$

Gamma-Ray Spectra

Gamma photons are monoenergetic. Therefore, gamma-ray spectra are made up of discrete lines, each characteristic of a particular nuclear transition. The ability to resolve such lines is limited by the detector employed (see Section 23.5), but with modern instrumentation very well defined spectra can be obtained, as shown in Figure 23.1.

The monoenergetic nature of gamma radiation makes its use very advantageous for chemical analysis. A gamma peak at a specific energy affords

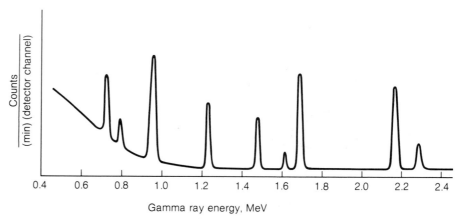

FIGURE 23.1   A gamma-ray spectrum

identification of the isotope of the element responsible for the peak. The excellent resolution (separation) of many important analytical peaks enables their quantitative measurement with minimum interference, and hence a quantitative determination of the element responsible for the peak.

Gamma-Ray Energies

Gamma-ray energies are commonly expressed in units of million electron-volts, MeV. An energy of 1 MeV is equal to 1.60×10^{-13} J. When a positron and electron meet and undergo annihilation, two gamma rays, each with an energy of 0.51 MeV, are produced, characteristic of positron emission. Gamma rays in the energy range of 0.5 to 3.0 MeV are commonly measured in chemical analysis.

23.3   Radioactive Decay

Decay Rate and Half-Life

The rate of decay of a radionuclide is first-order in the **number of nuclei** N. The mathematical expression for such a decay rate is the **rate equation**

$$-\frac{dN}{dt} = kN \tag{23.8}$$

where the **decay rate** is $-(dN/dt)$, and is negative because it represents a *loss* of the radionuclide, and k is the **rate constant**. Integration of the rate equation yields

$$\log \frac{N_0}{N} = \frac{kt}{2.303} \tag{23.9}$$

where N_0 is the number of nuclei initially, and N is the number of nuclei at time t. When decay has progressed to the point at which $N = \frac{1}{2}N_0$, time is symbolized as $t_{1/2}$, with the value

$$t_{1/2} = \frac{2.303 \log [N_0/(N_0/2)]}{k} = \frac{2.303 \log 2}{k} = \frac{0.693}{k} \qquad (23.10)$$

The term $t_{1/2}$ is the **half-life** of the radionuclide and is a constant for a specific mode of decay of the radionuclide. The integral form of the rate equation (Equation 23.9) with $t_{1/2}$ as a constant is obtained by substituting into it the expression for k in terms of $t_{1/2}$ (from Equation 23.10) to yield

$$\log \frac{N_0}{N} = \frac{0.301}{t_{1/2}} t \qquad \text{or} \qquad \log \frac{N}{N_0} = -\frac{0.301}{t_{1/2}} t \qquad (23.11)$$

Activity

The property of a quantity of radioactive material most readily measured in the laboratory is the **activity**, A, expressed in nuclear disintegrations, or counts registered by a measuring device, per unit time. Traditionally, activity has been expressed as counts per minute (cpm); the SI unit of activity is counts per second, called the **becquerel**. A counting device normally registers a constant fraction, less than 1, of the total disintegrations per unit time. This is because for most configurations not all ionizing radiation from a source can enter the detector that converts ionizing radiation to counts, and the conversion is not 100% efficient. This is not a problem, however, as long as the activity registered is directly proportional to the rate of decay of the radionuclide, $-(dN/dt)$. When these conditions are met, the rate equation in terms of activity is

$$\log \frac{A_0}{A} = \frac{0.301}{t_{1/2}} t \qquad (23.12)$$

where A_0 is the activity at time zero, and A is the activity at time t. In most applications ratios of decay rates provide sufficient information so that relative activities are all that need to be measured.

Illustration of Half-Life

The first-order nature of radioactive decay means that the activity at the end of each time span corresponding to one half-life is half of the activity at the beginning of that period of time. This is illustrated in Figure 23.2 for ^{76}As, half-life 27 h. Initially, the measured activity of the sample is 4.00×10^6 cpm. After one half-life has elapsed, the activity has decreased to 50% of the initial activity; after two half-lives have elapsed, the activity has decreased to 25% of its initial value, and so on.

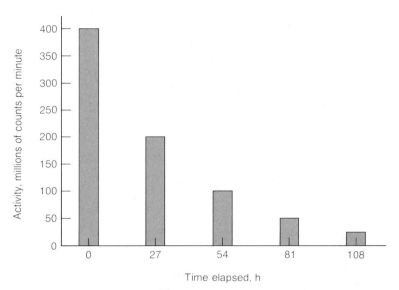

FIGURE 23.2   Activity of ^{76}As, half-life 27 h, as a function of time. Each division represents one half-life.

23.4   Decay Schemes of Radionuclides

It is useful to represent the mode by which radionuclides give off ionizing radiation by **decay schemes** that denote the radionuclide, its half-life, the radiation emitted, and the product. Two such decay schemes are shown in Figure 23.3. The first is that of ^{56}Mn, a radionuclide with a half-life of 2.58 h. It is seen that there are three initial decay paths, each involving the evolution of a beta particle, β^-. Two of these paths yield gamma emissions of 1.811 MeV and 2.110 MeV, respectively. The percentage figures below each mode of emission denote the percentage of radionuclides decaying by the path shown; these do not add up to exactly 100 because of other minor decay paths that are not shown. The final step in the decay of ^{56}Mn is a 0.847-MeV gamma emission; the stable product is ^{56}Fe. In order to determine the presence and quantity of ^{56}Mn, it would be possible to monitor any of the three gamma emissions of 0.847, 1.811, and 2.110 MeV. In modern practice all three gamma energies normally would be monitored simultaneously and corrections made by computer for background and interfering gamma emissions to give the best possible value for the quantity of the parent radionuclide.

The second decay scheme in Figure 23.3 shows the decay of a metastable (m) cobalt radionuclide, which first emits a very weak 0.025 MeV gamma. This scheme involves the emission of a positron, β^+. In this case the radionuclide could best be measured with the 0.80-MeV gamma.

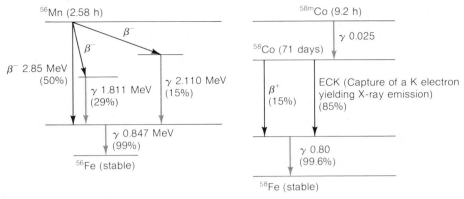

FIGURE 23.3 Decay schemes for two radionuclides

Measurement of Radioactivity

Detection of Radiation in Ionization Chambers

The most obvious means of measuring ionizing radiation is by the ions that such radiation produces. This is the principle behind the **Geiger counter** and the **proportional counter**. For both of these instruments the sensing device (transducer that produces an electrical signal) is an **ionization chamber** in which a strong potential gradient is applied between a cylinder and a wire located at its center (see Figure 23.4). Used as a Geiger counter, the metal cylinder contains argon gas at a low pressure of about 0.1 atm and a trace of methanol, ethanol, or Cl_2 gas. For Geiger counting, the tube is operated at an applied voltage of about 900 V, which gives a very high electrical field of approximately 20,000 V/cm near the anode. A beta particle entering the thin end window produces an ion pair in the Geiger tube. The electron from the ion pair is strongly accelerated by the high potential gradient around the wire anode producing a **cascade reaction** that yields about 10^9 electrons for each one

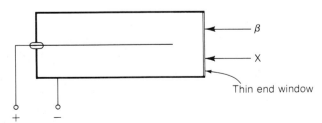

FIGURE 23.4 Ionization chamber for a Geiger counter or proportional counter

from the initial ion pair. This constitutes an electrical pulse, which is readily registered electronically by a **counter**. In the Geiger counting region the magnitude of the pulse produced is largely independent of the energy of the beta particle, and the counter is most useful for measuring emissions from pure β^--emitters.

A proportional counter tube contains a mixture of methane and argon gases at atmospheric pressure, and the voltage applied between the cathode and anode is in the range of 300 to 750 V. Under these conditions the magnitude of the electrical pulse depends upon the energy of the ionizing radiation, enabling the measurement of this energy. Proportional counting is most useful for beta particles and low-energy X-rays.

Scintillation Counting

Scintillation counters measure light produced in liquid or solid material by ionizing radiation. The scintillation phenomenon is adaptable to the measurement of alpha, beta, gamma, and X radiation. The luminescent intensity of the scintillating material is proportional to the energy of the ionizing radiation, enabling measurement of the energy. The most common scintillating solid for gamma-radiation detection is sodium iodide doped with a trace of thallium (NaI/Tl). A schematic of a scintillation counter using a crystal of NaI/Tl is shown in Figure 23.5.

For the measurement of very weak beta particles, especially those from the commonly analyzed ^{14}C and ^{3}H (tritium) isotopes, the sample is mixed directly with a scintillation liquid, such as 2,5-diphenyloazole in toluene. The very low energy betas from the radionuclides produce scintillation directly in

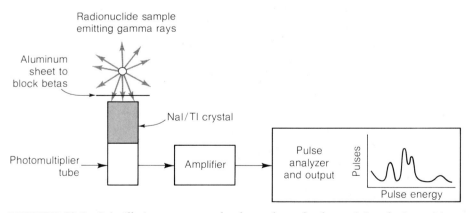

FIGURE 23.5   Scintillation counter and pulse analyzer for determining the intensities of gamma radiations at various energies

the liquid. This important measurement technique is called **liquid scintillation counting**.

Semiconductor Detectors

Scintillation detectors have very good counting efficiency but poor ability to separate (resolve) peaks of similar energy. More recently developed solid state semiconductor materials, particularly germanium or silicon drifted with lithium—Ge(Li), Si(Li)—have excellent resolution and reasonably good counting efficiencies. These detectors are now the ones of choice for taking gamma-ray spectra.

23.6   **Statistical Aspects of Radioactivity Measurement**

Random Nature of Nuclear Decay

Nuclear decay is a random event. Therefore, it is necessary to measure a minimum number of nuclear decay events to obtain a number of counts with the desired degree of precision. The statistical basis for handling a relatively large number of analytical results having only random error was discussed in Section 3.7. The mathematical treatment of the counting of radioactive emissions is similar.

Precision of Radioactivity Measurement

For the measurement of N counts from a radioactive source, the precision of the result is a function of the standard deviation σ, where

$$\sigma = \sqrt{N} \qquad (23.13)$$

According to statistical principles, a total of 68% of all values of a large series of measurements should lie within $\pm 1\sigma$ of the true value (discounting variables other than the inherent random nature of radioactivity). The relative standard deviation expressed as a fraction is

$$\text{Relative standard deviation} = \frac{\sqrt{N}}{N} \qquad (23.14)$$

Examination of this relationship shows that the higher the value of N, the lower the value of the relative standard deviation; a larger number of counts favors higher precision. Normally it is desired to measure sufficient counts such that there is 95% confidence that the value obtained lies within a specified percentage of the true value. The 95% confidence level corresponds to $\pm 2\sigma$, or the following in terms of relative standard deviation:

$$\frac{2\sqrt{N}}{N} \tag{23.15}$$

For example, to have a value of N within $\pm 3\%$ of the true value at the 95% confidence level, the number of counts to be taken is

$$0.03 = \frac{2\sqrt{N}}{N} \tag{23.16}$$

$$N = 4444 \text{ counts}$$

Totals of 10,000 and 40,000 counts would be required to have N within 2% and 1%, respectively, of the true value at the 95% confidence level.

23.7   **Neutron Activation Analysis**

Activation by Neutrons

Neutron activation analysis consists of producing radionuclides of an element by subjecting a sample containing the element to a flux of neutrons and measuring the radioactivity of the radioactive nuclei produced. For example, consider the neutron irradiation of aluminum, which has a 100% isotopic abundance of $^{27}_{13}\text{Al}$. This produces the radioactive isotope $^{28}_{13}\text{Al}$ by the nuclear reaction

$$^{27}_{13}\text{Al} + ^{1}_{0}n \rightarrow ^{28}_{13}\text{Al} + \gamma \tag{23.17}$$

This reaction involves the absorption of a neutron and the immediate ejection of a gamma photon (prompt gamma; it is not measured for analytical purposes, except in the specialized analytical technique of prompt gamma neutron activation analysis). This kind of nuclear reaction is so common that it is called an **n-gamma reaction** and is abbreviated by a notation of the type

$$^{27}_{13}\text{Al}(n, \gamma)^{28}_{13}\text{Al} \tag{23.18}$$

As discussed in Section 23.4, radionuclides decay by a multitude of paths. The radioactive $^{28}_{13}\text{Al}$ produced by the *n*-gamma reaction above decays according to the scheme shown in Figure 23.6. In terms of nuclear reactions, the activation and subsequent decay may be shown as

$$^{27}_{13}\text{Al}(n, \gamma)^{28}_{13}\text{Al} \xrightarrow[t_{1/2} = 2.3 \text{ min}]{\beta^- \ (2.85 \text{ MeV})} {}^{28}_{14}\text{Si}^* \xrightarrow{\gamma \ (1.78 \text{ MeV})} {}^{28}_{14}\text{Si (stable)} \tag{23.19}$$

Neutron activation is almost always performed with **thermal neutrons**, that is, those with a kinetic energy in thermal equilibrium with ambient surrounding matter. The number of radionuclides produced per second by neutron irradiation is a function of the number of atoms of the target nuclide in the

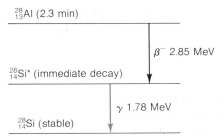

FIGURE 23.6   Decay scheme of $^{28}_{13}$Al produced by neutron activation. The symbol $^{28}_{14}$Si* represents an excited state that decays virtually instantaneously.

sample, the **reaction cross section** of the target nuclide (describing the probability that an incident neutron will react with a nucleus of the target nuclide), and the **neutron flux**, a measure of the number of bombarding neutrons per second per square centimeter. In the example just cited, $^{27}_{13}$Al is the target nuclide and $^{28}_{13}$Al is the radionuclide produced. As soon as radionuclides are produced, they begin to decay, and for a given neutron flux a point is eventually reached in which the decay rate equals the rate of production. When a radionuclide is very short lived, such as is the case with $^{28}_{13}$Al ($t_{1/2} = 2.3$ min), the point at which production and decay are in equilibrium occurs rather soon after the initiation of irradiation.

Reaction cross sections vary greatly among elements and among isotopes of specific elements, determining the amenability of various elements to neutron activation analysis. An element must have at least one isotope with a sufficiently large cross section and an adequate isotopic abundance to enable determination of the element by neutron activation analysis. The sensitivity of the analysis is also proportional to the available neutron flux. Adequate neutron fluxes are normally obtained from neutron fission reactors, the availability of which after World War II enabled development of neutron activation analysis as a practical analysis technique.

Use of Internal Standards

A number of variables, such as changing neutron flux, the presence in the sample matrix of an element with a very high neutron-capture cross section, and errors in timing irradiations and counting, make the determination of an element in a sample rather difficult unless an internal standard consisting of a known quantity of the analyte element is used. Typically the sample is irradiated in a container adjacent to another one containing a portion of the sample plus a known quantity of added analyte element. Thus the unspiked sample is irradiated under exactly the same conditions as the spiked sample so that variables such as those described above are the same for the sample and the sample plus standard. Computers are routinely used to reduce the rather complex data often obtained from neutron activation analysis.

Instrumental Neutron Activation Analysis

Before the development of high resolution solid state detectors for gamma rays (Section 23.5), neutron activation analysis frequently included chemical separation steps to isolate a particular radionuclide for interference-free counting. It is much more convenient to count the activation products of the analyte without a chemical separation, a routine process with modern detectors. This approach is called **instrumental neutron activation analysis**.

23.8   Radiolabeled Compounds

Uses of Radiolabeled Compounds

One of the more useful applications of radioactive materials in chemical analysis consists of employing compounds to which a radionuclide has been attached so that the compound can be followed through an analytical procedure. Such a compound is called a **radiolabeled** compound and the procedure is sometimes known as **tagging**. For example, oxalate ion can be prepared in which one of the carbon atoms (designated with an asterisk) is a radioactive $^{14}_{6}C$ isotope.

$$\overset{\overset{\displaystyle O}{\|}}{}\ \overset{\overset{\displaystyle O}{\|}}{}$$
$$^-O-C-C^*-O^-$$

When oxalate ion is used as an analytical reagent to precipitate metal cations, the course of the analysis can be followed at any stage by measuring the activity of the carbon-14. Typically, the completeness of precipitation with oxalate as a precipitant can be determined by measuring the activity in the precipitate and that remaining in solution. The solubility product of a compound can be determined by activity measurements if one of the ions is radiolabeled. In addition, radiolabeled species may be used to measure errors from adsorption, coprecipitation, and occlusion in gravimetric determinations and to devise means to avoid these errors.

Location of Radiolabeled Atom

The location of a radioactive atom on a compound can be crucial to the successful use of the radiolabeled compound. For example, acetic acid labeled with tritium (^3_1H) on the methyl group functions well as a radiolabeled compound, whereas tritium attached to the —OH group rapidly exchanges with solvent, losing the compound specificity of the label (Figure 23.7).

$$H^*\!\!-\!\!\underset{\underset{H}{|}}{\overset{\overset{H}{|}}{C}}\!\!-\!\!\overset{\overset{O}{\|}}{C}\!\!-\!\!OH \qquad\qquad H\!\!-\!\!\underset{\underset{H}{|}}{\overset{\overset{H}{|}}{C}}\!\!-\!\!\overset{\overset{O}{\|}}{C}\!\!-\!\!OH^*$$

Tritium attached to the —CH₃ group on acetic acid remains on the molecule enabling it to function as a radiolabeled compound.

Tritium substituted for the ionizable hydrogen is readily lost, so that the compound cannot function as a radiolabeled compound.

FIGURE 23.7 Importance of location of the radioactive atom in preparing radiolabeled compounds

Isotope Dilution Analysis

Many potential analytes are not amenable to quantitative isolation procedures, such as by precipitation. However, when it is possible to measure the yield of nonquantitative processes, these processes may be used for quantitative analysis. In some cases radiolabeled compounds can be used to measure yields of nonquantitative processes by **isotope dilution analysis**. With this technique a nonradiolabeled analyte compound is determined in a sample. A portion of the analyte compound is obtained or prepared having a known **specific activity** A_1 (activity per unit weight) and a quantity of the radiolabeled compound of weight W_1 is added to the sample. Subsequently, a small portion of the analyte compound is isolated in pure form and a determination is made of its specific activity A. It is necessary to isolate only a sufficient fraction of the analyte to weigh or otherwise determine the amount isolated accurately. The total amount of the nonlabeled analyte W is calculated by substitution into the following equation:

$$W = W_1\left(\frac{A_1}{A} - 1\right) \tag{23.20}$$

Isotope dilution analysis is especially useful for biochemical and other complex samples.

Radioimmunoassay

Radioimmunoassay is another very useful analytical technique that applies radiolabeled compounds. As with isotope dilution, a standard of the analyte compound is prepared in the radiolabeled form, usually employing carbon-14 as the radionuclide. This standard is reacted with an antibody specific for it that has previously been produced in the blood of an animal. The product of this reaction is a complex that is introduced into the sample. The unlabeled analyte compound in the sample displaces a portion of the radiolabeled compound from the complex. The degree of the displacement is a function of the amount of the unlabeled analyte compound in the sample. It follows that the quantity of the radiolabeled compound displaced from the complex is a measure of the unlabeled analyte compound. The quantity of radiolabeled compound displaced is determined by precipitating the complex and measuring the activity remaining in solution.

Programmed Summary of Chapter 23

The major terms and concepts introduced in this chapter are contained in this summary in a programmed format. To derive the most benefit from the summary, you should fill in the blanks for each question and then check the answers at the end of the book to see if your choices are correct.

Nuclear and radiochemical analysis covers a number of analytical techniques that involve the use, production, and measurement of (1) _____, also called (2) _____. Neutron activation analysis consists of making stable nuclei radioactive by exposure to (3) _____ and subsequent measurement of the activity produced. A major use of radionuclides is their application as (4) _____ to locate very small quantities of chemical species and follow them through physical and chemical processes. The loss of energy by emission of a particle or gamma ray from an unstable nucleus is called (5) _____. An alpha particle has a mass number of (6) _____ and a charge of (7) _____. A beta particle has a mass of (8) _____ and a charge of (9) _____. A positron can be regarded as a positively charged (10) _____ particle. The electromagnetic radiation that carries away excess energy from unstable radioactive nuclei is called (11) _____. The process of K capture is that by which a nucleus (12) _____ and emits an (13) _____ as a result. In the notation $^{236}_{92}U^*$ the superscript 236 is the (14) _____ of the nuclide, the subscript 92 is the (15) _____, and the asterisk shows that the nucleus is (16) _____. For radionuclides the number of disintegrations, or counts, registered by a measuring device per unit time is called the (17) _____. For a single radionuclide, the decay rate $-(dN/dt)$ is proportional to (18) _____. The integral form of this relationship in terms of activity A and time t is (19) _____. In the preceding equation $t_{1/2}$ is defined as the (20) _____, which is the time required for (21) _____. A decay scheme is a (22) _____ by which a radionuclide emits (23) _____ to eventually reach a state of (24) _____. For both the Geiger counter and proportional counter the sensing device is an (25) _____, in which a beta particle produces (26) _____ in the gas resulting in an (27) _____ that can be detected and registered by the counter circuitry. Scintillation counting depends upon the production of measurable (28) _____ by gamma rays in a suitable solid or liquid detector. Detectors that have a particularly good ability to resolve close-lying gamma-ray peaks are the (29) _____. Because of the random nature of radioactive decay, a larger number of counts favors (30) _____. Where N is the total number of counts from a radioactive source, the standard deviation σ is equal to (31) _____. Neutron activation analysis consists of producing (32) _____ of an element by exposing a sample containing the elements to (33) _____ and measuring the (34) _____ of the products. The abbreviation (n, γ) commonly associated with neutron activation analysis denotes a nuclear reaction in which (35) _____. The number of radionuclides produced by neutron activation increases with increased (36) _____ of neutrons and increased (37) _____ of the absorbing nucleus. A compound to which a radioactive atom is deliberately bound in order to detect small quantities of the compound is called a (38) _____ compound, and the process is sometimes called (39) _____. Use of a radioisotope to compensate for nonquantitative isolation of an isotope is called (40) _____. A technique with radiolabeled compounds that makes use of antibodies produced in living systems is called (41) _____.

Questions

1. A radionuclide with a particular mass number and nuclear charge emits a $_{-1}^{0}\beta$ particle. What can be said about the nature of the atom remaining?

2. What kind of radiation is emitted along with electron capture? Why is this type of radiation observed?

3. What tremendous advantage for chemical analysis is provided by the fact that gamma photons from a particular nuclear transition are monoenergetic?

4. What is the nature of radioactive decay in terms of its rate as a function of time?

5. What do decay schemes show about the nature of radioactive decay?

6. How do Geiger counter ionization chambers, solid scintillation counters, and liquid scintillation counters detect ionizing radiation?

7. Why is a relatively low activity sample usually "counted" for a long time in order to measure the activity?

8. What does the term (n, γ) stand for in nuclear reactions involved with neutron activation analysis?

9. What are two properties of the parent isotope of an element necessary to enable that element's determination by neutron activation analysis?

10. What is a radiolabeled compound?

11. What is the major advantage of isotope dilution analysis?

12. What does electron capture do to a nucleus? What relationship does the elemental product of electron capture bear to the original radionuclide? Since electron capture is the *capture* of an electron, how is the process manifested by *emission* of radiation?

13. What is the special significance of 0.51 MeV gammas?

14. In radioactive decay, what is the distinction between activity and specific activity?

15. The second decay scheme in Figure 23.3 shows an example of the equivalence of positron emission and electron capture. Explain.
16. What is the major advantage of solid state semiconductor detectors of the Ge(Li) and Si(Li) type?
17. In terms of what you know about the nature of nuclear decay, comment on the statement, "given initially exactly 10 nuclei of a radionuclide with a half-life of 38 s, there will be exactly 5 of the nuclei left after a time period of 38 s has elapsed."
18. What is the distinguishing feature of instrumental neutron activation analysis?

Problems

1. The initial activity of a radioactive sample was 32,000 cpm, and it was 1100 cpm after exactly 12 h. What is the half-life of the radionuclide being measured (assuming the presence of only one radionuclide) expressed to the nearest minute?
2. Cation M^{2+} and anion X^{2-} react to form the slightly soluble salt $MX(s)$. To a 100-mL flask was added 10 mmol of soluble Na_2X and a stoichiometrically insignificant ($\ll 10$ mmol) amount of highly radioactive Na_2X^*, in which the X^{2-} ion was radiolabeled. The solution was diluted to the mark and mixed. The activity per unit volume of this solution was measured and found to be 175,000 cpm/mL. A 50-mL portion of this solution was mixed with 100 mL of 0.100 M $M(NO_3)_2$ and the solid MX removed after equilibration. The specific activity of the remaining solution was 320 cpm/mL. From this information estimate the solubility product of $MX(s)$.
3. Radionuclide X was produced by exposure to neutrons in a nuclear reactor. Exactly 2 h after removal from the reactor, the activity of X was 259,000 cpm and 50 min later it was 194,000 cpm. Calculate the half-life of the radionuclide, the rate constant k, and the activity when it was first removed from the reactor.
4. From the data plotted below, calculate the half-life of the radionuclide being counted.

It may be helpful to plot the data in a different way.

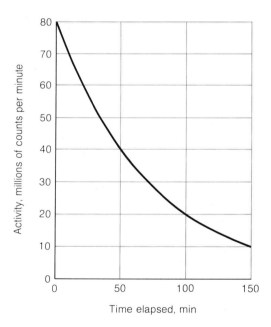

Time elapsed, min

5. Activity is normally considered to be "gone" when it has reached 0.1% of its initial level. After how many minutes could the activity from the radionuclide in the preceding example be considered to be gone? What would be another way of stating this rule in terms of half-life?
6. For this question, refer to Figure 23.3, the

decay scheme of ^{56}Mn. A sample of this radionuclide was found to emit 1.811 MeV gamma radiation at a rate of 5.3×10^5 s^{-1}. What was the rate of emission of 2.110 MeV gamma and 2.85 MeV β^-?

7. A sample of a radioisotope with a half-life of the order of years was counted for 1 min, yielding 532 counts during this period. Estimate the approximate length of time that the analyst should plan to count this sample to have a value within ±2% at the 95% confidence level.

8. A 564-mg sample of the solidified crust of sludge from a waste chemical dump was analyzed for trace element Z by neutron activation in a nuclear reactor. Adjacent to this sample was a standard solution containing 2.00 mg of Z. The gamma-emission characteristic of the trace element was counted for the unknown, giving 5324 cpm, and for the standard, giving 6235 cpm. The counts were taken consecutively over a time period that was short compared to the half-life. What was the content of Z in the sample in units of milligrams per kilogram?

24 The Analysis of Real Samples

L E A R N I N G   G O A L S

1. Overall scheme for chemical analysis.
2. The analytical approach.
3. Sampling procedures.
4. Reducing a sample to analyzable size.
5. Sample preservation.
6. Water in solid samples.
7. Dissolving a sample, including sample digestion and wet ashing.
8. Dry ashing of samples.
9. Combustion tube techniques.

24.1 The Analytical Approach

Planning Chemical Analysis

The practicing analytical chemist almost always works with another person or group to solve applied problems in chemistry or many other areas. Professionally the analytical chemist may be part of a large organization, such as a petrochemical company, or may work in an independent analytical laboratory. In such a capacity, the analytical chemist's role is much greater than simply receiving a sample ready for analysis and determining how much of an analyte is in it by methods such as those described in this text. Although many analyses, such as standard methods for the determination of water pollutants prescribed by the U.S. Environmental Protection Agency, have routine procedures, others require a great deal of thought, literature search, and even experimentation before they can be carried out.

In planning chemical analysis, decisions must be made regarding the degree of accuracy required (Section 3.2), permissible costs, length of time before results are required, and similar matters. An analysis accurate to within 2 or 3 parts per thousand with results required within a day after the request was made generally will cost far more than one accurate to within 2 or 3 percent with results not needed for several days. It is generally a waste of time and money to perform a chemical analysis to a degree of accuracy much greater than that required. It can be an equal waste of time to perform a "cheap" chemical analysis with accuracy insufficient to be useful.

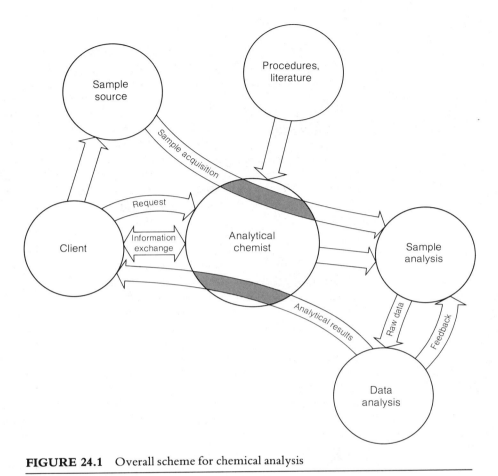

FIGURE 24.1   Overall scheme for chemical analysis

A General Scheme for Nonroutine Analysis

Figure 24.1 illustrates a general scheme for the analysis of a nonroutine sample. Normally the first step is a request or inquiry from an individual or group (client) wanting to have an analysis performed. This is followed by an exchange of information between the client and the analytical chemist. The latter may need additional information regarding the nature of the sample, the degree of accuracy required, anticipated levels of analyte in the sample, and other types of information that may not be given in the initial request. The client will want to know such things as costs, previous experience with the kind of sample involved, and how the sample is to be taken. The analytical chemist may need to consult books, manuals, or the scientific literature at this stage to learn if the requested analysis is feasible. Sometimes the end result of the discussions between the client and the analytical chemist is that the analysis cannot be done or cannot be done at an acceptable price.

When a decision is made to proceed with an analysis, consideration must be given to sampling. If the sample is already taken, it must be determined that it is representative of the material from which it was taken. If the sample is inadequate or has not been acquired, the client will need careful instructions for sampling, or the analytical chemist may do the sampling using proper procedures.

After all the information for the sample analysis has been obtained, the analysis itself is performed. The whole process may require elaborate sample preparation, such as digestion of a biological sample in hot oxidizing acids. Often interferences must be removed or measures taken to negate their detrimental effects. The actual determination of the analyte may be rather simple, such as the development of a colored species in solution and the measurement of the absorbance of that species. Sometimes a rather complex instrument is involved. Data obtained from the determination are subjected to data analysis, frequently by computer. In some sophisticated systems there is feedback from the data processing component of the system to the analytical instrument to change the measurement parameters so that a better determination can be obtained.

The end result of a chemical analysis is an answer with information about its precision and reliability. For a nonroutine sample, the analytical chemist carefully examines the results before passing them on to the client. If the results are judged inadequate, the determination may be repeated, sometimes going back as far as the sampling stage. In any case, when the results are reported, any cautions or constraints pertaining to them should be specified to the client.

Examples of the Analytical Approach

It would be misleading to represent the practice of analytical chemistry as an invariably interesting and exciting endeavor. However, many very interesting problems, large and small, are solved every day by the successful practice of analytical chemistry. Examples of these are described in the journal *Analytical Chemistry* (American Chemical Society, Washington, D.C.) in a series entitled "The Analytical Approach." Many of these articles have been collected in a single volume.* Some examples of the analytical approach are summarized in Table 24.1.

---

* Grasselli, Jeanette G. (ed.), *The Analytical Approach*. American Chemical Society, Washington, D.C., 1983.

TABLE 24.1   Examples of problems solved by the analytical approach

Problem	Solution	Analytical technique used
Piney-spruce off flavor in cold cereal	Resin used to bind layers of inner package lining were found to contain odorous terpenes, borneol, and fenchyl alcohol. New sources of nonodorous resin were found.	Gas chromatography was used to separate resin constituents, which were "sniffed" individually for odor. Odorous compounds were identified by mass spectrometry.
Presence of material causing turbidity in water solutions of manufactured thiamine (vitamin B_1)	The impurity was eliminated by better pH control of the synthesis reaction, which was monitored carefully by in-process analysis.	Foreign material was analyzed by thin-layer chromatography and ultraviolet spectrometry showing it not to be precursor material, thiothiamine. IR, NMR, and mass spectra proved material to be oxidation product, thiamine thiazalone.
Corrosion of steam turbine blades in a power plant	A condenser leaking impure water into the steam condensate was repaired and water purification procedures were improved.	An ion chromatograph was used to monitor sodium, chloride, and sulfate in boiler steam condensate, showing excessive levels of corrosive salts.
Corrosion and solids accumulation in the naphtha-handling system just downstream from the crude oil distillation tower in a petroleum refinery	A corrosion inhibitor with proven performance was substituted for the one causing deposits of solids.	Solids deposits were analyzed by solubility, elemental analysis, infrared spectrophotometry, and X-ray diffraction analysis, showing them to be a product of the corrosion inhibitor used.
Excessive levels of phosphorus, which damages auto catalytic converters, were present in gasoline from some turnpike service station pumps	Pump hoses from which phosphorus and other impurities were leached were replaced by brands resistant to leaching.	Phosphorous in gasoline was determined by the ASTM D3231 spectrophotometric procedure; leachable hose materials were measured by X-ray emission.
A method was needed to detect phenylcyclidine (PCP, "angel dust") in drug abusers weeks after use	Although drugs are rapidly cleared from body fluids, such as blood serum or urine, they can be detected in hair for many weeks.	Hair samples were analyzed for PCP using radioimmunoassay.
A need developed to trace the source of stolen uranium yellow cake, $(NH_4)_2[(UO_2)_2SO_4(OH)_4(H_2O)_n]$	The content of 22 trace elements was used as a means of finding the source of the yellow cake.	Uranium was extracted with tri-n-butyl phosphate and the trace element content in the remaining aqueous phase was determined by inductively coupled plasma atomic emission.

24.2   Sampling

Sampling and the Validity of Analytical Results

The validity of any analytical result depends absolutely upon having a sufficient quantity of a sample representative of the material being analyzed and stored in such a manner that the analytes are not affected adversely. If an analysis goes wrong at any point subsequent to sampling, it is possible to repeat some steps or start over again with a fresh portion of sample. However, if the sample is not valid, the whole analysis is a waste of time.

Sampling Procedures

Sampling procedures range from the very simple to the very complex. An ideal means of sampling occurs when a measurement can be taken directly on a material analyzed. An example of this would be when a glass electrode system is placed directly into a body of water and the pH registered on a pH meter. Even in this case depth and location must be considered. Another simple sampling procedure is provided by a single chip of paint found at the scene of a crime to be analyzed for elemental content by instrumental neutron activation analysis with no chemical processing involved. However, sampling becomes very complex in a case where a 100-car trainload of coal is to be analyzed for moisture, heat content, volatile matter, fixed carbon, ash, and sulfur. The actual analyses are simple and straightforward, but reducing thousands of tons of coal to a laboratory sample representative of the whole lot is a laborious and painstaking procedure.

In performing a chemical analysis it is always necessary to know and use the sampling procedures developed for the kind of material being analyzed. These must be obtained from procedures manuals or the literature. One of the most common types of analysis performed on an industrial scale is that of a large quantity of solid material that may range from granular size to large chunks. Suppose, for example, that the material to be analyzed is a crushed mineral ore as it enters a smelter via a conveyer belt. To obtain a sample representative of a day's run, a scoopful containing 500 g of the solid might be removed from the belt every 30 minutes over a 24-hour period. At the end of this period the analyst would have about 24 kg of solid that must be reduced to only a few grams. This final sample must (1) *be thoroughly mixed and uniform* and (2) *must be exactly representative of the larger quantity of material from which it was taken.* There are numerous ways to ensure that these criteria are met. For the mineral ore just described, **coning and quartering** can be used. First the substance to be analyzed is mixed thoroughly by shoveling it into a succession of cones, removing material from the edges of one cone and shoveling it into the center of the next (Figure 24.2). Next the quantity of the substance is reduced by flattening successive cones and removing alternate

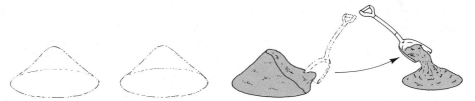

FIGURE 24.2   Coning, the successive shoveling of material from the edge of a cone to the center of the next, is used for uniform mixing as part of the sampling process.

FIGURE 24.3   Quartering involves flattening a cone, dividing it into quarters, and removing alternate quarters to produce a cone half the size of the preceding one. The process is repeated until a small enough portion of the sample is obtained.

quarters (Figure 24.3) to produce cones of smaller size until a manageably small portion of the sample is obtained.

Rolling and quartering is another means of mixing a sample and reducing its size. For a large quantity of material a tarpaulin can be used to place the sample on; for small quantities a piece of glazed paper works well. The material is first placed in a cone in the center of the sheet on which it is to be mixed. The cone is flattened and the material is mixed (rolled) by pulling successive corners of the sheet over the opposite corner. The rolling process is repeated about 100 times to ensure complete mixing, then the material is collected in the center of the sheet by raising all four corners of the sheet simultaneously. The pile of material is then flattened, divided into quarters, and opposite quarters are discarded. The material is then collected in the center of the sheet and the whole rolling and quartering process is repeated as many times as needed to reduce the sample to the desired size.

Examples of Sampling Methods

Given the vast number of materials subjected to chemical analysis, a large number of different sampling methods have been developed. Some of these are summarized in Table 24.2.

TABLE 24.2   Sampling methods for various kinds of substances

Substance	Sampling method
Fluid fertilizers with suspended material	Run the mixer in the storage tank for 15 min to produce a homogeneous suspension. Then lower a 500-mL Missouri sampling bottle [Figure 24.4(a)] to the bottom of the tank and raise it slowly as it fills.
Animal feed in a bag	Insert slotted tube and rod assembly [Figure 24.4(b)] diagonally into the bag lying horizontally and remove a core the length of the tube.
Fresh meat	Separate from bone and pass three times through a food chopper with 3-mm plate openings, mixing after each chopping step. These steps and subsequent analysis should be completed as rapidly as possible to avoid loss of moisture and decomposition.
Bulk lump lime in railroad cars	Collect at least 10 shovelfuls from different parts of each car. Crush to pass 5-cm opening. Mix and reduce to about 1 kg by quartering. Store in dry, sealed container. All of these steps must be performed as rapidly as possible to avoid uptake of water.
Purgeable halogenated hydrocarbons in water	Collect the water in a glass bottle of at least 25-mL volume just to overflowing in a manner that does not allow air bubbles to pass through the sample. If the water contains free or combined chlorine, 10 mg per 40 mL of fresh solid sodium thiosulfate preservative should be added to the bottle. Seal the bottle without entrapment of air bubbles and keep sealed and refrigerated until the analysis is performed.

Sample Preservation

Sample preservation is an important aspect of the sampling process. Samples are preserved to prevent such things as bacterial decomposition of biological samples, loss of water from hygroscopic materials, precipitation of metals from water samples, and loss of volatile analytes from water samples. Typically, water in which metal ions are to be determined is acidified immediately upon collection to prevent precipitation of analyte ions. Even if the solubility products of the metal ions are not likely to be exceeded, they can coprecipitate with iron(III) produced by air oxidation of iron(II) in water as follows:

$$4Fe(aq)^{2+} + O_2 + 10H_2O \rightarrow 4Fe(OH)_3(s) + 8H^+ \tag{24.1}$$

Bacterial action is an important cause of analyte loss in water samples. For example, through the action of bacteria, nitrate ion is reduced by organic matter (represented in general as $\{CH_2O\}$)

$$4NO_3^- + 5\{CH_2O\} + 4H^+ \rightarrow 2N_2(g) + 5CO_2(g) + 7H_2O \tag{24.2}$$

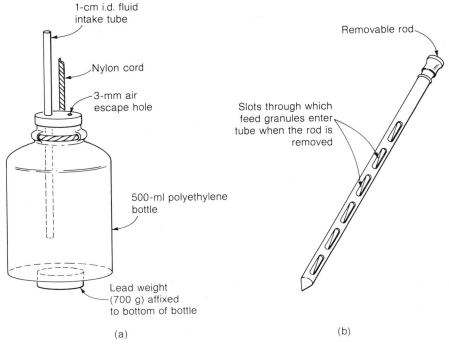

1-cm i.d. fluid
intake tube

Nylon cord

3-mm air
escape hole

500-ml polyethylene
bottle

Lead weight
(700 g) affixed
to bottom of bottle

(a)

Removable rod

Slots through which
feed granules enter
tube when the rod is
removed

(b)

FIGURE 24.4   Some sampling devices: (a) Missouri sampling bottle for fluid fertilizers with suspended material; (b) slotted tube and rod assembly for animal feed in a bag

To slow this kind of reaction, water samples are normally refrigerated from collection until the time of analysis. The addition of strong acid, mercury salts, or other bactericidal agents can prevent the loss of biodegradable analytes in water.

24.3   Water in Solid Samples

Water is one of the most common constituents of solid samples, and it must be considered when analyzing such a sample. In a solid sample, water may be held as readily evaporated surface moisture, by relatively strong physical attraction in pores and capillaries, and chemically as an integral part of the analyte or sample matrix. The various sources of water in solid samples are summarized in Table 24.3.

A common approach to taking account of water in solid samples is to dry the sample in a manner that is presumed to get rid of adsorbed water. Drying

TABLE 24.3   Sources of water in solid samples

Name of water source	Nature of water in sample
Water of constitution	Not in the sample per se, but produced upon strong heating—such as $Ca(OH)_2 \rightarrow CaO + H_2O$
Water of crystallization	Part of the crystal matrix of the compound—such as $CuSO_4 \cdot 5H_2O$
Adsorbed water	Condensed onto the surface of particles
Capillary or pore water	Sometimes called sorbed water, this water is contained in pores, capillaries, and interstices of primarily colloidal material, such as starch, clay, or silica gel. These materials may contain up to 20% water while appearing to be dry.
Occluded water	The water is contained in microscopic pockets in crystalline solids. The water is the solvent for the mother liquor of the crystal from which the solid was precipitated.
Solid solution	Some solids, such as natural glasses, may contain up to several percent water in solution in the solid.

at 105° C for 1 hour is commonly done. Higher temperatures can be employed if it has been shown that they eliminate more strongly held water in a reproducible manner. In some cases the sample may be heated strongly enough to drive off even the most strongly held water, and the water evolved is measured. In other cases it is best to let the sample equilibrate with water in an atmosphere with regulated humidity. This gives the sample a reproducible water content and it can be analyzed on that basis.

24.4   Sample Dissolution

Getting a Sample into Solution

Although a number of chemical analyses are performed directly on solids and gases, normally a solid must be gotten into aqueous solution prior to analysis. This section deals with the sometimes difficult task of dissolving a solid sample.

In rare cases a sample is dissolved by simply mixing it with water; this can be done with industrial soda ash, for example. Strong mineral acids are commonly used for dissolving a number of different kinds of samples. Hot, concentrated hydrochloric acid works well to dissolve many metal oxides. When heated, hydrochloric acid loses HCl gas to produce a constant-boiling, approximately 6 M HCl solution that boils at 110° C that dissolves some metal oxide samples. Most metal samples are dissolved by concentrated nitric acid. The oxidizing ability of this acid enables it to dissolve some metals that do not displace H_2 from H^+ solutions. Whereas copper metal will not dissolve in hydrochloric acid, for example, it dissolves in concentrated nitric acid because of the reaction

$$Cu + 4HNO_3 \rightarrow Cu^{2+} + 2NO_3^{2-} + 2NO_2(g) + 2H_2O \qquad (24.3)$$

Caution: When HNO_3 is used as an oxidizing agent, toxic NO_2 gas is evolved. Inhalation of this gas is to be strictly avoided, and operations in which it may be produced must be performed in a hood.

Sulfuric acid has a very high boiling point of $340°C$, a temperature that results in the dissolution of many kinds of samples, such as metals, alloys, and biological materials. In addition to its acidic strength, sulfuric acid is a dehydrating agent and oxidant, properties that may enhance its ability to dissolve some kinds of samples. Concentrated perchloric acid, $HClO_4$, is a very strongly oxidizing acid that can be used to dissolve samples as intractable as stainless steel. Though a rather mild reagent at room temperature, this acid forms a 72.4% $HClO_4$ solution boiling at a constant $203°C$ when heated. *Hot perchloric acid is a potentially explosive reagent in the presence of oxidizable materials. It must be used with extreme caution in special hoods that provide for the collection of perchloric acid vapors that might otherwise accumulate to explosive levels in the hood system.*

Oxidizing mixtures of acids are often more effective in dissolving oxidizable samples than are the pure acids. Historically a mixture of one volume of concentrated nitric acid and three volumes of hydrochloric acid, called *aqua regia*, has been used to dissolve materials as chemically resistant as platinum metal. The addition of nitric acid to perchloric acid yields a reagent with better properties for oxidizing and dissolving organic samples than perchloric acid alone. *Both of these mixtures are very hazardous.*

Hydrofluoric acid, HF, is used to dissolve silicate minerals. The silicon is lost as volatile silicon tetrafluoride, SiF_4. The remaining HF, which interferes with many chemical analyses, is driven off by heating after the addition of high-boiling perchloric or sulfuric acid. *Hydrofluoric acid and HF vapor are both very toxic and react with glass; they must be handled in special apparatus.*

Reaction of a sample with an oxidizing acid is not a simple dissolution process because oxidation and a complete change in the sample matrix is involved. For example, if the sample matrix is an organic material, the organic matter is completely destroyed. In such cases the process of placing the sample in solution is more properly called **sample digestion** or **wet ashing**.

Sample Fusion with Fluxes

Some samples, notably those composed of mineral oxides and silicates, are very resistant to aqueous reagents and are best dissolved in molten salt mixtures called **fluxes**. Fluxes that are used for sample dissolution are normally liquids at 300 to 1000° C; such high temperatures contribute to their effectiveness in dissolving samples. Fluxes fall into the three general categories *of acidic, basic,* or *oxidizing*. Potassium pyrosulfate, $K_2S_2O_7$, is a typical acid flux; when heated, it decomposes to highly acidic sulfur trioxide

$$K_2S_2O_7 \xrightarrow[300°C]{} K_2SO_4 + SO_3 \tag{24.4}$$

Sodium carbonate, Na_2CO_3, is a typical basic flux. Sodium peroxide, Na_2O_2, is a strongly basic oxidizing flux.

A powdered sample is normally mixed with a flux in a tenfold excess in a crucible. The crucible and its contents are heated until the flux melts and dissolves the sample. Just before the flux melt solidifies, the crucible is rotated to distribute the melt around the edges, from which the solidified flux can easily be dislodged.

24.5   Dry Ashing Samples

Organic samples to be subjected to elemental analysis must first be treated to oxidize the organic matter. As described in Section 24.4, this can be done with an oxidizing mineral acid—for example, perchloric acid—by a procedure known as digestion or wet ashing. Another means of oxidizing an organic sample matrix is **dry ashing**, in which the sample is heated in a crucible with exposure to oxygen. This is typically accomplished in three steps involving progressively higher temperatures and exposure to oxygen in the air. The first step is **drying**, in which the sample is warmed to about 100° C to drive off water. Following this step it may be advisable to crush the sample into a more compact mass to prevent mechanical loss during later stages. The next step is **charring**, in which the sample is heated to a sufficiently high temperature with the lid on the crucible to drive off volatile matter and leave a carbon residue. Last comes the **ashing** itself, in which the residue is heated strongly in contact with air to oxidize the carbon. At no time should the sample be allowed to flame, which carries off parts of the sample as volatile material or ash. Dry ashing is commonly carried out in a small high-temperature furnace called a **muffle furnace**. With care a burner flame can be used as a source of heat.

Conventional dry ashing carries with it the potential for a great deal of sample loss. Convection currents swirling around the crucible may entrain very fine sample particles and carry them away. The oxidation often requires the application of red heat, which causes the evaporation of even some inorganic constituents. Therefore, high-temperature dry ashing should not be used except on samples where it has been proven to be effective.

Low-temperature ashing makes use of chemically reactive oxygen species to ash organic samples at slightly above room temperature. The sample is held in an evacuated chamber into which O_2 is bled through a radiofrequency field. Species such as atomic oxygen, O, react with carbon and organically bound H, S, and N to leave the mineral constituents in a largely unaltered form. The long times required and the low number of samples that can be handled are major disadvantages of low-temperature ashing.

24.6   Combustion Tubes

Collection of Volatile Combustion Products

Loss of volatile components during ashing has been mentioned as a source of error in the analysis of organic compounds. This loss can be turned to an advantage by heating the sample in a glass or quartz **combustion tube** and collecting the volatile components given off in a trap. During this procedure a carrier gas flows over the heated sample. The gas may act as an oxidizing agent if the sample is to be ashed. The elements commonly determined by combustion tube methods are halogens, hydrogen, carbon, nitrogen, oxygen, and sulfur.

The Pregl Method

Combustion tube methods are commonly used for the elemental analysis of organic samples. The best known such procedure is the **Pregl method**, which can be adapted to the determination of carbon, hydrogen, the halogens, and sulfur. With this method the sample is contained over a hot platinum catalyst and burned in an oxygen stream. Halogens in the sample are converted to the elemental form (X_2) and to the hydrogen halides (HX). The gas stream is passed through a basic carbonate solution to retain the acidic hydrogen halides; the solution also contains SO_3^{2-} ion to reduce elemental halogens and oxyhalides to halide ions, X^-. The halide ions are then determined by standard methods. Sulfur in the Pregl process is oxidized to SO_2 and SO_3, which are collected in a solution of hydrogen peroxide, H_2O_2. The hydrogen peroxide oxidizes SO_2 to H_2SO_4, which is titrated by standard base. Carbon is converted to CO_2, absorbed in a trap packed with a base coated on a solid support (Ascarite), and the weight gain of the trap used to measure the quantity of carbon dioxide evolved. Hydrogen goes to H_2O, is absorbed in a desiccant in a trap, and determined by the weight change of the trap.

The Dumas Method

The **Dumas method** is used to determine nitrogen. Oxidation of the sample over hot CuO yields N_2, along with CO_2 and H_2O. The CO_2 and H_2O are retained in a trap of concentrated KOH solution, and the volume of N_2 is measured. The **Unterzaucher method** measures oxygen by sample pyrolysis over carbon in a stream of H_2; this converts oxygen to CO, which is measured by its reduction of I_2O_5 to I_2, followed by titration of the I_2 liberated.

Programmed Summary of Chapter 24

The major terms and concepts introduced in this chapter are contained in this summary in a programmed format. To derive the most benefit from the summary, you should fill in the blanks for each question and then check the answers at the end of the book to see if your choices are correct.

The validity of any analytical result depends upon having a sufficient quantity of a sample that is (1) _____ of the material being analyzed and preserved in such a manner that (2) _____. An ideal means of sampling occurs when a measurement can be taken (3) _____. Two means of reducing the quantity of a large sample to a small sample representative of the whole are (4) _____ and (5) _____. A slotted tube and rod assembly can be used for sampling (6) _____. An interference caused by precipitation of iron(III) hydroxide in a sample is (7) _____ of metal ions. Six "types" of water in a sample are (8) _____. An advantage that sulfuric acid has for getting sample material into solution when heating is required is its high (9) _____. A very strongly oxidizing (and dangerous) acid in the hot, concentrated form is (10) _____. An acid used to dissolve samples in silicate mineral matrixes is (11) _____. The process of putting a sample in solution that involves oxidation of organic matter is called (12) _____. In regard to sample processing, a molten salt mixture is called a (13) _____. Oxidation of an organic sample matrix without liquid reagents is called (14) _____. Use of chemically reactive oxygen species produced in a radiofrequency field to oxidize organic sample matrixes at slightly above room temperature is called (15) _____. The Pregl method is used for the determination of (16) _____ in organic samples. The Dumas method is used to determine (17) _____. Both of these methods make use of (18) _____, in which the sample is oxidized.

Questions

1. The percentage of sulfur in the coal contained in a 100-car trainload of coal is to be determined. Using material in this chapter, discuss the sampling and analysis procedures to be used. Assume that the coal has been crushed to 2.5 cm in diameter or less.

2. In addition to the analytical result, such as the percentage of an analyte in a sample, what other information should be included in the answer for a chemical analysis?

3. What are two major criteria of the small final sample derived from a large quantity of material to be analyzed?

4. What are some of the detrimental things that proper sample preservation avoids?

5. Why are water samples normally refrigerated prior to analysis?

6. How strongly is water normally held in a sample?

25 Suggested Laboratory Determinations

25.1 Use Care!

But Have Confidence

A chemistry student who is genuinely afraid of chemicals may have difficulty in performing efficiently and may even have a greater likelihood of accidents with chemicals than does the chemist who handles them confidently. The performance of laboratory procedures requires pouring, weighing, mixing, and heating chemicals that are dangerous if improperly handled. Many chemicals can cause chemical burns, explode, or be toxic, such as by irritating respiratory passages. The good chemist or well-prepared chemistry student knows how to do various operations with chemicals in a safe manner so that accidents or excessive exposure are not likely to occur. This is the first line of defense against any unpleasant incidents with chemicals. But even in the best of circumstances, humans are fallible, laboratory apparatus breaks, and chemical systems do things that you may not have anticipated. Therefore, there is a second line of defense consisting of safety glasses, laboratory jackets, eyewash fountains, safety showers, and fire blankets. So, given

1. A knowledge of proper safe laboratory procedures
2. Knowledge of the hazards of specific chemicals and laboratory procedures
3. Proper protective glasses, clothing, and laboratory safety features

you are safer in the chemistry laboratory than in crossing any busy streets leading to it.

Safety Precautions

In most chemistry laboratory courses, the student is given a list of safety precautions and asked to sign a form saying that these precautions have been read and will be heeded. This is for the protection of both the student and the institution. Keep this list and review it from time to time. It will make you a better laboratory worker and may save you from injury.

Precautions with chemicals and procedures are noted in the laboratory procedures contained in this chapter. This does not guarantee that all hazards have been covered, however. The types of equipment and practices vary from laboratory to laboratory. Therefore, the laboratory instructor will give you written and verbal precautions pertaining to potentially hazardous chemicals and procedures. Take heed! If you don't completely understand, ask! You will work more confidently and get better results if you understand the nature of the hazards involved.

Disposal of Waste Chemicals

Everyone has heard of hazardous wastes, waste chemicals whose improper disposal can cause pollution problems or even direct hazards to humans. You will be instructed on the proper collection of waste chemicals so that something potentially dangerous does not get flushed down the drain or carried out with wastepaper. Follow these instructions carefully. The hazardous by-products of a quantitative analysis laboratory are small in quantity, but everyone wants to prevent even a slight addition of pollutant chemicals to the environment.

25.2   **Preparation and Standardization of 0.1 *M* Sodium Hydroxide**

An accurately standardized solution of 0.1 *M* NaOH is commonly used for titrating weak and strong acids. This determination describes how to prepare the NaOH solution so that it is free of carbonate, which can give erroneous end points in some titrations, and how to standardize the NaOH with potassium hydrogen phthalate, KHP. *Text chapter references for this section are Chapter 5, Sections 5.1 to 5.8.*

1. Prepare carbon dioxide–free water by boiling about 1 L of distilled or deionized water in a beaker or conical flask for about 5 min. Allow the boiled water to cool covered with a watchglass or a piece of aluminum foil. After cooling, transfer the water to a suitable polyethylene or rubber-stoppered borosilicate glass flask in which the NaOH is to be contained (avoid glass stoppers because base solutions cause them to "freeze" in place).
2. Using a 10-mL graduated cylinder, add 5.5 to 6.0 mL of 50% NaOH to the cooled CO_2-free water, replace the stopper, and mix thoroughly by inverting the bottle 20 times with the stopper held in place. *Note:* The solution of 50% NaOH in water is used as a source of NaOH because carbonate precipitates from it as Na_2CO_3 and thus does not contaminate the 0.1 *M* standard solution of NaOH. Normally the 50% NaOH will be prepared in advance and contained in an apparatus that enables siphoning the liquid from the container so that no solid Na_2CO_3 is entrained with it. *Warning: The 50% NaOH solution is very*

damaging to skin and eyes. Under no circumstances should this solution be pipetted by mouth. In the event of accidental contact, laboratory first-aid measures for contact with strong caustic must be followed.

3. Prepare to standardize the NaOH solution with primary standard acid, potassium hydrogen phthalate, KHP, which reacts with OH^- according to the reaction

$$\underset{\substack{\text{KHP, KHC}_8\text{H}_4\text{O}_4, \\ \text{fw} = 204.23}}{}$$

KHP, KHC$_8$H$_4$O$_4$,
fw = 204.23

An approximately 5-g quantity of KHP in an open weighing bottle should be dried for 1 hour in a drying oven at $110°\,C$ (the laboratory instructor may provide you with dried primary standard KHP). Measuring to the nearest 0.1 mg with an analytical balance, weigh out by difference three 0.8- to 0.9-mg portions of the KHP, and transfer each quantitatively to a 250-mL conical flask. Add 40 mL of water to each flask and dissolve the KHP.

4. Titrate the KHP with the 0.1 *M* NaOH to standardize the NaOH. Add 4 drops of phenolphthalein indicator to one of the KHP solutions and titrate with NaOH to the appearance of a faint pink color of phenolphthalein that persists throughout the solution for at least 20 seconds (the end point of this titration may suggest the use of a drop or two more or less of the phenolphthalein for subsequent titrations). Record the volume of sodium hydroxide required. Add phenolphthalein to the two remaining solutions of KHP and repeat the titration for each.

5. Calculate the molarity (normality) of the sodium hydroxide for each of the titrations. This value should lie between 0.085 and 0.115 *M*. If the NaOH solution is too dilute, add some more 50% NaOH; if it is too concentrated, dilute it with carbon dioxide–free water. If the relative range of the three values

$$\text{Relative range} = \frac{\text{highest value} - \text{lowest value}}{\text{mean value}} \times 1000$$

exceeds 6 parts per thousand (ppt) you may want to run additional titrations (consult your laboratory instructor).

6. Correctly label the bottle of standard NaOH with its molarity, date, and number of the page of the laboratory notebook where the titration data are recorded.

25.3   **Preparation and Standardization of HCl with Standard NaOH**

1. Prepare HCl solution, approximately 0.1 M, by adding 10.5 mL of concentrated ($\cong$ 12 M) hydrochloric acid to 1 L of deionized or distilled water in a storage bottle. Stopper the bottle and mix well. *Caution: Concentrated HCl gives off strong fumes of HCl gas, is damaging to eye tissue, and deadly to cotton cloth, where accidental spills are manifested by holes in blue jeans following the first washing after contact.*

2. Titrate the HCl with standardized NaOH. Using a volumetric pipet, place 25-mL aliquots of the HCl in each of three 25-mL conical flasks and add 4 drops of phenolphthalein indicator to each. Titrate each with the standardized 0.1 M NaOH (prepared in Determination 25.2) until the faint pink of phenolphthalein indicator persists for at least 20 seconds.

3. Calculate the molarity (normality) of each of the HCl solutions. If the range of the results (see Determination 25.2, step 5) exceeds 6 ppt, consult the laboratory instructor about doing additional titrations.

4. Correctly label the bottle of standard HCl with its molarity, date, and number of the page of the laboratory notebook where the titration data are recorded.

25.4   **Standardization of HCl with Primary Standard Na$_2$CO$_3$**

Primary standard Na$_2$CO$_3$ can be used for direct standardization of HCl solution (see Determination 25.3 for the preparation of 0.1 M standard HCl). This is done by titrating Na$_2$CO$_3$ to NaCl and CO$_2$ gas with the HCl titrant. The overall ionic reaction is

$$CO_3^{2-} + 2H^+ \rightarrow CO_2(g) + H_2O$$

Since two H$^+$ ions are required for each CO$_3^{2-}$ ion, the equivalent weight of Na$_2$CO$_3$ (53.0 g/eq) is half the formula weight (106.0 g/mol). Much of the CO$_2$ produced tends to remain dissolved in solution during the titration, adversely affecting the observation of the end point. Removal of this CO$_2$ by boiling just prior to the end point enables a sharp end point color change.

1. Dry primary standard sodium carbonate (analytical reagent grade may be used if a correction is made for impurities) at 110°C in a drying oven for at least 2 hours (this is best done with the assistance of the laboratory instructor between laboratory periods).

2. Calculate the weight of Na$_2$CO$_3$ required to react with 40 mL (an optimum volume for titration) of 0.1 M HCl. Remember that the equivalent weight of Na$_2$CO$_3$ is half its formula weight. Note that the weight calculated is not large; a small error in weighing such a small

quantity of a chemical or small losses in transferring the chemical lead to large relative errors. Therefore, 10 times as much Na_2CO_3 will be placed in solution, and a 1/10 aliquot of the solution taken for each titration.

3. On an analytical balance, weigh by difference 10 times the weight of Na_2CO_3 required for each titration, add the solid to a 250-mL volumetric flask, dissolve in distilled or deionized water, and make up the solution to the mark with thorough mixing.

4. With a volumetric pipet, withdraw a 25-mL aliquot of the sodium carbonate solution for titration in a 150- or 250-mL conical flask. Add 4 drops of methyl orange indicator solution, a drop or two more if the yellow color is insufficiently distinct.

5. Titrate with 0.1 M HCl until the first distinct change of indicator color from yellow to orange has occurred.

6. Boil the solution for 30 seconds to expel dissolved CO_2, cool, and observe the color. The color should have reverted back to yellow; if it is too faint, add three more drops of indicator.

7. Continue the titration until a distinct change of indicator color from yellow to red occurs.

8. Repeat steps 4 through 6 until at least three satisfactory titrations have been obtained (for example, those in which the end point volume was not exceeded prior to boiling).

9. Calculate the molarity of the HCl solution for each of the satisfactory titrations. If the range of the results (see Determination 25.2, step 5) exceeds 6 ppt, consult the laboratory instructor about doing additional titrations.

25.5   Titration of Unknown Acids with Standard Base

The base prepared and standardized in Determination 25.2 can be used for the titration of unknown acids. Important examples of such acids are the following:

1. Pure solid weak organic acids for which the equivalent weights can be determined by acid–base titration.

2. Mixtures of potassium hydrogen phthalate (see Determination 25.2, step 3) with inert solid. These are sold as commercial solid "unknowns," which the student can analyze to determine the percentage of KHP.

3. Solutions of acids, such as vinegar.

Detailed instructions will not be given for the determination of acid in the various possible kinds of unknown samples. The instructor may choose to have you determine the acid in a commercial unknown, in an unknown prepared by the instructor, or in a commercial product such as vinegar or industrial waste acid (for example, steel pickling liquor, an acid solution used to remove corrosion from steel surfaces). General instructions are given here for the titration of acid; the instructor will give you specific details about handling the sample and obtaining the accurately measured, proper size of sample for titration.

1. Measure accurately three portions of the sample for titration. For a solid sample, these portions must be taken from the predried material and weighed to the nearest 0.1 mg on an analytical balance. For a liquid sample, the appropriate amount must be withdrawn with a volumetric pipet. The instructor will tell you the amounts of sample to take, or give you the range of analytical results expected so that you can calculate the appropriate amounts to take. In the case of synthetic samples (for example, an accurately measured volume of an acid of known concentration), you may receive the sample in a volumetric flask, in which the sample is diluted to the mark and a measured aliquot is taken for titration.
2. To each of the three measured portions of the sample dissolved and contained in about 25 mL of solution in 150- or 250-mL conical flasks, add 4 drops of phenolphthalein indicator.
3. Titrate each of the sample portions with standard NaOH to the phenolphthalein end point.
4. Calculate the requested analytical result (for example, percent KHP in a solid, equivalent weight of a pure solid acid, percentage of acetic acid in vinegar). If the range of these results is excessive, or if the volumes of titrant required were too low or too high (less than 20 mL, introducing too much relative volume error, or greater than 50 mL, requiring that a 50-mL buret be refilled during the course of the titration), consideration should be given to doing additional titrations.

25.6   Titration of a Mixture of HCl and H_3PO_4 with Standard NaOH

Using Indicators

When a mixture of HCl and H_3PO_4 is titrated, a titration curve with two breaks (two end points) is obtained. The volume of base required to reach the first end point is always greater than the volume required in going from the first to the second end point. After studying Chapter 5, you should be able to explain why this is the case and to give the reactions that occur. It is possible to observe the first end point from the color change in methyl orange indicator and the second from the color change in thymolphthalein indicator.

1. Receive the unknown in a 250-mL volumetric flask containing a mixture of HCl and H_3PO_4. Dilute the contents of the flask to the mark and mix by inverting the stoppered flask 20 times.
2. Pipet a 25-mL aliquot of the acid mixture into each of two 150- or 250-mL conical flasks, add 4 drops of methyl orange indicator, and titrate until the color matches that of a dilute solution of NaH_2PO_4 containing a like amount of indicator.
3. Repeat the preceding step, substituting thymolphthalein indicator for methyl orange. The solution is colorless up to the end point and turns a permanent blue color at the end point.
4. Calculate the millimoles of HCl and millimoles of H_3PO_4 in each of the three aliquots in which these values were determined. Report the total millimoles of each acid in the sample you received.

25.7   Titration of a Mixture of HCl and H_3PO_4 with Standard NaOH

Using a pH Meter

The titration of a mixture of HCl and H_3PO_4 using standard base was described in the preceding determination. The actual titration curve with its two breaks is readily plotted by following the titration with a pH meter (see Chapter 5, Section 5.3, for basic information about pH meter measurements). The end point volumes are readily determined by graphical methods (see Chapter 5, Section 5.4).

1. Place a magnetic stirring bar in a 400-mL beaker.
2. To the beaker add 25 mL of the acid mixture and 150 mL of distilled or deionized water and place the beaker on the magnetic stirrer.
3. Immerse the tip of the pH electrode in the solution to be titrated. (The pH meter should have first been calibrated with a standard buffer so that it gives the correct pH values.) Set the meter to read pH. Turn up the control on the magnetic stirrer on which the beaker is resting to give a moderate stirring rate without excessive turbulence or splashing.
4. Add NaOH and record pH as a function of millimeters of added NaOH. Readings should be taken at 0.5-mL intervals around the two end points and at 2- to 3-mL intervals elsewhere. In addition to recording data in the laboratory notebook, plot it on graph paper so that the shape of the emerging titration curve can be observed. *It is easy to consume too much time doing this titration.* About 15 min should be required to complete a titration curve after the apparatus is set up.
5. Connect the points on the titration curve with a smooth line. A French curve can be employed for this purpose.

6. Determine the volume of NaOH at the first and second end points. Report the total number of millimoles of HCl and of H_3PO_4 in the sample you received.

Suggested variation: A mixture of H_3PO_4 and NaH_2PO_4 gives a titration curve with a shape very similar to that obtained with the HCl–H_3PO_4 mixture, except that the relative volumes at the end points are different. Your instructor may choose to give you such a sample and ask for the millimoles of H_3PO_4 and NaH_2PO_4, as well as an explanation of the reactions involved. The course of the titration can be followed either with indicators (Determination 25.6) or with a pH meter (Determination 25.7).

25.8   **Titration of Unknown Sodium Carbonate with HCl**

The percentage of Na_2CO_3 in a solid also containing inert solid material can be determined by titration with standard HCl. The reaction and principles are exactly those described in Determination 25.4 for the standardization of HCl with primary standard Na_2CO_3. If a solid sample containing Na_2CO_3 is not available, your instructor may choose to provide a sample of dissolved Na_2CO_3 in a volumetric flask.

1. Receive the sample and process it (drying, weighing, dissolving, taking solution aliquots, etc.) as directed by the instructor to obtain three portions of sample dissolved in a minimum of 25 mL of water and contained in a 150- or 250-mL conical flask.
2. Add 4 drops of stock methyl orange indicator solution to each of the portions of sodium carbonate to be titrated.
3. Titrate with 0.1 *M* HCl until the first distinct change of indicator color from yellow to orange has occurred.
4. Boil the solution for 30 seconds to expel dissolved CO_2, cool, and observe the color. The color should have reverted back to yellow; if it is too faint, add three more drops of indicator.
5. Continue the titration until a distinct change of indicator color from yellow to red has occurred.
6. Calculate the percentage of Na_2CO_3 in the sample.

25.9   **The Mohr Precipitation Titration of Chloride**

Chloride is readily determined by titration with standard $AgNO_3$ solution to precipitate solid AgCl. The net ionic reaction is

$$Ag^+ + Cl^- \rightarrow AgCl(s)$$

The Mohr method, described in Chapter 9, Section 9.6, enables determination of the end point by formation of red Ag_2CrO_4 in excess Ag^+ when a low concentration of chromate ion is present.

Because of the expense of silver nitrate reagent solution, you may be supplied this reagent in a standardized form. The procedure is written for a sample supplied in solution but is readily modified for use with solid samples.

1. Receive the sample in a 250-mL volumetric flask and dilute it quantitatively with mixing.
2. With a volumetric pipet transfer 25 mL of the diluted sample to a 150- or 250-mL conical flask. Add 25 mL of water and 2.0 mL of 0.10 M K_2CrO_4.
3. Titrate with standard $AgNO_3$. *Caution: $AgNO_3$ solution discolors the skin, although the discoloration leaves with time.* The contents of the flask should be swirled vigorously with each addition of titrant. Observation of the color at the point where the $AgNO_3$ contacts the solution provides an indication of the color that will be observed at the end point. The end point should be taken as the first permanent red-orange coloration. This end point is a little tricky and is best observed as a *change* in color that occurs upon the addition of one or two drops of titrant solution and that persists with mixing.
4. For at least three titrations with acceptable end points, calculate the millimoles of chloride in the sample you received, or other results the laboratory instructor requests. (You may have been given a solid sample or asked to assume that your sample was from a solid of a specified weight.)

25.10   Titration of a Mixture of Iodide and Chloride

The titration of a mixture of chloride and iodide with silver nitrate gives a titration curve with two breaks as discussed in Section 9.3 and illustrated in Figure 9.3. The log[Ag^+] function used to plot this titration curve is readily measured potentiometrically (see Chapters 12 and 13) by the potential of a silver electrode versus a reference electrode, both in contact with the solution. The equation relating this potential E to log[Ag^+] at 25° C is

$$E = E^0 + 59.1 \log[Ag^+]$$

where E^0 is a constant for the electrode system employed, and E is in units of millivolts (mV). A pH meter that can be set to read potential in mV is commonly used to measure the potential.

1. Obtain the iodide–chloride mixture sample in a 250-mL volumetric flask and dilute to the mark with mixing.
2. Place a 50-mL aliquot of the sample in a 400-mL beaker containing a magnetic stirring bar and set up for stirring. Place the beaker on the

magnetic stirrer, immerse the ends of the electrodes in the solution, and adjust the stirring to a moderate rate without splashing or excessive turbulence.

3. Titrate with $AgNO_3$ solution, such that titrant is added in 0.1- or 0.2-mL intervals in the immediate vicinity of the first (iodide) equivalence point, 0.5-mL intervals over the second equivalence point and 2- to 5-mL intervals elsewhere on the titration curve. Record the data and simultaneously plot the data points on graph paper so that the locations of the end points will be more obvious. You may be advised to run a first titration rapidly by adding titrant in 2-mL increments and plotting the data to get the approximate locations of the end points.

4. Draw a smooth-line plot that best fits the data points. A French curve and straightedge may be employed for this purpose.

5. Determine the two end point volumes. Refer back to Figure 9.3 before doing this and take special note of the unique shape of the first "break" in the titration curve.

6. Report the millimoles of I^- and of Cl^- in your sample.

25.11   Gravimetric Determination of Nickel by Homogeneous Precipitation
of Nickel(II) Bisdimethylglyoximate

In Chapter 8, Section 8.7, the advantages of homogeneous precipitation are discussed. In Section 8.10 the use of dimethylglyoxime to precipitate nickel(II) from solution is discussed, and the reaction is given in Equation 8.15.

Nickel can be precipitated homogeneously by the *in situ* generation of dimethylglyoxime by the oximation of biacetyl as shown by the following reaction:

$$H_3C-\overset{\overset{O}{\|}}{C}-\overset{\overset{O}{\|}}{C}-CH_3 + 2NH_2OH \rightarrow \underset{HO}{\overset{H_3C}{\underset{N}{\overset{|}{C}}}}\overset{CH_3}{\underset{N}{\overset{|}{C}}}_{OH} + 2H_2O$$

| Biacetyl | Hydroxylamine | Dimethylglyoxime |

This reaction occurs slowly and throughout the solution (that is, homogeneously). When a significant concentration of dissolved nickel is present, nickel(II) bisdimethylglyoximate $[Ni(C_8H_{14}N_4O_4)$, fw = 288.91] precipitates in a very pure form that is readily filtered and weighed.

It should be noted that dimethylglyoxime is a virtually specific reagent for nickel. Palladium(II) likewise forms a precipitate with this reagent, but it is most unlikely to be present as an impurity. Either iron or cobalt in a large excess over nickel may interfere, but this interference is eliminated with suitable masking agents (that is, species that bind to interfering species and prevent their adversely affecting an analytical reaction). For example, tartaric acid can be used as a masking agent to prevent interference by iron.

1. Dry three clean, numbered glass filtering crucibles to constant weight at 140°C. This procedure consists of alternate heating in a drying oven, cooling in a desiccator, and weighing until the weight of each filtering crucible agrees to within 0.3 mg (or other value specified by the instructor) of its preceding weight.

2. Weigh out about 3 g of hydroxylamine hydrochloride on a top-loading balance, dissolve in 300 mL of water, and adjust the pH to 7.5 ± 0.1 with 6 M ammonia and 6 M acetic acid, using a pH meter and electrode. This will be referred to as the hydroxylamine solution.

3. As directed by the laboratory instructor, measure out three portions of unknown into separate 250-mL beakers. (The sample may be a solid to be weighed or a liquid of which aliquots are to be taken.) The sample should be dissolved in approximately 25 mL of water.

4. Add 20 mL of biacetyl solution (1.0% in ethanol solvent) and adjust the pH to 7.5 ± 0.1 with 6 M ammonia and 6 M acetic acid.

5. Add 100 mL of hydroxylamine solution, stir, and adjust the final volume to approximately 175 mL.
   Note: If the determination must be interrupted before filtration, this is a good place to stop. Cover the beakers with plastic wrap and put them in a location where they will not be disturbed (that is, your laboratory drawer, if you remember to open and close it carefully).

6. Allow the mixed solutions containing analyte to stand in the beakers at room temperature for at least one hour (or between laboratory periods as noted above).

7. Heat the beakers and contents on a water bath at approximately 80°C for one-half hour.
   Warning: Do not exceed 90°C; the reaction will go too fast. The red nickel(II) bisdimethylglyoximate precipitate will appear as a layer floating on the liquid.

8. Allow the precipitates to cool and collect each by filtration with separate glass filter crucibles prepared in step 1.

9. Dry the precipitate in each crucible at 110°C, cool in a desiccator, and weigh. If time permits, repeat this procedure until a constant weight is obtained.

10. Calculate the percentage of nickel in the sample, or other result as requested by your laboratory instructor. Note the formula weight of nickel(II) dimethylglyoximate given in the introduction to this determination. You may want to verify it from the atomic weights of the constituent elements and the formula of the precipitate. What is the value of the gravimetric factor that you might use for your calculation?

25.12   **Gravimetric Determination of Magnesium as the**

Ammonium Phosphate

Magnesium can be precipitated from solution as magnesium ammonium phosphate hexahydrate according to the reaction

$$Mg^{2+} + NH_4^+ + HPO_4^{2-} + OH^- + 5H_2O \rightarrow MgNH_4PO_4 \cdot 6H_2O(s)$$

The product may be weighed after filtration and washing or converted to magnesium pyrophosphate by ignition in a muffle furnace at $1050 \pm 50°\,C$:

$$2MgNH_4PO_4 \cdot 6H_2O \xrightarrow[heat]{} Mg_2P_2O_7 + 2NH_3(g) + 13H_2O(g)$$

1. According to directions given by the instructor, prepare a solution of the sample containing not more than 0.15 g of magnesium in a total volume of 200 mL and a pH of 4 to 7.
2. Add 7 mL of concentrated HCl solution, stir, then add 3 to 5 drops of methyl red indicator.
3. Add 10 mL of ammonium phosphate solution (freshly prepared by dissolving 35 g of $(NH_4)_2HPO_4$ in water and diluting to 100 mL).
4. Precipitate the magnesium ammonium hexahydrate salt as follows: Add concentrated NH_3 solution with vigorous stirring until the indicator color turns yellow. (Stirring should be done with a magnetic stirrer or with a glass rod tipped with a rubber policeman; scratching the sides of the beaker with a glass rod will result in formation of hard-to-remove precipitate at the point of contact on the beaker walls.) The stirring should be continued for 5 minutes, with addition of concentrated ammonia solution dropwise as necessary to maintain the yellow color.
5. After completion of at least 5 minutes of stirring, add 7 mL of concentrated ammonia in excess and mix.
6. Allow the solution to stand for at least 3 hours, and preferably overnight.
7. Collect the precipitate by filtration through a medium-porosity-frit glass filtering crucible that has been washed with 95% ethanol followed with diethyl ether and dried to a constant weight in a desiccator.
   Caution: Diethyl ether is a highly flammable organic liquid. Exposure of the liquid or vapor to flame or spark is to be strictly avoided. It is to be used in a well-ventilated area and inhalation must be minimized. Use and disposal of this reagent must be in strict accordance with directions from the laboratory instructor.
8. Wash the precipitate with small batches of 0.8 M ammonia until the filtrate gives a negative test for the presence of chloride ion (no turbidity when 2 to 4 mL of the wash liquid are acidified by nitric acid and several drops of $AgNO_3$ solution added).

9. Remove surface water from the precipitate by washing with three 10-mL portions of 95% ethanol.

10. Wash the precipitate with four 5-mL portions of anhydrous diethyl ether (*note precautions with this reagent in step 7*).

11. Evaporate the ether from the precipitate by drawing air through the filtering crucible for 10 minutes, using the laboratory vacuum system provided.

12. Wipe any liquid film from the surface of the filtering crucible with a lint-free laboratory tissue (such as Kim–Wipe).

13. Allow the crucible and contents to dry in a desiccator for 30 minutes.

14. Weigh the crucible and contents; subtract the weight of the crucible to get that of the $MgNH_4PO_4 \cdot 6H_2O$ product.

15. Calculate the magnesium content as directed by the laboratory instructor.

25.13   Determination of Water Hardness in an Unknown

Water hardness is a measure of the concentration of Ca^{2+} and Mg^{2+} in the water. Hard water causes a curdy precipitate with soap (infamous bathtub ring), clogging of water pipes as a result of the precipitation of calcium carbonate, and scale in water heaters. However, some evidence suggests that a moderate degree of hardness in water is beneficial to health. Water hardness must be known accurately for water treatment processes and is one of the most commonly determined water quality parameters.

Water hardness is readily determined by complexometric titration with EDTA (see Chapter 10). Largely as a result of water treatment plant practices, water hardness is frequently expressed as milligrams of $CaCO_3$ per liter of water.

In this experiment the unknown hardness in a sample of water will be determined as the sum of $[Ca^{2+}] + [Mg^{2+}]$. The Ca^{2+} concentration in both solutions will be determined by titration with EDTA employing Eriochrome black T (see Table 10.6) as an indicator.

1. Prepare standard EDTA titrant, 0.01 M, by drying the dihydrate $(Na_2H_2Y \cdot 2H_2O)$ at 80°C, cooling, weighing 1.9 g to the nearest 0.1 mg, dissolving, and diluting to 500 mL in a volumetric flask. Calculate the exact molarity of the EDTA standard, and label the standard container with that value, the date of preparation, and the notebook pages on which the preparation is described.

2. For titration of an unknown water hardness sample, receive the sample in a flask; it may be distributed in a volumetric flask requiring dilution to the mark prior to analysis.

3. Transfer three 25-mL aliquots to each of three 250-mL conical flasks,

add about 75 mL of water and 3 drops of concentrated HCl to each flask and boil gently for 2 min to remove CO_2; allow to cool.

4. Add 3 drops of methyl red indicator and neutralize the acid by adding 0.5 M NaOH dropwise until the indicator changes color.

5. Buffer the solutions by adding 2 mL of pH 10 ammonia–ammonium chloride buffer.

6. Add magnesium required for the proper functioning of the indicator by adding 1 mL of magnesium-EDTA (1×10^{-3} M MgY^{2-}) solution to each sample.

7. Add 4 drops of Eriochrome black T indicator to each portion of analyte immediately prior to its titration.

8. Titrate with standard EDTA to a color change from red to pure blue. *Note: Observation of this end point is often not easy and may require several practice runs.*

9. Repeat steps 7 and 8 for each of the remaining samples.

10. Calculate the water hardness in the original sample in units of molarity of Ca^{2+}; optionally, the hardness can be calculated in units of milligrams per liter of $CaCO_3$ (a standard engineering and water treatment unit).

25.14   Complexometric Determination of Calcium in Water Equilibrated with CO_2 and $CaCO_3$

Solid $CaCO_3$ reacts with CO_2 dissolved in water to go into solution as Ca^{2+} ion and HCO_3^- ions. This reaction is a function of K_{sp} of $CaCO_3$, K_{a1} and K_{a2} of $CO_2(aq)$ (as discussed in Chapter 7, Section 7.6), and the Henry's law solubility of gaseous CO_2 (see Section 7.12). You should try to write the reaction and check it with your laboratory instructor. It is the reaction by which $CaCO_3$ as limestone is dissolved by CO_2-rich groundwater to form cavities and caves in limestone formations.

This experiment involves bubbling pure CO_2 gas through water in contact with $CaCO_3$, so that the water is saturated with both CO_2 gas (at almost 1 atm pressure) and with $CaCO_3$. This will probably be done for the whole laboratory in a single experimental setup. The water saturated with $CaCO_3$ and CO_2 is then filtered to remove any suspended $CaCO_3$ and titrated complexometrically with EDTA to determine the Ca^{2+} concentration, which is also equal to the solubility S of $CaCO_3$ in equilibrium with pure CO_2. Another aliquot of the water is boiled to drive off CO_2 and precipitate solid $CaCO_3$ (can you write the reaction?), filtered, and titrated with EDTA to determine the new concentration of Ca^{2+}.

1. Receive a sample consisting of water that has been equilibrated with solid $CaCO_3$ and with CO_2 gas.

2. Filter the sample to remove suspended $CaCO_3$; keep the sample stored in a stoppered or plastic-wrap-covered flask to prevent loss of CO_2 to the atmosphere.

3. Transfer three 10-mL aliquots of filtered sample to each of three 250-mL conical flasks, add about 90 mL of water and 3 drops of concentrated HCl. Stir until evolution of CO_2 gas subsides. Add additional HCl dropwise followed by stirring (swirling) until no significant fizzing occurs when a drop of acid is added.

4. Continue with the determination, following steps 4 through 9 of Determination 25.13 for titration with EDTA.

5. Calculate $[Ca^{2+}]$ in the solutions titrated.

6. After steps 1 through 5 have been completed to your satisfaction, heat the remaining filtered solution saturated with solid $CaCO_3$ and CO_2 gas to a point of barely boiling (incipient boiling) and stir it at that temperature for 5 min. Bubbles of CO_2 gas should be evolved, and a precipitate of $CaCO_3$ should be visible.

7. Cool the solution to room temperature in contact with the atmosphere. If the solution has to be left between laboratory periods, it should be sealed with plastic wrap after cooling.

8. Decant (pour off) the solution from the precipitated $CaCO_3$, and filter through fine filter paper to remove suspended $CaCO_3$.

9. Determine the calcium ion concentration in three 25-mL aliquots of the filtered solution as instructed by steps 3 through 9 of Determination 25.13.

10. Calculate the concentration of Ca^{2+} in the boiled, filtered solution.

Explain the results obtained in this experiment on the basis of chemical reactions and solubility and acid–base equilibria. If the results do not conform to your calculations, attempt to offer explanations as to why they do not. You may want to compare your results and thoughts with other students and with the laboratory instructor.

25.15   Oxidation-Reduction Determination of Ferrous Ammonium Sulfate

Unknown by Titration with Potassium Dichromate

Oxidation-reduction titrations are covered in Chapter 12. In this determination, iron in a mixture of ferrous ammonium sulfate, $Fe(NH_4)_2(SO_4)_2 \cdot 6H_2O$, and unreactive material is determined by oxidation of iron(II) to iron(III) with Cr(VI). Prior to titration the iron is all placed in the +2 oxidation state by reduction with stannous chloride. The excess $SnCl_2$ is eliminated by reaction with $HgCl_2$. Standardization of the potassium dichromate titrant solution is accomplished by using it to titrate a known amount of iron(II) prepared by dissolving a weighed quantity of primary standard iron wire in acid.

Preparation of Standard $K_2Cr_2O_7$ Solution

1. In preparation for making up the standard titrant solution, dry a sufficient quantity of $K_2Cr_2O_7$ at 150 to 160° C for 2 hours. Store the dried reagent in a desiccator. In order to save time and the chemical, this may have been done for the class by the instructor in advance.
2. Weigh a 2.452-g portion of the $K_2Cr_2O_7$ reagent on an analytical balance, dissolve in a 500-mL volumetric flask, dilute to volume, and mix. The reagent can be transferred to a glass-stoppered bottle for storage.
4. With a knowledge of the formula weight of the potassium dichromate, and the fact that its reduction product is Cr^{3+}, calculate the solution normality. A more exact value of the normality will be obtained by standardization with iron wire.

Standardization of $K_2Cr_2O_7$ with Iron Wire

5. In each of three 500-mL conical flasks, place a 0.20- to 0.25-g portion of reagent-grade iron wire weighed to the nearest 0.1 mg.
6. *This step must be performed in a laboratory hood to collect HCl vapors.* To each conical flask containing iron wire, add 10 mL of concentrated HCl. Cover each flask with a watchglass. If the iron wire does not dissolve, or dissolves too slowly, heat the flask contents mildly. Replace evaporated HCl to restore a volume of 5 to 10 mL, if necessary.
7. Reduce the iron(III) to iron(II) with $SnCl_2$ solution. (This reagent is 0.67 M $SnCl_2$–6 M HCl prepared by dissolving 150 g $SnCl_2 \cdot 2H_2O$ in 500 mL of concentrated HCl, and making up to 1 L with water. Oxidation of $SnCl_2$ in storage is prevented by adding several pieces of mossy tin to the reagent bottle.) To each individual flask, add $SnCl_2$ solution dropwise with swirling to a point 2 drops beyond the disappearance of the yellow Fe(III) chloride color.
8. This step should be done immediately before the titration. Dilute the contents of one of the flasks to 150 mL and rapidly add 15 mL of 5% $HgCl_2$ while swirling the contents of the flask. *Avoid contact of the $HgCl_2$ solution with the skin.* (The mercury(II) chloride solution is prepared by dissolving 50 g of $HgCl_2$ in 950 mL of water.) Cover the flask with a watchglass and wait 5 min, *but not appreciably longer*, before proceeding to the next step.
9. Add 15 mL of concentrated H_3PO_4 to the flask, and cool the flask externally in a stream of tap water.

10. Add 2 drops of sodium diphenylamine sulfonate indicator. (This reagent can be prepared by dissolving 0.25 g of diphenylamine sulfonate in 100 mL of water. Alternatively, 0.32 g of the barium salt can be dissolved in 100 mL of water, 0.50 g of Na_2SO_4 added, and the $BaSO_4$ precipitate produced removed by filtration.)
11. Immediately titrate with the $K_2Cr_2O_7$ solution. The end point is manifested by a color change from gray-green to the first perceptible tinge of violet that persists for 10 s of swirling.
12. Repeat steps 8 through 11 for each of the remaining flasks of dissolved standard iron wire.
13. Because of a significant end point error, it is best to *calculate the iron titer* of the standard dichromate solution; this solution's iron titer is the number of milligrams of iron(II) oxidized by each milliliter of standard dichromate solution.

Determination of Percentage Iron in Ferrous Ammonium Sulfate Samples

14. Weigh out three portions of ferrous ammonium sulfate unknown, and place in each of three 500-mL conical flasks. Your laboratory instructor will suggest the appropriate range of weights. Alternatively, solution samples may be provided.
15. Dissolve the sample in 10 mL of concentrated HCl plus the minimum amount of water required to complete dissolution.
16. Titrate by going through steps 7 through 12 above.
17. Calculate the percentage of iron in the sample, using the iron titer of the solution in the calculation.

25.16 Potentiometric Titration of Iron(II)

The oxidation of iron(II) by cerium(IV) reagent is represented by the reaction

$$Ce^{4+} + Fe^{2+} \rightarrow Ce^{3+} + Fe^{3+}$$

This titration can be followed potentiometrically at a platinum electrode (see Chapter 13). The potential E of a platinum electrode is a function of the relationships

$$E(\text{mV at } 25°\text{C}) = E^0_{Fe} - 59.1 \log \frac{[Fe^{2+}]}{[Fe^{3+}]} = E^0_{Ce} - 59.1 \log \frac{[Ce^{3+}]}{[Ce^{4+}]}$$

where the E^0's are for the following half-reactions:

$$Fe^{3+} + e^- \rightleftarrows Fe^{2+} \qquad E^0_{Fe}$$
$$C^{4+} + e^- \rightleftarrows Ce^{3+} \qquad E^0_{Ce}$$

The ratios $[Fe^{2+}]/[Fe^{3+}]$ and $[Ce^{3+}]/[Ce^{4+}]$ change over several orders of magnitude at the equivalence point, causing E to shift sharply at the equivalence point. The titration curve consists of a plot of E versus volume of titrant; the end point is manifested as a sharp break in the titration curve.

1. Prepare 0.1 M cerium(IV) titrant solution. Make up a solution of sulfuric acid by adding 15 mL of concentrated sulfuric acid to 250 mL of water. (*Caution: Concentrated sulfuric acid is one of the more dangerous of the common laboratory reagents because of is extreme acidity and dehydrating action. In case of contact with the skin, flush with water immediately and take first-aid measures as directed by the laboratory instructor.*) To the solution of dilute H_2SO_4, add approximately 32 g of ceric ammonium sulfate, $Ce(NH_4)_4(SO_4)_4 \cdot 2H_2O$, and stir until the maximum amount of solid has dissolved (a small amount of the salt may fail to dissolve). If the solution is not clear, particularly after standing overnight, it should be filtered through a sintered glass filter crucible.

2. Dissolve three portions of iron wire, reagent grade, each 0.20 to 0.25 g, weighted to the nearest 0.1 mg. Each portion of iron wire should be placed in a 400-mL beaker, 10 mL of concentrated HCl added to each beaker, and each covered with a watchglass. Place the beakers in the hood and gently apply heat until the iron wire is dissolved. Rinse the consensed liquid from the bottom of the watchglass back into the beaker it was used to cover to recover volatile iron compounds. To each beaker add an additional 10 mL of concentrated HCl.

Perform steps 3 through 6 on separate iron solutions before proceeding to the next solution.

3. Prereduce the iron by first heating the iron solution mildly and adding stannous chloride solution (0.67 M $SnCl_2$–6 M HCl, see Determination 25.15, step 7) dropwise until the yellow iron(III) color just disappears, then add 2 drops excess.

4. Dilute the contents of the beaker to 150 mL and rapidly add 15 mL of 5% $HgCl_2$ while swirling the contents of the beaker. *Avoid contact of the $HgCl_2$ solution with the skin.* (Preparation of the mercury(II) chloride solution is explained in Determination 25.15, step 8.) Cover the beaker with a watchglass and wait 5 min, *but not appreciably longer,* before proceeding to the next step.

5. Add 10 mL of concentrated H_3PO_4 and set up the solution for potentiometric titration with magnetic stirring as directed by the laboratory instructor.

6. Titrate with the cerium(IV) solution, taking potential measurements

at intervals of 0.5 mL or less around the end point, and at 2- to 3-mL intervals elsewhere on the titration curve.

7. Repeat steps 3 through 6 with each of the remaining solutions.

8. Determine the end point volume graphically and calculate the molarity of the cerium(IV) titrant.

9. Determine the concentration of iron in an unknown sample. The directions apply to the determination of iron in a commercial ferrous ammonium sulfate unknown, but may be modified for other types of unknowns. To the nearest 0.1 mg, weigh out ferrous ammonium sulfate unknown of an approximate weight and number of portions stipulated by the laboratory instructor.

10. Dissolve the sample in 10 mL of concentrated HCl plus the minimum amount of water required to complete dissolution.

11. Go through the titration procedure in steps 3 through 7 of this determination.

12. Calculate the percentage of iron in the sample.

25.17   Solvent Extraction of Iron(III) from an HCl Medium

One of the most well known solvent extraction procedures is that of iron(III) from aqueous HCl solution into an organic solvent. The iron(III) is thought to be extracted as the $HFeCl_4$ species. The organic solvents that work well are those that contain electron-donor oxygen atoms. Classically, diethyl ether has been used, but several esters, including butyl acetate, ethyl acetate, pentyl acetate, and isobutylacetate offer advantages, such as greater safety and lower temperature rise during extraction.

For this extraction to work, the iron must be in the +3 oxidation state. One of the most crucial experimental parameters is the concentration of HCl in the aqueous phase. It has been found that 6 M HCl is the optimum concentration.

1. Receive from the instructor, or prepare according to instructions, an iron(III) solution containing 6 to 10 g of iron per liter of 6 M HCl solution. Such solutions are conveniently and accurately prepared from hydrated iron(III) sulfate salt. With a pipet, take a 25 mL aliquot of the solution for extraction.

2. Place the 25 mL aliquot of iron solution in a 125 mL separatory funnel. Add 25 mL of ethyl acetate and extract by shaking for 3 to 4 minutes. Beware of excessive heat generated during the extraction. Pressure buildup in the separatory funnel, which occurs particularly during the early parts of the extraction, may be relieved by inverting the funnel and momentarily opening the stopcock.

3. Repeat the extraction with two more fresh 25 mL portions of ethyl acetate. Combine all three of the organic extracts.

4. Discard the extracted sample solution from the separatory funnel, rinse the funnel with water, and add to it the combined ethyl acetate extracts.

5. Add 25 mL of water to the funnel to back-extract (strip) the iron from the organic solvent (the reverse extraction occurs because the water does not have enough HCl needed to form $HFeCl_4$).

6. Repeat the stripping of iron from the organic solvent with a fresh 25 mL portion of water, and combine the aqueous solutions used for back-extraction.

7. Determine the iron content of the aqueous solutions by titration with dichromate or other means specified by the instructor.

25.18   Determination of Total Equivalents of Cation in Water

by Cation Exchange

The total number of equivalents of cation in a water sample is equal to the sum of the number of moles of each cation times its charge (for example, a mole of Na^+ contributes 1 equivalent, a mole of Ca^{2+} contributes 2 equivalents, and a mole of Al^{3+} contributes 3 equivalents). A strong acid cation exchanger in the H^+ ion form can be used to exchange cations for H^+ and the latter titrated with base to give the number of equivalents of cation in the sample. The reaction of the cation exchanger is shown below for Ca^{2+} ion:

$$Ca^{2+} + 2H^{+\,-}Cat(solid) \rightarrow 2H^+ + Ca^{2+}\{^-Cat(solid)\}_2 \tag{25.1}$$

Alkalinity, such as that from HCO_3^- ion or OH^- ion, is subtracted from the total equivalents of cation by this method. This is seen from the example of HCO_3^- alkalinity associated with calcium cation illustrated by the exchange reaction below:

$$Ca^{2+} + 2HCO_3^- + 2H^{+\,-}Cat(solid) \rightarrow 2H_2O + 2CO_2 + Ca^{2+}\{^-Cat(solid)\}_2 \tag{25.2}$$

This reaction shows uptake of the cation by the ion exchanger without any release of titratable H^+ ion.

1. Place the strong acid cation exchanger in the H^+ ion form in a 14- to 16-mm inside diameter glass chromatographic column to produce a column of ion exchanger beads 25 to 30 cm long. The cation exchanger will be provided as a pre-prepared water slurry; alternatively, you will be given instructions for properly conditioning the cation exchanger.

2. Rinse the column of cation exchanger with about 200 mL of distilled water over an approximately 20-minute time period. Do not allow the

level of the water to drop below the surface of the cation exchanger beads during and at the end of the rinsing procedure.

3. Pass a 50 mL aliquot of the analyte solution through the column over a 15-minute period and discard the effluent.

4. Pass two successive 100 mL portions of the sample over the column, allowing about 30 minutes for each, and collect these portions separately for analysis. Flush the column with 100 mL of distilled water and discard the water.

5. Titrate each of the 100 mL portions of the sample with 0.02 M standardized base and calculate the equivalents of cation in each.

Answers to Programmed Summaries

Chapter 1

1. qualitative analysis   **2.** quantitative analysis   **3.** to obtain a representative sample   **4.** sample
5. an analyte   **6.** to get the sample into solution   **7.** to remove interferences   **8.** nondestructive
9. wet chemical   **10.** chemical   **11.** classical   **12.** instrumental   **13.** physical   **14.** gravimetric
15. titrimetric   **16.** calibration curve   **17.** standard solutions   **18.** calibration   **19.** stability   **20.** purity
21. weighability   **22.** solubility   **23.** mass   **24.** liter   **25.** degree Celsius   **26.** kilo   **27.** centi
28. milli   **29.** mole   **30.** millimoles   **31.** organic chemistry   **32.** aromatic   **33.** moiety
34. functional groups   **35.** polymer   **36.** the calculation of the quantities of chemical species reacting with each
other and the quantities of the products of these reactions   **37.** molarity, M **38.** analytical concentration
39. equivalents   **40.** titer   **41.** parts per million   **42.** milligrams per liter
43. micrograms per liter   **44.** an equilibrium constant expression   **45.** equilibrium constant
46. thermodynamic equilibrium constant   **47.** $K_w = [H^+][OH^-]$ **48.** pH   **49.** $K_{sp} = [Ag^+]^2[CrO_4^{2-}]$
50. solubility product   **51.** complexation   **52.** $K_f = [CuNH_3^+]/[Cu^{2+}][NH_3]$ **53.** formation constant
54. an oxidation-reduction reaction   **55.** distribution coefficient or partition coefficient
56. a liquid immiscible with water and a solid ion-exchange resin

Chapter 2

1. mass   **2.** weight   **3.** knife-edge   **4.** rider   **5.** 0.1 mg, 0.0001 g   **6.** constant load   **7.** arrest or
release the beam and pan   **8.** add or remove weights   **9.** buoyancy   **10.** an auxiliary balance   **11.** top
loading   **12.** taring feature   **13.** those with inside-fitting lids and those with outside-fitting lids   **14.** weighing
by difference   **15.** lid, base, plate, and desiccant   **16.** P_4O_{10} **17.** drying to constant weight   **18.** filter
paper, glass frit, unglazed porcelain, and asbestos mat   **19.** ashless   **20.** decantation of the solution, washing
the precipitate, and removing residual particles with a rubber policeman   **21.** vacuum   **22.** 250° C   **23.** 250 to
1200° C   **24.** the pipet, the buret, and the volumetric flask   **25.** volumetric (transfer)   **26.** Mohr
27. syringe pipet   **28.** resistance to chemical attack, little tendency to freeze, and no requirement for lubrication
29. the bottom of the meniscus   **30.** parallax   **31.** buret-reading card   **32.** piece of black tape on a white card
33. deliver   **34.** contain   **35.** detergent solution and a brush   **36.** dichromate cleaning solution
37. sodium dichromate   **38.** sulfuric acid   **39.** calibrated   **40.** mass of water

Chapter 3

1. precision   **2.** determinate   **3.** absolute uncertainty   **4.** relative uncertainty   **5.** true value
6. first uncertain one   **7.** 4.94   **8.** 4.93   **9.** 4.94   **10.** 4.96   **11.** 370.2   **12.** 114   **13.** mantissa
14. three   **15.** five   **16.** exponential notation   **17.** $\bar{x}$ **18.** $(x_1 + x_2 + x_3 + x_4)/4$ **19.** μ **20.** $\bar{x} - \mu$
21. $[(\bar{x} - \mu)/\mu] \times 1000$ **22.** $x_3 - x_2$ **23.** the deviation of x_i from the mean d_i **24.** $\sum |d_i|/n$
25. $\sqrt{\sum d_i^2/(n-1)}$ and $\sqrt{\dfrac{\sum x_i^2 - (\sum x_i)^2/n}{n-1}}$ **26.** Gaussian distribution   **27.** 68, 95, and 99.7%

28. confidence level   **29.** $\bar{x} \pm ts/\sqrt{n}$ **30.** $\dfrac{\bar{x} - \bar{y}}{s}\sqrt{\dfrac{mn}{m + n}}$ **31.** different   **32.** Q test

33. $|x_o - x_n|/(x_h - x_l)$ **34.** $>Q_{crit}$ **35.** independent variable   **36.** dependent variable   **37.** the entire plot fills most of the available space   **38.** plot only a portion of the data full scale   **39.** best fit   **40.** method of least squares   **41.** $y = mx + b$

Chapter 4

1. a substance capable of donating a proton   **2.** a substance capable of accepting a proton   **3.** amphiprotic
4. conjugate base   **5.** conjugate acid   **6.** strong   **7.** leveling solvent   **8.** differentiating solvent
9. ion product   **10.** K_w **11.** $H_2O \rightleftarrows H^+ + OH^-$ **12.** $[HB^+][OH^-]/[B]$ **13.** K_b **14.** K_w/K_b
15. OH^- **16.** hydrolysis   **17.** buffer   **18.** pH   **19.** buffer capacity   **20.** exactly 1

Chapter 5

1. acid-base titration   **2.** equivalence point   **3.** end point   **4.** strong   **5.** titration curve   **6.** initial point, region from the initial point to the equivalence point, equivalence point, and region beyond the equivalence point
7. differential titration curve   **8.** second-derivative plot   **9.** higher   **10.** acid-base indicator
11. indicator pH range   **12.** ± 1 **13.** phthalein, sulfonphthalein, azo   **14.** ± 0.5 to 1   **15.** [(end point volume–equivalence point volume)/equivalence point volume] $\times 100$ **16.** initially, from the initial point to the equivalence point, at the equivalence point, and beyond the equivalence point   **17.** K_a **18.** greater than
19. less than   **20.** be separated by several orders of magnitude   **21.** be less than approximately 1×10^{-10}
22. two breaks

Chapter 6

1. acidity or basicity   **2.** solubility   **3.** acidity or basicity   **4.** self-dissociation   **5.** aprotic   **6.** inert
7. autoprotolysis   **8.** amphiprotic   **9.** $HSolv + HSolv \rightleftarrows H_2Solv^+ + Solv^-$ **10.** self-dissociation constant K_s
11. water, methanol, and ethanol   **12.** acidic solvents   **13.** basic solvents   **14.** boiling point, dielectric constant, and self-dissociation constant   **15.** K_a/K_s **16.** self-dissociation   **17.** basic   **18.** acidic
19. phenol   **20.** pyridine   **21.** leveling solvent   **22.** differentiating solvent   **23.** separate oppositely charged ions   **24.** ion pairs   **25.** decrease   **26.** formation of additional charged species
27. ethylenediamine   **28.** acetic acid   **29.** perchloric acid, $HClO_4$ **30.** a titration curve prepared with a glass electrode used to measure pH   **31.** methyl violet and crystal violet

Chapter 7

1. form a precipitate containing the analyte and weigh the precipitate or a modified form of it, form a precipitate containing the analyte and analyze it by nongravimetric means, remove interferences by precipitation
2. $AgCl$, Hg_2Cl_2, $PbCl_2$ **3.** Hg_2SO_4, $HgSO_4$, Ag_2SO_4, $PbSO_4$, $CaSO_4$, $SrSO_4$, $BaSO_4$ **4.** $4S^3$ **5.** $108S^5$
6. activities   **7.** activity coefficient   **8.** ionic strength   **9.** $\frac{1}{2}(m_1Z_1 + m_2Z_2 + m_3Z_3 + m_4Z_4 + \cdots)$

10. mean activity coefficient   **11.** individual activity coefficients   **12.** $-\log f_X = \dfrac{0.5085 Z_X \sqrt{\mu}}{1 + 3.281 \times 10^{-3} a_X \sqrt{\mu}}$

13. $25°C$ **14.** the charge of the ion   **15.** a_X **16.** diameter of the hydrated ion   **17.** picometers
18. the ion concentrations, and, therefore, the total ionic strength are very low for a slightly soluble salt equilibrated with pure water   **19.** higher   **20.** activity coefficients   **21.** lower   **22.** multiplied   **23.** complexing agents (ligands)   **24.** $Ag(NH_3)_2^+$ **25.** $[AgCl(aq)]$ **26.** $AgCl(aq)$ **27.** $Pb(OH)_3^-$

Chapter 8

1. formation of a precipitate, evolution of a volatile product, extraction, and electrodeposition   2. AgCl
3. $2Fe(OH)_3 \cdot xH_2O \xrightarrow{\text{heat, } 1000°C} Fe_2O_3(s) + (2x + 3)H_2O(g)$ 4. mother liquor   5. very low solubility, no contamination by impurities that cannot be removed, readily separable from mother liquor, possible to convert to material of well-defined chemical composition   6. colloidal particles and crystalline particles   7. $(Q - S)/S$, where S is the solubility of the precipitate at equilibrium in the precipitating medium and Q is its dissolved concentration in solution at a particular time and place in the medium   8. nucleation and particle growth
9. elevated temperature, dilute solutions, slow addition of precipitating reagent, gradual elevation of pH when the precipitating reagent is the conjugate base of a weak acid   10. coagulation or agglomeration   11. heating, stirring, and electrolyte   12. primary adsorption   13. counter-ion   14. electrical double layer
15. digestion   16. coprecipitation   17. peptization   18. volatile electrolyte   19. inclusion
20. occlusion   21. $(H_2N)_2CO + 3H_2O \rightarrow CO_2(g) + 2NH_4^+ + 2OH^-$ 22. drying, removal of volatile electrolytes, conversion to a weighable form, burning off filter paper   23. change in composition
24. $Fe^{3+} > Cu^{2+} > Fe^{2+} > Cd^{2+} > Ag^+$ 25. OH^- and S^{2-} 26. high molecular weight of precipitates and some degree of selectivity   27. dimethylglyoxime   28. nickel   29. 8-hydroxyquinolone
30. tetraphenylarsonium chloride   31. calcium chloride and magnesium perchlorate
32. a solid support material covered with Na_2CO_3 or Na_2CO_3 plus $NaOH$ 33. burning the sample in pure oxygen and collecting the CO_2 and H_2O, respectively   34. to estimate the oxygen content of metals
35. mass of product, conversion to moles of product, conversion to moles of analyte, conversion to mass of analyte

36. gravimetric factor   37. $\text{percent analyte} = \dfrac{(\text{mass of product})(\text{gravimetric factor})}{\text{mass of sample}} \times 100$

Chapter 9

1. solubility   2. excess analyte, equivalence point, and excess reagent   3. $2(4K_{sp})^{1/3}$ 4. midway on the titration curve break   5. I^- as $AgI(s)$ 6. "corner" at the top of the first break   7. initiation of AgCl precipitation
8. turbidity, formation of a colored precipitate, formation of a colored complex ion, adsorption of a dye onto the precipitate, and potentiometric methods   9. precipitate in the analyte solution upon addition of reagent
10. red Ag_2CrO_4 11. CrO_4^{2-} concentration below that required for Ag_2CrO_4 formation exactly at the equivalence point and the necessity of forming a finite amount of Ag_2CrO_4 before the end point is reached   12. an indicator blank
13. $FeSCN^{2+}$ 14. adsorption   15. adsorption of dye ions to charged colloidal precipitate particles

Chapter 10

1. ligand   2. central metal ion   3. complex ion, also called metal complex or simply complex
4. hexammineiron(II) ion   5. bond simultaneously in two or more places   6. pentadentate   7. stepwise
8. overall formation constant   9. K_3 and β_3 10. decreases   11. CN^-, SCN^-, Cl^-, Br^-, and I^-
12. 1:1   13. aminopolycarboxylic acids

14.
$$\begin{matrix} & H & O & \\ & | & \| & \\ - & C & - C & -OH \\ & | & & \\ & H & & \end{matrix}$$

15. ethylenediaminetetraacetic acid   16. EDTA   17. $\dfrac{K_{a1}K_{a2}K_{a3}}{[H^+]^3 + K_{a1}[H^+]^2 + K_{a1}K_{a2}[H^+] + K_{a1}K_{a2}K_{a3}}$

18. conditional formation constant   19. $K\delta\alpha_{Y^{4-}}$ 20. metallochromic indicator   21. Mg^{2+}
22. indicator transition range   23. back-titration   24. alkalimetric titration   25. water hardness

Chapter 11

1. $+2$, -2, and $+6$ **2.** gains   **3.** loses   **4.** $MnO_4^- + 8H^+ + 5e^- \rightleftarrows Mn^{2+} + 4H_2O$ **5.** electrochemical cell
6. liquid junction   **7.** salt bridge   **8.** hydrogen electrode   **9.** $Pt\,|\,H_2,\,H^+\,\|\,Fe^{3+},\,Fe^{2+}\,|\,Pt$
10. standard hydrogen electrode   **11.** exactly 0   **12.** reductions   **13.** standard electrode potential
14. exactly 1   **15.** Nernst equation   **16.** $MnO_4^- + 8H^+ + 5e^- \rightleftarrows Mn^{2+} + 4H_2O$ **17.** constant
18. exactly 1   **19.** concentration   **20.** calomel electrode   **21.** silver–silver chloride   **22.** E_M^0 and K_{sp} of MX
23. to the right   **24.** $nE^0/0.0591$ **25.** liquid junction potential, internal cell resistance, and overvoltage
26. indicator electrode   **27.** reference electrode

Chapter 12

1. potentiometric titration   **2.** analyte   **3.** Nernst equations or E^0's for both the analyte and titrant
4. $MnO_4^- + 5Fe^{2+} + 8H^+ \rightarrow Mn^{2+} + 5Fe^{3+} + 4H_2O$ **5.** $(3E_{Ox}^0 + E_{Red}^0)/4$ **6.** $[H^+]$ **7.** the concentrations of
the species involved in the titration reaction at the equivalence point   **8.** it has two S-shaped breaks
9. upward (sharply so at the two equivalence points)   **10.** the color of MnO_4^- ion, starch, and $FeSCN^{2+}$ complex ion
11. a soluble dye that undergoes reversible oxidation-reduction and has different colors in the oxidized and reduced
forms   **12.** 0.118 V   **13.** $KMnO_4$, cerium(IV), $K_2Cr_2O_7$, and I_2 (as I_3^-)   **14.** an auxiliary reductant
15. amalgamated zinc   **16.** silver-coated copper   **17.** chloride ion   **18.** peroxides, ammonium
peroxodisulfate, and sodium bismuthate   **19.** are readily oxidized by oxygen in air   **20.** MnO_2 **21.** oxalate
22. $2MnO_4^- + 5H_2C_2O_4 + 6H^+ \rightarrow 2Mn^{2+} + 10CO_2 + 8H_2O$ **23.** six   **24.** two Cr^{3+} ions   **25.** dichromate
26. iron(II)   **27.** iodine   **28.** I_3^- **29.** starch   **30.** elemental iodine   **31.** thiosulfate ion
32. dissolved oxygen in water   **33.** iron(II)

Chapter 13

1. potentiometry   **2.** $E = E_a + (2.303RT/zF)\log a_X z$ **3.** an indicator electrode   **4.** first-order, metal-metal
cation; second-order, metal-metal complex in solution; metal amalgam, inert electrode; solid membrane; liquid
membrane   **5.** EDTA titrations of metal ions   **6.** hydrated   **7.** H^+ ions   **8.** internal reference electrode
9. internal reference electrode potential, junction potential, asymmetry potential, boundary potential, and external
reference electrode potential   **10.** pH   **11.** alkaline error   **12.** selectivity constant   **13.** calibration
14. Na^+, Li^+, K^+, NH_4^+, Rb^+, Cs^+, and Ag^+ **15.** no electrode is totally specific—that is, without interferences—
for any particular ion   **16.** direct electrical connection to the inside surface of the solid membrane   **17.** silver
18. organic-soluble ion exchanger bound to the analyte ion, dissolved in an organic solvent, and soaked into a filter
disk composing the membrane   **19.** selectivity constant   **20.** the ratio of the charges of analyte and interfering ions
21. gas-permeable membrane   **22.** the solubility of the LaF_3 membrane   **23.** 4%   **24.** standard addition
25. calomel and silver–silver chloride electrodes   **26.** a glass and a silver–silver chloride electrode
27. acid-base, precipitation, oxidation-reduction, and complexation   **28.** Gran plots

Chapter 14

1. microelectrodes   **2.** electroactive species   **3.** current-voltage   **4.** current   **5.** potential versus a reference
electrode   **6.** working or indicating   **7.** auxiliary electrode   **8.** working   **9.** reference
10. voltammogram   **11.** polarogram   **12.** conduct electricity well   **13.** electrostatic forces
14. residual current   **15.** limiting current   **16.** the rate at which analyte ions can diffuse to the electrode surface
17. $i = \frac{1}{2}i_d$ **18.** reversible   **19.** the half-wave potential   **20.** the diffusion current   **21.** concentration of
analyte   **22.** polarographic maxima   **23.** pulse of an additional 20 to 100 mV   **24.** peak   **25.** three
26. several minutes   **27.** rapid removal of analyte   **28.** $1 \times 10^{-9}\ M$ **29.** amperometric titration
30. potential   **31.** current   **32.** Karl Fischer   **33.** water

Chapter 15

1. quantity of electricity   2. weight   3. an electrolysis   4. weight and electrical charge   5. metals
6. complexed with ligands   7. current density   8. hydrogen gas   9. cathode depolarizer   10. lead
11. lead dioxide, PbO_2 12. metal impurities from solution   13. potential of a working electrode $vs.$ a
reference electrode   14. total charge C 15. current-time plot   16. mediator   17. secondary analytical
reaction   18. turning the current on and off   19. isolated by a glass frit from the rest of the solution
20. Ag^+ ion   21. $Ag \rightarrow Ag^+ + e^-$ 22. these are not needed   23. 100% current efficiency

Chapter 16

1. phases   2. multistage operation   3. thin layer   4. adsorbent surface of a solid and a layer of liquid coated on
a solid   5. stationary   6. mobile   7. coprecipitation   8. inclusion and occlusion   9. peptization
10. oxides and hydroxides   11. sulfide   12. organic   13. ion pairs   14. organic extractants
15. $K = [A_{org}]/[A_{aq}]$ 16. continuous   17. countercurrent fractionation   18. organic extractants
19. $(n\text{-}C_4H_9)_4N^+$ and $(C_6H_5)_4As^+$ 20. tri-n-butyl phosphate   21. nickel   22. separate an extracted substance
from the solvent into which it was extracted   23. remove impurities from the organic solvent back into aqueous
solution   24. ion exchange   25. chelating   26. ion exchange capacity

Chapter 17

1. toward a stationary phase and a mobile phase that moves relative to the stationary phase   2. adsorption, partition,
ion exchange, molecular exclusion, and affinity   3. paper and thin-layer   4. the walls of a narrow-bore column
5. capillary chromatography   6. eluent   7. elution   8. eluate   9. peak   10. retention time
11. retention volume   12. zone broadening   13. plate theory   14. theoretical plates   15. the height
equivalent of a theoretical plate   16. rate theory   17. dynamic   18. longitudinal   19. eddy
20. van Deemter   21. resolution   22. 16   23. $\sqrt{N}$ 24. peak tailing   25. column overload
26. mass spectrometer   27. retention times   28. detector   29. peak area   30. area normalization

Chapter 18

1. liquid and gas   2. column and planar   3. a flat surface, such as a microscope slide   4. the solvent front
5. solute zone or "spot"   6. sulfuric acid, iodine, ninhydrin, and fluorescent materials   7. silica and alumina
8. binder   9. retardation factor, R_f 10. water   11. paraffin oil   12. silicone oil   13. reverse phase
14. two-dimensional paper chromatography   15. silica or alumina   16. adsorb to the solid phase and displace
solute from it   17. ion-exchanging functional group   18. two different ions   19. $-SO_3^-$ 20. $-NR_4^+$
21. sizes   22. pores of a well-defined size   23. void volume, V_0 24. structural group   25. much smaller
26. pressures   27. isocratic   28. gradient elution   29. reverse phase chromatography
30. ion chromatography

Chapter 19

1. gas-liquid chromatography   2. gas-solid chromatography   3. gas-liquid chromatography   4. injector,
column, detector, and amplifier-recorder   5. argon, helium, hydrogen, and nitrogen   6. measure carrier gas flow
rate   7. packed and capillary columns   8. liquid phase   9. solid support   10. gas-sampling valve and
septum   11. capillary   12. temperature programming   13. flame ionization detector   14. electron-capture
detector   15. the time required for an unretained peak, usually air, to go through the column   16. add to the
sample to see if the peak height of a suspected compound, and only that compound, increases   17. an internal
standard compound added in known quantity to the sample

Chapter 20

1. electromagnetic radiation   **2.** wavelength   **3.** frequency   **4.** refractive index, n **5.** photons
6. energy   **7.** Planck's constant   **8.** excited state   **9.** electronically   **10.** singlet   **11.** triplet
12. fluorescence   **13.** $T = P/P_0$ **14.** $100/\% \ T$ **15.** $A = abC$ **16.** absorption spectrum   **17.** peak
18. source, monochromator, sample holder, and detector   **19.** tungsten lamp   **20.** diffraction grating
21. wavelength range   **22.** nominal wavelength   **23.** bandwidth   **24.** cuvettes   **25.** emits electrons
26. electrical current   **27.** photomultiplier   **28.** dynodes   **29.** double-beam   **30.** 0.2 to 0.8
31. several orders of magnitude   **32.** gaseous analyte atoms   **33.** detailed multiple-peaked
34. several sharp lines of definite wavelengths   **35.** Beer's law

Chapter 21

1. several species in equilibrium in solution   **2.** MnO_4^- ion   **3.** CuY^{2-} **4.** 260 nm   **5.** neutrality
6. low polarity   **7.** acetylacetone   **8.** react or interact with each other   **9.** additive   **10.** molar
absorptivities   **11.** absorptivity value   **12.** analyte, titrant, or product   **13.** equilibrium constants
14. mole ratio plot   **15.** curved   **16.** sharp break   **17.** luminescence   **18.** chemiluminescence   **19.** NO_2^*
20. NO and O_3 **21.** fluorescence   **22.** intensity of fluorescence as a function of wavelength of exciting radiation
23. emission intensity as a function of wavelength at a constant excitation wavelength

Chapter 22

1. emission   **2.** absorption   **3.** elemental atoms   **4.** gas   **5.** resonance energy, E **6.** excited
7. atomic fluorescence   **8.** bandwidths   **9.** gas   **10.** very specific   **11.** hollow cathode
12. the same element   **13.** atomizer   **14.** flame   **15.** flameless   **16.** graphite   **17.** passage of an
electrical current through it   **18.** reducing the mercury to the elemental state with a solution of stannous chloride
19. sweeping it from solution in a stream of air   **20.** an electrical discharge between graphite or metal electrodes
21. qualitative determination   **22.** flame atomic emission spectrophotometry   **23.** flame temperature
24. relative proportions of fuel and oxidant   **25.** maximum formation of free atoms in the flame
26. maximum tendency of the atom to reach an excited state   **27.** ions in the flame   **28.** an ionized gas, or plasma
29. radiofrequency signal   **30.** 7000 K   **31.** relatively low detection limits, applicability to a large number of
elements, absence of interfering reactions multielement capability, and linearity of response over several orders of
magnitude of analyte concentration   **32.** Doppler broadening   **33.** rapid motion   **34.** pressure broadening
35. collisions   **36.** cooler outer   **37.** hotter inner   **38.** negative deviation from linearity at higher analyte
concentrations as shown in Figure 22.9   **39.** self-reversal   **40.** spectral interference   **41.** matrix

Chapter 23

1. radioactive nuclei   **2.** radionuclides   **3.** neutrons   **4.** tracers   **5.** radioactive decay   **6.** 4   **7.** +2
8. virtually zero   **9.** -1 **10.** beta   **11.** gamma radiation   **12.** takes an electron from the K or L shell
of an atom   **13.** X-ray   **14.** mass number   **15.** atomic number   **16.** in an excited state
17. activity   **18.** the number of radioactive nuclei, N **19.** $\log(A_0/A) = (0.301/t_{1/2})t$ **20.** half-life
21. half of the radioactive nuclei to decay   **22.** model of pathway   **23.** radiation
24. stability   **25.** ionization chamber   **26.** ions or ion pairs   **27.** electrical pulse   **28.** flashes of light
29. semiconductor detectors   **30.** higher precision   **31.** $\sqrt{N}$ **32.** radionuclides   **33.** neutrons
34. radioactivity   **35.** a nucleus absorbs a neutron and immediately emits a gamma ray, usually producing
a radionuclide   **36.** flux   **37.** reaction cross section   **38.** radiolabeled   **39.** tagging
40. isotope dilution analysis   **41.** radioimmunoassay

Chapter 24

1. representative   2. any analytes are not adversely affected   3. directly on a material to be analyzed
4. coning and quartering   5. rolling and quartering   6. animal feed in a bag   7. coprecipitation
8. water of constitution, water of crystallization, adsorbed water, capillary or pore water, occluded water, and water in solid solution   9. boiling point   10. perchloric acid   11. hydrofluoric acid
12. digestion or wet ashing   13. flux   14. dry ashing   15. low-temperature ashing
16. carbon, hydrogen, the halogens, and sulfur   17. nitrogen   18. combustion tubes

Answers to Selected Questions

Chapter 1

1. A particular constituent of a sample is said to be *determined*.
4. A-4, B-3, C-1, D-2
5. Gravimetric (mass) and volumetric or titrimetric (volume)
7.

2-methylpropane   2-nitronaphthalene

10. $K = \dfrac{[Cr^{3+}]^2\,[Fe^{3+}]^6}{[Cr_2O_7^{2-}][Fe^{2+}]^6\,[H^+]^{14}}$

Chapter 2

3. Because the balance has a constant load
6. Knife edges
9. Setting the balance to zero with the empty weighing container on the pan
11. To dry the atmosphere in the desiccator

Chapter 3

2. Expressing the volume as 37.3 mL implies that the volume was read to only the nearest 0.1 mL, whereas volumes on such a buret can and should be read to 0.01 mL
4. $\frac{1}{5}X$ to $2X$
5. (A) 3.5031, (B) -3.12, (C) 5.473, (D) -2.21896
8. (A) $w = x_{highest} - x_{lowest}$, (B) $\dfrac{w}{\bar{x}}$, (C) $d_i = x_i - \bar{x}$, (D) $\bar{d} = \dfrac{\Sigma|d_i|}{n}$
11. Gaussian distribution
13. Both give wider confidence limits implying less knowledge of the central value
14. To decide if a discordant result should be discarded
17. (A) High precision
18. (B) Agreement with the true value of the mean of a large number of analyses of a sample of known concentration

Chapter 4

2. An amphiprotic solvent
4. $HX + HX \rightleftharpoons H_2X^+ + X^-$, $K = [H_2X^+][X^-]$
5. (A), s; (B), w; (C), s; (D), w; (E), w; (F), s.
7. $K_a K_b = K_w$
10. Buffer capacity $= \dfrac{\Delta \text{ equivalents acid or base added per L}}{|\Delta pH|}$
12. (B) NH_4^+ ion acts as a weak acid

Chapter 5

1. A titrant is a standard solution normally added by a buret to react with analyte in a titrimetric determination.
3. Exactly half-way up the steeply rising portion of the titration curve (pH 7)
5. For the weak acid the end point pH is greater than 7, whereas for the strong acid the end point pH is 7.
7. The primary factors are differences in the relative intensities of the indicator colors and the ability of the human eye to respond to different colors.
9. The break for the acid with K_a of 1×10^{-8} is of marginal magnitude. Titration of any weaker acid would be virtually impossible.
11. The buffer regions are the relatively flat parts of the plot prior to the first break or between breaks, where the pH changes only by small amounts with added acid or base.
15. (A)
16. (C)

Chapter 6

1. $H_2S + H_2S \rightleftharpoons H_3S^+ + HS^-$
3. Ethylenediamine would function as a basic solvent because of the electron-donor pairs on each of the two N atoms.
4. Sodium ethoxide, $Na^+ C_2H_5O^-$
7. There is relatively little tendency for the reaction $NH_4^+ + C_2H_5OH \rightleftharpoons NH_3 + C_2H_5OH_2^+$ to proceed to the right, so that this reaction does not compete strongly with the reaction of NH_4^+ with a base B added to the solvent, $NH_4^+ + B \rightleftharpoons NH_3 + HB^+$.
9. Acetic acid, HOAc, is an acidic solvent. Therefore, it has a relatively strong proton-donor property and tends to react with all bases above a certain strength according to the reaction $B + HOAc \rightleftharpoons HB^+ + OAc^-$, where OAc^- is the strongest base that can exist in acetic acid.
11. (E)
13. (B)

Chapter 7

1. A precipitate is formed by a stoichiometric reaction from the analyte, and the precipitate is either weighed directly or in a modified (such as ignited) form.
3. $K_{sp} = (mS)^m (xS)^x$
5. When a complex ion, such as $AgCl_2^-$ in the case of $AgCl(s)$ or $Pb(OH)_3^-$ in the case of $Pb(OH)_2$ (s), is formed
7. Because CN^- ion readily forms its conjugate acid form, HCN, whereas Cl^- ion does not

Chapter 8

1. Precipitation, evolution of a volatile substance, extraction, or electrodeposition
3. Nucleation increases geometrically and particle growth linearly with increasing degree of supersaturation.
5. A(2), B(1), C(3), D(4)
7. The initial small decrease in mass is due to loss of water in drying; the three abrupt decreases in mass indicate changes in composition, such as loss of strongly bound water of hydration or loss of CO_2.
9. A-4, B-3, C-1, D-2
11. Conversion to moles of product, conversion to moles of analyte, and conversion to mass of analyte
13. (C)
15. (B)

Chapter 9

1. A-3, B-4, C-1, D-2
4. A-6, B-3, C-1, D-5
6. The blank is run by adding Ag^+ reagent to a suspension of solid $CaCO_3$ in a dilute CrO_4^{2-} medium to measure the volume of titrant that must be added after the equivalence point to initiate precipitation of Ag_2CrO_4 and to produce a visible quantity of that brick-red precipitate.
7. A-2, B-4, C-1, D-3
10. (A)

Chapter 10

1. Solvation is the binding of electron-donor solvent molecules to a solute metal ion; hydration is the special case when the solvent is water. These phenomena are complexation in the sense that the solvent molecules bind to a metal ion by donor electron pairs.
3. The maximum number of electron-pair donor groups that the metal ion can accommodate
5. (A) Tricyanocopper(II) ion, (B) dihydroxiron(III) ion, (C) hexamminemanganese(II) sulfate, (D) zinc hydroxide
7. Chelation is a special case of complexation in which the ligand bonds to the central metal ion in at least two sites.
10. High pH deprotonates ligands that are conjugate bases of weak acid (which favors complexation) and tends to precipitate metal ions (which competes with complexation).
12. Aminopolycarboxylic acids containing acetic acid groups bonded to N
13. Five rings of five atoms each
16. To keep a metal ion in solution during titration with a chelating titrant
18. A dye that is an organic complexing agent and that has distinctly different colors when complexed and when not complexed with a metal ion
20. (C)

Chapter 11

1. (A) +6, (B) +2, (C) +2, (D) +5, (E) +2, (F) +5, (G) +7, (H) +2/3
3. The number of electrons gained by the oxidizing agent species must equal the number of electrons lost by the reducing agent species.
6. (A) A galvanic or voltaic cell, (B) an electrolytic cell

8. (A) By listing the metal electrode and its associated solution on the left, (B) by listing the metal electrode and its associated solution on the right, (C) by a single vertical line, (D) by a double vertical line

10. They are the same, including sign.

13. (A) Concentrations, (B) pressure in atmospheres, (C) exactly 1, (D) exactly 1

16. $Ag_2C_2O_4(s) + 2e^- \rightleftarrows 2Ag(s) + C_2O_4^{2-}$ E^0

$$E = E^0 - \frac{0.0591}{2} \log[C_2O_4^{2-}]$$

18. $E_{cplx}^0 = E^0 - 0.0591 \log \beta_2$

19. Oxidizing agents, $Cr^{3+} < Ti^{3+} < Fe^{3+} < Br_2 < Cl_2$; reducing agents, $Cl^- < Br^- < Fe^{2+} < Ti^{2+} < Cr^{2+}$

21. That it goes to the right

23. $E = E^0 - \dfrac{0.0591}{2} \log \dfrac{[Br^-]^2[Ti^{3+}]}{[Ti^{2+}]^2}$

25. $\log K = \dfrac{nE^0}{0.0591}$

27. Different mobilities of positive and negative ions

30. To indicate species concentrations or concentration ratios

32. The filling solution is saturated with KCl.

Chapter 12

1. The value of E in the solution is monitored by a Pt electrode *vs.* a reference electrode, such as the SCE. The data may be displayed as a plot of E *vs.* volume of titrant. At the equivalence point E increases sharply for an oxidizing titrant, and decreases sharply for a reducing titrant.

2. (A) $M^{3+} + Z^+ \rightarrow M^{2+} + Z^{2+}$, (B) the initial value of E cannot be calculated exactly because the initial value of Z^{2+} concentration is not known, (C) the value of E half-way to the equivalence point is equal to E_Z^0, (D) the value of E at the equivalence point is $\dfrac{E_M^0 + E_Z^0}{2}$, in a two-fold excess of M^{3+} titrant $E = E_M^0$.

5. Increasing $[H^+]$ should increase the oxidizing strength of $Cr_2O_7^{2-}$ according to Le Chatelier's principle.

7. A-2, B-4, C-1, D-3

9. To reduce an analyte species to a single, known oxidation state

11. The reaction is catalyzed by the Mn^{2+} product, little of which is present initially.

14. I_2 is only slightly soluble in water. The I_3^- complex ion is quite soluble.

17. Fading of the red-brown color of I_3^- to the light yellow of a very dilute solution of the complex.

Chapter 13

2. A-2, B-4, C-1, D-3

3. (A) An indicator electrode, (B) a reference electrode

5. The electrode may be employed as a dropping amalgam electrode with a continually renewed surface.

7. To serve as a selective site for the binding of ions to which the electrode gives a Nernstian potential response

9. At the interface of the external reference electrode with the analyte solution

11. It is a hygroscopic layer containing chemically bound water molecules and is capable of binding H^+ ion by cation exchange.

12. A-2, B-4, C-1, D-3

15. Equilibrating the solid with pure water and measuring the pH of the water
17. Liquid membrane electrode for calcium and solid membrane electrode for fluoride
19. It should contain both Cl^- and ClO_4^- ions, such as a solution of $NaCl$ and $NaClO_4$.
21. The key component of a gas-selective electrode is a gas-permeable membrane selective for analyte gases that allows the gases to penetrate to an indicator electrode, such as a glass electrode.
23. In direct potentiometry, log of concentration is a function of E, so that small variations in E result in relatively large variations in measured concentration.

Chapter 14

1. A-3, B-1, C-4, D-2
3. Because of the discontinuities of the falling mercury drops
6. E^0 applies to reduction to the solid metal, E_A^0 applies to reduction to the metal in an amalgam (solution in mercury metal).
8. $[Cd^{2+}]_0 = 0$ when $i = i_d$
10. $E_{1/2}$ for qualitative analysis and i_d for quantitative analysis
11. To eliminate polarographic maxima
13. Differential pulsed polarography has about 3 orders of magnitude lower detection limit than DC polarography and is capable of resolving waves with half-wave potentials that are much closer.
15. Peak-shaped
18. Either the analyte or titrant, or both, must produce a reversible voltammogram so that if both oxidized and reduced forms are present, there is not an inflection point between the anodic and cathodic sections of the voltammogram.
20. A constant small current is maintained between the electrodes, and the voltage necessary to maintain the current is measured. With the appearance or disappearance of a reversible oxidation-reduction couple at the end point, the potential either decreases or increases markedly.

Chapter 15

3. Lead as PbO_2
4. Deposition at low current density and with stirring from a complexed metal at an empirically determined optimum temperature
6. The high overpotential for the evolution of H_2 at a mercury surface prevents interference from the half-reaction $2H_2O + 2e^- \rightarrow 2OH^- + H_2(g)$.
8. Mt^+ could be oxidized to Mt^{2+} at the anode with Me^+ as the mediator species. For the half-reactions
$$Mt^{2+} + e^- \rightleftharpoons Mt^+ \quad E_{Mt}^0$$
$$Me^{3+} + 2e^- \rightleftharpoons Me^+ \quad E_{Me}^0$$
E_{Me}^0 must be significantly greater than E_{Mt}^0.
10. It is a reaction with analyte, at some distance from the working electrode, of titrant species generated at the working electrode.
12. The faraday, which is 96,485 coulombs/mole of electrons

Chapter 16

1. Chromatography and mass spectrometry
3. A-3, B-5, C-1, D-4, E-2
5. Inclusion or occlusion of impurities, coprecipitation, peptization of colloidal precipitates
7. Normally only neutral species are extracted into organic solvents.

9. Five extractions with $\frac{1}{5} V_{org}$ each
11. A fresh batch of each of the solvents involved
14. $(nC_4H_9)_4N^+$ and $(C_6H_5)_4As^+$
16. They often function as colorimetric reagents enabling determination of the analyte by absorption of light.
17. Buildup of pressure from organic vapor, emulsion formation, entraining water droplets in the organic phase
21. When a particular ion strongly predominates in both the liquid and solid phases

Chapter 17

1. Plant pigments
3. A-3, B-1, C-5, D-4, E-2
4. Affinity chromatography
5. A plot of quantity or concentration of solute emerging from a column as a function of time or eluent volume
8. Bands
10. The partition coefficient may vary with solute concentration, equilibrium of solute partition between mobile and stationary phases may not be as rapid as assumed, longitudinal diffusion is not dealt with, and the dimensions of the two phases are not considered.
12. The height equivalent to a theoretical plate
13. For best chromatographic separations the particles of solid phase should be quite small, but this requires very high pressures.
17. Its ability to separate solutes

Chapter 18

1. Planar and column
3. The solvent front is the leading edge of the solvent; a solute zone is a "spot" to which a particular solute has migrated.
5. By hydrogen bonding
6. Its ability to retain solute
9. By replacing the water in the paper by paraffin oil or silicone oil
11. Separation in one direction with a particular solvent followed by separation at a right angle by a solvent with appreciably different separation properties
14. Systematic change of solvent from one of low eluent strength to one of high eluent strength to elute the more strongly retained peaks sooner and reduce total separation time.
16. High-molecular-weight solutes
18. The rare earth ions
20. To determine void volume in molecular exclusion chromatography

Chapter 19

2. The most common criterion is selectivity, which is the ability to separate similar compounds.
4. To remove impurities from carrier gas
6. As solid support materials in packed columns
7. Low volatility, chemically unreactive, and proper solvent characteristics for solutes to be separated
10. Increasing column temperature decreases the resolution, but results in faster separations.
13. A-2, B-2,4, C-1, D-3
14. The electron-capture detector
17. When the peaks are very sharp

Chapter 20

1. Electromagnetic radiation has a dual nature because it behaves as both particles and waves.
3. $Ca + heat \rightarrow Ca^*$ (where Ca^* is an electronically excited calcium atom)
   $Ca^* \rightarrow Ca + h\nu$
6. The absorptivity a (which may vary with wavelength) and the path length b must remain the same.
8. The glow that ceases immediately illustrates fluorescence, whereas that which persists is phosphorescence.
11. Light photons hitting a light-sensitive cathode eject electrons that are attracted to the anode, thus resulting in a small electrical current proportional to light intensity.
13. Because infrared spectra normally have a number of peaks characteristic of bonds and groups of atoms in a compound

Chapter 21

1. Detection limits down to about 1×10^{-7} M can be achieved with accuracies of 0.1 to 0.3% achievable with special care.
3. Complexation or chelation of the metal
5. An isosbestic point is commonly encountered when a series of spectra are taken for solutions having the same analytical concentration but under different conditions (e.g., of pH) such that the analyte is convertible between two forms having different spectra in a 1/1 ratio.
7. A plot of absorbance $vs.$ ratio of ligand to metal
8. Chemiluminescence is excited by a chemical reaction, fluorescence by a photon.
11. Because the formulas of chromium(VI) species change with concentration (and with pH)
12. Persulfate is used to oxidize manganese to the highly colored permanganate form.
14. $A_{total} = A_X + A_Y + A_Z = a_X C_X + a_Y C_Y + a_Z C_Z$ (where $b = 1$ cm)
16. Omission of a dilution correction for a relatively dilute titrant would result in a curved plot and an erroneous extrapolation to the end point; the omission would have a negligible effect for a more concentrated titrant.
19. Since fluorescence (ideally only from the analyte) is the only source of light at the wavelength measured, it is like measuring the fluorescence light against a dark background when a monochromator is used to eliminate interfering light of other wavelengths. Furthermore, fluorescence is measured at right angles to the exciting radiation.

Chapter 22

1. A-3, B-4, C-1, D-2
3. A large number of elements can be determined with detection limits in the parts per billion range in favorable cases. In general, atomic spectroscopy is rather simple in application and is amenable to routine and automated analysis.
5. A device that converts atoms of analyte in the form of ions or molecules in solution to gas-phase analyte atoms
7. The graphite furnace has detection limits as much as 1000 times lower for some important metals.
9. (A) Continuous; (B) band; (C) line
10. Calcium, lithium, potassium, and sodium
12. By using a fuel-rich mixture for the flame
15. Self-reversal

Chapter 23

1. It has the same mass number, but a nuclear charge one unit higher than that of the parent radionuclide. Therefore, it is an isotope of the next element higher in the periodic table.
3. The gamma line from a particular nuclear transition is very sharp and discrete, enabling excellent qualitative analysis.
5. The parent and intermediate radionuclides and their half-lives, the stable radionuclide product, the particles and gamma rays given off, and their energies and percentage of the decay scheme (if less than 100%)
7. Decay is a random event, and the precision is increased by taking a larger total number of counts.
9. Sufficient abundance of that isotope and an adequate cross section for the absorption of neutrons by the isotope
11. It enables determining the amount of an analyte without a quantitative separation of the entire quantity of the analyte.

14. Activity applies to the total sample, whereas specific activity is per unit quantity of the sample.
18. Instrumental neutron activation analysis does not involve chemical separations.

Chapter 24

2. Information about precision and reliability and any cautions or constraints pertaining to the result
4. Biological decomposition of biological samples, loss of water from hygroscopic materials, precipitation of metals from water samples, loss of volatile analytes from water samples
6. Water may simply be present as readily vaporized surface moisture or as strongly held water of constitution, which is released only by chemical decomposition of the sample material. There are varying degrees of water retention between these two extremes.

Answers to
Selected Problems

Chapter 1

1. $3 \times 10^{-8}\%$
4. $[Cl^-] = 2.5 \times 10^{-4}\ M$
7. 4.09 mmol
10. 2.693 g $BaSO_4$
13. $2.78 \times 10^{-4}\ M$
16. $C_{HAC} = 2.20 \times 10^{-4}\ M$, $[HAC] = 1.66 \times 10^{-4}\ M$; $[H^+] = [Ac^-] = 5.40 \times 10^{-5}\ M$
19. 18.00 mg Na_2CO_3/mL HCl
21. 2.22 mol/L
24. $[NO_3^-] = 1.16 \times 10^{-4}$ mol/L
27. $[H^+] = 3.09 \times 10^{-13}\ M$, pH = 12.5
31. $3.91 \times 10^{-13}\ M$
34. $[Fe^{3+}] = 1.85 \times 10^{-4}\ M$
37. $K_d = 121$

Chapter 2

1. $M_c = 1.3289$ g
5. The corrected mass of the water is 24.9297 g; the true volume of water is 25.00 mL.
9. 10 mL, +0.04; 20 mL, +0.01; 30 mL, −0.14; 40 mL, −0.01; 50 mL, −0.01
11. $V_c = 25.10$ mL
12. 0.43%

Chapter 3

1. 303.90
4. 1.006×10^3
5. 8.287
7. 21.46
8. 0.15786
10. $\bar{x} = 5.11\%$; absolute error = −0.09%; relative error = 17 ppt; $w = 0.15\%$; $d_1 = 0.07\%$, $d_2 = -0.08\%$, $d_3 = 0.00\%$, $d_4 = -0.04\%$, $d_5 = 0.04\%$, $\bar{d} = 0.046\%$
13. $s = 0.060$
17. 87.4 ± 1.0 mg/L at the 90% confidence level

19. There is a 95% probability that the two means are the same.
21. The discordant value may be retained at the 90% confidence level.

Chapter 4

1. pH = 7.16
4. pH = 4.92
5. pH = 4.86
8. $[H^+] = 8.74 \times 10^{-3}\ M$, pH = 2.059
10. $[H^+] = 9.53 \times 10^{-12}$, pH = 11.021
11. 0.0340 moles of CO_2
13. 0.0576 mol $\times L^{-1}$
15. $[H^+] = 2.11 \times 10^{-8}\ M$, pH = 7.68
17. (A)

Chapter 5

1. −0.65%
3. A pH range of 1.18 from $pK_a - 0.70$ to $pK_a + 0.48$
5. Volume NaOH titrant with corresponding pH in parentheses immediately following: 0 (1.00), 10 (1.48), 19 (2.59), 20 (4.05), 21 (4.72), 30 (6.80), 39 (8.08), 40 (9.66), 41 (11.22).
10. (E)
11. (B)
12. (E)

Chapter 6

1. 2.3%
3. Neutral when $[CH_3OH_2^+] = [CH_3O^-] = (K_s)^{1/2}$

Chapter 7

1. $K_{sp} = 3.5 \times 10^{-11}$
4. $S = 7.57 \times 10^{-7}\ M$
5. $S = 1.33 \times 10^{-8}\ M$

10. $S = 5.62 \times 10^{-9} M$

13. An equation that can be derived and solved for S is
$S^2 - 9.76 \times 10^{-7} S^{1/2} - 4.47 \times 10^{-9} = 0$. Solving by
Newton's approximation or other methods gives
$S = 1.24 \times 10^{-4} M$.

15. 0.578

17. $S = 7.99 \times 10^{-2} M$

19. $S = 5.35 \times 10^{-7} M$

22. (B)

Chapter 8

1. 0.263 g H_2O, 1.287 g CO_2, 0.820 g CaO

3. 68.86% C, 4.95% H

6. 1.24% U_3O_8

9. 16.88% K

11. 19.3% Fe_2O_3

Chapter 9

1. 1.24 mmol I^-

4. 42.0% KI

5. 28.72% NaCl

7. $[CrO_4^{2-}] = 2.5 M$

10. $AgCl(s) + SCN^- \rightleftharpoons AgSCN\ (s) + Cl^-$
The value of K is 165.

11. $[Cl^-] = 1.00 \times 10^{-3} M$, $[SCN^-] = 6.1 \times 10^{-6} M$

12. $Ag_2CrO_4\ (s) + 2Cl^- \rightleftharpoons AgCl(s) + CrO_4^{2-}$,

$$K = \frac{[CrO_4^{2-}]}{[Cl^-]^2} = 3.89 \times 10^7,$$

$[CrO_4^{2-}] = 5.00 \times 10^{-4} M$, $[Cl^-] = 3.59 \times 10^{-6} M$

15. (B)

Chapter 10

1. (A) 1.6×10^{13}, (B) 8.1, (C) 1.3×10^{15}

3. $K' = 1.29 \times 10^8$

5. $[Ag^+] = 6.3 \times 10^{-10} M$

6. $\alpha_Y4- = 0.24$

7. $K' = K\alpha_Y4- = 1.2 \times 10^8$

8. $\delta = 2.51 \times 10^{-5}$

11. 7.64% Ni

13. 870 mg/L $CaCO_3$

Chapter 11

1. The half-reaction is
$Br^- + 3H_2O \rightleftharpoons BrO_3^- - 6H^+ + 6e^-$ and $E = 1.38$ V.

4. $[Cl^-] = 2.9\ M$ (The actual value is slightly over

4 M; the discrepancy comes from approximating
activities with concentrations.)

7. $K_{sp} = 5.16 \times 10^{-13}$

10. $E = -0.027$ V vs. SHE

11. $E = -0.159$ V vs. SHE

13. $E = 0.91$ V

16. $K = 3.6 \times 10^{21}$

17. $K = 1.6 \times 10^9$

Chapter 12

1. $E^0 = 1.58$ V vs. SHE

2. $E = 0.99$ V vs. SHE

4. $E_{eq} = 1.36$ V vs. SHE

6. The indicator transition range is 0.070 V.

7. Percent unoxidized = 1.8×10^{-4}%

8. 10.0 mg/L of dissolved oxygen

12. 322 mg O_2/L

Chapter 13

1. (A) $E = 0.516$ V, (B) $E = 0.387$ V, (C) $E = 0.270$ V

2. $E = 1.91$ V

4. $E_a = -0.018$ V

6. $\Delta E/\Delta pH = -0.0560$ V/pH unit compared to the
theoretical value of -0.0591 V/pH unit

8. $4.6 \times 10^{-6} M$

10. Percent relative concentration error = 3%

11. Error in $E = 2.4$ mV

Chapter 14

2. i_d is 12 μA and $E_{1/2}$ is -0.413 V

4. $[Cd^{2+}] = 6.6 \times 10^{-4} M$

6. Moles M(Hg) = 6×10^{-10} moles

Chapter 15

1. 0.85 milliequivalents, 0.42 millimoles, and
8.5×10^{-3} mmol/mL

3. The value of $E_{Me} - E_{Mt}$ must be at least 0.030 V.

4. Current density = 0.123 A/cm^2

7. Millimoles of allyl alcohol = 0.880 mmol

Chapter 16

2. 49.4 mL

4. Wt. sample = 22.70 mg

5. Distribution coefficient = 0.627

7. 5.5×10^{-4} mmol/mL

Chapter 17

1. $W = 20$ sec, $W_{1/2} = 11$ sec, $\sigma = 5$ sec
2. H-100 μm and $N = 1.15 \times 10^4$ theoretical plates
3. Mass of Y = 4.5 μg
8. Resolution = 2.32
9. Air, 8.5%, A, 20.8%; B, 16.2%, (C) 28.5%; (D), 17.1%; (E), 9.0.

Chapter 18

2. $R_f = 0.80$
4. 591 cm^3
6. $K_{av} = 1.829$; such a value exceeding 1 shows adsorption of the solute by the solid phase.

Chapter 19

2. 12.6
3. 93.0
6. Concentration phenol in original sample = 12.4 μg
7. 35.4 mL/min

Chapter 20

2. 2.38×10^5 J
3. %T = 53.4%, $A = 0.272$
5. 6.5 mg/L
7. $\varepsilon = a \times 2.393 \times 10^5$ mg/mol, ε in units of $L \times mol^{-1} \times cm^{-1}$

Chapter 21

1. Slope = 218 $L \times mol^{-1} \times cm^{-1}$ molar absorptivity since the path length is 1 cm

3. $[X] = 5.37 \times 10^{-5}$ M and $[Y] = 1.72 \times 10^{-4}$ M
4. Dissolve the sample in acid and dilute such that a 5% manganese sample would yield a solution about 4×10^{-3} M in manganese. Treat the sample with ammonium persulfate to develop the permanganate color and dilute 1:2 to place the absorbance in a suitable range.
5. From the information that HA is 89.5% dissociated at an analytical concentration of 1.00×10^{-3} mol/L, the value of K_a can be calculated ([HA] = 1.05×10^{-4} and [A$^-$] = 8.95×10^{-4}). Knowing that the absorbance due entirely to A$^-$ is 0.0325 in a 1-cm cell enables calculation of the molar absorptivity of A$^-$. Absorbance values can be calculated for a range of analytical concentrations of HA and plotted $vs.$ the analytical concentration, giving a curved line.
9. The end point volume is 38.1 mL.

Chapter 22

1. 52 μg/L
2. The cadmium concentration in the original sample is 4.96 μg/L.
7. The detection limit is about 20 μg/L.
8. About 0.05 μg

Chapter 23

1. 148 minutes
2. $K_{sp} = 6.09 \times 10^{-6}$
4. Half-life is 50 minutes
6. Rate for the 2.110 MeV γ is 2.7×10^5 s^{-1}; rate for the 2.85 MeV γ is 9.1×10^5 s^{-1}.

Appendix

The values given in Tables A-G are from a variety of sources, including compilations and original research papers. The most prominent of these sources are listed below, with the tables to which they pertain in brackets.

Sillén, L. G., and A. E. Martell. *Stability Constants of Metal-Ion Complexes.* Special Publication No. 17. London: Chemical Society, 1964. [Tables A, B, C, and D]

Kortüm, G., W. Vogel, and K. Andrussow. Dissociation constants of organic acids in aqueous solution. *Pure and Applied Chemistry* 1:187–536 (1960). [Tables A and B]

Feitknecht, W., and P. Schindler. Solubility constants of metal oxides, metal hydroxides . . . in aqueous solution. *Pure and Applied Chemistry* 6:130–199 (1963). [Table C]

Yatsimirskiĭ, K. Y., and V. P. Vasil'ev. *Instability Constants of Complex Compounds.* Translated by D. A. Paterson, Oxford: Pergamon Press, 1960. [Tables C and D]

Charlot, G., D. Bezier, and J. Courtot. *Selected Constants, Oxydo-Reduction Potentials.* Oxford: Pergamon Press, 1958. [Table F]

Bishop, E. In *Comprehensive Analytical Chemistry*, eds. C. L. Wilson and D. W. Wilson. Amsterdam: Elsevier Publishing Company, 1960. Volume IB, pp. 151–184. [Tables A and B]

Page, C. H., and P. Vigoureux, eds. *The International System of Units (SI).* National Bureau of Standards Special Publication 330, 1972 ed. Washington: U.S. Government Printing Office, April 1972. 42 pp. [Table G]

TABLE A   Dissociation constants of some acids in water at 25° C

Acid		K_a	Acid		K_a
Acetic	K_1	1.75×10^{-5}	Nitrilotriacetic	K_1	2.18×10^{-2}
Arsenic	K_1	5.6×10^{-3}	acid (NTA)	K_2	1.12×10^{-3}
	K_2	1.2×10^{-7}		K_3	5.25×10^{-11}
	K_3	3×10^{-12}	Nitrous	K_1	5×10^{-4}
Arsenious	K_1	1.4×10^{-9}	Oxalic	K_1	5.36×10^{-2}
Benzoic	K_1	6.14×10^{-5}		K_2	5.42×10^{-5}
Boric	K_1	5.9×10^{-10}	Phenol	K_1	1.1×10^{-10}
Carbonic	$K_1{}^a$	4.45×10^{-7}	Phosphoric	K_1	7.1×10^{-3}
	K_2	4.69×10^{-11}	(ortho)	K_2	6.17×10^{-8}
Chloroacetic	K_1	1.36×10^{-3}		K_3	4.4×10^{-13}
Citric	K_1	7.4×10^{-4}	Phosphorous^b	K_1	1.00×10^{-2}
	K_2	1.7×10^{-5}		K_2	2.6×10^{-7}
	K_3	3.9×10^{-7}	o-Phthalic	K_1	1.1×10^{-3}
Ethylene dinitrilo-	K_1	1.0×10^{-2}		K_2	3.9×10^{-6}
tetraacetic (EDTA)	K_2	2.1×10^{-3}	Salicylic	K_1	1.0×10^{-3}
	K_3	6.9×10^{-7}		K_2	4×10^{-14}
	K_4	5.5×10^{-11}	Sulfamic	K_1	1.0×10^{-1}
Formic	K_1	1.8×10^{-4}	Sulfuric	K_1	1.1×10^{-2}
Hydrocyanic	K_1	2.1×10^{-9}	Sulfurous	K_1	1.7×10^{-2}
Hydrofluoric	K_1	6×10^{-4}		K_2	6.3×10^{-8}
Hydrogen sulfide	K_1	5.7×10^{-8}	Tartaric	K_1	9.2×10^{-4}
	K_2	1.2×10^{-15}		K_2	4.3×10^{-5}
Hypochlorous	K_1	2.8×10^{-8}	Thiocyanic	K_1	1.4×10^{-1}
Iodic	K_1	1.8×10^{-1}			

^{a} Apparent constant based on $C_{H_2CO_3} = [CO_2] + [H_2CO_3]$.
^{b} Only two of the H's in H_3PO_3 are ionizable.

TABLE B   Dissociation constants of some bases in water at 25° C

Base	K_b	Base	K_b
2-Amino-2-(hydroxymethyl)-1,3-propanediol	1.2×10^{-6}	Hydrazine	9.8×10^{-7}
		Hydroxylamine	9.6×10^{-9}
		Lead hydroxide	1.2×10^{-4}
Ammonia	1.76×10^{-5}	Piperidine	1.3×10^{-3}
Aniline	4.2×10^{-10}	Pyridine	1.5×10^{-9}
Diethylamine	1.3×10^{-3}	Silver hydroxide	6.0×10^{-5}
Ethanolamine	3.18×10^{-5}		
Hexamethylene-tetramine	1×10^{-9}		

TABLE C Solubility products of some common electrolytes in water at 25° C

Substance	Formula	K_{sp}
Aluminum hydroxide (amorphous)	$Al(OH)_3$	6×10^{-32}
Barium carbonate	$BaCO_3$	5.5×10^{-10}
Barium chromate	$BaCrO_4$	1.2×10^{-10}
Barium oxalate	BaC_2O_4	1.7×10^{-7}
Barium sulfate	$BaSO_4$	1.2×10^{-10}
Bismuth sulfide	Bi_2S_3	$\sim 10^{-97}$
Cadmium sulfide	CdS	$\sim 10^{-27}$
Calcium carbonate	$CaCO_3$	4.47×10^{-9}
Calcium fluoride	CaF_2	4.9×10^{-11}
Calcium hydroxide	$Ca(OH)_2$	3.7×10^{-6}
Calcium oxalate	CaC_2O_4	2.3×10^{-9}
Calcium phosphate	$Ca_3(PO_4)_2$	$\sim 10^{-26}$
Calcium sulfate	$CaSO_4$	2.6×10^{-5}
Chromium(III) hydroxide	$Cr(OH)_3$	$\sim 10^{-30}$
Cobalt(II) hydroxide (pink, inactive),	$Co(OH)_2$	4×10^{-16}
Cobalt(III) hydroxide	$Co(OH)_3$	$\sim 10^{-43}$
Cobalt(II) sulfide	CoS	$\sim 10^{-23}$
Copper(I) chloride	$CuCl$	1.2×10^{-6}
Copper(II) hydroxide (inactive)	$Cu(OH)_2$	2×10^{-19}
Copper(I) iodide	CuI	5.0×10^{-12}
Copper(II) sulfide	CuS	$\sim 10^{-36}$
Iron(II) hydroxide (inactive)	$Fe(OH)_2$	8×10^{-6}
Iron(III) hydroxide (amorphous, inactive)	$Fe(OH)_3$	8×10^{-40}
Iron(II) sulfide	FeS	$\sim 10^{-19}$
Lanthanum iodate	$La(IO_3)_3$	6.2×10^{-12}
Lead carbonate	$PbCO_3$	1×10^{-13a}
Lead chloride	$PbCl_2$	1.6×10^{-5}
Lead chromate	$PbCrO_4$	1.8×10^{-14}
Lead hydroxide	$Pb(OH)_2$	$\sim 10^{-20}$
Lead iodide	PbI_2	7.1×10^{-9}
Lead sulfate	$PbSO_4$	1.7×10^{-8}
Lead sulfide	PbS	2.5×10^{-27}
Magnesium carbonate	$MgCO_3$	1×10^{-5}
Magnesium fluoride	MgF_2	6.6×10^{-9}
Magnesium hydroxide	$Mg(OH)_2$	1.8×10^{-11}
Manganese(II) carbonate	$MnCO_3$	1.8×10^{-11}
Manganese(II) hydroxide	$Mn(OH)_2$	1×10^{-13}
Manganese(II) sulfide	MnS	$\sim 10^{-13}$
Mercury(I) chloride	Hg_2Cl_2	1.3×10^{-18}
Mercury(II) hydroxide	$Hg(OH)_2$	$\sim 10^{-26}$
Mercury(II) sulfate	Hg_2SO_4	6.8×10^{-7}
Mercury(I) sulfide	Hg_2S	$\sim 10^{-47}$
Mercury(II) sulfide (black)	HgS	$\sim 10^{-52}$
Nickel hydroxide (inactive)	$Ni(OH)_2$	6×10^{-18}
Nickel sulfide	NiS	$\sim 10^{-21}$
Potassium tetraphenylboron	$KB(C_6H_5)_4$	3.2×10^{-8}
Silver arsenate	Ag_3AsO_4	1×10^{-22b}
Silver bromide	$AgBr$	5.2×10^{-13}
Silver carbonate	Ag_2CO_3	8.1×10^{-12}

TABLE C   **(Continued)**

Substance	Formula	K_{sp}
Silver chloride	AgCl	1.82×10^{-10}
Silver chromate	Ag_2CrO_4	1.29×10^{-12}
Silver cyanide	AgCN	7.2×10^{-11}
Silver dicyanoargentate	$Ag[Ag(CN)_2]$	1.3×10^{-12}
Silver hydroxide	AgOH	2.0×10^{-8}
Silver iodide	AgI	1.00×10^{-16}
Silver sulfate	Ag_2SO_4	2.1×10^{-5}
Silver sulfide	AgS	6×10^{-50}
Silver thiocyanate	AgSCN	1.1×10^{-12}
Strontium carbonate	$SrCO_3$	1.1×10^{-10}
Strontium sulfate	$SrSO_4$	3.2×10^{-7}
Zinc carbonate	$ZnCO_3$	1.4×10^{-11}
Zinc hydroxide (amorphous)	$Zn(OH)_2$	2.5×10^{-16}
Zinc sulfide	ZnS	$\sim 10^{-24}$

[a] At 18° C.
[b] At 20° C.

TABLE D   **Stability constants of some metal ion complexes in water**

Unless it is otherwise indicated, the values are for 25° C and ionic strength μ of 0.1. The value given to the right of the formula of a complex is the *overall* stability constant of that complex. Where known, the *stepwise* stability constants are given as log K values beneath the formula.

Ammonia	
$Ag(NH_2)_2^+$	1.59×10^7
3.31, 3.89	
$Cd(NH_3)_4^{2+}$	5.8×10^6
2.56, 2.01, 1.35, 0.84	
$Cu(NH_3)_4^{2+}$	2.0×10^{13}
4.30, 3.70, 3.00, 2.30	
$Hg(NH_3)_4^{2+}$	2×10^{19}
8.8, 8.7, 1.0, 0.8	
$Ni(NH_3)_4^{2+}$	4.2×10^7
2.71, 2.16, 1.64, 1.11	
$Zn(NH_3)_4^{2+}$	8.0×10^8
2.23, 2.30, 2.36, 2.01	
Chloride	
$AgCl_2^-$	1.9×10^5
3.31, 1.97	
$HgCl_4^{2-}$ $(\mu = 0.5)$	1.2×10^{15}
6.74, 6.48, 0.85, 1.05	
Iodide	
CdI_4^{2-}	2.3×10^5
2.40, 1.26, 1.0, 0.7	
HgI_4^{2-} $(\mu = 0.5)$	7.2×10^{29}
12.87, 10.95, 3.67, 2.37	

TABLE D   (Continued)

Unless it is otherwise indicated, the values are for 25° C and ionic strength μ of 0.1. The value given to the right of the formula of a complex is the *overall* stability constant of that complex. Where known, the *stepwise* stability constants are given as log K values beneath the formula.

Thiocyanate	1.07×10^3
$\quad$ Fe(SCN)$^{2+}$	
Cyanide	
$\quad$ Ag(CN)$_2^-$	1.1×10^{21}
$\quad$ Cu(CN)$_2^-$	1×10^{16}
$\quad$ Fe(CN)$_6^{4-}$	1×10^{24}
$\quad$ Fe(CN)$_6^{3-}$	1×10^{31}
$\quad$ Hg(CN)$_4^{2-}$	3.2×10^{41}
$\qquad$ 18.0, 16.70, 3.83, 2.98	
$\quad$ Ni(CN)$_4^{2-}$	1×10^{22}
Ethylenedinitrilotetraacetate($=$Y^{4-})	
$\quad$ (values at 20° C)	
$\quad$ CaY^{2-}	5.0×10^{10}
$\quad$ CdY^{2-}	2.9×10^{16}
$\quad$ CuY^{2-}	6.3×10^{18}
$\quad$ FeY^{2-}	2.1×10^{14}
$\quad$ FeY$^-$	1×10^{25}
$\quad$ MgY^{2-}	4.9×10^8
$\quad$ ZnY^{2-}	3.2×10^{16}
Eriochrome Black T ($=$D^{3-})	
$\quad$ (values at 20° C)	
$\quad$ CaD$^-$	1.9×10^5
$\quad$ MgD$^-$	7.4×10^6
$\quad$ ZnD$^-$	2×10^{12}

TABLE E   Common acid-base indicators

Common name	pK_{In}	pH visual transition interval	Color[a] Acidic	Basic
Thymol blue	1.65	1.2–2.8	Red	Yellow
Methyl yellow	3.1	2.4–4.0	Red	Yellow
Bromophenol blue	4.2	3.0–4.6	Yellow	Blue
Methyl orange	3.46	3.1–4.4	Red	Orange
Bromocresol green	4.66	3.8–5.4	Yellow	Blue
Methyl red	5.0	4.2–6.3	Red	Yellow
Chlorophenol red	6.2	4.8–6.4	Yellow	Red
Bromocresol purple	6.4	5.2–6.8	Yellow	Purple
Bromothymol blue	7.10	6.2–7.6	Yellow	Blue
Cresol red	8.4	7.2–8.8	Yellow	Red
Phenol red	7.81	6.8–8.4	Yellow	Red
Thymol blue	8.90	8.0–9.6	Yellow	Blue
Phenolphthalein	8.7	8.3–10.0	Colorless	Pink
Thymolphthalein	9.2	9.3–10.5	Colorless	Blue

[a] Colors in aqueous solution at lower and upper pH limits of the visual transition interval, respectively.

TABLE F Standard electrode potentials at 25° C

Half-reaction equation	E^0, V	Solution conditions for formal potentials
$S_2O_8^{2-} + 2e \rightleftharpoons 2SO_4^{2-}$	+2.01	
$H_2O_2 + 2H^+ + 2e \rightleftharpoons 2H_2O$	+1.78	
$MnO_4^- + 4H^+ + 3e \rightleftharpoons MnO_2(s) + 2H_2O$	+1.69	
$Ce^{4+} + e \rightleftharpoons Ce^{3+}$	+1.70	(in 1 M HClO$_4$)
	+1.61	1F HNO$_3$
	+1.44	1F H$_2$SO$_4$
	+1.28	1F HCl
$NaBiO_3(s) + 6H^+ + 2e \rightleftharpoons Na^+ + Bi^{3+} + 3H_2O$	~+1.6	
$MnO_4^- + 8H^+ + 5e \rightleftharpoons Mn^{2+} + 4H_2O$	+1.51	
$2BrO_3^- + 12H^+ + 10e \rightleftharpoons Br_2 + 6H_2O$	~+1.5	
$PbO_2(s) + 4H^+ + 2e \rightleftharpoons Pb^{2+} + 2H_2O$	+1.46	
$Cl_2 + 2e \rightleftharpoons 2Cl^-$	+1.359	
$Cr_2O_7^{2-} + 14H^+ + 6e \rightleftharpoons 2Cr^{3+} + 7H_2O$	+1.33	
$Tl^{3+} + 2e \rightleftharpoons Tl^+$	+1.28	
$MnO_2(s) + 4H^+ + 2e \rightleftharpoons Mn^{2+} + 2H_2O$	+1.23	
$O_2(g) + 4H^+ + 4e \rightleftharpoons 2H_2O$	+1.229	
$ClO_4^- + 2H^+ + 2e \rightleftharpoons ClO_3^- + H_2O$	+1.19	
$2IO_3^- + 12H^+ + 10e \rightleftharpoons I_2(s) + 6H_2O$	+1.19	
$Br_2 + 2e \rightleftharpoons 2Br^-$	+1.087	
$VO_2^+ + 2H^+ + e \rightleftharpoons VO^{2+} + H_2O$	+0.9994	
$HNO_2 + H^+ + e \rightleftharpoons NO(g) + H_2O$	+0.99	
$NO_3^- + 3H^+ + 2e \rightleftharpoons HNO_2 + H_2O$	+0.94	
$2Hg^{2+} + 2e \rightleftharpoons Hg_2^{2+}$	+0.907	
$Cu^{2+} + I^- + e \rightleftharpoons CuI(s)$	+0.86	
$Ag^+ + e \rightleftharpoons Ag$	+0.7994	
$Hg_2^{2+} + 2e \rightleftharpoons 2Hg$	+0.792	
$Fe^{3+} + e \rightleftharpoons Fe^{2+}$	+0.771	
Benzoquinone + 2H$^+$ + $e \rightleftharpoons$ hydroquinone	+0.6994	
$O_2(g) + 2H^+ + 2e \rightleftharpoons H_2O_2$	+0.69	
$MnO_4^- + e \rightleftharpoons MnO_4^{2-}$	+0.6	
$MnO_4^- + 2H_2O + 3e \rightleftharpoons MnO_2(s) + 4OH^-$	+0.57	
$H_3AsO_4 + 2H^+ + 2e \rightleftharpoons H_3AsO_3 + H_2O$	+0.559	
$I_3^- + 2e \rightleftharpoons 3I^-$	+0.545	
$I_2(s) + 2e \rightleftharpoons 2I^-$	+0.536	
$Ag_2CrO_4(s) + 2e \rightleftharpoons 2Ag + CrO_4^{2-}$	+0.447	
$Fe(CN)_6^{3-} + e \rightleftharpoons Fe(CN)_6^{4-}$	+0.356	
$VO^{2+} + 2H^+ + e \rightleftharpoons V^{3+} + H_2O$	+0.337	
$Cu^{2+} + 2e \rightleftharpoons Cu$	+0.337	
$Hg_2Cl_2 + 2e \rightleftharpoons 2Hg + 2Cl^-$	+0.2680	
	+0.3337	(in 0.1 M KCl)
	+0.2801	(in 1 M KCl)
	+0.2412	(in Sat'd KCl)
$AgCl + e \rightleftharpoons Ag + Cl^-$	+0.2224	
$SbO^+ + 2H^+ + 3e \rightleftharpoons Sb + H_2O$	+0.21	
$SO_4^{2-} + 4H^+ + 2e \rightleftharpoons H_2SO_3 + H_2O$	+0.17	
$Cu^{2+} + e \rightleftharpoons Cu^+$	+0.153	
$Sn^{4+} + 2e \rightleftharpoons Sn^{2+}$	+0.15	

TABLE F (Continued)

Half-reaction equation	E^0, V	Solution conditions for formal potentials
$S + 2H^+ + 2e \rightleftharpoons H_2S$	+0.14	
$Hg_2Br_2(s) + 2e \rightleftharpoons 2Hg + 2Br^-$	+0.1392	
$S_4O_6^{2-} + 2e \rightleftharpoons 2S_2O_3^{2-}$	+0.09	
$AgBr(s) + e \rightleftharpoons Ag + Br^-$	+0.071	
$2H^+ + 2e \rightleftharpoons H_2(g)$	±0.0000	
$Pb^{2+} + 2e \rightleftharpoons Pb$	−0.126	
$Sn^{2+} + 2e \rightleftharpoons Sn$	−0.140	
$AgI(s) + e \rightleftharpoons Ag + I^-$	−0.152	
$Ni^{2+} + 2e \rightleftharpoons Ni$	−0.23	
$V^{3+} + e \rightleftharpoons V^{2+}$	−0.255	
$Co^{2+} + 2e \rightleftharpoons Co$	−0.28	
$Tl^+ + e \rightleftharpoons Tl$	−0.336	
$Ti^{3+} + e \rightleftharpoons Ti^{2+}$	−0.37	
$Cd^{2+} + 2e \rightleftharpoons Cd$	−0.402	
$Cr^{3+} + e \rightleftharpoons Cr^{2+}$	−0.41	
$Fe^{2+} + 2e \rightleftharpoons Fe$	−0.440	
$Cr^{3+} + 3e \rightleftharpoons Cr$	−0.74	
$Zn^{2+} + 2e \rightleftharpoons Zn$	−0.7628	
$Mn^{2+} + 2e \rightleftharpoons Mn$	−1.190	
$Al^{3+} + 3e \rightleftharpoons Al$	−1.66	
$Mg^{2+} + 2e \rightleftharpoons Mg$	−2.37	
$Na^+ + e \rightleftharpoons Na$	−2.713	
$Ca^{2+} + 2e \rightleftharpoons Ca$	−2.87	
$K^+ + e \rightleftharpoons K$	−2.925	
$Li^+ + e \rightleftharpoons Li$	−3.03	

TABLE G The International System of Units (SI)

Physical quantity	Name of unit	Symbol of unit	Expression in SI base units
Length	meter	m	[base]
Mass	kilogram	kg	[base]
Time	second	s	[base]
Current, electric	ampere	A	[base]
Temperature, thermodynamic	kelvin	K	[base]
Amount of substance	mole	mol	[base]
Luminous intensity	candela	cd	[base]
Frequency	hertz	Hz	s^{-1}
Force	newton	N	$m \cdot kg \cdot s^{-2}$
Pressure	pascal	Pa	$m^{-1} \cdot kg \cdot s^{-2}$
Energy, work, quantity of heat	joule	J	$m^2 \cdot kg \cdot s^{-2}$
Power, radiant flux	watt	W	$m^2 \cdot kg \cdot s^{-3}$
Quantity of electricity, electric charge	coulomb	C	$s \cdot A$

TABLE G (Continued)

Physical quantity	Name of unit	Symbol of unit	Expression in SI base units
Electrical potential, potential difference, electromotive force	volt	V	$m^2 \cdot kg \cdot s^{-3} \cdot A^{-1}$
Electrical resistance	ohm	Ω	$m^2 \cdot kg \cdot s^{-3} \cdot A^{-2}$
Electrical conductance	siemens	S	$m^{-2} \cdot kg^{-1} \cdot s^3 \cdot A^2$
Electrical capacitance	farad	F	$m^{-2} \cdot kg^{-1} \cdot s^4 \cdot A^2$
Magnetic flux	weber	Wb	$m^2 \cdot kg \cdot s^{-2} \cdot A^{-1}$
Magnetic flux density	tesla	T	$kg \cdot s^{-2} \cdot A^{-1}$
Inductance	henry	H	$m^2 \cdot kg \cdot s^{-2} \cdot A^{-2}$
Activity (radioactive)	becquerel	Bq	s^{-1}

NOTE: The General Conference of Weights and Measures has adopted this practical system of units of measurement. Listed are the SI base units and also some SI derived units having special names. For the derived units, the expression combines base units and centered dots.

SI prefixes

Multiple or Submultiple	Prefix	Symbol
10^{12}	tera	T
10^9	giga	G
10^6	mega	M
10^3	kilo	k
10^2	hecto	h
10	deka	da
10^{-1}	deci	d
10^{-2}	centi	c
10^{-3}	milli	m
10^{-6}	micro	μ
10^{-9}	nano	n
10^{-12}	pico	p
10^{-15}	femto	f
10^{18}	atto	a

NOTE: Terms formed from prefixes listed, joined to the unit names, provide the multiples and submultiples in the International System. For example, the unit name *meter* with the prefix *kilo* added produces *kilometer*, meaning "1000 meters."

Newton's Approximation Method for the Solution of Polynomials

Newton's approximation method is useful for solving any equation is one unknown, where an approximate solution is known (which is generally true of solution equilibrium problems). The method is especially useful for solving higher-degree polynomial equations, such as cubic equations. The generalized form of such an equation is

$$a_n X^n + a_{n-1} X^{n-1} + \cdots + a_1 X + a_0 = 0$$

where X is the unknown, n is the maximum power of X, and the a's are numerical coefficients. According to Newton's approximation method, if X_0 is an approximate value of the desired root of X, X_1 is a better value where

$$X_1 = X_0 - \frac{f(X_0)}{f'(X_0)}$$

and where

$$f(X_0) = a_n X^n + a_{n-1} X^{n-1} + \cdots + a_1 X + a_0$$

$$f'(X_0) = n a_n X^{n-1} + (n-1) a_{n-1} X^{n-2} + \cdots + a_1$$

After solving for X_1 in one iteration, X_1 then becomes X_0 for the next iteration. The solution is reached when X_1 and X_0 are equal to each other to the desired number of significant figures. For an example, see the solution to a cubic equation at the end of Section 7.5 in Chapter 7.

Index

Photo Credits

This page constitutes an extension of the copyright page.